Related Titles

Brennan, J.G., Grandison, A.S. (eds.)

Food Processing Handbook
Second Edition

2012
ISBN: 978-3-527-32468-2 (Also available in a variety of electronic formats)

Norton, J.E., Fryer, P., Norton, I.

Formulation Engineering of Foods

2013
ISBN: 978-0-470-67290-7 (Also available in a variety of electronic formats)

Dunford, N. (ed.)

Food and Industrial Bioproducts and Bioprocessing

2012
ISBN: 978-0-813-82105-4 (Also available in a variety of electronic formats)

Bhandari, B.R., Roos, Y. (eds.)

Food Materials Science and Engineering

2012
ISBN: 978-1-405-19922-3 (Also available in a variety of electronic formats)

Tanja Wüstenberg

Cellulose and Cellulose Derivatives in the Food Industry

Fundamentals and Applications

WILEY-VCH
Verlag GmbH & Co. KGaA

The Author

Dipl.-Ing. Tanja Wüstenberg
25421 Pinneberg
Germany

Library of Congress Card No.: applied for

British Library Cataloguing-in-Publication Data

A catalogue record for this book is available from the British Library.

Bibliographic information published by the Deutsche Nationalbibliothek

The Deutsche Nationalbibliothek lists this publication in the Deutsche Nationalbibliografie; detailed bibliographic data are available on the Internet at http://dnb.d-nb.de.

Print ISBN: 978-3-527-33758-3
ePDF ISBN: 978-3-527-68295-9
ePub ISBN: 978-3-527-68296-6
Mobi ISBN: 978-3-527-68294-2
oBook ISBN: 978-3-527-68293-5

Cover Design Adam-Design, Weinheim, Germany
Typesetting Thomson Digital, Noida, India
Printing and Binding Markono Print Media Pte Ltd, Singapore

Printed on acid-free paper

Contents

Preface

Cellulose derivatives belong to the complex group of food hydrocolloids. Despite their close chemical relationship with the frequently used starches and their high technological capacity, cellulose-based thickeners and stabilizers play only a minor role in terms of quantity in the food industry. Owing to the high availability of native cellulose as a renewable resource and the multifunctional properties of its derivatives, cellulose derivatives have attracted steadily increasing interest and growing demand in industrial applications for some years.

The aim of this book is to collect the available traditional and current knowledge about this product group, which for a long time has been rather underrated, and make it accessible to a wide readership in an improved form. The background is both theoretically and practically illuminated to generate ideas for product development, to facilitate technological implementation with respect to handling, and to ensure successful use.

The focus is on the application in food products. Limited insight into other regulated non-food areas such as pharmaceuticals and cosmetics, as well as in technical applications, is intended to show the general feasibility and to provide inspiration for innovation. In particular, Chapters 6 (Ethylcellulose) and 7 (Hydroxypropylcellulose) examine whether the solvents, plasticizers, and stabilizers used are appropriate and permitted for use in foods or products that come in contact with food.

The information in this book – especially the recommendations, instructions, and guideline formulations – is provided by the author and the companies involved solely as a service and is of a general nature. Any suggestions serve only to assist the reader and users in the exploration of possible applications. The author and the aforementioned manufacturers make no guarantee as to the accuracy or completeness of the information for a specific application case. They assume no responsibility for the results arising from the practical implementation of such information, advice, and support. Each user is solely responsible for determining whether such products are suitable for the intended use, and for obtaining the necessary approvals and permits.

I would like to thank the companies FMC Biopolymer and Ashland Inc. for providing very valuable technical support in the realization of this cellulosic project. Very special thanks go to my former colleagues from Wolff Cellulosics and Dow Company. Through our many years of inspiring collaboration my passion for cellulose derivatives has always been encouraged, which has made this book possible. In particular, I would like to thank my family and my partner for their patience and motivation to negotiate this long-term project successfully.

Tanja Wüstenberg

List of Abbreviations

Abbreviation	Meaning
1 Å	The ångström is a unit of length, but it is not an SI- unit. It is used in chemistry and in crystallography to describe the bond lengths in molecules, distances between atoms in the crystal structures, and magnitudes of atomic radii. 1 Å = 100 pm = 0.1 nm = 10^{-4} µm = 10^{-7} mm = 10^{-10} m
ADI	Acceptable daily intake. Limit for the daily intake of a substance in the human diet that was evaluated to be harmless
AfG	Soft drinks (free of alcohol)
ASTM	American Society for Testing and Materials. ASTM International is a globally recognized leader in the development and delivery of international voluntary consensus standards. Today, some 12,000 ASTM standards are used around the world to improve product quality, enhance safety, facilitate market access and trade, and build consumer confidence.
D.E. or DE	Dextrose equivalent. The D.E. serves to characterize polysaccharide mixtures and designates the percentage of reducing sugars in the dry matter (DM). It corresponds to the amount of glucose (= dextrose) in 100 g dry matter that would have the same reducing power. The D.E. is a measure of how much polymer degradation has occurred. Products with low D.E. have a high content of polysaccharides and a low proportion of low-molecular-weight sugars. Products with high D.E. are mainly composed of short-chain sugars. D.E. of starch = 0; D.E. of glucose = 100; D.E. of maltodextrins = 3–20 (depending on degree of hydrolysis); D.E. of glucose syrups > 20
DM	Dry matter
DP	Degree of polymerization. The DP is a measure of the chain length of molecules. It indicates the number of single units (monomers) in a chain.

Abbreviation	Meaning
DS	Degree of substitution. The degree of the average substitution; also referred to as the degree of etherification. The DS is a measure of the number of etherified hydroxyl groups of an anhydroglucose unit (AGU). DS = [mol substituent per mol AGU]
EUPh	*Pharmacopoea Europaea* pharmacopoeia of the European Union
FDA (also USFDA)	Food and Drug Administration: U.S. bureau of the security and surveillance of food, tobacco products, food supplements and medicines.
FP	Fruit preparation
FSMP	Food for special medical purposes and special diets.
HLB	HLB = hydrophilic–lipophilic balance. According to Griffin, the HLB value is defined as a number between 1 and 20 for the characterization of emulsifiers. The lower the number the more lipophilic is the substance. The formula is: $\text{HLB} = 20 \times (1 - M_L/M)$ where M_L is the lipophilic portion of the molecule and M is the molecular weight of the entire molecule. The factor 20 is a scaling factor freely chosen by Griffin.
HP	Hydroxypropyl ($CH_2CHOHCH_3$)
IEP	Isoelectric point. The IEP is the pH at which the net charge of a protein is 'zero'. The number of positively charged groups and the number of negatively charged groups is the same, so that the protein shows the lowest mobility in an electrically charged field. At the IEP, the sensitivity of a protein to heat and acid addition is highest.
IFT	Interfacial tension
LBG	Locust bean gum – E 410. A polysaccharide from mannose and galactose used to thicken and stabilize food products
MDM	Mechanically deboned meat
Me	Methyl (CH_3)
MS	Molar substitution. The MS is the average number of moles of substituted groups per glucose monomer
MSG	Monosodium glutamate (E 621) – used as flavour enhancer
MSNF	Milk solids non-fat
MW	Molecular weight – the weight (mass) of a molecule
MWh	Megawatt hour. A watt-hour (Wh) is a unit of energy. It corresponds to the energy that is required from a machine of one watt in an hour.
NDC	Non-dairy cream
NFDM	Non-fat dry milk
QS - qs	*Quantum satis.* This means, in relation to the dosage of substances, the amount necessary to achieve the desired effect
rpm	Revolutions per minute – unit for the speed of stirring equipment
RTD	Ready-to-drink

Abbreviation	Meaning
RTU	Ready-to-use
SMP	Skimmed milk powder
T	Temperature
TSPC	Texturized soy protein concentrate. TSCP, with its firm bite, is used as meat replacement
USPh	Pharmacopoeia of the United States of America
WPC	Whey protein concentrate

Units

Symbol	Unit	Description
$[\eta]$	mPa s	Intrinsic viscosity
η_S	—	Specific viscosity
η	1 mPa s = 1 cP	Dynamic viscosity 1 Poise = 100 cP = 0.1 Pas = 100 mPa s
ν	$1\ mm^2\ s^{-1}$	Kinematic viscosity ($1\ mm^2\ s^{-1} = 1$ cSt; St = Stoke)
γ	$N\ m^{-1}$	Interfacial tension
γ_S	$dynes\ cm^{-2}$	Surface tension

1
General Overview of Food Hydrocolloids

1.1
Introduction to the World of Hydrocolloids

The term 'hydrocolloid' is derived from the Greek *hydro* 'water' and *kolla* 'glue'.

Hydrocolloids are colloidal substances with an affinity for water. From a chemical point of view, they are macromolecular hydrophilic substances. Some of them are water soluble and form colloidal solutions others are only able to swell in water and can be dispersed by means of shear forces. Hydrocolloids produce viscous solutions, pseudo-gels, or gels in water. The heterogeneous group consists of polysaccharides and proteins.

Hydrocolloids are used in technical and regulated applications to thicken and to stabilize formulations. In processed foods, they are ubiquitous – no other group of ingredients contributes more to viscosity, texture, and body like hydrocolloids do.

Hydrocolloids are not really emulsifiers because, mostly, they do not have the characteristic linkage of lipophilic and hydrophilic groups in the molecular structure. The molecules are too big and complex in size and therefore are not flexible enough to cover the interfaces being formed during homogenization of oil–water mixtures fast enough to create a long-term stable emulsion with sufficiently small droplet diameter. However, these thickeners can stabilize emulsions by increasing the viscosity of the water surface or by interaction with surface-active substances. Some hydrocolloids like gum Arabic or non-ionic products such as methylcellulose (MC), HPMC (hydroxypropylcellulose), or propylene glycol alginate (PGA) reduce the surface tension and exhibit limited emulsifying properties [1].

In accordance to their origin and way of manufacturing, hydrocolloids can be classified in four different groups:

1) hydrocolloids purely isolated from plants (without chemical modification);
2) hydrocolloids obtained by fermentation;
3) plant-derived hydrocolloids that are chemically modified;
4) hydrocolloids from animals.

Cellulose and Cellulose Derivatives in the Food Industry: Fundamentals and Applications, First Edition.
Tanja Wüstenberg.

According to their botanical origin and their function in the plant organism, naturally occurring vegetable hydrocolloids can be divided into [1]:

- exudates (protective colloids being deposited on wounds):
  acacia gum/gum Arabic, tragacanth, karaya gum, ghatti gum;
- seed flours (reserve polysaccharides):
  guar gum, locust bean gum, tara gum, tamarind seed gum;
- extracts from land plants and marine algae (scaffolding substances):
  pectins, agar, alginate, carrageenan, starches, cellulose, furcelleran, larch gum.

Additionally, there are [1]:

- microbial or bacterial polysaccharides:
  xanthan, dextran, curdlan, scleroglucan, gellan, pullulan;
- modified polysaccharides:
  propylene glycol alginate, amidated pectin, modified starches, cellulose derivatives;
- proteins of animal origin:
  gelatine, caseinates.

Figure 1.1 presents an overview of globally used food hydrocolloids. There are also other substances available and in use, but several of them are restricted to local use and, depending on availability and legislation, are not in industrially-produced applications. Please always check the relevant legislation before using one of these stabilizers, gelling agents, or thickeners.

The individual substances are described in subsequent sections. Information on their origin, manufacturing, structure, properties, and handling are provided. For the most used products there are overview tables; Tables 1.1–1.9 below give a quick orientation.

The individual cellulose-based substances are then described in more detail. Overview tables for selected cellulose derivatives are given in Section 1.6 (Tables 1.10–1.14).

1.2 Plant Extracts

1.2.1 Agar

Raw Material, Harvesting, and Manufacturing

Agar is a structure-building component of the cell wall of red algae (Rhodophyceae). *Gelidium*, *Gracilaria*, and *Pterocladia* species especially serve as a source of raw materials. The main producing countries are Japan, United States of

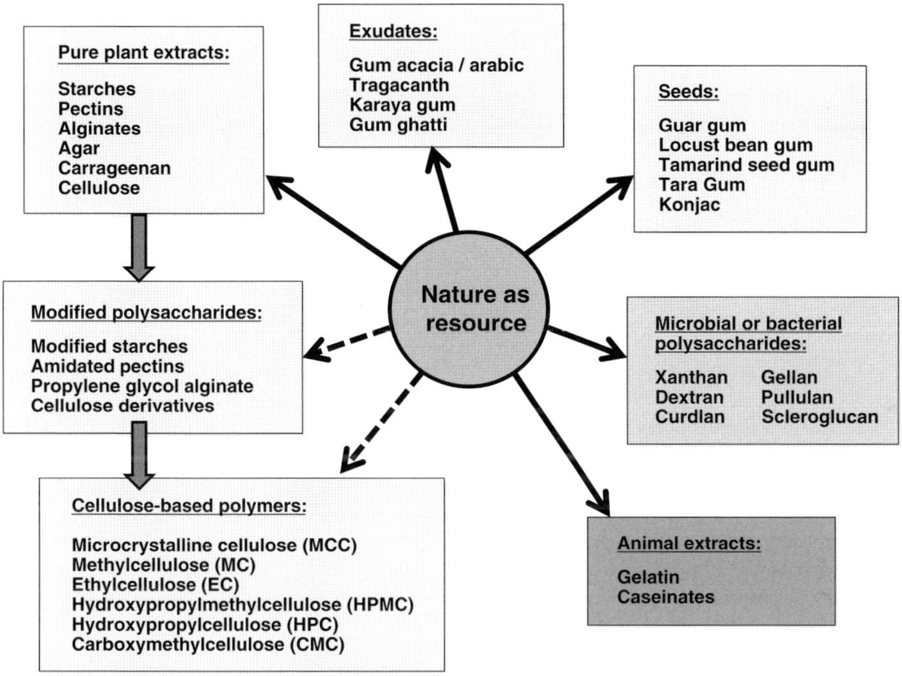

Figure 1.1 Overview of food hydrocolloids used globally.

America (California), Chile, and Spain, on whose rocky shores they occur. Agar was discovered in 1658 in Japan.

The red algae are harvested and extracted under pressure with hot water (100–130 °C (212–266 °F)) at pH 5–6. The extract is purified by filtration or centrifugation and subsequently bleached with calcium hypochlorite. To isolate the agar, the extract is frozen and, after thawing, the remaining gelatinous residue is dried. More recently, the water is squeezed out by means of a high-pressure press, and the remaining water is removed by drying. The annual production involves about 55 000 MT of dried seaweed to manufacture 7500 MT of agar [1,2].

Chemical Structure

Agar is a heterogeneous polysaccharide composed of the monomeric substances D-galactose and (3 → 6)-anhydro-L-galactose (Table 1.1). The main component, which is responsible for the strong gelling ability, is so-called agarose. Additionally, a small amount of a slightly acidic polysaccharide, the non- or only weakly gelling agaropectin, is present. Agaropectin also contains sulfate ester, glucuronic, and pyruvic acid groups. Agarose is a neutral, linear galactan. The D-galactose and 3,6-anhydro-L-galactose monomers are linked alternately by α-(1 → 3)- and β-(1 → 4) bonds with each other. Owing to the high anhydrogalactose portion and the absence of sulfate ester groups, which strengthen the hydrophobic character, agar is a good gelling agent that is independent of cations [1].

Table 1.1 Characteristics of agar.

	Agar
E-No.	E 406
Origin	Seaweed extract
Chemical composition	• Galactose and anhydro-galactose; • low sulfate content (<4.5%, mostly 1.5–2.5%)
Nutritional value (in 100 g) – metabolism	1425 kJ (340 kcal); slow resorption
Fibre content	2.2% (42% protein, 36% carbohydrates)
Toxicology	No health concerns, consumption generally 1–2 g per day (mild laxative effect at 6 g per day); no ADI value defined; considered as food in Asia, not as additive
Solubility at low temperature (H_2O)	Insoluble
Appearance of an aqueous solution	Opaque, yellowish
Viscosity of solution in water	Low, insoluble in cold water
Impact of heat on viscosity in water (pH 7)	Soluble at $T > 80\,°C$ (176 °F)
Viscosity development in water at pH 7 ($T = 0–100\,°C$)	Forms a thermoreversible gel after dissolution and cooling to 35 °C (95 °F) (gel melts at $T > 85\,°C$ (185 °F))
Shear stability	High
Thickening effect	High (gel formation)
pH stability	Medium, hydrolysis by cooking in acidic system
Decomposition	Combination of heat + low pH (below pH 4)
Film formation	High
Emulsion stabilization	No
Gelation	Thermoreversible gelation after heating to $T > 80\,°C$ (176 °F) and cooling to 35 °C (95 °F); also gels in saturated sugar solutions; gelation temperature is independent of sugar concentration
Gel strength and gel stability	High gel strength, gels are highly heat-stable
Gel transparency	Low
Tendency for gel synaeresis	High
Impact of electrolytes (cations +, 2+, 3+)	No
Reaction with Ca^{2+} ions	No
Protein activity	Low/no
Crystallization control	No
Synergistic effects with other hydrocolloids	+ LBG or guar: less synaeresis, more elasticity; + 10% LBG: maximum gel strength (+8%); + 10% konjac: maximum gel strength
Other synergistic effects	Enhanced gel strength with high sugar concentrations (>60%) for some agar types
Negative interactions	• Tannic acid can inhibit the gelling process; • gum karaya reduces gel strength of agar gels; • proton scavengers like potassium iodide, urea, guanidine, sodium thiocyanate, and so on block the gelling process and prevent agarose gel formation
Dosage level in foods	Low, typical 0.5–2%, gelation already at 0.2%

Gelation of agarose from aqueous solutions is assumed to occur because of the association of the molecular chains into double helices, which then aggregate to form a network capable of immobilizing the water [3].

The sulfate content of agar is below 4.5% (mostly between 1.5–2.5%) which is very low compared to that of carrageenans [2].

Solubility, Viscosity, and Gelation

Agar is insoluble in cold water. Complete dissolution occurs at temperatures above 80 °C (176 °F). Upon cooling to 30–40 °C (86–104 °F), a thermoreversible gel is formed that melts again when heating above 76–92 °C (169–197 °F), depending on seaweed species and methyl ester content. This transition occurs at higher temperature with increasing agar concentration and decreasing cooling rate. Another vey important property of agar is gel hysteresis – the difference between the gelling and the melting temperature – which is around 50–60 °C (90–108 °F). This hysteresis of agar types with low ester content is much larger than for many other gelling agents (e.g. κ-carrageenan with 15–20 °C (27–36 °F)). Agar is a neutral hydrocolloid whose solubility and gelation is not affected by the addition or presence of electrolytes. Agar does not require cations to gel. The agar gelation process is totally reversible. The gel melts on heating and resets on cooling. This cycle can be repeated many times without significant change in the mechanical properties of the gel, provided the agar is not used at very low pH conditions (<4) or with oxidizing agents. Agar consists of a macromolecular network of agarose molecules associated through hydrogen bonds. Consequently, the presence of proton scavengers such as potassium iodide, urea, guanidine, or sodium thiocyanate block the gelling process by impeding the formation of hydrogen bonds and hence prevent agarose gel formation. Moreover, agar is resistant against enzymes used in food technology [2].

Properties

The ability to form reversible gels by simply cooling hot, aqueous solutions is the most important property of agar. The greatest advantages for agar in different food applications derive from the characteristic firm texture and heat tolerance of the gels, stability under acidic conditions, high solubility in concentrated sugar solutions, and limited reactivity with other food components. Agar can be hydrolysed by acid at high temperature, so for pH values below 5 it is recommended to lower the pH just before cooling to form the gel. The level of use is low and typically between 0.5% and 2%. The threshold for gelation is 0.2%. With locust bean gum (LBG) or guar gum, agar gels exhibit less synaeresis and more elasticity. The maximum gel strength (+8%) of agar is obtained with 10% LBG galactomannan or 10% konjac glucomannan in the mixture. Gel strength is enhanced by high sugar concentrations (>60%). Gum karaya reduces agar gel strength. The presence of tannic acid or pentadiagaloyl glucose found in fruits like quince, apples, and plums can inhibit the gelling process because of their function as proton scavenger. This effect can be overcome by adding low levels of glycerol [2].

Applications

Agar is used in canned meat, fish, and poultry products. It is suitable for use in water- and milk-based desserts, any kind of aspics, and artificial caviar. Sugar-based coatings, glazes, and icings can be formulated with low to high sugar concentrations and are never sticky. Agar is added to confectionary, including jelly beans, nougat, candy fillings, piping gels, jams, and jellies. It acts as texture stabilizer in ice cream and as fining agent to clarify wine, juice, and vinegar. Besides food, it is used in microbiology in high volumes for the preparation of culture media.

Toxicology and Regulatory Affairs

Since time immemorial, seaweed serves the people as part of its food. Long-term experiences in using agar in animal nutrition and clinical results in using as laxative are available. Extensive toxicological studies did not justify any health concerns. Agar has GRAS (generally recognized as safe) status from US FDA. No ADI (allowable daily intake) value has been defined. In Europe, agar is approved as food additive with unrestricted dosage ('quantum saris') and classified as E 406. In Asia it is has been considered to be a food ingredient, not an additive, for centuries with an estimated daily consumption of 1–2 g. The formed gels slowly pass through the stomach, increasing satiety by increasing the feeling of fullness. The increased viscosity moderates glucose and fat uptake and controls cholesterol.

1.2.2 Alginates and PGA

Raw Material Origin, Harvesting, and Quality

Alginate was first described in 1881. The name is now used as the general term for the range of alginic acid salts that can be applied in foods. Alginate occurs in the cell walls and intercellular spaces of brown algae (Phaeophyceae). The alginate molecules provide both flexibility and strength to the plants adapted to the special conditions in the sea. Main sources of raw material are species of *Macrocystis, Laminaria, Lessonia, Eklonia, Ascophyllum,* and *Durvillaea.* Brown algae need clean water at 4–18 °C (39–64 °F) and sunlight to produce alginate by photosynthesis.

Alginic acid is extracted directly from brown seaweed through a long and slow process. This is the free acid form of alginate and the water-insoluble intermediate in the commercial manufacture of alginates. The concentration of alginic acid in seaweed is about 15–40% of the dry matter. To produce alginic acid from, for example, *Macrocystis* species the seaweed is cleaned, shredded, and washed after harvesting. The wet or dried algae are extracted by alkaline treatment. The extracts are purified by filtration and the alginic acid is isolated by precipitation with calcium ions or mineral acid. Subsequently, the partially soluble alginic acid is converted into water-soluble alginate products (an exception is Ca alginate) by neutralization with alkali and incorporation of different inorganic salts ($NaCO_3$, K_2CO_3, NH_4CO_3, $CaCO_3$,). Propylene glycol alginate (PGA) is produced by adding propylene oxide under pressure to a partially neutralized alginic acid. The resulting products are dried and milled. The global annual production of alginate is estimated to be 38 000 MT [1,2].

Chemical Structure

Alginate is a family of unbranched binary copolymers of (1 → 4)-linked β-D-mannuronic acid (M) and α-L-guluronic acid (G) residues of widely varying composition and sequential structure (Table 1.2). Alginates are salts of alginic acid, a linear polyuronide, with the cations of Na, K, NH_4, or Ca. Depending on the type of algae, the alginic acid is composed of mannuronic acid and guluronic acid in ratios of M : G = 0.4–1.9 : 1. In addition to alternating polymers (M-G-M-G), block polymers also occur in the alginate macromolecule and are constructed only of mannuronic acid (M-M-M-M) or guluronic acid (G-G-G-G). In the pure GGGG- or MM-blocks there is a kind of folded structure, which has an essential role in gelation. In particular, the GG-blocks form a regular zigzag structure. The degree of polymerization for commercial products is in the range 50–3000, corresponding to molecular weights of 10–600 kDa. Propylene glycol alginate is an ester of alginic acid whose uronic acid units are partially (40–85%) esterified with propylene glycol. The remaining carboxyl groups occur partially in free form and partially as sodium salt. The average molecular weight is between 32 and 200 kDa [1–3].

Solubility, Viscosity, and Gelation

Alginic acid and calcium alginate are insoluble in water, while the alkali metal (sodium, potassium) and ammonium salts are easily soluble. Propylene glycol alginate is water-soluble down to pH 2.5, with stable solutions in the range pH 2.5–8. Typically, the alginate powder is dry-blended with particles, such as sugar, or suspended in a hydrophobic solvent, for example vegetable oil to prevent lump formation. The dissolution rate depends on shear force, ionic level with respect to water hardness and solvent temperature. Viscosity in water is correlated to the molecular lengths of the alginate molecules, but is also several influenced by ions present, sugar, polyols, and alcohols. Aqueous solutions have a high viscosity, even at low concentrations, and are very stable when heated in a neutral pH range. In acidic systems at pH <5, the viscosity decreases strongly. Calcium ions greatly increase the viscosity and lead to gelation by incorporation in the zigzag structure of the GG-blocks (egg-box model), which then form three-dimensional networks with other blocks. MM-blocks and MG-blocks will not participate in the junction zones but form so-called elastic segments in the gel network. Viscous solutions of alginate exhibit shear-thinning properties caused by parallel arrangement of the linear molecule chains. Alginate solutions (1–2%) form irreversible gels in the presence of Ca^{2+} ions at room temperature. The gel strength depends on the amount of MM- or GG-block polymers, which cause different bonds. Owing to the high sensitivity to calcium and acids, buffers and sequestrants,

Table 1.2 Characteristics of alginates and PGA.

	Alginates and PGA
E-No.	E 400–E 405
Origin	Seaweed extract/alginic acid derivatives
Chemical composition	• Salt of a linear polyuronide with cations (Na, K, NH_4, Ca); • mannuronic acid (M), guluronic acid (G) (ratio M : G = 4–19 : 10)
Nutritional value (in 100 g) – metabolism	0 kJ (0 kcal); no resorption
Fibre content	100% soluble fibre
Toxicology	No health concerns for salts of alginic acid (therefore no ADI for E 400–404); considered as food in Asia, not as additive; but ADI for PGA (25 mg kg^{-1} BW per day)
Solubility at low temperature (H_2O)	100% Soluble (except for pure alginic acid and calcium alginate)
Appearance of an aqueous solution	Opaque, yellowish
Viscosity of solution in water	High (low for alginic acid and calcium alginate)
Impact of heat on viscosity in water (pH 7)	Heat-stable gels
Viscosity development in water at pH 7 (T = 0–100 °C)	• Ca^{2+}-free: high-viscosity solution; • gel formation in presence of calcium
Shear stability	Pseudoplastic (shear-thinning)
Thickening effect	High
pH stability	Medium (pH 5–9)
Decomposition	• Acid (pH < 5) and alkaline (pH > 9) conditions; • oxidation by free radicals; • enzymes
Film formation	High
Emulsion stabilization	High
Gelation	Cold formation of thermo-irreversible gels by acid and multivalent cations (except magnesium)
Gel strength and gel stability	High strength + high stability during cooking, baking, cooling, freezing
Gel transparency	Low
Tendency for gel synaeresis	Medium
Impact of electrolytes (cations +, 2+, 3+)	High
Reaction with Ca^{2+} ions	Gel formation
Protein activity	Low/no
Crystallization control	Freeze–thaw stable gels possible
Synergistic effects with other hydrocolloids	• Gum karaya: interactions with sodium alginate (modified solutions properties); • HM pectin: strong thermoreversible gels are formed at low solid levels over a wide pH range
Other synergistic effects	—
Negative interactions	—
Dosage level in foods	Medium (typical 0.7–2%)

which slowly release calcium and hydrogen ions, are necessary for the preparation of gels. Suitable substances are slightly soluble calcium salts, polyphosphates, and glucono-δ-lactone [2].

Properties

The most important property of alginates is the ability to form heat-stable gels by cold processing in presence of acids and all multivalent cations except magnesium. Calcium is the most often used gelling cation. Gel formation is controlled by controlling the release of calcium or acid into an alginate solution. Both acid and calcium alginate gels are thermo-irreversible and will form over a wide temperature range. A thermoreversible gel can be made under acidic conditions below pH 4.0, preferably around pH 3.4, by using a combination of alginate and HM pectin. In general, alginate gels exhibit synaeresis over storage time. Synergistic effects of alginates with other hydrocolloids are rare. There are interactions between gum karaya and sodium alginate (modified solutions properties). Alginate forms strong thermoreversible gels with HM pectin at low solid levels over a wide pH range [2].

Applications

In the food industry, alginates are used as emulsifiers, gelling agents, coating agents, and thickeners – and also for organic products. They are applied in diet and light products, baked goods, frozen foods, mayonnaise, salad dressings, dessert jellies, ice cream, mousses, foams, processed cheese, in meat and canned vegetables, and soups. The gums are suitable for encapsulation of prebiotics, flavours, and functional food oil. Alginates are also often used in so-called molecular gastronomy for the production of fish eggs and caviar imitations or in artificial mozzarella. In Asia, bird's nests or shark fins are made of it.

PGA has lipophilic and hydrophilic groups in the molecules and therefore combines the properties of a real emulsifier with those of a hydrocolloid. The ester is less sensitive to calcium ions than alginic acid and does not flocculate at low pH. In particular, PGA is used to thicken and stabilize acidic and Ca^{2+}-rich foods. The emulsifying effect is advantageous in producing stable foams, for example in beer (allowed limit $100\,mg\,l^{-1}$). In addition, it is approved for use in sauces for fish products ($20\,g\,kg^{-1}$ product) [1].

Toxicology and Regulatory Affairs

In some regions of the world, seaweed has been part of human nutrition for centuries. Alginates are neither resorbed nor metabolized in the intestinal tract. They can reduce the resorption rate of mineral cations like calcium and iron. Numerous toxicological studies have found no negative effect on human health. There is no specified ADI. Alginates are approved for use in foods in numerous countries. Specifications for alginic acid, sodium, potassium, calcium, and ammonium alginate salts and propylene glycol alginate (PGA) are listed in the USA Food Chemical Codex (FCC) and are also considered as GRAS in accordance with US Food and Drug Regulations (CFR 21). In Europe, alginic acid, its Na, K, NH_4, and Ca salts and PGA are listed as food additives with E number 400–405 [2].

In contrast to the pure alginic acid salts, PGA has an ADI of 25 mg kg^{-1} body weight (BW) per day, because the effect of propylene glycol, formed during metabolism, on the human organism has not been fully clarified [1].

1.2.3
Carrageenan

Raw Material Origin, Manufacturing, and Quality

For centuries, red seaweeds (family Rhodophyceae) have been harvested and used in foods in the Far East and Europe. They all contain naturally occurring polysaccharides that act as structural substances that fill the voids within the cellulose structure of the plant. Beside agar and furcelleran, carrageenans are a very important group of hydrocolloids extracted from red seaweed of *Chrondrus, Gigartina,* and *Euchema* species growing close to the coasts of the North Atlantic and Pacific Ocean. Different manufacturing processes are used to make carrageenan extract and semi-refined carrageenan, also known as processed *Euchema* seaweed (PES), PNG, SRC, or ARC. The primary process difference between 'extract' and 'PES' is that the extract process solubilizes the carrageenan and removes the solids, whereas the PES process leaves the carrageenan within the seaweed cellulosic structural matrix. To manufacture refined carrageenan, the harvested seaweed is dried and washed with water to remove sand and stones. Then it is extracted with alkaline water under pressure at 100–140 °C (212–284 °F). The specific alkali, for example sodium, potassium, or calcium hydroxide, is selected depending upon the carrageenan salt to be produced. Subsequently, the raw extracts are purified by filtration, treatment with activated carbon, and centrifugation, and finally concentrated. The concentrated solutions are then precipitated with isopropyl alcohol to give a fibrous mass that is squeezed to remove the alcohol and dried. An alternative recovery process is used for κ-carrageenan whereby the solution is extruded into a concentrated solution of potassium chloride and a fibrous mass is formed. The precipitated gel mass is dewatered under pressure and may be frozen and thawed to increase dehydration before further drying and grinding. PES manufacture includes, after appropriate washing, a soaking in potassium hydroxide solution (with excess potassium cations to prevent carrageenan solubility) before chopping and bleaching to enhance the colour of the finished powder. After washing, the drying, grinding, and blending steps are the same as those for the extract carrageenan process. The microbiological requirements for PES used in foods are identical to those for carrageenan. Only requirements for the so-called alkali-modified flour (AMF) are less strict as this product is targeted for pet food and non-food applications [1,2].

The global annual production of carrageenan is about 50 000 MT, of which about 70–80% is used in the food industry.

Chemical Structure

Carrageenan is the name for a family of salts of sulfated galactans with a sulfate content of 18–40% (Table 1.3). They are all high-molecular-weight linear hydrophilic polysaccharides consisting of repeating disaccharide units of galactose and 3,6-anhydro-galactose, both sulfated and non-sulfated, joined by alternating α-(1 $\rightarrow$ 3)- and β-(1 $\rightarrow$ 4)-glycosidic links. The main carrageenan fractions are the κ-, λ-, and ι-carrageenans (kappa, lambda, and iota, respectively), which differ in number and position of sulfate ester groups and in their content of 3,6-anhydro-galactose units. In the natural state in live seaweed, unmodified κ- and ι-carrageenan contain about 30% μ- and ν-carrageenan (mu- and nu-c, respectively), which are the precursor structures in carrageenan and which are converted into κ- and ι-carrageenan by alkaline modification (reducing these precursors to less than 5%). λ-Carrageenan occurs naturally in seaweed and can be transformed into ϑ-carrageenan (theta). κ-Carrageenan consists of a chain of alternating galactose and anhydro-galactose-units where the galactose residues carry a sulfate ester group at the 4-position. Its sulfate ester content is approximately 22% and its 3,6-anhydro-galactose content is about 33%. ι-Carrageenan has an additional sulfate ester group at the 2-position on the anhydro-galactose residue and therefore has a higher sulfate ester content of 32%, along with 26% anhydro-galactose. λ-Carrageenan is an alternating chain of galactose units linked in β-(1 $\rightarrow$ 4) and α-(1 $\rightarrow$ 3) mode. The galactose at the 4-position is mainly present as 2,6-disulfate, and the galactose at the 3-position mainly occurs as 2-sulfate. It contains approximately 37% sulfate ester with little or no 2,3-anhydro-galactose. PES differs from traditional clarified carrageenan extracts in that it contains 8–15% acid insoluble matter (AIM) compared to 2% maximum for an extract. AIM makes up the structural network of plant cellulosic and proteinaceous material, maintaining its integrity during the PES process. Consequently, in applications using PES, enough energy must be applied during processing to break down the AIM structure and release the carrageenan. Hydration and solubility profiles are different for carrageenan and PES. For applications requiring clear solutions or gels, only carrageenan extract can be used. The average molecular weight of carrageenan and PES is between 200 and 800 kDa, but it can be as high as 1.5 million daltons (MDa). κ-Carrageenan is the most used of these seaweed extracts. λ-Carrageenan is the least utilized; its production is very expensive because special plants need to be selected. Most commercial products labelled 'λ-carrageenan' are blends of non-gelling unmodified κ-types and λ-types [1,2].

Solubility, Viscosity, and Gelation

The solubility of carrageenans depends on their structure, the applied temperature, and the presence of cations. All carrageenans are soluble in hot water, but – with the exception of λ-grades, only the sodium salts of κ- and ι-carrageenan are soluble in cold water. Potassium and calcium salts of κ- and ι-types are completely soluble at 50–60 °C (122–140 °F). All carrageenans are soluble in hot milk. In cold milk, only λ-carrageenan has solubility, producing a thickening effect via protein interactions, an effect that is enhanced by phosphates.

Table 1.3 Characteristics of κ-, ι-, and λ-carrageenan.

	κ-, ι-, and λ-carrageenan
E-No.	E 407
Origin	Seaweed extract
Chemical composition	• Family of salts of sulphated galactans (high sulfate content 18–40%); • D-galactose, D-(3 → 6)-anhydro-galactose, sulfate ester groups
Nutritional value (in 100 g) – metabolism	0 kJ (0 kcal); no resorption
Fibre content	100% Soluble fibre
Toxicology	Global approval as food additives; viscosity: minimum 5 mPa s as 1.5% solution at 75 °C (167 °F) (average MW of 100 kDa); Europe: ADI (75 mg kg^{-1} BW per day) and MW limit of maximum 5% below 50 kDa
Solubility at low temperature (H_2O)	κ-C: only Na^+ salts are soluble, limited swelling of K^+ and Ca^{2+} salts; ι-C: only Na^+ salts are soluble, Ca^{2+} salt gives thixotropic swollen particles; λ-C: all salts are soluble in cold water
Appearance of an aqueous solution	Clear for pure extracts, opaque-turbid for PES
Viscosity of solution in water	Medium to high (hot solutions set to gels when cooling)
Impact of heat on viscosity in water (pH 7)	All κ-, ι-, and λ-C types are completely soluble in water and milk at $T = 80\,°C$ (176 °F)
Viscosity development in water at pH 7 ($T = 0–100\,°C$)	Complete solubility in water (or milk) at $T = 80\,°C$ (176 °F), viscosity is reduced by further heating (fully reversible at pH 7–9), solutions of κ- and ι-types set to gels upon cooling, gels are stable at room temperature, gels melt by heating and re-set upon cooling without loss of gel strength or texture in neutral conditions
Shear stability	Solutions are pseudoplastic (reversible shear-thinning); κ-C-gels break when sheared (irreversible); ι-C-gels break by shear, but recover and re-gel after shear stress is removed
Thickening effect	High
pH stability	Medium (pH 5.5–9)
Decomposition	κ-C + ι-C: hydrolysis by acid (accelerated by heat, low pH value, and time); however, gels are acid-stable; λ-C: hydrolysis in acidic systems
Film formation	High
Emulsion stabilization	High
Gelation	κ-C: strongest gels with K^+; ι-C: strongest gels with Ca^{2+}; λ-C: no gelation with cations (but gelling with very high salt concentrations)
Gel strength and gel stability	κ-C: firm brittle gels with strong synaeresis, not freeze–thaw stable, hysteresis 10–20 °C (18–36 °F);

Table 1.3 (Continued)

	κ-, ι-, and λ-carrageenan
	ι-C: soft elastic texture, no synaeresis, freeze–thaw stable; λ-C: no gels, solutions are freeze–thaw stable
Gel transparency	High
Tendency for gel synaeresis	κ-C-gels: strong synaeresis; ι-C-gels: no synaeresis
Impact of electrolytes (cations +, 2+, 3+)	κ-C: strong gelation with monovalent potassium ions; ι-C: strong gelation with divalent calcium ions; λ-C: no impact
Reaction with Ca^{2+} ions	Gelation with ι-carrageenan
Protein activity	κ-C. forms weak gels with κ-casein in milk to stabilize neutral dairy products and particles in suspensions; ι-C. and λ-C. have a strong protein interaction in acid
Crystallization control	Gels of ι-C. and solutions of λ-C. are freeze–thaw stable, gels of κ-C. exhibit synaeresis
Synergistic effects with other hydrocolloids	κ-Carrageenan forms synergistic gels (increased gel strength and elasticity, reduced synaeresis) with galactomannans (e.g. LBG) and konjac glucomannan; synergism between starch and ι-carrageenan
Other synergistic effects	Solubility in 50% sugar solutions: κ-C. is hot-soluble, ι-C. is insoluble, λ-C. is soluble; solubility in 10% salt solutions: κ-C. is insoluble, ι-C. and λ-C. are hot-soluble
Negative interactions	—
Dosage level in foods	Low to medium (typical 0.02–3%)

In principle, when a carrageenan dispersion is heated, there is no significant particle swelling or hydration until the temperature exceeds about 40–60 °C (104–142 °F). As the particles hydrate, the viscosity increases as the swollen particles offer more resistance to flow. Further heating to 75–80 °C (167–176 °F) produces a drop in viscosity. On cooling the solution shows a marked increase in viscosity followed by gelation below 40–50 °C (104–122 °F).

Carrageenan solutions are highly viscous and viscosity is reduced with increasing temperature. The viscosity drop is reversible when heat is applied in neutral systems (pH 7–9). At room temperature viscosity is stable over a wide pH range. Acid and pH values below 4.5–5.5 lead to hydrolysis of the carrageenan solutions with heat accelerating the degradation, but the gels of κ- and ι-types are acid-stable. On this account, carrageenan should be added to acidic products at the last moment or acid should be added to the food immediately before filling in order to minimize polymer breakdown.

Pure solutions show pseudoplasticity or shear-thinning when pumped or stirred.

All carrageenans hydrate at high temperatures, with κ- and ι-types in particular exhibiting a low fluid viscosity in both water and milk. On cooling to below the gel point, which is between 30 and 70 °C (86–158 °F), these κ- and ι-carrageenans set to form a range of gel textures depending on the cations and other ingredients present. In the presence of potassium (K^+), κ-carrageenan forms firm, brittle gels that exhibit synaeresis and are not freeze–thaw stable. ι-Carrageenan gives soft, elastic syneresis-free gels with calcium (Ca^{2+}). κ-Carrageenan gels break when sheared and this effect is permanent. ι-Type gels break by shear but recover and re-gel after shear stress is removed, indicating thixotropic behaviour, but with a longer time to recover fully than, for example, xanthan gum. Both gels are stable at room temperature but can be remelted by heating to 5–30 °C (9–54 °F) above the gelling temperature. On cooling, the system will re-gel without loss of gel strength or change in texture under neutral conditions. Sugar and salt affect the solubility of carrageenans. In 50% sugar solution, κ-types are hot-soluble, ι-types are insoluble, and λ-types are soluble. In 10% salt solutions, κ-carrageenan is insoluble and ι- and λ-types are hot-soluble. Salts and sugar have a strong effect on the hydration temperature of carrageenan and on its subsequent setting and remelting temperatures. For example, 2% salt (NaCl) can increase the hydration temperature of the sodium salt of κ-carrageenan from 40 °C (104 °F) to 55 °C (131 °F) or higher in meat brines. At concentrations of 4% NaCl or above, hydration is fully prevented. The presence of high solids, in confectionary for example, effectively concentrates the carrageenan and cations on the aqueous phase so that gelation may occur at 80–85 °C (176–185 °F) or higher. To ensure full hydration and prevent lumping, carrageenan particles are premixed with sugar (5–10 times amount) or pre-dispersed in oil, sugar syrup, alcohol, or a salt solution [2].

Properties

The ability of κ- and ι-carrageenans to form gels in the presence of K^+ and Ca^{2+} ions forms the basis of many applications in food. Gelation is caused by the relatively high content of hydrophobic anhydro-galactose residues at low to medium content of sulfate ester groups. The gelation mechanism consists of two steps. Gel-I is an intermediate step including formation of zones of single helices by ionic interaction and spiral-like association of different molecules. In the final gel-II state, the single helices arrange with adjacent ones to form double or triple helical structures. Calcium ions induce divalent bridging between adjacent chains in ι-types, while potassium ions counter sulfate charges without sterically hindering close approach and double helix formation. Hydrogen bonding occurs between different helices.

κ-Carrageenan forms synergistic gels with galactomannans such as locust bean and tara gum, as well as glucomannans such as konjac flour. These gels have an increased gel strength and elasticity, while synaeresis is reduced. The optimum blend ratio is 60 : 40 to 70 : 30 of κ-carrageenan to mannan gum.

Of note is the interaction of κ-carrageenan with κ-caseins of milk at around neutral pH. This protein and milk reactivity is due to electrostatic attractive forces between the negatively charged hydrocolloid with the positively charged regions of κ-caseins. Already at very low concentrations of 0.01%, κ-carrageenan forms a weak gel of fluid character. Additionally, κ-carrageenan builds structure by interacting directly with positively charged amino acids and indirectly, via divalent cations, with negatively charged amino acids in the proteins at the surface of the casein micelles. Very low levels (0.015–0.025%) are sufficient to prevent whey separation during processing and storage of ice cream, milk shakes, cream cheese, and dairy desserts. This 'liquid gel network' can also stabilize particles such as cocoa in chocolate milk and other suspensions at only 0.02% without increasing viscosity significantly and without phase separation or sedimentation [2].

Applications

Carrageenans are used as gelling and thickening agents for water-based jellies and for dairy products such as desserts, puddings, ice cream, mousses, whipping creams, cheese spreads, cream cheese, infant formula, and neutral milk drinks (e.g. calcium-enriched milk, nutritional beverages, chocolate milk, and milk shakes). They are applied in soups, sauces, and salad dressings, especially in transparent low-viscosity products with herbs/spices. Carrageenan is used successfully in meat products such as cooked ham where it stabilizes the injected water. The clarification of beer and fining of wine are other applications.

Toxicology and Regulatory Affairs

Carrageenan and semi-refined carrageenan (PES) have a long history of use and are permitted globally for use in food. Carrageenan is not assimilated by the human body, providing only fibre with no nutritional value.

In the European Union, carrageenans are approved as E 407 (and E 407a for PES) as food additives with unrestricted level (quantum satis). In the USA, the US Food and Drug Administration makes no distinction between carrageenan and PES and both are regulated as carrageenan.

In the past, carrageenan was rated as harmless by the independent Expert Committee of FAO/WHO and by the US FDA. But, more recently, in animals, ulceration and changes in the immune system have been observed with degraded carrageenan. When carrageenan is heated at high temperatures and at low pH, degraded carrageenan (so-called polygeenan with MW of 10–20 kDa) is produced. The longer the heating, the more the molecule length of carrageenan is reduced. Therefore, an ADI of 75 mg carrageenan per kg of body weight per day has been set. Additionally, JECFA (Joint FAO/WHO Expert Committee on Food Additives) and SCF (Scientific Committee of Food, in Europe) have limited the viscosity of a 1.5% solution to a minimum of 5 mPa s at 75 °C (167 °F), which corresponds to an average molecular weight of 100 kDa. A limit for the MW is set with a maximum of 5% below 50 kDa. Commercial carrageenan and PES normally have MW in the range 200–800 kDa. However, it is also found that the presence of associated cations prevents carrageenan hydrolysis during digestion and that normal food processes do not significantly increase the proportion of low-molecular-weight material [2].

There are claims that carrageenan hinders the absorption of minerals such as potassium and can cause allergies.

1.2.4 Pectins

Raw Material Origin, Quality, and Manufacturing

Pectin was discovered as a gelling agent in 1820. It is widely distributed in the tissue of land plants, where it acts in combination with cellulose as an intercellular matrix substance (cell walls, middle lamellae). For commercial use, pectin is primarily extracted from apple pomace (15% pectin) and citrus peel of lime, lemon, or orange, which contain about 30%. Sugar beet pulp with 20% pectin is used to a lesser extent. All raw materials are by-products from other industries of fruit juice or sugar production. Most pectin is manufactured be extraction with hot aqueous acid solution at 50–90 °C (122–194 °F) for 3–12 h at pH 1–3. Insoluble protopectin is degraded to soluble pectin by this method. The raw extract is filtrated. Potentially available starch is enzymatically degraded and thus removed. The purified extract is concentrated and drum- or spray-dried. Another option is to precipitate the pectin by alcohol. The precipitation method results in highly pure products and is preferably used today. By modifications during extraction process and controlled partial de-esterification, pectins can be produced with different degrees of esterification and polymerization. Amidated pectin is typically obtained by amidating conventional pectin with ammonia in an alcoholic suspension. More recent technology includes methyl de-esterification with biocatalysts.

The annual production of more than 45 000 MT is mainly for food use [1,2].

Chemical Structure

The substance class of pectins occurs in various structures (Table 1.4). Plant species, environmental conditions, and plant maturity have an impact. Common to all pectins is linear polysaccharides whose main component is the α-D-galacturonic acid (pK_a 3.5) as a monomer. According to a European directive, commercial pectin is classified as a substance that contains at least 65% galacturonic acid. Pectin is a polyuronide consisting of a linear backbone of α-(1 → 4)-linked galacturonic acid units. The carboxy groups (COOH) are partially esterified with methanol (to give the $-COOCH_3$ group). The secondary hydroxyl groups may be partially esterified with acetate depending on the origin of the pectin. Covalently bound side chains of neutral sugars (xylose, β-(1 → 4)-D-galactans, α-(1 → 5)-L-arabenes) may be included in the macromolecule. Rhamnose units, linked with galacturonic acid units at the 1- and 2-positions, are integrated into the backbone. The basically quite stretched pectin molecule thereby has a kink.

When extracting pectin for commercial use, most of the neutral sugar side chains are removed. Thus, commercial pectin is often referred to as the homogalacturonic backbone. By weight, typically more than 70% of these pectins is galacturonic acid, and, depending on pectin quality and origin, up to 75% of the galacturonan groups are methyl esterified. According to the

Table 1.4 Characteristics of pectins.

	Pectins
E-No.	E 440
Origin	Fruit extract
Chemical composition	• Linear polyuronide, partially esterified with methanol and acetate; • backbone of galacturonic acid
Nutritional value (in 100 g) – metabolism	0 kJ (0 kcal); no resorption
Fibre content	100% soluble fibre
Toxicology	No health concerns, ADI value 'not specified'; considered as natural component of foods
Solubility at low temperature (H_2O)	HMP: cold water soluble; LMP: only soluble as Na or K salt
Appearance of an aqueous solution	Opaque, yellowish
Viscosity of solution in water	Depends on concentration (low at 0.5%, high at 3–4%)
Impact of heat on viscosity in water (pH 7)	Irreversible reduction of viscosity and gelation power due to molecular degradation
Viscosity development in water at pH 7 ($T = 0–100\,°C$)	• Ca^{2+}-free: high-viscosity solution; • gel formation in presence of calcium
Shear stability	High, but reversible shear-thinning
Thickening effect	Low – thickening only after gelation
pH stability	Low (HMP: 2.5–4.5, LMP: 2.5–5.5)
Decomposition	• Acid or alkali (only stable at pH 3.5–5); • high temperatures; • many enzymes
Film formation	High
Emulsion stabilization	High for pectins with high acetyl content
Gelation	HMP: with sugar + H^+ LMP (amidated or non-amidated): with Ca^{2+} ions
Gel strength and gel stability	HMP: – no gelation at $pH > 3.5$; firm thermo-irreversible gels at $pH < 3.5$ LMP: • non-amidated: spreadable, thermoreversible; • Amidated: semi-firm to spreadable, reversible
Gel transparency	High
Tendency for gel synaeresis	Low
Impact of electrolytes (cations +, 2+, 3+)	+ Metal ions (2+/3+): precipitation or gelation
Reaction with Ca^{2+} ions	HMP: no reaction; LMP: gel formation
Protein activity	High for HMP (solubilizing effect around IEP + heat protection effect at acidic pH)
Crystallization control	Freeze–thaw stable gels are possible
Synergistic effects with other hydrocolloids	Alginate forms strong thermoreversible gels with HM pectin ($pH < 4$) at low solid levels and with LM pectin at $pH < 2.8$
Other synergistic effects	Crosslinking of pectin molecules with positively charged molecules (chitosan, poly-L-lysine)
Negative interactions	Precipitation with gelatine and CMC (only HMP)
Dosage level in foods	Medium (0.5–5%)

number of carboxyl groups being esterified with methanol, the class of pectins is divided into:

- high-esterified pectins (HM pectin) with a methoxyl content >7% and degree of esterification >50% (typically 55–75%);
- low-esterified pectins (LM pectins) with a methoxyl content <7% and degree of esterification <50% (typically 20–45%).

Additionally, amidated LM pectin can be prepared by treating pectin with ammonia during processing to convert some of the C6 methyl ester groups into amide groups. Values for amidated LM pectin are 30% degree of methyl esterification and 20% degree of amidation. Another way of differentiating pectins is by the degree of acetylation.

The molecular weight of commercial products is between 30 and 300 kDa [1,2].

Solubility, Viscosity, and Gelation

In general, pectin is soluble in water or forms a colloidal solution. It is insoluble in alcohol and most organic solvents. Powdered pectin tends to form lumps when added directly to water. As concentrated pectin exhibits non-Newtonian behaviour and the viscosity decreases with increasing rates of shear, a high-speed mixer is useful. By applying high shear forces, solutions of up to 10% can be prepared; without such forces it is difficult to obtain smooth pectin solutions above 3%, owing to the high viscosity impeding dispersion. Preparation of dry blends with sugar (ratio of sugar to pectin = 5 : 1) or pre-wetting in liquids in which pectin is insoluble are alternative hydration methods. Pectins form both soluble and insoluble salts. Sodium pectinates are more soluble than pectic acids, which, in turn, are more soluble than calcium pectinates. LM pectins have a lower solubility – they are only soluble as sodium or potassium salt. Sodium, potassium, and other monovalent ions are bound electrostatically to the pectin backbone, whereas divalent cations are involved in gelation of pectin, visualized as the so-called egg-box model. As a general rule, high levels of soluble solids and salts decrease pectin solubility. The addition of sequestering agents such as pyro- or ortho-phosphates improves the hydration of pectin under such adverse conditions. Pectin creates viscous solutions. The viscosity depends on concentration and increases greatly above the concentration at which the molecules entangle. A weak solution (<0.5%) exhibits almost Newtonian behaviour, whereas concentrated solutions are shear-thinning. The viscosity also depends on pectin type, solvent, pH value, temperature, and presence of salts. Polyvalent ions like calcium generally increase viscosity because of crosslinking of the polymer chains.

Pectin has optimal stability at pH 3.5–4.0, the natural pH of fruit preserves, and slowly degrades outside this range. In neutral and alkaline conditions, the homogalacturonic backbone depolymerizes by β-elimination, a process in which the backbone glycosidic bonds at the C4 position of methylated galacturonic acid units are cleaved. As a consequence, HM pectin is more vulnerable to β-elimination than LM grades. At elevated temperature, depolymerization at the pectin backbone starts at pH 5, although pectin is more robust at room temperature. In acidic conditions (pH <3), methyl ester and acetyl groups are cleaved and the neutral sugars are hydrolysed even at low temperatures. At elevated temperatures, the degradation reactions accelerate. Evidently, high pH values and high temperatures are unfavourable parameters for pectin stability [2].

Properties

Pectin is a weak acid with polyelectrolyte behaviour. At neutral pH, pectin is negatively charged and occurs as polyanion. Because of the negative charge of pectins, they react with positively charged polymers, including proteins. This feature is utilized in acidified milk drinks.

The most important property of pectin is its ability to gel. To induce gelation, high-esterified pectins require low water activity due to high sugar levels (55–85% soluble solids) and an acidic pH of 2.5–3.8. Gelation results from diminished electrostatic repulsion and is caused by hydrogen bonding between non-dissociated carboxyl groups and secondary alcohol groups together with hydrophobic interactions between methyl ester groups. The degree of methyl esterification is fundamental for gelling HM pectin. Increased esterification degrees and low pH enhance the ability to gel. For this reason further groupings have been established. The sub-groups include ultra-rapid set, rapid set, medium rapid set, slow set, and extra-slow set pectin with the degree of methyl esterification ranging from 74% to 77% for ultra-rapid set types to 58–60% for extra slow set pectins. For a given pH, setting time increases and setting temperature decreases with decreasing degree of methyl esterification. Gels of HM pectin are formed at 25–90 °C (77–194 °F) and are thermo-irreversible. Gels are of a firm texture at pH below 3.5. Above pH 3.5, no is gel formed, but viscosity is provided.

Low-esterified pectins gel over a broader pH range of 2.5–6 and at lower concentrations of sugar (5–65%) but require the presence of cations, which in terms of food is generally calcium. Magnesium ions do not initiate gelation. Since the calcium reacts with de-esterified galacturonic acid units, a low degree of methyl esterification enhances the ability to gel and the more calcium sensitive the pectin becomes. Low pH favours gelation of LM pectin, but gelling is possible up to pH 6. Gels of non-amidated LM pectin are formed at a soluble solids range of 25–55% and 40–100 °C (104–212 °F). These gels are, depending on the calcium level in the system, thermoreversible. The gels are soft and spreadable above pH 3.5. Firmness increases when pH is lowered (<3.5).

Amidated LM pectin gels at lower pectin concentration and lower soluble solids content (5–65%). Gelation requires less calcium than for non-amidated LM pectin. Amidated products are less sensitive to high calcium levels (no precipitation). Gels are formed at 30–70 °C (86–158 °F) and are always thermoreversible. Gels above pH 3.5 are spreadable and similar to gelled non-amidated LM pectin. Below pH 3.5, gels are semi-firm and similar to HM pectin but more rubbery.

Pectic acid has a degree of esterification below 5% and gels like LM pectin. It precipitates at high pH values and high levels of polyvalent cations as salts (pectates).

The presence of acetyl groups may prevent pectin gelling, but makes pectin a useful emulsifier. The acetyl groups enhance the hydrophobic nature of the molecule and provide a surface-active character. Thus, pectin is able to act as interfacial agent in the stabilization of oil–water or air–water systems.

Generally, pectin does not show any synergy with other hydrocolloids, but its gel strength might be influenced by addition of other substances. It is worth mentioning that HM pectin forms strong thermoreversible gels with high-polyguluronic acid alginate at low solid levels and pH <4. LM pectin gels with these alginates at pH <2.8. More interactions can be found between pectin and positively charged molecules such as chitosan and poly-L-lysine, which are able to crosslink pectin molecules. Such interactions may find use in films [2].

Applications

Pectin is used as gelling, thickening, and stabilizing agent in foods and, to a lesser extent, in pharmaceuticals. Basically it controls moisture content and helps to create the desired texture. HM pectin gives structure, bite, and bake-stability to acidic jams, jellies, and confectionary with high sugar content. It is used as non-gelling stabilizer in acidified dairy or soya drinks and in fruit beverages. In bread and frozen dough, HM pectin improves volume, retains moisture, softness, and freeze–thaw stability.

LM pectin thickens fruit-based ripples and toppings with pumpable consistency. Other applications are fruit preparations for yogurt and cold-setting milk desserts. In low-fat products, LM pectin acts as fat substitute in sauces, dressing, ice cream, processed meat and cheese products, and spreads. Amidated LM pectin types find use as gelling and foaming agents in non-fruit confectionary (marshmallows, toffee, and peppermint). Amidated pectin gels are more robust than gels of non-amidated pectin types. Furthermore, pectin may be used to purify and concentrate protein solutions by exploiting the electrostatic repulsion effect at pH values above the isoelectric point [2].

Commercial products often contain sugar to adjust gel strength and buffering salts to adjust setting time.

Toxicology and Regulatory Affairs

Pectin is a soluble dietary fibre and a natural part of the human diet, but does not contribute significantly to nutrition. The daily intake of pectin from fruits and vegetables can be estimated to be around 5 g assuming a consumption of approximately 500 g fruits and vegetables per day. In human digestion, pectin binds to cholesterol in the gastrointestinal tract, slows glucose absorption by trapping carbohydrates and thus attenuates blood glucose response. Consumption of pectin has been shown to reduce blood cholesterol levels. The mechanism appears to be an increase of viscosity in the intestinal tract, leading to a reduced absorption of cholesterol from bile or food. In the large intestine and colon, microorganisms degrade pectin and liberate short-chain fatty acids that have a positive influence on health (prebiotic effect). Improved mineral absorption, reduction of gastric emptying time, and anti-diarrhoea effects are additional favourable physiological effects reported.

Pectin is regarded as a safe food additive and has been given an acceptable daily intake (ADI) of 'not specified' by the committees of FAO/WHO and European Union. Furthermore, the US FDA accords pectin 'generally recognized as safe' (GRAS) status. Notably, the Codex Alimentarius and US specifications include both amidated and non-amidated pectins in one class. EU legislation distinguishes between non-amidated and amidated pectins. Nevertheless, both types are labelled as E 440 and permitted uses are identical, except for organic foods, where only non-amidated pectin is permitted [2].

1.2.5 Native and Modified Starches

Raw Material Origin and Manufacturing

Starch is widely used in plants as a reserve carbohydrate in various organs. The starch granules formed in the amyloplasts of plants consist of concentric or eccentric layers of different density and have different sizes (2–150 μm), size distributions, and shape. The quantitatively most important raw materials are corn, waxy maize, potato, cassava, wheat, rice, waxy rice, and tapioca. In addition, starches are extracted from peas, lentils, sweet potatoes, sago palms, lotus roots, mung beans, water chestnuts, sorghum, arrowroot, kouzu, taro, and edible canna.

The manufacture of a starch depends on the location of the reserve material in the specific plant. In some cases, for example potato, the starch granules occur free in the cells, so that their isolation is a relatively simple process. The plant material is crushed, the starch is slurried with water, separated from the suspension (starch milk), and dried.

In other cases, for example cereals and corn, the starch is stored in the endosperm incorporated into a protein matrix and, thus, extraction is more difficult.

Maize, for example, is soaked for about 48 h in warm water (50 °C (122 °F)). To break up the protein matrix, the soaking water contains SO_2 (about 0.2%). This speeds up the process and increases the starch yield. Subsequently, the corn kernels are crushed and the germs that are rich in fat are separated by flotation. Gluten (protein) and starch are separated in hydrocyclones due to their density difference. The starch is washed and dried.

Modified starch is the most used food hydrocolloid, with a volume of more than 1.5 million tons. The global volume of native and modified starches is estimated to be about 80 million tons (78% corn, 11% tapioca, 7% wheat, and 4% potato).

Chemical Structure of Native Starches

Starches are long-chain polymers of α-D-glucopyranose units with the general molecular formula $(C_6H_{10}O_5)_n$ (Table 1.5). Starch is a mixture of two glucans – amylose and amylopectin – the content of which depends on the plant source. While the average amylose content is 20–30%, there are so-called amylo-varieties with 50–80% amylose. Normal starches contain 70–80% amylopectin, but the so-called waxy varieties have 99% (only 1% amylose) due to breeding measures. Amylose can be extracted from starch by, for example, crystallization

Table 1.5 Characteristics of native and modified starches.

	Native starches	Modified starches
E-No.	—	E 1400 Following
Origin	Seed extracts (germs, roots)	
Chemical composition	α-D-Glucose	
Nutritional value (in 100 g) – metabolism	1507 kJ (360 kcal); fast resorption	1750 kJ (418 kcal); fast resorption
Fibre content	1% (~86% glucose)	0% (~100% glucose)
Toxicology	No health concerns; no ADI specified; GMO – freedom of raw material is an issue	
Solubility at low temperature (H_2O)	Cook-up starches are insoluble; development of cold-water soluble instant starches	
Appearance of an aqueous solution	Turbid suspension that sediments at room temperature, slurry needs cooking to dissolve	
Viscosity of solution in water	Low at room temperature (only pre-gelatinized native and modified starches are cold water soluble)	
Impact of heat on viscosity in water (pH 7)	Swelling + gelling (at $T > 60\,°C$ (140 °F)); formation of heat-stable gels	
Viscosity development in water at pH 7 ($T = 0–100\,°C$)	Swelling + gelling (at $T > 60\,°C$ (140 °F)); formation of heat-stable gels	
Shear stability	Low after gelation, irreversible viscosity loss	Shear-stable starches are available
Thickening effect	High	
pH stability	Low after gelation	
Decomposition	• Cooking in acidic or alkaline systems ($6 > pH > 8$); • oxidation by free radicals; • high shear forces; • enzymes	
Film formation	High	
Emulsion stabilization	No – only by thickening	Emulsifying starches are available
Gelation	Upon heating	
Gel strength and gel stability	High gel strength + high stability during cooking, baking, cooling	
Gel transparency	Low	
Tendency for gel synaeresis	High	Medium
Impact of electrolytes (cations +, 2+, 3+)	Low	
Reaction with Ca^{2+} ions	No	
Protein activity	Low/no	
Crystallization control	No – syneresis	Medium
Synergistic effects with other hydrocolloids	With gelatine: improved gelation, with guar: viscosity increase, with MC: reduced boiling, improved freeze–thaw-stability	
Other synergistic effects	Sugar increases the temperature of gelation	
Negative interactions	Formation of acrylamide: heating (>120 °C (248 °F)) in presence of the amino acid asparagine (promoted by fructose and glucose)	
Dosage level in foods	High (typical 1–5%)	

from a starch dispersion, usually in the presence of salts ($MgSO_4$) or polar organic compounds (butanol or low molecular weight fatty acids), which form complexes with amylose and can be fractionated precipitated. About 70% of the mass of a starch grain is considered to be amorphous (disordered regions) and 30% to be crystalline (high degree of orientation). Amylose is water soluble and made of helical, but unbranched chains of 1000–10 000 glucose residues with $\alpha(1 \rightarrow 4)$ linkage. The molecular size depends on the source of raw material. The length is about 0.5 μm, the molecular weight about 0.15–1.5 million daltons (MDa). The water-insoluble amylopectin consists of branched chains of glucose residues in $(1 \rightarrow 4)$- and $(1 \rightarrow 6)$-α-glycosidic bonds. On average, there is a branch point every 15–30 glucose residues. A phosphate residue occurs on about every 400th glucose residue. The structure is supposed to be a double helical with a parallel arrangement (globules about 0.05 μm in diameter). The crystallinity of starch is caused mainly by amylopectin. The molecular weight is very high and in the range 10–700 MDa.

Native corn starch has 25–28% amylose with a DP (degree of polymerization) of 2000 and 72–75% amylopectin with a DP of 2 million. Native tapioca starch consists of 17% amylose (DP of 5000) and 83% amylopectin (DP 3 million) [2].

Modification

The properties of native starch and various starch fractions (amylose, amylopectin) can be improved by physical or chemical modifications and adjusted to defined uses. Some modifications include several conversion processes (e.g. acetylated oxidized starch). A modification can be made on the native starch granule or after gelatinization. For starches, the functional properties change already when only a few new groups are incorporated into the molecule. The aim of modification is an increased tolerance and stability of the starch against the effects of temperature (heat/cold), shear forces, and changes in pH (acid addition). For regulatory reasons, the declaration is significantly different between chemically and physically modified products. Physically modified starches are handled like native starches because they are 'only' treated with physical methods (heat, pressure, water). Instant or pregelatinized native starches are an example. In the chemical modification of cook-up and instant starches, there is a distinction between 'stabilization' and 'crosslinking'. Stabilization aims to improve the storage stability. The shelf life is extended by introduction of foreign groups such as acetyl, phosphate, or hydroxypropyl with placeholder function. Products exhibit a delayed retrogradation, increased freeze–thaw stability, and enhanced mouthfeel. Examples are starch acetates and hydroxypropylated starches. The crosslinking process with adipates and phosphates increases the process tolerance. The stability to heat, shear forces, and acids is significantly improved. Crosslinked starches are characterized by a short texture. The denser the network, the more process-tolerant the resulting starch will be and the more difficult it will be to gel this specific starch. Stabilization and crosslinking can be combined to produce highly storage- and process-stable starch products (e.g. acetylated distarch adipate and hydroxypropylated distarch phosphate).

Starch-Based Products

There are several starch derivatives available. The most important are explained below.

By acidic hydrolysis of starches, dextrins and maltodextrins are produced that are used as adhesives and fat substitutes.

Starches are partially damaged by mechanical forces and pressure, for example by an extrusion process, used to improve cold dispersibility and ease of handling in bakery applications.

Thin-cooking starches result after partial hydrolysis with acid and are applied in protective films.

Starch ethers like hydroxyethyl or hydroxypropyl derivatives are formed by treatment of 30–40% starch suspensions with ethylene or propylene oxide in a strongly alkaline system (pH 11–13) with alkali metal hydroxides. The substitution varies between 0.1 and 1 mol of alkyl groups per mole of glucose. By introduction of hydroxyalkyl groups, often in combination with light crosslinking, the swelling capacity, solubility, and freeze–thaw stability are significantly improved, while the gelatinization temperature is reduced. Frozen foods and heat-sterilized canned foods are preferred applications. By reaction of starch with monochloroacetic acid in an alkaline medium, carboxymethyl starches are manufactured that are very soluble in cold water or ethanol and form gels.

Starch monophosphates are produced by dry-heating of starch with alkali orthophosphates or alkali tripolyphosphates to 120–175 °C (248–347 °F). Esters of starch are prepared by reaction with organic acids (e.g. acetic acid, higher fatty acids, (C_6–C_{26}), succinic acid, adipic acid, and citric acid), suitable activated derivatives of the acids (such as vinyl acetate), or by heating starch with the acids or their salts. The thickening ability of the esters is better than that of native starch. They have high freeze–thaw stability. Starch esters are applied for thickening and stabilization of baked goods, dried soups and sauces, desserts, preserves, margarine, and canned food. They form protective coatings on dried fruits and can encapsulate flavours.

Crosslinked starch results from the reaction of starch with bifunctional or polyfunctional reagents. Suitable substances are special phosphates or mixed anhydrides of acetic acid and dicarboxylic acids like adipic acid. Depending on the degree of crosslinking, the gelatinization temperature is increased, while the swelling capability is reduced. The stability to shear forces and extremes of pH is improved. Crosslinked starches are used in applications that require large process stability.

Oxidized starch is produced by reaction of starch with an alkaline hypochlorite solution at temperatures below the gelation point. Hydrolysis and oxidation lead to products having carboxyl groups (on average 1 to 25–50 glucose residues). In contrast to a thin-boiling product, an oxidized starch exhibits neither retrogradation nor gel formation. It is used as a low-viscosity filling material in salad dressings and mayonnaise.

Emulsifying starches or OSA (= starch sodium octenyl succinate) starches result from reaction of starch with sodium octenylsuccinate. They are readily soluble in cold liquids and can stabilize multiphase mixtures such emulsions and foams.

Cyclodextrins (CDs) are a class of compounds belonging to the cyclic oligosaccharides. They are degradation products of starch in the shape of a ring and consist of α-(1 $\rightarrow$ 4)-glycosidically linked glucose molecules (6–8 monomers). β-Cyclodextrins with seven rings are used most often. The hydrophilic groups are directed to the outside, the hydrophobic groups to the inside. The result is a toroidal structure with a central cavity. Active ingredients can be microencapsulated by inclusion and be protected or 'neutralized' when having an unpleasant smell or taste.

Solubility, Viscosity, and Gelation

Native untreated starch, so-called cooking starch, is insoluble in cold water and will only form opaque suspensions with a strong tendency to sediment. When this suspension is heated, it passes through the gelling point of the starch, which is the temperature at which it starts to swell. As the starch granules swell they absorb water until they reach their maximum size. At this point the maximum thickening viscosity is achieved. As heating continues, the starch granules cannot swell any further and so they start to break down. If this process is continued long enough all the granules break down and the amylose and amylopectin polymers are released into the solution. When this solution cools, it tends to set to give a gel-like structure [2].

Only when boiling in water does starch forms colloidal solutions, which form a gel during cooling. This process is called gelatinization. By pre-gelatinization and subsequent drying (on drums or spraying), so-called instant starches are produced that swell and hydrate even at cold temperatures.

Starches form high-viscosity solutions.

Properties

Upon heating, starches can physically bind many times their own weight of water, as they swell and gelatinize. When the aqueous suspensions are heated, the starch swells at 47–57 °C (116–134 °F), the layers burst, and at 55–87 °C (131–188 °F) a starch paste is produced. Potato starch gelatinizes at 62.5 °C (144.5 °F), wheat starch at 67.5 °C (153.5 °F). Depending on the type of starch, the pastes may have various consistencies and thickening effects. The starch gels are stable at room temperature; however, they tend to separate water (synaeresis) during cold storage and freezing. The course of gelatinization depends on three factors – botanical origin of the starch, temperature, and water content of the suspension. Dried starch (1–3% water) can be heated to 180 °C (356 °F) without major changes occurring. Starch containing about 60% water is already completely gelatinized at 70 °C (158 °F).

Starch and starch-rich products exhibit retrogradation, which refers to the recovery of previously gelatinized starch. This re-crystallization is mainly caused by the amylose of the starch because, in contrast to amylopectin, it is not composed of a three-dimensional network of glucose and therefore the water is not bound so strongly. Amylopectin, however, forms stable gels with no tendency to retrogradation. Retrogradation is the primary reason for the staling of bread and pastries. In the flour, the starch present partially releases the physically bound liquid and transitions into the crystalline state. The consistency becomes dry and rubbery. Retrogradation of the starch is strongly favoured by cool temperatures. Between −8 and +8 °C (17–46 °F) the ageing is reinforced by a factor of three. The recrystallization can be reduced by addition of fats, emulsifiers, and thickening agents, such as guar gum, CMC (carboxymethylcellulose), HPMC, or xanthan gum.

Starch (native and modified) is detected by iodine with the iodine–starch-reaction. In the presence of starch, an inclusion compound of iodine–starch is formed that is visible due to a dark blue or blue–violet to black colour.

Applications

Traditionally, starches are used as thickening and gelling agents in all kinds of foods.

They control the rheology in soups and sauces, spreads, dressings, and condiments. Starches give texture and body to dairy products like yogurt and pudding. They improve the stability of fillings and dough for bakery products. In confectionary, starches contribute to a jelly consistency in moulded products and can replace gelatine, and in chewy confections they give the appropriate gel texture and stretch. Emulsifying starches are used in flavour-oil-emulsions for application in beverages.

Toxicology and Regulatory Affairs

In the human body, the ingested starch is first degraded to glucose. Starch digestion starts already in the mouth by enzymes (α- and β-amylases). Glucose that is not transferred into energy is used in the liver to compose glycogen ('animal starch') and be stored as reserve substance. Starch delivers bout 3.6–4.2 $kcal\,g^{-1}$ of energy. So-called resistant starches are available that are digested much more slowly and only partially, making less energy contribution.

Native and modified starches have wide approval from the US FDA and European Union (EU). Most producers offer different product programs for Europe and USA because local legislation may be different.

According to European food law, several starches are not permitted in the EU. In general, modified starches are considered as food additives to be used for technological purposes in unrestricted dosage ('quantum satis').

According to current knowledge, the possibly carcinogenic acrylamide is formed by overheating of starches – especially during baking, frying, roasting, and grilling – in the presence of the amino acid asparagine. Acrylamide formation is promoted by sugars such as fructose and glucose. In the manufacture of French fries it can be formed due to high temperatures. Particularly high levels of acrylamide occur when potato- and cereal-containing foods are heated dry above 180 °C (356 °F). Although acrylamide formation starts already at 120 °C (248 °F), it rapidly increases at 170–180 °C (338–356 °F).

1.2.6 Furcellaran

Raw Material Origin, Harvesting, and Manufacturing

Furcellaran is produced from the red algae *Furcellaria fastigiata*, where it is the structural substance. The seaweed grows close the coasts of Denmark and Canada. Production started in 1943, when Europe was cut off from seaweed supply in World War II. The biggest volume is manufactured in Denmark (approx. 1000–1500 MT). The seaweed is harvested, washed, and treated with alkali to remove proteins and colouring substances. The furcelleran is extracted with hot water under pressure. The raw extract is purified by centrifugation or filtration. The extract is concentrated under vacuum. Then furcelleran is precipitated in the form of gelatinous threads by injection of the concentrated extract into a cold potassium chloride solution (1–1.5% KCl). The precipitated threads are further concentrated by freezing, pressing, or centrifuging and are finally dried. The resulting commercial product is a potassium salt that contains 8–15% free KCl. It is a colourless to yellowish, practically odourless powder with a slightly salty taste.

Chemical Structure

Furcellaran consists of D-galactose (45–54%), 3,6-anhydro-D-galactose (30–35%), and sulfates of both monosaccharides (15–20%). All commercially available products are potassium salts of furcellaran. In composition and structure, it is very similar to κ-carrageenan; therefore, it is included in the group of carrageenans. The essential difference is that κ-carrageenan has one sulfate ester residue on two sugar residues while furcellaran has one sulfate residue on three to four sugar residues. This means furcellaran has a lower content of sulfate ester groups than κ-carrageenan. The galactose- and anhydro-galactose-units of furcellaran are alternately linked by $\beta(1 \rightarrow 4)$- and α-$(1 \rightarrow 3)$-bonds as in κ-carrageenan. The polysaccharide chain can be branched. The average molecular weight is 20–80 kDa.

Solubility and Viscosity

Furcellaran can be dispersed in water to a homogeneous suspension. Full hydration requires heating to 75–80 °C (167–176 °F). During cooling, a gel is formed. Neutral solutions (pH about 7) can be sterilized without loss of viscosity. During heating in acidic systems of pH <5, the viscosity is irreversibly reduced by hydrolysis and the capability for gel formation is lost. Monovalent ions such as K^+, NH_4^+, Rb^+, and Cs^+ cause a strong increase of gel strength; Ca^+ has a lower effect. There is no gelation with Na^+. The addition of sugar has a positive influence on the gel strength.

Properties and Applications

The properties of furcellaran are very similar to those of κ-carrageenan. It is supposed to have a cleaner taste in puddings, jellies, and chocolate milk than carrageenan [1].

Furcellaran forms very good gels with milk and is added to neutral dairy desserts and beverages such as pudding and cocoa drinks. It is also very suitable for cake icings, fruit preparations, jams, and marmalades because the gel is quickly formed in the presence of sugar and is highly stabile in acidic systems. In jams, furcellaran has an advantage over pectin due to being able to form stable gels at sugar concentrations below 50–60%. The amount of polysaccharide in gels is low, 0.2–0.5%, depending on sugar content and desired gel strength. To prevent hydrolysis during heating in low pH systems, furcellaran is added as aqueous solution (2–5%) as final step to the hot and ready cooked fruit–sugar mixture.

Other application areas are reformed meat products and fillings for meat products. It is also used for protein separation during the final filtration of beer.

Regulatory Affairs

Furcellaran is misleadingly called 'Danish agar'. In Europe, it is classified as carrageenan with E 407. There is no defined ADI.

1.2.7
Larch Gum

Raw Material and Manufacturing

Larch gum is a component of the heartwood of all kinds of larches (*Larix* species). Larches contain 5–35% (on a dry matter basis) of a water-soluble arabinogalactan, a polysaccharide consisting of galactose and arabinose. The substance is manufactured by crushing the wood and countercurrent extraction with water or diluted acids. Accompanying substances such as phenols, terpenes, and ferrum are removed by treatment of the hot gum solution with active magnesium oxide. The purification can also be done by treatment of the solution with activated carbon, ion exchangers, electrodialysis, or repeated re-precipitation with ethanol. The highly purified extract is dried on drums. The resulting white to light-brown powder is tasteless and odourless [1].

Chemical Composition

Purified larch gum is a mixture of neutral polysaccharides of different molecular size. The ratio of the building blocks galactose and arabinose varies greatly – depending on type and age of the tree, the location within the tree, and the general ratio of high- and low-molecular-weight components. The ratio of galactose to arabinose is between 4 : 1 and 23 : 1. The macromolecule consists of a main chain of β-(1 → 3) linked galactose units that has side chains in α-(1 → 6) and β-(1 → 4) linkage of galactoses, arabinoses, and mixtures of both monosaccharides.

The degree of branching is high. The molecular weight is 50–70 kDa on average. The molecule is more or less spherical [1].

Solubility and Viscosity

Larch gum is easily and completely soluble in cold and in hot water. Up to 60% of the gum can be dissolved at room temperature. It exceeds the solubility of gum acacia. Even the addition of alcohol, as much as 70%, does not precipitate the polysaccharide from aqueous solution if the alcohol is added slowly under stirring.

The viscosity of highly concentrated larch gum solutions is relatively low. The viscosity of a 10% solution at 20 °C (68 °F) is below 2 mPa s. The viscosity of a 30% solution at pH 4 is 7.8 mPa s and changes only slightly to 8.1 mPa s on increasing the pH to 11. A 40% gum solution has a viscosity of 24 mPa s. The thickening effect is stable over a wide pH range (1.5–10.5) and is not sensitive to addition of electrolytes. Newtonian flow behaviour is observed at concentrations as high as 50%. At concentration above 60%, the liquid solution turns into a thick paste. These rheological properties are caused by the highly branched molecular structure and the relatively low molecular weight. Viscosities of larch gum solutions decrease with increasing temperature, but return to their original values after cooling. There is no depolymerization or irreversible breakdown of ordered structure at elevated temperature [3].

Properties and Application

Larch gum is surface active. The decrease in surface tension produced by the gum is greatest at low pH, and the addition of electrolytes enhances surface activity, with monovalent cations being more effective than divalent ones [3].

Larch arabinogalactan is a potential substitute for gum Arabic; since it is readily dispersible in water it gives low-viscosity solutions at high concentrations and has the same emulsifying properties. Options are dietetic low-sugar products because larch gum solutions show the same physical properties as sugar solutions – except for the sweetening effect and the energy content. Further applications are emulsions of flavour oils and carrier substance for highly-effective additives such as sweeteners.

For pharmaceutical applications, larch arabinogalactan has been discussed and examined as a novel immune modulator for reducing intestinal problems, improving immune defence against colds, and effectiveness in cancer tumour therapy.

1.3 Seed Flours

1.3.1 Guar Gum

Raw Material Origin, Harvesting, and Quality

Non-ionic galactomannans act as reserve carbohydrates and are found as storage polysaccharides in the cell walls of various seeds. Guar gum is one example for this type of seed gum. It is manufactured from the endosperm of the seeds of the guar plant (*Cyamopsis tetragonolobus*, family of Leguminosae), or cluster bean. The plant is a drought-resistant annual legume that has been cultivated in India for centuries; 80% of the globally produced guar is from India. Guar pods are 3–12 cm long and contain 5–12 seeds. Depending on weather conditions, the annual guar crop fluctuates, but in average 500 000 MT of guar seeds are available. Approximately 55 000 MT per year is used as food-grade guar gum [2].

The endosperm (32–36%) is separated from the other components of the seed (20–22% hull, 43–47% germ) by milling procedures [1].

Guar gum is a yellow to light grey powder with a slight smell and taste of beans. Its solutions are opaque to turbid.

Chemical Structure

Guar gum contains 15% water (max.), % protein (max.), 7% acid insoluble residue (fibre) (max.), less than 1.5% ash, and a minimum of 75% polysaccharides (galactomannans). Guar gum is a neutral polysaccharides consisting of D-mannose and D-galactose in a molecular ratio of 1.6 : 1 (Table 1.6). The molecule consists of a linear backbone of β-(1 → 4)-linked mannose units. Single terminal D-galactose units are linked to the main backbone by α-(1 → 6)-glycosidic bonds to the 4,6-mannose units. Statistically, nearly every second mannose unit is linked to a galactose unit. Enzymatic hydrolysis, though, has shown that the galactose residues are distributed irregularly within the mannose chain. There are zones without branches and zones where every mannose building block carries a galactose residue (mostly double or triple arrangements). The average molecular weight is about 220 kDa [1,2].

Solubility, Viscosity, and Properties

Owing to its high degree of branching, guar gum has good solubility in cold water. There are high-viscosity solutions at low concentrations, which exhibit pseudoplasticity above 0.5%. Water retention capability is high. Solutions of guar gum are freeze–thaw stable. Monovalent cations have no effect, while an addition of di- or trivalent cations leads to a viscosity increase and polyols reduce viscosity. Viscosity is irreversibly reduced when guar is heated at neutral pH to above 90 °C (194 °F) or under moderate heat in acidic systems at a pH below 3.5, because the galactomannan molecule is decomposed. Cold viscosity is stable from pH 2 to 10. Borate ions cause gelation of guar gum solutions. Addition of guar gum to starch, xanthan, or CMC has a synergistic effect on viscosity of the blend. Addition of guar gum to gelling polysaccharides such as agar increases the strength and elasticity of the gels [1].

Functions and Applications

Guar gum is used as binder and thickener in ice cream, dairy desserts, soups, sauces, meat, and bakery products. There it controls texture, influences crystallization, and prevents synaeresis and phase separation. In baked goods, it prolongs shelf life and retards starch retrogradation. It is used as a processing aid in canned foods. In healthy and dietary products, guar contributes to calorie

Table 1.6 Characteristics of guar gum.

	Guar gum
E-No.	E 412
Origin	Extract of endosperm of seed (Leguminosae)
Chemical composition	• Medium-galactose galactomannan; • mannose + Galactose (ratio M : G = 1.6 : 1)
Nutritional value (in 100 g) – Metabolism	292 kJ (70 kcal); slow resorption
Fibre content	Approx. 80% (contains 10% protein)
Toxicology	No health concerns, no ADI value defined; considered as food in Asia, not as additive
Solubility at low temperature (H_2O)	Highly soluble
Appearance of an aqueous solution	Opaque, grey, cloudy
Viscosity of solution in water	High in cold water, lower in hot water
Impact of heat on viscosity in water (pH 7)	Viscosity decrease
Viscosity development in water at pH 7 (T = 0–100 °C)	Irreversible viscosity loss at $T > 90$ °C (194 °F)
Shear stability	Pseudoplastic > 0.5% concentration, shear thinning
Thickening effect	High
pH stability	High (pH 2–10)
Decomposition	• With heat $T > 90$ °C (194 °F); • heating at pH < 3.5; • strong acids or alkali, strong oxidizing agents; • specific enzymes
Film formation	Low
Emulsion stabilization	Medium
Gelation	No (only with borate ions)
Gel strength and gel stability	—
Gel transparency	—
Tendency for gel synaeresis	—
Impact of electrolytes (cations +, 2+, 3+)	Addition of cations 2+ or 3+: viscosity increase
Reaction with Ca^{2+} ions	No (only viscosity increase)
Protein activity	No
Crystallization control	High
Synergistic effects with other hydrocolloids	+ Starch/xanthan/CMC: viscosity increase; + gelling polysaccharides (e.g. agar): increased gel strength and elasticity
Other synergistic effects	Gelation with borate ions
Negative interactions	Viscosity reduction with polyols
Dosage level in foods	Low to medium (0.05–2%, mostly 0.2–0.5%)

reduction and fibre enrichment. In beverages, the gum maintains turbidity (clouding effect) and solid suspension.

Toxicology and Regulatory Affairs

Animal studies have not shown negative interactions. Guar gum is a globally approved food additive (as E 412 in Europe). There is no ADI defined. In some countries, guar is considered as food. The pods are cooked as vegetable for human consumption and are used as cattle feed. During passage through the digestive tract, the β-(1 → 4)-linkage of the main backbone is not cleaved, the α-(1 → 6)-linkages of the side chains are partially cleaved. Therefore, the water retention capability of guar is maintained. The gum is a soluble fibre with a strong retarding effect on the speed of glucose resorption. A positive effect on cholesterol level in the blood serum has been observed [1].

There are strongly depolymerized guar gum products, sold as food ingredients on the market, with molecular weights about 20 kDa. However, these do not meet the specifications of E 412 in Europe and have no E-number [2].

1.3.2
Locust Bean Gum (Carob)

Raw Material and Manufacturing

Carob gum or locust bean gum (LBG) is produced by milling the endosperm of the seeds of the carob tree (*Ceratonia siliqua*, family Leguminosae). The trees grow around the Mediterranean Sea and in California; 70–75% are found in Italy, Spain, Portugal, and Greece. It takes 8–10 years before the trees bear fruit called pods or St. John's bread. The pods are 10–20 cm long. The carob seeds or kernels are 8–15% of the pod's weight. Under normal weather conditions, about 280 000–300 000 tonnes of pods are collected manually each year. The pods are broken using a kibbling process, during which the seeds are released. The kibbled de-seeded pods are known as 'pulp' and finely ground pulp, also called carob powder, which is used as animal feed for all kinds of livestock, as a remedy against diarrhoea in children, and as raw material for alcohol production. Roasted pulp is also used as replacement for cocoa. The seeds consist of 30–33% hulls, 23–25% germs, and 42–46% endosperm. The hulls are either carbonized by hot sulfuric acid, followed by intensive washing with water, or largely removed by roasting in a rotating furnace at about 550 °C (1022 °F), after which residual hull fragments are rubbed off mechanically. The treated remnants of the seeds are then split and sifted. The recovered endosperm halves are dried and ground into a fine, off-white powder of desired particle size. The average annual production of food-grade gum is about 11 000 MT [1,2].

Chemical Structure

Carob gum contains (max.) 15% water (max.), 7% protein (max.), 4% acid insoluble residue (fibre) (max.), less than 1.2% ash, and a minimum of 75% polysaccharides (galactomannans). It is a neutral polysaccharide consisting of D-mannose and D-galactose in a molecular ratio between 73 : 27 and 86 : 14, with an average of 3.5 : 1 (Table 1.7). The molecule consists of a linear backbone of β-(1 → 4)-linked mannose units. Statistically nearly every fourth mannose molecule is bond to a galactose residue with an α-(1 → 6)-linkage. However, enzymatic hydrolysis has shown that the galactose residues are distributed irregularly within the mannose chain. There are zones with unsubstituted mannose blocks

Table 1.7 Characteristics of locust bean/carob gum.

	Locust bean gum/carob gum
E-No.	E 410
Origin	Extract of endosperm of seed (Leguminosae)
Chemical composition	• Low-galactose galactomannan; • mannose + galactose (ratio M : G = 3.5 : 1)
Nutritional value (in 100 g) – metabolism	252 kJ (60 kcal); slow resorption
Fibre content	Approx. 80% (contains 7% protein)
Toxicology	No health concerns, no ADI value defined
Solubility at low temperature (H_2O)	Limited soluble
Appearance of an aqueous solution	Opaque, grey, cloudy
Viscosity of solution in water	Low in cold water, high after heating + cooling
Impact of heat on viscosity in water (pH 7)	Viscosity increase of cold processed solutions, no heat thinning
Viscosity development in water at pH 7 ($T = 0–100\,°C$)	High viscosity after heating and cooling
Shear stability	Pseudoplasticity (shear thinning)
Thickening effect	High, after heating to 80 °C (176 °F)
pH stability	High (pH 3–11)
Decomposition	• Heating at pH < 3.5; • strong acids or alkali, strong oxidizing agents; • specific enzymes
Film formation	Low
Emulsion stabilization	Medium
Gelation	Only with xanthan (optimum ratio at pH 7 = 1 : 1) or κ-carrageenan
Gel strength and gel stability	Stable thermoreversible gels with xanthan
Gel transparency	Low
Tendency for gel synaeresis	No synaeresis in gels with xanthan or κ-carrageenan
Impact of electrolytes (cations +, 2+, 3+)	Low
Reaction with Ca^{2+} ions	No
Protein activity	No
Crystallization control	High
Synergistic effects with other hydrocolloids	+ Xanthan: viscosity increase + gel formation + κ-carrageenan: increased gel formation (higher gel strength + elasticity, syneresis-free) + other gel formers (agar, alginate): increased gel strength and elasticity
Other synergistic effects	—
Negative interactions	—
Dosage level in foods	Low to medium (0.1–2%, mostly 0.2–0.5%)

and zones where two adjacent mannose units carry a galactose residue. The average molecular weight is about 310 kDa [1].

Solubility and Viscosity

Owing to its reduced amount of side chains compared to guar gum, LBG is not fully soluble in cold water. When heated to above 80 °C (176 °F), it is totally soluble and forms pseudoplastic solutions. Solutions heated to this temperature and cooled have a much higher viscosity than cold-prepared solutions. The viscosity is stable in a range of pH 3–11. Mono- and divalent cations only have a slight impact; the addition of anions usually has no effect.

Properties

The properties of LBG correspond substantially to those of guar gum so that both gums are used alternatives. Certain hot water-soluble fractions of LBG self-associate to form weak three-dimensional gel networks, especially after a freezing process.

LBG has the very important characteristic of being able to act synergistically with κ-carrageenan and xanthan to form gels that are elastic, very cohesive, and relatively free of synaeresis. LBG interacts more strongly with other polysaccharides than does guar gum. Evidence is accumulating that the interaction effects of LBG with hydrocolloids like cellulose, xanthan, κ-carrageenan, and agarose are due to direct molecular association between the two polymer types. The superior interactive properties of LBG are clearly linked to its lower degree of galactose substitution [2,3].

Applications

LBG is used in the same applications as guar gum and in all products where the synergistic gelation gives additional benefits (e.g. ice cream, sorbets, chilled dairy desserts). Water gels are made with LBG to be heated and form solutions to accompany prepared meat, fish and other seafood products. The low energy and high fibre content makes it suitable for healthy foods and products intended for diabetics.

Toxicology and Regulatory Affairs

Clinical trials with commercial LBG products did not show negative results on the human digestive tract and no allergenic reactions. In addition, animal studies did not show negative interactions. LBG is a globally approved as food additive (as E 410 in Europe). There is no ADI defined [1].

1.3.3 Tara Gum

Raw Material and Manufacturing

Tara gum, also known as Peruvian carob bean gum, is obtained by grinding the endosperm of the seeds of the South American tara tree *Caesalpinia spinosa*, which belongs to the Leguminosae family. The trees grow at up to 3000 m above sea level in Peru and Bolivia and tolerate dry climate and poor soil conditions. The fruit is a flat oblong reddish pod that contains 4–7 large round black seeds composed of endosperm (22% by weight), germ (40%), and hull (38%). Mature pods are usually harvested by hand and typically sun dried before processing.

Based on seed weight, the yield of high-grade tara gum is only 21–22%.

However, the EU purity criteria specify a maximum limit for protein content of only 3.5% (N% × 5.7). This is significantly lower than the permitted maximum limits for carob/LBG (7%) or guar gum (10%). Thus, tara gum is clearly the purest seed gum. An estimated volume of 1500–2000 MT per year is available globally and is mainly used in the food industry. The purification principle is the same as for carob and locust bean gum (LBG) [2].

Chemical Structure

Tara gum is a neutral polysaccharide, a galactomannan, of high molecular weight. It consists of a linear main chain of β-(1 → 4)-linked D-mannose units that has side chains of α-(1 → 6)-linked galactose units. Statistically, each third mannose unit carries one galactose (M : G = 3 : 1) [1].

Solubility, Viscosity, and Stability

Tara gum is partially soluble in cold water (80% at 25 °C/77 °F) and it is completely soluble in hot water. The aqueous solutions are neutral. A solution of tara gum is less viscous than a guar gum solution of the same concentration, but more viscous than a solution of LBG. Generally, tara gum presents a viscosity of around 5500 cps (1% in water). Furthermore, tara gum shows an intermediate acid stability between LBG and guar gum. It resists the depolymerization effect of organic acids down to a pH of 3.5. This gum is also stable to high temperature heat treatment, up to 145 °C in a continuous process plant.

Properties

Tara gum is a white to yellowish powder. The properties are similar to those of guar and LBG. The addition of 60–70% sugar to a tara gum solution leads to weak gel formation. Xanthan causes a gelation of tara gum, too, but the gels are weaker than those of xanthan with LBG. More synergistic effects are found with agar and carrageenan, where tara gum supports gel formation.

Applications

Tara gums function mainly as a thickener and stabilizer in several food products. Blends with modified and unmodified starches can be produced that have enhanced stabilization and emulsification properties and are used in convenience foods, such as ice cream.

Toxicology and Regulatory Affairs

Tara gum is not digested by enzymes in the human body. Animal studies have not shown negative effects. In the USA, it is approved as a food additive by the

Food Chemicals Codex (FCC). In Europe, it is approved as hydrocolloid E 417. There is no specified ADI.

1.3.4 Tamarind Seed Gum

Raw Material and Manufacturing

The raw material for tamarind gum is the flour of tamarind seed. These seeds grow on the tamarind tree (*Tamarindus indicus*, family of Leguminosae), one of the most important and popular trees in India. Tamarind seed flour is extracted by cooking it for 30–40 min with boiling water (30–40 vol.%) under stirring. The water contains 0.2% organic acid. The extract is left overnight and insoluble material is removed by filtration. 0.5% Kieselguhr is added to the remaining liquid and pressure filtration performed. The filtrate is dried under vacuum and precipitated by ethanol. Tamarind gum is a creamy-white powder without taste and odour [1].

It is the only seed xyloglucan that is exploited commercially [3].

Chemical Composition

Besides polysaccharides, tamarind gum contains about 15% water, 3% protein, 1% fibre, and small amounts of phosphate and oil. Building blocks of the polysaccharide are D-glucose, D-galactose, and D-xylose in the ratio 3 : 1 : 2. L-Arabinose is also present in smaller portions. The macromolecule consists of a linear main chain of β-(1 → 4) linked glucose units. Side chains bond at the 6-position are xyloses, arabinoses, and galactose with different chain lengths. The average molecular weight is 50–115 kDa.

Properties

The characteristic property of tamarind gum is its capability to form stable gels with sugar at 0.5–1.5% gum concentration in water. These gels are stable over a wide pH range. The sugar concentration required to form a specific gel strength is lower than for pectin. The gels do not tend to show phase separation (synaeresis). With 45–70% sugar, the gum forms heat- and pH-stable gels with very high gel strength. Sugar can be partially replaced by alcohol (up to 20%). The maximum gel strength for a gel with 1% gum and 50% sugar is at pH 2.8.

Application

Owing to this gelation behaviour, tamarind gum can replace pectin in fruit preparations, jams, and marmalades. Another application field is ice cream and mayonnaise-type products.

Toxicology and Regulatory Affairs

Tamarind gum is used traditionally in Japan to produce jams, jellies, dressing, and ice cream. Tests have not shown negative effects on animals.

In Europe and the USA the use of tamarind gum has lapsed and it does not currently have clearance for use food use [3].

1.3.5 Konjac Gum

Raw Material Origin and Manufacturing

Konjac flour was known in China as long ago as 200 BC. It yields a high molecular weight viscous polysaccharide – konjac glucomannan. The flour is obtained from the tubers of less than ten species of *Amorphophallus*. Most species have a glucomannan concentration of more than 50% (dry matter). The size and shape of tubers depend on age and cultivar (diameters of 15–20 cm are frequent). Mainly, 3-year-old tubers 0.3–1.5 kg in weight are washed and peeled before being sliced into chips. The characteristic smell of konjac may be washed out along with other alcohol-soluble ingredients by a 10–50% alcohol solution to give a cleaner odour and taste. Processing steps are dry and semi-wet milling to increase the concentration of glucomannan. The polysaccharide precipitates when organic solvent is added [2].

Chemical Structure

Konjac glucomannan is mainly composed of a backbone of β-(1 → 4)-linked D-glucopyranose and β-D-mannopyranose sugars in a random order. The molecular ratio of mannose to glucose is on average 1.6 : 1 and is random, though small repeats of three to five mannose blocks are frequent. Randomly spread acetyl groups occur every 10–19 units (3–6%) on carbon atoms C2, C3, or C6 of mannopyranose. Methylation and carboxylation, as well as single sugar substituents or longer side chains, have been documented. Although derivatization improves chain flexibility, these products are not permitted for food use. The average molecular weight is 1 MDa. Commercial products range between 200 000 and 2 million daltons [2].

Solubility and Viscosity

Konjac flours even solubilize in ice water. Hydration is accelerated by raising the temperature and speed of agitation. Viscosity and solubility can be hindered by addition of competing solutes, low-molecular-weight ingredients or maltodextrin, branched dextrins, and salts. Alcohols, such as ethanol or isopropyl alcohol, also reduce solubility and can eventually precipitate the glucomannan chains out of solution. Konjac flours have a remarkable absorption capacity and can absorb 15–20 g of water or 1–2 g of oil per g. Increased acetylation levels reduce the water absorbency and viscosity. The gum is degraded by prolonged exposure above 80 °C (176 °F), especially in acidic media. Konjac solutions are sensitive to shear-thinning. High-purity galactomannans are difficult to solubilize. Konjac exhibits high viscosity solutions at 1%, but it is much less viscous than guar or xanthan gum at lower concentrations. Extreme values of konjac for molecular weight and viscosity could be linked to the presence of microgel particles, so-called clusters, produced during processing. Native konjac material, crude flour, is still bound to the original protein membrane, which must swell first before the konjac molecules are released into solution. Viscosity is positively correlated with glucomannan content and degradation of chain length reduces it. Solutions of native grades exhibit non-Newtonian flow behaviour and shear-dependant viscosity. Preservatives should be used to prolong the shelf life of gel and stock solutions [2].

Properties

The less-substituted konjac glucomannan has a stronger interaction effect than locust bean gum because the unsubstituted mannan backbone segments are critical for forming interchain associations [3].

Eliminating acetyl groups produces deacetylated konjac glucomannan, which is able to build random junction zones through hydrogen bonds to form an irreversible and extremely stable gel with similar elastic recovery to polyacrylamide gels. Acidic deacetylation is possible, but alkali reactions are preferred to limit hydrolysis. In some cases, temperature-stable gels may be obtained without deacetylation. Non-modified konjac flours show different gelling properties depending on pH, concentration of other ingredients, and presence of synergistic hydrocolloids, such as semi-refined, κ- or ι-carrageenan, or xanthan gum. With those, konjac forms elastic, cohesive gels. Maximum peak viscosity is achieved with a blend of 75–90% konjac and 10–25% carrageenan. For milk gels with ι-carrageenan, the maximum gel strength is obtained with a 20 : 80 to 40 : 60 blend of konjac flour to carrageenan. In brines, konjac is only used in low levels to boost carrageenan gel strength and reduce thickening power. A blend of xanthan (50%) and konjac (50%) with a total gum concentration of 0.02% gives a very elastic thermoreversible gel. Xanthan shows more affinity for glucomannan than the galactomannans (LBG, tara gum). A konjac–xanthan blend is more influenced by salts than a LBG–xanthan blend. With 0.04 $mol\,l^{-1}$ NaCl, xanthan gels konjac glucomannan already at 42 °C (107 °F). Konjac improves the functional properties of starch (viscosity, freeze–thaw stability) and reduces synaeresis. Gels of 1% konjac and 9% starch are thermo-irreversible and very acid stable. Konjac improves the gel strength of agar when added in a ratio of 10% to 90% agar. Long-chain konjac glucomannan can influence gellan gum gelation promoted by salts. High concentrations of konjac glucomannan (2–5%) may gel without any additive. These gels are not clear and do not exhibit a specific fusion temperature. Glucomannan has affinity for iron, zinc, calcium, and magnesium salts – high amounts of glucomannan can reduce the mineral intake because of physical entrapment. Lyotropic salts alter the gel properties.

Some grades show suspension activity. Sheared konjac gels have an oily creamy mouthfeel [2].

Applications

Konjac is used for shelf-life extension in bakery products at 0.1–0.5%. It provides adhesion for coatings and binding in complex matrixes such as restructured meat or vegetable products, pasta, and desserts. Often, konjac is blended with xanthan, carrageenan, or starches. The gum is used to thicken fluids for patients with dysphagia [2].

It is used as vegan substitute for gelatine. Konjac root powder is used as an ingredient in vegan alternatives to seafood products. It can be incorporated into animal product-free versions of scallops, fish, prawns, crab, and shrimps.

Toxicology and Regulatory Affairs

Konjac glucomannan is sensitive to enzymes such as β-D-glucanase and β-D-mannanase and thus bacterial fermentation is able to occur in the intestines.

Konjac lowers cholesterol level more effectively than other gums and it regulates the blood sugar level. It synergistically acts with stanols and sterols for low-density lipoprotein (LDL) reduction. Konjac acts as prebiotic in the human colon. It has low energy content and reduces appetite. It is a globally accepted gelling agent and thickener. It has a substantial record as a common healthy ingredient in Asia. In Europe it is approved as E 425 with a maximum intake of $10\,g\,kg^{-1}$ food [2].

1.4 Exudates

1.4.1 Acacia Gum/Gum Arabic

Raw Material Origin, Quality, and Purification

Acacia gum, also known as gum Arabic, is a natural gum exudate obtained from acacia trees in the 'African sub-Saharan zone' from 12 different countries, from Senegal to Ethiopia, the so-called gum belt. It has been known since antiquity and used for thousands of years as food additive and ingredient. The FAO/WHO has defined acacia gum as the dried exudate obtained from the stems and branches of *Acacia senegal* or *Acacia seyal* (family Leguminosae). The gum occurs as resinous mass naturally or is secreted through artificial cuts made from December to May. One tree produces only about 400 g of gum per year. Within a few weeks, the exudate dries to tubers and sheds in sunlight. These are collected, cleaned of the rest of bark and, depending on colour, sorted into different quality classes. Raw gum has several vegetable and mineral impurities and fluctuating bacterial contamination. Using dry purification steps such as sieving only, the level of impurities can be reduced but microbial contamination cannot be improved and, most of the time, raw gum does not meet international requirements for foods. Consequently, dry methods of purification have been substituted by purification in aqueous solutions, which is much more efficient. The gum is fully dissolved in water and all impurities are removed by a cascade of filtration steps giving levels of insoluble matter in the finished products as low as 0.02%. Bacteriological contamination is reduced by pasteurization and the gum syrup is concentrated to 25–35% and dried. During solubilization, purification, and drying the thermal conditions are critical. Acacia gum contains proteins that are very important for the emulsifying properties but are sensitive to heat denaturation. Roller drying gives gum powders with good hydration properties, but less emulsifying capability due to the drastic thermal treatment. By spray-drying, highly functional fine powders are obtained. Recently, multistage drying processes have been developed to produce agglomerated gum particles with unique dissolution in water without lump formation and with a maximum solubility of 45–50%. The fine-to-coarse powders are off-white to yellowish. Exudates of other plants (e.g. *Combretum* species) are falsely offered as gum Arabic. The consumption of acacia gum from *A. seyal,* restricted to technical applications in the past, is now about the same as from *A. senegal.* With more than 60 000 MT, it is the third largest hydrocolloid additive used by industry [1,2].

Chemical Structure

Acacia gum is a highly branched neutral to slightly acidic complex polysaccharide (Table 1.8). Building blocks of the arabinogalactan macromolecule are D-galactose, L-arabinose, L-rhamnose, and D-glucuronic acid in a molecular ratio of about 3 : 3 : 1 : 1. The structure consists of a linear backbone of β-(1 → 3) linked galactose units that has side chains of α- or β-(1 → 6) linked galactose units with a terminal glucuronic acid group. The acid groups are partially neutralized by calcium, magnesium, sodium, or potassium ions and occur partially in a free-acid form. The side chains are further branched with rhamnose or arabinose units. The arabinogalactan chains are attached to a protein skeleton forming an arabino-galactoprotein (AGP). Depending on source, acacia gum has different compositions. A simple test to identify the gum source is the specific rotation (*A. senegal* = −30°; *A. seyal* = +51°). For further structural information, sugar composition is determined by HPLC after acidic gum hydrolysis. The nitrogen content of *A. senegal* (0.29%) is about double that of *A. seyal* (0.14%). The lower content of protein combined with a lower protein availability explains why *A. seyal* is less efficient for emulsion stabilization and, consequently, it is not used for emulsions where long-term stability is required. The molecular weight varies within wide limits and is on average 350 kDa [1,2].

Solubility and Viscosity

Acacia gum powder hydrates readily in cold water and concentrations up to 40–50% can be handled easily. Compared to other water-soluble polysaccharides with a similar molecular weight, acacia gum exhibits very low viscosity in water. To reach a viscosity of 3000–5000 mPa s (Brookfield), only 1% of guar or xanthan gum is necessary, but acacia gum has to be dissolved at concentrations of 40–45%. Rheological behaviour of acacia gum solutions is Newtonian up to 25% and then becomes pseudoplastic. Highly concentrated solutions exhibit only a slight shear-thinning. The aqueous solutions have an acidic character (pK_a 4.5–5.5).

Table 1.8 Characteristics of acacia gum/gum Arabic.

	Acacia gum – gum Arabic
E-No.	E 414
Origin	Exudate of a tree (Acacia)
Chemical composition	• Highly-branched arabinogalactan; • galactose, arabinose, rhamnose, glucuronic acid in a molecular ratio of 3 : 3 : 1 : 1
Nutritional value (in 100 g) – Metabolism	837 kJ (200 kcal); slow resorption
Fibre content	High (contains 80% starch/sugar)
Toxicology	No health concerns, no ADI value defined
Solubility at low temperature (H_2O)	100% cold-water soluble (solutions possible with 40–50% concentration)
Appearance of an aqueous solution	Grey-white to yellowish
Viscosity of solution in water	Very low in cold water (natural pH about 4–5), viscosity is further reduced upon heating
Impact of heat on viscosity in water (pH 7)	Slight viscosity decrease – fully reversible at pH 4–9
Viscosity development in water at pH 7 (T = 0–100 °C)	Slight viscosity decrease upon heating
Shear stability	Newtonian ≤ 25%, pseudoplastic > 25%
Thickening effect	Low, very high concentrations necessary
pH stability	Medium (pH 4–9)
Decomposition	By strong acids or alkali (4 > pH > 9), but extremely heat- and enzyme-stable
Film formation	High
Emulsion stabilization	High
Gelation	—
Gel strength and gel stability	—
Gel transparency	—
Tendency for gel synaeresis	—
Impact of electrolytes (cations +, 2+, 3+)	Precipitation + viscosity reduction (e.g. $CaCl_2$)
Reaction with Ca^{2+} ions	Precipitation + viscosity reduction (e.g. $CaCl_2$)
Protein activity	No
Crystallization control	Low
Synergistic effects with other hydrocolloids	—
Other synergistic effects	• Interaction with sugar and sugar alcohols; • complex formation with sodium citrate (viscosity increase)
Negative interactions	• Coacervate formation with gelatine + precipitation; • electrolytes reduce solution viscosity
Dosage level in foods	High (in most cases >5%)

The presence of uronic acids in acid and salt forms makes the viscosity of solutions dependent on pH value and electrolyte content. However, it also gives acacia gum buffer properties so that solutions are more pH stable after moderate addition of acids or bases. Viscosity has a broad maximum at pH 4.5–7 and is stable in a pH range of 4–9. Degradation occurs in stronger acidic or alkaline conditions, whereby viscosity decreases irreversibly (50% loss at pH 2 or pH 12). The addition of electrolytes results in a decrease in viscosity. An exception is sodium citrate, which increases solution viscosity due to complex formation. Acacia gum can be dissolved in hot ethylene glycol or glycerol as well as in aqueous ethanol solutions of up to 60 vol.% alcohol. Viscosity decreases upon ageing, but this effect can be minimized by addition of preservatives such as benzoic acid (0.2%). Normal mechanical treatment has no effect on viscosity. Prolonged exposure to ultrasonic vibration or ultraviolet irradiation reduces viscosity. The highly branched structure makes it highly resistant to hydrolysis in medium-acidic systems and to degradation under extreme thermal conditions and by enzymes [1–3].

Properties

The highly branched compact arabinogalactan structure is responsible for the low viscosity of solutions while the central protein fraction provides good emulsification properties, especially when using grades from *A. senegal*. The emulsifying power is lost upon heating, which causes some precipitation of proteinaceous material. It is the high molecular mass protein-rich fraction that preferentially adsorbs at the oil–water interface. In confectionary products, acacia gum in high concentrations (>40%) interacts with sugar, also present in high amounts, thereby preventing its crystallization and giving smooth products of uniform texture. A similar strong interaction occurs between the gum and polyols (sorbitol, mannitol) that replace sugar in dietetic hard candies. Corn starch and maltodextrins, which are ingredients of chewy or hard sweets, are prevented from aggregating. The gum (1%) offers a long-lasting, cohesive chew in chewy confectionary. Furthermore, it stabilizes the fats in caramels and toffees. The film-forming ability is high [2,3].

Applications

Acacia gum is used as an emulsifier and stabilizer in a range of confectionary products. It stabilizes flavoured oil emulsions and is often found in beverages. In tablets, it is used as binder. It improves the quality of soft and hard coatings for food, nutraceutical, and pharmaceutical products. Acacia gum is used for effective encapsulation – as emulsifier as part of the matrix or as coacervate partner in membrane encapsulation.

It brings benefits to bakery products in terms of processing, texture, and shelf life due to its moisture regulation and film-forming properties. The gum is used as traditional oenological additive to protect red and white wine against destabilization (10–30 g per 100 l). In health foods it is a source of soluble fibre with prebiotic properties and lowers the glycemic index by reducing the speed of sugar resorption [2,3].

Toxicology and Regulatory Affairs

Acacia gum resists digestion in the stomach and the small intestine. However, it is slowly fermented by the microflora colonizing the gastrointestinal system. Resorption occurs in the large bowel after synthesizing short-chain fatty acids. The gum has prebiotic function and promotes selectively the beneficial bacteria from the lactic acid group. The gut tolerance is high; there is no gas production and no laxative effect. It has a hypoglycaemic effect and reduces GI at concentrations of 3–6%. It is a non-carcinogenic polysaccharide.

Acacia gum is approved by the US FDA and has GRAS status. In Europe, it is classified as E 414 as food additive for technological purposes in unrestricted dosage ('quantum satis'). The calorific value is $2\,\text{kcal}\,\text{g}^{-1}$ (Europe). The gum is globally approved for pharmaceutical applications and it is listed in the US National Formulary and European Pharmacopeia [2,3].

1.4.2
Tragacanth

Raw Material Origin, Harvesting, and Quality

Tragacanth is mentioned in the Bible. It is the exudate from the trunk and branches of the shrubby representatives of the species *Astragalus* (family Leguminosae). The two main commercially exploited species are grown predominantly in Iran, Syria, and Turkey. Approximately 1000 MT per annum where used in 1980, then demand fell to about 300 MT. It was replaced by new xanthan grades and the Gulf War made tragacanth supply erratic.

Longitudinal incisions are made with a knife into the lower stem of the shrubs in May or June. Collection of the exudate occurs in the hot summer months for the high-quality ribbon grades and later for the flake grades. Collected material is hand-sorted according to colour and viscosity. It is exported mainly to Europe and USA where further processing steps – such as grinding down to 150 μm particle sizes and selective sieving – are carried out. Blends are made to ensure consistent quality of this natural product [2].

Unground tragacanth is a white to yellowish, transparent, horn-like material 0.5–2.5 mm thick. Milled products are white or off-white powders.

Chemical Structure

Gum tragacanth can be described as a complex, acidic, highly branched, heterogeneous hydrophilic polysaccharide. The molecular weight is about 840 kDa. Tragacanth is a mixture of two components present as a mixed calcium, magnesium, and potassium salt. The first component is tragacanthic acid, commonly known as bassorin, a water-swellable polymer (60–70%).

The second component is a water-soluble arabinogalactan polysaccharide, known as tragacanthin (30–40%). It is made of 75% L-arabinose, 10% D-galactose, and 10% D-galacturonic acid which is esterified with methanol. The structure is believed to consist of a core of repeating $(1 \rightarrow 6)$-linked D-galactose units to which highly branched chains of L-arabinofuranose residues are attached. Traces of starch and cellulosic materials are present and the proteins (3–4%) contribute to the emulsification properties of the gum.

The water-swellable tragacanthic acid yields D-xylose, L-fucose, D-galactose, L-rhamnose, and D-galacturonic acid. The viscous property of the gum is largely due to this component having a high molecular weight and rod-like molecular shape as it has a $(1 \rightarrow 4)$-linked D-galactose backbone with short side chains of D-xylose, L-fucose, or D-galactose connected by $(1 \rightarrow 3)$-linkages [2].

The tragacanthic acid fraction occurs associated with Ca^{2+}, Mg^{2+}, and K^{+} [3].

Solubility and Viscosity

Gum tragacanth hydrates in hot or cold water to form viscous solutions. Tragacanthin dissolves in water to form a colloidal solution, while bassorin forms a gel. Gelation properties of this fraction are improved by Ca^{2+} ions. Viscosity depends on the grade and is regarded as the main indicator for quality. Viscosity is fully developed after 24 h at 25 °C (77 °F). The hydration time can be reduced as the initial water temperature is increased (e.g.: 2 h at 50 °C (122 °F)). Tragacanth solutions are pseudoplastic at concentrations as low as 0.5%. Aqueous dispersions containing more than 2% of gum form thick pastes that have the texture of soft gels [2,3].

Properties

The gum is stable over the pH range 2.5–10; the highest stability is between pH 4 and 8. The pH of a 1% solution is acidic (pH 4.5–6). Viscosity decreases by only 30% as the pH falls to 2. This unusual stability under acidic conditions (~20% viscosity loss only when 1% gum is stored in 1% acetic acid for 3 weeks) was the reason for the extensive use of gum tragacanth in foods such as salad dressing and acidified sauces, where maintenance of viscosity at low pH is required. Heating leads to a reversible decrease in viscosity by about 25–50% during heating from 25 to 90 °C (77–194 °F) – upon cooling the initial thickening effect is almost fully recovered. Permanent viscosity loss will occur during prolonged heating at elevated temperature or if heat is applied during hydration. Then an irreversible viscosity loss (about one-third) occurs that is ascribed to autohydrolysis. Gum tragacanth is an effective bifunctional emulsifier that lowers the interfacial tension between oil and water. The minimum surface and interfacial tension is reached at a gum concentration of only 0.25%. Tragacanth is compatible with most hydrocolloids, proteins, and fats used in food preparation. However, interaction with acacia gum results in a viscosity reduction and improved emulsifying properties of the mixture. This reaction is exploited to produce pourable emulsions with long shelf life. Pure gum solutions are unusually stable to microbial attack and have long shelf lives without loss of viscosity [2,3].

Applications

Many of the traditional tragacanth applications now use more cost-effective additives such as xanthan gum, but it is still used where the viscosity properties, emulsifying ability, and acid stability are superior to other food hydrocolloids [2].

Tragacanth is used as filler, stabilizer, emulsifier, thickener, or gelling agent in icings, confectionary, salad dressings, sauces, ice creams and sherbets, bakery fillings, and oil emulsions. In beverages, it suspends the pulp, prevents settling of fruit particles, and gives the desired body.

Toxicology and Regulatory Affairs

The US FCC classifies tragacanth as GRAS. There is no ADI specified from WHO/FAO. The Scientific Committee of Food (SCF) in Europe lists it in Annex I of the 'Directive for food additives other than colours and sweeteners' – with unrestricted dosage (quantum satis). Although allergenic responses have been reported from animal studies, it is not listed in the European allergen list (Annex IIIa) [2].

1.4.3
Karaya Gum

Raw Material Origin, Harvesting, and Quality

Gum karaya, also known as sterculia gum, India gum, or Indian tragacanth, is a dried tree exudate obtained from certain species of *Sterculia* (*Sterculia urens* and other species, family Sterculiaceae) or from *Cochlospermum* species (family Bixaceae). The gum is obtained by making incisions into the trunks. The gum exudes within 24 h in large irregular lumps or tears that dry in the hot climate. Each tapping yields between 1 to 5 kg of gum and each tree can be tapped about five times during its life time. The Indian crop is harvested between April and June, before the monsoon season. The crop in Senegal is collected in September to January or from March to June. The gum is collected in bigger volumes. Then bark and foreign matter (BFM) are removed by hand. Lumps are broken down into smaller pieces. Sorting occurs on the basis of gum colour and residual BFM. Premium quality superior no. 1 is used for food and pharmaceutical applications and may only contain 0.5% BFM. It gives light-coloured solutions and high viscosities. As BFM increases, the solution colour becomes darker. Technical grades contain more than 3% BFM. Further processing is carried out in Europe and USA. The gum is mechanically ground to particle sizes below 110 μm. Fibres, bark, and other foreign material are removed by aspiration and density-table separation. Blending is an option to obtain products with consistent colourizing and thickening effect [2].

The demand for highly-purified gum karaya has decreased from approximately 6000 MT globally in the early 1980s to 3000 MT in 2010. The shift has been towards more cost-effective gums. India is the biggest producer (50–70%), with the remaining volume produced in North Africa. Some 75–80% of the crop is imported by the USA.

Chemical Structure

Gum karaya is a complex, branched, acid polysaccharide composed of the building blocks D-galactose, L-rhamnose, D-galacturonic, and L-glucuronic acid. The gum naturally occurs with Ca^{2+} and Mg^{2+} ions associated with the uronic acid groups. The amount of uronic acid is about 37%. The L-rhamnose content is much higher compared to other exudate gums – this can be used for identification. The monosaccharides are partially acetylated. The molecule contains 8–14% acetyl groups (dry matter basis), depending on karaya source and its age. Free acetic acid is split off on ageing. Increased temperature and humidity and fine particle size increase the rate of acetic acid formation. The structure of karaya contains a central chain of galactose, rhamnose, and galacturonic residues with side chains of glucuronic acid. The exudate occurs as a mixed calcium and magnesium salt. Approximately 1% proteinaceous material has been detected. The average molecular weight is extremely high and is about 9.5 MDa; it can be up to 16 MDa [2].

Karaya can vary widely in its chemical composition, corresponding to which are several different compositions that are summarized below.

Solubility and Viscosity

Because of the presence of acetyl groups and the strong crosslinking, gum karaya is the least soluble of the commercial exudates. It swells in water, to many times its original volume, without dissolution. It does not dissolve in water to give a clear solution but absorbs cold water rapidly to form viscous colloidal dispersions at low concentrations (<0.02% in cold water, 0.06% in hot). At high concentrations of 3–5%, the gum forms thick, soft gel-like pastes that have spreadable quality and are ideal for suspending particles. Karaya solutions exhibit true plasticity at 2–3% as a consequence of the high degree of ordered structure and its large, heavily branched molecular network, which absorbs and immobilizes water very strongly and therefore offers considerable resistance to flow. Hydration depends on particle size – finer gums hydrate much more rapidly than coarser gums. Application of high shear to grainy dispersions from coarse powders will give smooth solutions with reduced viscosity. Heating increases solubility, due to a change in the polymer conformation, but results in a permanent viscosity loss. A maximum of up to 4% gum can be hydrated in water. By heating under pressure, colloidal solutions with 20% concentration can be produced. The viscosity decreases with increasing age of the powder. Grinding to fine particles reduces viscosity, too. High storage humidity or high storage temperatures also contribute to viscosity loss. Boiling gum solutions for more than 2 min leads to an irreversible viscosity reduction. The solutions show thixotropic flow behaviour. The viscosity depends strongly on pH. The pH of a 1% solution is about 4.6. Above pH 7, alkali irreversibly transforms the characteristics of short-bodied karaya solution into a ropy, stringy mucilage with a ninefold increase in viscosity. This has been ascribed to irreversible deacetylation of the karaya molecule. The maximum viscosity occurs at pH 5–7. At pH 2–5 and above 11, there is a strong viscosity decrease. Karaya forms viscous solutions in 60 vol.% alcohol, but is insoluble in higher concentrations of alcohol. The viscosity of karaya solutions remains constant for several days. Since these solutions are subject to microbial attack, preservatives are recommended. Depending on the food formulation benzoic or sorbic acid, methyl and propyl *para*-hydroxybenzoate, glycerol, and propylene glycol are commonly used [2,3].

Properties

Karaya has an acetic odour and taste. Its powder is light grey to pinkish grey. Cost is based on colour and purity. There is no distinct correlation between viscosity and grade. Where viscosity is important, the powdered karaya should be used within six months after processing, because its viscosity decreases with age.

Karaya is compatible with most other gums, as well as proteins and carbohydrates.

Interactions with sodium alginate have been reported to modify solution properties. Karaya reduces the gel strength of agar gels. Pyrilamine maleate, a strong hydrotrope and antihistaminic, and karaya are incompatible. Karaya dispersions lose viscosity when certain strong electrolytes (e.g. $CaCl_2$) are added in small amounts. Alkali and alkaline conditions make karaya solutions very ropy.

Applications

Karaya has similar physical characteristics to gum tragacanth and can replace it in some applications. The elimination of acetic acid restricts its use due to sensory reasons (odour, taste). The relatively good acid stability and strong water binding capability makes it suitable as a thickener and emulsifier in foods, as a bulk laxative, and adhesive for dentures or colostomy bags in medical and pharmaceutical applications.

Gum karaya is used as suspending aid and emulsifier in cold-processed dressings and sauces. Other application fields are aerated dairy desserts, cream, beer, and meringues where it stabilizes the foam. The gum prevents whey separation in cheese spreads, acts as binder in meat products, extends shelf life of baked goods, prevents synaeresis in gels, and controls ice crystal growth in sherbets, sorbets, and ice lollies. It is added as fibre source for dietary supplements – often in combination with guar.

Toxicology and Regulatory Affairs

Karaya gum is digested only to a minor extent, if at all [3]. It is classified in the USA as 'generally recognized as safe' (GRAS). The WHO/FAO did not specify an ADI. Within the EU, karaya is listed as E 416. In the past, the SCF of the EU set an ADI of 12.5 mg per kg body weight per day and allocated it to Annex IV of the 'Directive for food additives other than colours and sweeteners'. This legislation restricts karaya use to special food groups such as chewing gum, dietary food supplements, desserts, emulsified sauces, cereal and potato based snacks, egg-based liqueurs, bakery filling, toppings, and coatings [2].

1.4.4
Ghatti Gum

Raw Material Origin, Harvesting, and Quality

Gum ghatti is the dried exudate from the bark of the tree *Anogeissus latifolia* of the family Combretaceae. The tree is quite large and found in large numbers in the dry, deciduous forests of India and Sri Lanka. The crude gum has a glossy fracture and occurs in rounded tears, which are normally less than 1 cm in diameter. The bigger pieces are the size of a hazelnut. The colour of the gum exudates varies from off-white to dark brown. The lighter the colour, the better the quality. Crude ghatti has up to 15% total impurities at maximum 15% water. Since ghatti and karaya are found in the same geographic areas, the harvesting and grading are similar. The best crops are picked outside of the monsoon season, and the largest crop is harvested in April. After picking the gum, it is dried in the sun for several days. It is then hand sorted according to colour and impurities into various grades. There are also white commercial powder grades available, which are highly purified and spray-dried.

Chemical Structure

Ghatti gum is a water-soluble calcium–magnesium salt of a complex acidic polysaccharide. The monomers are L-arabinose (47.6%), D-galactose (28.6%), D-mannose (9.5%), D-xylose (4.8%), and D-glucuronic acid (9.5%). The molecular ratio is 10 : 6 : 2 : 1 : 2. Traces of rhamnose are found, too. The sugar components are partially acetylated (5.5% on dry matter). Three different structural elements are detected. The macromolecule consists of several main chains from β-(1 → 6) linked galactoses bonded to arabinose at the 3-position or to (1 → 4) linked glucuronic acid, as well as α-(1 → 2) linked mannoses bonded to glucuronic acid at the (1 → 4) position. In the 3-position to galactoses, there are side chains with arabinose and xylose units. The average molecular weight is 270 kDa. The molecules may have an overall rod shape in solution [1,3].

Solubility, Viscosity, and Properties

Gum Ghatti is 80–90% soluble in water. It can be dispersed in hot or cold water to form colloidal dispersions. Because of this limited solubility it is impossible to attain the concentrations reached with acacia gum. The viscosity of dispersion is very variable from one batch of gum to another – for example, the viscosity of a 5% dispersion of the best grade of gum ghatti varies between 30 and 400 mPa s. This has been ascribed to the presence of two different fractions in the gum, one soluble in cold water, the other forming a dispersible gel. Viscosity in water is controlled by the proportion of the latter, which is 10–30 times more viscous than the soluble fraction. The gelling component occurs in amounts of 8–25% and can be dissolved to some extent (~20% of the total) by stirring the gum at 92 °C (197 °F) for about 2 h, but complete dissolution requires maceration. The two fractions of the heterogeneous gum molecules differ in chemical and physical properties. Both fractions are largely in salt form with less than 10% of free acid.

The significant difference lies in the proportion of the different cations present: the soluble fraction contains about 62% Ca^{2+}, 32% K^+, 5% Mg^{2+} and 1% Na^+. In contrast, the dispersible gel fraction has a large Ca^{2+} content (>90%), 4% Mg^{2+}, 2% K^+, and no Na^+. The higher viscosity and lower solubility of the gelling fraction are due to aggregation, through interaction between the divalent Ca^{2+} ion and anionic groups on different molecules. Removal of Ca^{2+}, by precipitation with sodium carbonate, reduces the viscosity of gum dispersions irreversibly. This effect is not reversed on subsequent addition of Ca^{2+} which, conversely to alginates and other gums, further decreases viscosity. The effects of other cations, and of pH, on viscosity of gum ghatti dispersions differ little from those observed with acacia gum. The addition of sodium salts decreases viscosity. Ghatti dispersions are not affected by small amounts of acid or alkali since the gum acts as a buffer and reverts to its normal pH of about 4.8. The finer the particle size of the powdered gum, the faster it will swell and reach its maximum viscosity. If the gum is coarser than 150 mesh, solutions will be grainy and require a longer hydration time. Gum ghatti will not form a true gel. The dispersions are slightly coloured due to traces of pigments. The viscosity is a function of pH value. The optimum of viscosity is at pH 5–8, but stability is found from pH 3 to 11. At 3–5% gum concentration, a highly viscous (100–500 mPa s) dispersion of uniform smoothness and texture results. The dispersions are non-Newtonian – viscosity increases geometrically with concentration. Solution viscosity increases with time (about 10% every 7 days). The molecule is more resistant to glycosidic fission than acacia gum. Its viscosity is greater than that of acacia gum, but less than that of gum karaya [3].

The gum tastes mild and it is almost odourless. The colour of commercial powders varies from tawny to dark brown. The lightest colour has the least impurities and the greatest effectiveness in most applications. Ghatti is compatible with other plant hydrocolloids as well as carbohydrates, most proteins, and some fats. Viscosity loss is noted below pH 3 and above pH 11. A higher apparent viscosity is observed in an aqueous ghatti gum dispersion containing 25% ethanol. Dry films are relatively soluble and brittle. Ghatti has good emulsifying properties in particular for oil-in-water emulsions. To prevent bacterial attack in solutions, the gum is preserved with methyl and propyl *para*-hydroxybenzoate (0.15–0.02%), glycerine, or propylene glycol.

Applications and Regulatory Affairs

Gum ghatti has GRAS status from US FDA. In Europe it is not classified as a food additive. The world tonnage is relatively low. In general, applications are the same as for acacia gum. The stabilization of suspensions and emulsions is the most important use. In food, gum ghatti stabilizes butter-containing table syrups. It is used at about 0.4% in combination with 0.08% lecithin. Ghatti modifies the refractive index of table syrup until the syrup becomes clear. In pharmaceuticals and cosmetics, the gum is used to prepare stable, powdered, oil-soluble vitamins. It acts as a stabilizer in oil-in-water emulsions and in X-ray suspensions with barium sulfate.

1.5 Bacterial Polysaccharides

1.5.1 Xanthan

Raw Material and Manufacturing

Xanthan gum was first discovered in the 1960s and commercialized in the 1970s. It is produced by aerobic fermentation of glucose or sucrose solutions with pure cultures of *Xanthomonas campestris* or *phaseoli*. Owing to the high viscosity of the formed xanthan gum, the sugar concentration of the substrate may not exceed 5%. Commercial production is carried out batchwise by submerged fermentation with strong agitation. Besides the carbohydrates, the sterile medium contains a nitrogen source (yeasts) as well as nutrient and buffer salts like magnesium sulfate and other trace minerals. After initial inoculation with the selected strain, fermentation is continued for approximately 3 days at 30 °C (86 °F). At the end of the fermentation, the broth undergoes sterilization treatment to eliminate any viable microorganisms. The cellular components are removed by centrifugation. The xanthan gum is then recovered by precipitation with isopropyl alcohol. After separation of the fibres by centrifugation or filtration, they are dried and milled before packaging. Annual production is approximately 50 000 MT, 50% of which is used in food applications.

Xanthan gum is sold as an off-white powder with a moisture content of about 11% and ash content of 6–9% [2].

Chemical Structure

Xanthan is a long-chain polysaccharide with a high number of trisaccharide side-chains. The building blocks D-glucose, D-mannose, and D-glucuronic acid occur in a molecular ratio of about 3 : 3 : 2 (Table 1.9). The primary structure of xanthan consists of a cellulosic backbone of β-(1 → 4) linked D-glucose units substituted on alternate glucose residues with a trisaccharide side chain. The trisaccharide chain is composed of two mannose units separated by a glucuronic acid. Approximately half the terminal mannose units are linked to a pyruvate group (about 3.5%) and the non-terminal residue usually carries an acetyl group (about 4.7%). The carboxyl groups on the side chains render the molecules anionic. Xanthan gum has an average molecular weight of about 2000 kDa with a narrow-molecular-weight distribution compared to most polysaccharides [2].

Solubility, Viscosity, and Properties

Xanthan gum is a fast-hydrating water-soluble hydrocolloid that can be dissolved at room temperature. For efficient hydration, the individual gum particles must be well dispersed in the solvent. Hydration time is reduced with increased mixing speed. As the particle size increases, xanthan becomes easier to disperse but slower to hydrate. Generally, high ionic strength (>1–2%) or high solid content slows down hydration. With high-speed mixing, hydration takes between 15

Table 1.9 Characteristics of xanthan gum.

	Xanthan gum
E-No.	E 415
Origin	Bacterial polysaccharide from fermentation (*Xanthomonas campestris*)
Chemical composition	• Glucose • mannose • glucuronic acid
Nutritional value (in 100 g) – metabolism	210 kJ (50 kcal); slight resorption
Fibre content	Approx. 100%
Toxicology	No ADI value defined
Solubility at low temperature (H_2O)	High – 100%
Appearance of an aqueous solution	Opaque
Viscosity of solution in water	Very high, low salt contents (<0.5%) increase viscosity and stability
Impact of heat on viscosity in water (pH 7)	Viscosity decrease (but 100% reversible below 100 °C (212 °F))
Viscosity development in water at pH 7 (T = 0–100 °C)	Slight viscosity decrease + conformational transition
Shear stability	High – pseudoplastic with flow limit
Thickening effect	High
pH stability	High (2 > pH > 10)
Decomposition	Extreme acid- and enzyme-stable
Film formation	High
Emulsion stabilization	High
Gelation	No – only in combination with LBG
Gel strength and gel stability	Gel of xanthan and LBG (50 : 50 or 60 : 40) at minimum 0.3% at low salt content and neutral pH: high gel strength, gels are highly heat-stable
Gel transparency	Low
Tendency for gel synaeresis	Low (syneresis-free)
Impact of electrolytes (cations +, 2+, 3+)	No
Reaction with Ca^{2+} ions	No
Protein activity	No – precipitation possible in milk
Crystallization control	High
Synergistic effects with other hydrocolloids	+ LBG or Cassia: viscosity increase + gel formation; + Guar or tara gum: viscosity increase; + konjac: gel formation
Other synergistic effects	Addition of maximum 0.5% NaCl improves stability
Negative interactions	—
Dosage level in foods	Low, typical 0.05–0.5%

and 30 min and with a low-speed mixer it takes up to 1 h. In solution, the side chains of xanthan wrap around the cellulose-like backbone, thereby protecting it. It is thought that this is responsible for the excellent stability of xanthan gum under adverse conditions. Its solutions undergo a conformational transition during heating, which is believed to be associated with the change from a rigid ordered state at low temperature to a more flexible, disordered state at high temperatures. The temperature of conformational transition depends on ionic strength and composition of the xanthan gum molecule, especially its acid group content. Typically, for low-level salt food applications, it occurs above 90 °C (194 °F). The presence of low salt amounts helps to maintain the rigid ordered conformation of xanthan gum, which in turn causes a relative insensitivity of the viscosity to additional salt and elevated temperature. Pure solutions exhibit Newtonian flow behaviour at very low shear rates, followed by a pseudoplastic region as shear rate increases and, finally, an upper Newtonian viscosity at very high shear rates. Xanthan solutions have the visco-elastic properties of a weak gel, giving effective suspending characteristics in liquid foods such as dressings, sauces, or cakes before baking. The gum has the ability to develop extremely high viscosities even at low concentrations. Xanthan solutions exhibit exceptional stability during heating, even in the presence of salts and/or acids. As temperature increases, solution viscosity decreases but recovers almost completely upon cooling. Sterilization temperatures reduce cold viscosity by 10–20%. Solutions are independent of pH over a wide range (pH 3–10). Xanthan is compatible with most organic acids, and it is more stable than other thickeners. It can be hydrated directly in an acidic solution, but better results are achieved by preparing the gum solution first and then adding the acid. As salts reduce hydration speed, it is recommended to hydrate xanthan in water before salt addition. Once hydrated, xanthan tolerates up to 20–30% salt without viscosity change. Generally, the gum is not soluble in organic solvents, although it will hydrate directly in glycerol at 65 °C (149 °F). After hydration, a xanthan solution tolerates up to 50% ethanol or isopropanol without precipitation. It is compatible with all other food hydrocolloids and it is very resistant to enzymes (e.g. amylases, proteases, pectinases, and cellulases). Xanthan gum has strong synergies with galactomannans such as locust bean gum (LBG), guar gum, tara gum, cassia gum, and glucomannans such as konjac gum. In solution the galactose-free (smooth) regions of the mannose backbone form associations with the ordered xanthan helices, which result in a synergistic viscosity increase in the case of guar or gelation in the case of LBG. Konjac forms strong elastic gels after heating and cooling in mixtures with xanthan gum. Cassia gum exhibits similar behaviour to LBG and tara gum has synergy intermediate to that of guar and LBG [2].

Applications

The largest application area for xanthan is dressings and sauces. In cakes, it prevents sedimentation of fruit or chocolate pieces before and during baking. It increases volume, retards staling, and prolongs freshness. Xanthan is added to dairy desserts to support gel formers and reduce synaeresis. Instant dry mix products such as beverages, soups, desserts, and low-calorie products achieve consistent particle distribution and body by xanthan.

Toxicology and Regulatory Affairs

Xanthan has had US FDA approval since 1969. In Europe, it has been approved since 1974 as food additive E 415 with a non-specified acceptable daily intake (ADI). All animal studies, biochemical, and toxicological examinations support its harmlessness. Xanthan is only slightly digested in the human intestinal tract (0.5 $kcal\,g^{-1}$).

1.5.2 Others

Gellan

Raw Material Gellan gum is an extracellular polysaccharide produced through fermentation by the microorganism *Sphingomonas elodea* (previously identified as *Pseudomonas elodea*, but later reclassified). All the different forms of gellan gum are made from the same basic fermentation process. Large, sterile fermentation vessels are used to allow the bacteria to convert simple sugars like D-glucose and other nutrients into this polysaccharide. Once the cells have been killed and separated from the fermentation broth, the solution is further treated to produce four different types of gellan gum:

1) high-acyl unclarified gellan gum is made by precipitation with organic solvents like alcohols;
2) high-acyl clarified gum is obtained after clarification and subsequent precipitation;
3) low-acyl unclarified gum is not manufactured commercially at present;
4) low-acyl clarified gum, a de-esterified product, is made by clarification, subsequent removal of acyl groups by treatment with alkali (deacetylation at 80 °C (176 °F) at pH ~10 for 10 min), and final precipitation [3].

Chemical Composition Gellan is a linear, anionic heteropolysaccharide with a straight chain consisting of the building blocks D-glucose, L-rhamnose, and D-glucuronic acid in a molecular ratio of 1.5 : 1 : 1. The chain is made of a tetrasaccharide chemical repeat unit in which β-(1 → 4)-linked glucose, glucuronic acid, glucose, and rhamnose in α-(1 → 3) linkage are bonded together. The native product is partially esterified: the (1 → 3)-linked glucose residue contains a C2 linked L-glycerate and about 50% C6-linked acetate substituents. In its native or high-acyl form, two acyl substituents – acetate and glycerate – are present. Both substituents are located on the same glucose residue and, on average, there is one glycerate per repeat and one acetate per every two repeating unit. In low-acyl gellan gum, the acyl groups are absent. According to X-ray diffraction studies, the molecule is supposed to adopt a double or threefold helical structure after heating and cooling [1–3].

Solubility, Viscosity, Gelation, and Properties The presence or absence of acyl groups on the gellan gum backbone has a profound effect on its physical, chemical, and functional properties. Therefore, both types, low-acyl and high-acyl, are discussed separately. In general, gellan gum is dispersible in cold water and fully soluble in hot water. The hydration temperature of low-acyl gellan gum is sensitive to the ionic environment and is particularly sensitive to divalent cations. The gum itself contains divalent cations and will only hydrate in cold, deionized water. Hydration is further inhibited by the divalent ions present in most water supplies. This effect makes low-acyl gellan gum easy to disperse in cold water without forming lumps. Subsequently, it can be hydrated by adding sequestrants or chelators, such as citrates and phosphates, to control the divalent ions, heat or a combination of both. Hence, its hydration temperature can be controlled effectively. Without sequestrants, low-acyl gellan gum requires a temperature above 75 °C (167 °F) to fully hydrate in soft water. However, it can be hydrated in cold, soft water using 0.12% sodium citrate. The pH value of solution also affects the hydration characteristics of low-acyl gum. At pH values above the pK_a of gellan gum (about pH 3.6), the gum is in a form that allows easy dissolution. If the solution pH is below 3.6, the gum will exist in a predominantly acid form that is not completely soluble. When formulating acidic products, the acid should be added after gum hydration. Dissolved sugars have an effect on gum hydration. Hydration is possible with up to 25% dissolved sugar. At higher sugar levels, the gum must be heated to be fully solubilized. Another option is to hydrate the gum in a low-sugar environment first and add the remaining sugar solids later. High-acyl gellan gum swells in deionized water, creating a consistency like a swollen starch paste. Low levels of sodium ions inhibit this swelling behaviour. The addition of sodium salts is a useful strategy to improve gum dispersion and minimize viscosity during processing. Heat is required to fully hydrate high-acyl gellan gum. It hydrates at 70–80 °C (158–176 °F), even with high ion concentrations. In contrast to low-acyl gum, the calcium effect on high-acyl gum is small and sequestrants do not facilitate hydration. Both gum types can be dispersed directly in milk and will hydrate during normal heat processing. In acidic systems, the pH must be above 4 for good hydration. Gels can be formed over a broad range of conditions. Critical parameters are pH value, temperature, sugar solids, and cation content. Gels occur at low concentrations of 0.05–0.3% and upon cooling to 30–45 °C (86–113 °F). High-acyl gellan gum forms soft elastic gels whose strengths increase in the presence of mono- or divalent cations. The gels of low-acyl gum are very firm and brittle and similar to agar gels. Sugar has a plasticizing effect on the gels, which are thermoreversible, stable at pH 3.5–8, and not attacked by enzymes. When the gellan gum concentration is high enough, the network structure becomes a demoldable gel. At lower concentrations, the gum molecules still associate and form a long-range network; the system remains very fluid and forms a so-called fluid gel. Fluid gels exhibit a highly pseudoplastic flow property – the viscosity decreases with increasing shear. While low in viscosity, however, these gels have a high elastic modulus, which imparts suspension properties to the system. In fact, fluid gels including gellan gum have a yield stress that keeps particles in suspension.

Blends of high-acyl and low-acyl gellan gum are commonly used to produce a 'gel within a gel' and create flexibility in formulation development. Gellan is also mixed with other gelling and non-gelling hydrocolloids to modify rheology, stability, heat performance, or setting temperatures. Xanthan or CMC are added to low-acyl gellan gum to produce softer gels with a smooth texture. Low-acyl gum is used to increase the firmness of soft, elastic gels such as ι-carrageenan, gelatine, or combinations of LBG and xanthan gum. Gellan alters the setting and melting properties of other gelling agents. For example, gellan gum can be used to improve the heat stability of gelatine gels and to raise the set temperature of gelatine-based confections and dessert gels. Additionally, gummy candies can be made gelatine-free by using a mixture of carrageenan and gellan gum. Gellan strongly reduces the setting time of starch confectionary and increases the heat stability of candies, preventing them from sticking together when exposed to warm environment [3].

Application Gellan gum is used as thickener, binder, film-former, and stabilizer in a wide range of food applications. It stabilizes water-based gels such as desserts and drinking jellies. A lot of Asian food products contain the gum as alternative to agar. It is used in bakery formulations, where it does not increase the batter viscosity when cold, but hydrates during baking. It improves bake-stability of fillings with low to high soluble solids content. Gellan gum replaces gelatine in cultured dairy products such as yogurt and sour cream in vegan, Kosher or Halal nutrition. It is typically added in low concentrations (maximum 0.1%) to raw milk prior to homogenization. In beverages, gellan gum reduces cloud and pulp settling. It is suitable in systems with low milk amounts, low-quality milk protein, or heat-damaged proteins such as those found in spray-dried milk powders where carrageenan comes to its functional limit. Gellan gum gives a short texture to gelled confectionary, improves gelatine gummy candies, reduces stickiness, and shortens production time of demoldable products. Other applications are low-calorie (sugar-free) jams in which pectin is not functional, fruit preparations for yogurt, sauces, no-fat salad dressings with herbs, and films and adhesion systems [3].

Toxicology and Regulatory Affairs Gellan gum was first used and approved in 1988 in Japan, where microbial polysaccharides are considered as naturals materials. Gellan gum is now approved for use in foods, cosmetics, pharmaceuticals, and non-food products in USA, Canada, Australia, Europe, and many other countries of Latin America, South America, and Asia. The FDA has approved it for food in the USA. There is no ADI specified. In Europe it is classified as food additive E 418 [3].

Pullulan

Raw Material and Manufacturing The manufacturing process of pullulan was developed in Japan in 1976. Pullulan is an extracellular glucan produced from starch syrup and sugar as substrate through fermentation by the fungus *Aureobasidium pullulans*, commonly called 'Black yeast'. With help of nitrogen sources and different salts, the cultivation is carried out at 30 °C (86 °F) under agitation. After 100 h, the yield is greater than 70%. To purify pullulan, the microbial cells are first removed by filtration. After decolouration with activated carbon, the filtrate is subjected to an ion-exchange process to remove the contaminating salts and proteins and to obtain a purified solution. After concentration, the solution is drum dried and pulverized to produce a fine powder [2].

Various bacteria such as *Klebsiella* species possess enzymes, the so-called pullulanases, with which they cleave pullulan and can make it available as a carbon source.

Chemical Composition Pullulan is a linear polysaccharide, a homoglucan, consisting of maltotriose building blocks. Three of the glucose units of maltotriose are linked through α-(1 $\rightarrow$ 4) glycosidic bonds while the subsequent maltotriose units are linked by α-(1 $\rightarrow$ 6) bonds. The molecular weight of the polymers is between 10 and 400 kDa.

Solubility and Viscosity Pullulan easily dissolves in cold or hot water to form a stable, viscous solution that does not gel. It is insoluble in organic solvents, with the exception of dimethylformamide and dimethyl sulfoxide. Pullulan can be modified with ethers or esters, so that, depending on the degree of substitution, it may be insoluble in water and soluble in other organic solvents such as acetone or ethanol. Pullulan solutions have a relatively low viscosity, like gum Arabic, when compared with other hydrocolloids. The solutions are Newtonian fluids with a surface tension similar to that of water (74 dyne cm^{-2}). Viscosity is essentially unaffected by pH over a wide range from pH 2 to 11 and is stable in the presence of most metal ions. With some ions, for example borate, complex formation with hydroxyl groups is observed, which leads to a strong increase of viscosity. However, even under such conditions, pullulan does not gel [2].

Properties Pullulan is a tasteless and odourless white powder. Its solutions are stable over a wide range of pH and it is relatively stable to heat. A significant feature of pullulan is its high stability to sodium chloride. Heating pullulan in 30% sodium chloride solution at 100 °C (212 °F) for 6 h produces no noticeable change in viscosity.

It is highly adhesive when dissolved in water and it has remarkable binding properties – it shows superior adhesiveness on wood compared to cornstarch and modified cellulose. Pullulan readily forms films that are edible, transparent, oil-impermeable, anti-static, readily soluble in water, and have low oxygen permeability. The powder is directly compressible with heat and pressure in the presence of moisture. Pullulan is a biodegradable polymer. It is easily metabolized by many microorganisms found in nature to give carbon dioxide and water. But it is largely undegraded by human digestive enzymes and can be used as low-calorie food additive. Pullulan is a soluble dietary fibre with prebiotic function [2].

It can replace gelatine in coatings.

Application Pullulan is used as binder, thickener, glazing, and coating agent in food (30 vol.%). Other applications are films (30%), capsules (30%), and cosmetics and drugs (10%). Edible packaging films of pullulan are common in Japan. Pullulan is used in the coating of food supplement products. Fresh breeze strips are made with pullulan. Instant beverages, creams, icings, frostings, soy sauces, other sauces, desserts, confectionary, and processed meat products are made with the addition of pullulan (0.2–3%).

Toxicology and Regulatory Affairs In the USA, the FDA approved the GRAS (generally recognized as safe) notification for pullulan in 2002. Pullulan was recently approved in the EU as food additive E 1204 for use in capsules, tablets, and films under directive 2006/52/EC. It is also permitted in several Asian countries, in Russia, and in some South American countries. No ADI has been specified [2].

Dextran

Raw Material Dextran is produced in an extracellular process by fermentation of the disaccharide sucrose by enzymes of bacteria of the genus *Leuconostoc* or *Streptococcus*. Sucrose is the carbon source; hydrolysed yeast is the nitrogen source. The pH value is about 6.5–7. After fermentation, the culture medium is cleaned of microorganisms. The solubilized dextrans are precipitated by addition of organic solvents such as methanol, ethanol, 2-propanol, or acetone. By re-precipitation, entrained contaminants are removed. The resulting product is a white powder.

In nature, dextran serves as reserve materials for yeasts and bacteria.

Chemical Composition Dextrans are high molecular weight, highly branched, neutral biopolysaccharides. Since the polymers are composed only of glucose units, they belong to the homoglycans. Some 95% of the glucose monomers are located in the main chain, an α-(1 $\rightarrow$ 6) glucan. The glycosidic bonds to the side chains occur as α-(1 $\rightarrow$ 4) or α-(1 $\rightarrow$ 3), rarely as α-(1 $\rightarrow$ 2) linkages. Natural dextrans have molecular masses of 10–50 000 kDa.

Solubility and Viscosity Dextrans are readily soluble in water or glycerine at room temperature. Sterile aqueous solutions are very stable at pH 4–7 at room temperature. Solubility and viscosity depend on the molecular mass and on the structure of the dextran. Highly viscous, slimy liquids can be formed.

Application The colloid osmotic pressure of a dextran solution (6% of a molecular weight of 75 kDa) corresponds to that of human blood and, therefore, dextrans are used in medicine as a substitute for blood plasma. Low molecular weight dextrans act as platelet aggregation inhibitors. Other uses are as a stabilizer in freeze-drying processes, in blood purification, and for X-ray contrast liquids. Dextrans act as carriers for gel chromatography, as molecular sieves in chemical analysis, as tablet binder, and as protective colloid for cells to prevent freezing damage. The tendency to form films makes dextrans suitable for protective coatings. Examples are improved solubility and flavour fixing for coffee and tea powders.

Toxicology Dextrans are decomposed in the human digestive tract and transferred into energy. As blood plasma expander, dextrans are excreted quickly by the kidneys. Highly branched substances can cause allergies in humans. There is no defined ADI value [1].

Curdlan

Raw Material The polysaccharide curdlan is produced in an extracellular procedure by the non-pathogenic bacteria *Agrobacterium biobar* and mutants of *Alcaligenes faecalis* var. *myxogenes*, a group of microorganisms occurring in the soil. The product is manufactured by conventional fermentation processes. It is an odourless white powder.

Chemical Composition Curdlan is a linear β-(1 → 3)-glucan (99%), a high-molecular-weight polymer of glucose with the general formula $(C_6H_{10}O_5)_n$. It contains hardly any branches. It can be partially esterified with succinic acid and occur as succino-glucan. The amount of esterified polysaccharide and degree of esterification depend on the bacteria producing it. Curdlan manufactured technically for industrial use contains only very small amounts of succinic acid. The average molecular weight is between 40 000 and 70 000 Da.

Solubility and Viscosity Curdlan is insoluble in water up to 54 °C (129 °F). Upon heating its aqueous suspensions above this temperature, it starts to swell and forms strong elastic gels. The gel strength is between that of agar and gelatine. Curdlan is soluble in aqueous alkaline solutions.

Properties Curdlan forms irreversible gels in water during heating. The strength of the gels increases with increasing temperature. It reaches a plateau at 80 °C (176 °F) and starts increasing again at 100 °C (212 °F). The gels exhibit synaeresis that strongly depends on the temperature of storage. Curdlan forms water-insoluble films with strength between those of cellulose and amylose. The films are edible, biodegradable, and impermeable to oxygen [1].

Application and Regulatory Affairs Curdlan has numerous applications as a gelling agent in the food, construction, and pharmaceutical industries. It is suitable to replace gelatine and agar in production of jellies, desserts, and confectionary. Curdlan can be used as thickener and binder in dietetic foods (e.g. in salad dressings, desserts, pasta). For long-life products, it can deliver edible coatings.

It has been approved as a food additive by the US FDA. In Japan, there is food approval, too.

Scleroglucan

Raw Material and Manufacturing Scleroglucan is the extracellular secretion of the filamentous fungus *Sclerotium, inter alia* the species *Sclerotium rolfsii* and *Sclerotium glucanicum.* The polysaccharide is produced by aerobic submerse fermentation in conventional fermentation tanks with controlled aeration. D-Glucose serves as carbon source and corn steep liquor or nitrate as nitrogen source. After fermentation, the fungi spawn is removed by filtration. The polysaccharide is precipitated with an organic solvent, filtrated or centrifuged, dried, and milled [1].

Chemical Composition Scleroglucan is a non-ionic neutral homoglucan consisting of D-glucose as building block. The β-(1 → 3)-glucan has one β-(1 → 6)-glucose residue as side chain every three main residues. Its molecular weight is very high (2 MDa).

Solubility and Viscosity Scleroglucan is easily soluble in cold and in hot water. It produces aqueous solutions with a very high viscosity. Owing to its non-ionic nature, acids and alkalis do not affect scleroglucan over a wide pH range (2.5–12). Mono-, di-, and trivalent cations have no impact on solution viscosity. Unlike most natural and synthetic gums, high temperature has little effect on the viscosity of a scleroglucan solution. Below 10 °C (50 °F), the solutions form a soft gel that can be eliminated by shaking or heating. Solutions of scleroglucan may be sterilized by heating them at 121 °C (250 °F) for 20 h without affecting their viscosity.

Application Besides the thermal stability, solutions of scleroglucan exhibit pseudoplastic behaviour with a high yield value, resulting in solutions of high-suspending power with good pouring properties. Because of its high yield value, it is extremely effective in holding particles in suspension, in static as well as in dynamic conditions, without any risk of sedimentation. Scleroglucan has a good emulsifying capability and acts as foam stabilizer. It is compatible, without synergism, with most other thickening hydrocolloids. It is also compatible with most widely used surfactants such as sulfates, sulfonates, and quaternary ammonium salts. Scleroglucan remains soluble in mixtures containing up to 50% of polyols and glycols.

Its remarkable rheological properties and stability over a wide range of pH values, salinities, and temperatures make scleroglucan suitable for several applications.

In Europe it is mainly used in cosmetic products for skin care and sun-care, in shower gels, body washes, shampoos, conditioners, shaving foam, make-up, mascara, and eyeliners.

1.6
Overview Tables for the Most Important Cellulose Derivatives

Tables 1.10–1.14 give an overview of characteristics and selected properties of the most important cellulose derivatives (MCC, MC, HPMC, EC, HPC, and CMC). The idea is to present the main features at a glance to simplify developmental work and check whether the substance is generally suitable for a specific application. The details for each characteristic can be studied in the respective chapter.

Table 1.10 Characteristics of powdered and microcrystalline cellulose (MCC).

	Powdered cellulose and colloidal MCC
E-No.	E 460 (i), E 460 (ii)
Origin	Wood pulp or cotton linters
Chemical composition	Linear molecule of ß-D-glucose with an increased amount of crystalline sections; colloidal MCC consists additionally of cellulose gum (CMC), alginates, guar or xanthan gum
Nutritional value (in 100 g) – metabolism	Powdered cellulose: 0 kJ (0 kcal), no resorption; MCC: depends on added hydrocolloid
Fibre content	Cellulose is 100% soluble dietary fibre
Toxicology	No health concerns, no ADI value defined, but laxative effects at 12–15 g day^{-1}
Solubility at low temperature (H_2O)	Low, only swelling (0.24% water-soluble components in powdered MCC)
Appearance of an aqueous solution	Opaque
Viscosity of solution in water	Formation of a dispersion and an extreme thixotropic stable 3D-gel after application of high shear forces
Impact of heat on viscosity in water (pH 7)	Stable viscosity (no heat-thinning)
Viscosity development in water at pH 7 ($T = 0$–100 °C (32–212 °F))	100% Cold water dispersible; stable viscosity upon heating
Shear stability	Shear forces activate the formation of a dispersion and thixotropic 3D-gel; viscosity of the gel is reduced by shear, but recovers to almost the initial value after a rest-time
Thickening effect	High – after shear treatment
pH stability	Medium (from pH 3.8–9); there are colloidal MCC types with added protective hydrocolloids (e.g. CMC) available to prevent flocculation
Decomposition	• By enzymes (cellulases); • flocculation by cooking in acidic systems
Film formation	Low

Table 1.10 (*Continued*)

	Powdered cellulose and colloidal MCC
Emulsion stabilization	Support of emulsifiers
Gelation	After application of high shear forces
Gel strength and gel stability	Gelation after shear treatment; spreadable heat-stable gels
Gel transparency	Opaque gels
Tendency for synaeresis	Low
Impact of electrolytes (cations +, 2+, 3+)	Low
Reaction with Ca^{2+} ions	Low
Protein activity	No for powdered MCC; added protective colloids in colloidal MCC may deliver protein activity
Crystallization control	Low
Synergistic effects with other hydrocolloids	Neutral inert character; microcrystalline cellulose does not impact other hydrocolloids because it does not consume water
Other synergistic effects	No for MCC, only possible due to interactions of the co-processed hydrocolloid
Negative interactions	At acidic pH values (<4.5) and without protective colloid, MCC flocculates to larger particle aggregates
Dosage level in foods	Medium to high, 0.1–5%

Table 1.11 Characteristics of methylcellulose (MC) and hydroxypropylmethylcellulose (HPMC).

	Methylcellulose	Hydroxypropylmethylcellulose
E-No.	E 461	E 464
Origin	Wood pulp or cotton linters	
Chemical composition	Linear molecule of ß-D-glucose with:	
	Uncharged hydrophobic (CH_3) substituents	Uncharged hydrophilic ($CH_2CHOHCH_3$) and hydrophobic (CH_3) substituents
Nutritional value (in 100 g) – metabolism	0 kJ (0 kcal); no resorption	
Fibre content	93.5 g per 100 g product (100% soluble fibre)	
Toxicology	No health concerns, but laxative effects at 12–15 g per day; no ADI value defined, GRAS status of MC; long-chain HPMC channels undesired substances 'out'	
Solubility at low temperature (H_2O)	High, 100%	
Appearance of an aqueous solution	Water-clear, transparent	
Viscosity of solution in water	Low to high (5–250 000 mPa s at 2%)	
Impact of heat on viscosity in water (pH 7)	1. Viscosity decrease; 2. viscosity increase; 3. gelation – flocculation possible; 4. viscous solution upon cooling	

(*continued*)

Table 1.11 (*Continued*)

	Methylcellulose	Hydroxypropylmethylcellulose
Viscosity development in water at pH 7 (T=0–100 °C (32–212 °F))	Reversible viscosity change, thickening and gelation	
Shear stability	Shear-thinning with re-thickening after a rest time, pseudo-plastic flow behaviour of aqueous solutions, gels are extremely thixotropic	
Thickening effect	Low to high	
pH stability	High (from pH 2–12)	
Decomposition	By enzymes (cellulases)	
Film formation	High	
Surface activity	High, good foam generator	Very high, excellent foamer
Emulsion stabilization	Support of emulsifiers	
Gelation	Reversible gelation upon heating, melting-back upon cooling	
Gel strength and gel stability	• High gel strength; • stable, when kept hot	• Soft to semi-firm gels; • stable, when kept hot
Gel transparency	Opaque gels	
Tendency for synaeresis	Slightly	Very low
Impact of electrolytes (cations +, 2+, 3+)	Low; high concentrations could cause reversible flocculation	
Reaction with Ca^{2+} ions	Low	
Protein activity	Low/no	
Crystallization control	High	
Synergistic effects with other hydrocolloids	No real synergistic effects	
Other synergistic effects	—	
Negative interactions	Possible with some preservatives in liquid systems (e.g. beverages with benzoate)	
Dosage level in foods	Low, 0.05–0.8%	

Table 1.12 Characteristics of ethylcellulose (EC).

	Ethylcellulose – EC
E-No.	E 462
Origin	Wood pulp or cotton linters
Chemical composition	Linear molecule of ß-D-glucose with uncharged ethyl groups as substituents (CH_2CH_3)
Nutritional value (in 100 g) – metabolism	kJ (0 kcal); no resorption
Fibre content	>99% Soluble dietary fibre in dry matter
Toxicology	No ADI value defined, but laxative effects at 12–15 g day^{-1}
Solubility at low temperature (H_2O)	Depending on DS: water-solubility at DS 0.8–1.3; only organo- and oil-soluble at DS 2.2–2.8 (commercial grades)
Appearance of an aqueous solution	Commercial grades are not water-soluble
Viscosity of solution	7–100 mPa s at 5% in 80 : 20 toluene–ethanol

Table 1.12 (*Continued*)

	Ethylcellulose – EC
Impact of heat on viscosity	Reversible viscosity decrease upon heating; no gelation
Viscosity development in oil/organic solvents at pH 7 ($T = 0$–100 °C (32–212 °F))	Reversible viscosity change/reduction; no gelation
Shear stability	High
Thickening effect	Low to high
pH stability	Very high (from pH 1–14); can be treated with hot concentrated alkalis, concentrated salt solutions and diluted acidic solutions
Decomposition	By enzymes (cellulases); highest stability of all cellulose ethers
Melting point	152–162 °C (305–324 °F)
Film formation	High
Emulsion stabilization	High
Gelation	Controlled gelation possible by hot-melting of EC (3–5%) at $T \geq 180$ °C (356 °F) in specific vegetable oils and subsequent cooling
Gel strength and gel stability	Rigid unelastic, but stable gels
Gel transparency	Transparent to opaque
Tendency for synaeresis	Low
Impact of electrolytes (cations +, 2+, 3+)	Low
Reaction with Ca^{2+} ions	Low
Protein activity	No
Crystallization control	Depending on DS
Synergistic effects with other hydrocolloids	No real synergistic effects
Other synergistic effects	—
Negative interactions	—
Dosage level in foods	Medium (0.5–5%)

Table 1.13 Characteristics of hydroxypropylcellulose (HPC).

	Hydroxypropylcellulose – HPC
E-No.	E 463
Origin	Wood pulp or cotton linters
Chemical composition	Linear molecule of ß-D-glucose with uncharged hydrophilic ($CH_2CHOHCH_3$) substituents
Nutritional value (in 100 g) – metabolism	0 kJ (0 kcal); no resorption
Fibre content	97% Soluble dietary fibre
Toxicology	No ADI value defined, but laxative effects at 12–15 g day^{-1}
Solubility at low temperature (H_2O)	$T = 0$–38 °C (32–100 °F): high, 100%; $T > 40$–45 °C (104–113 °F): insoluble in water (precipitation of dissolved HPC)
Appearance of an aqueous solution	Water-clear, transparent

(*continued*)

Table 1.13 (*Continued*)

	Hydroxypropylcellulose – HPC
Viscosity of solution in water	Low (10% = 200 mPa s) to high (1% = 3000 mPa s)
Impact of heat on viscosity in water (pH 7)	Strong viscosity decrease until the point of flocculation/ precipitation is achieved (reversible heat-thinning)
Viscosity development in water at pH 7 (T = 0–100 °C (32–212 °F))	100% Cold water solubility, reversible viscosity decrease upon heating up to 38 °C (100 °F); flocculation, precipitation, phase separation, significant viscosity decrease; re-dissolution upon cooling below 38 °C (100 °F)
Shear stability	Pseudoplastic flow behaviour at high shear forces, solutions of low-viscosity grades are shear-stable at moderate shear treatment
Thickening effect	Low to high
pH stability	High (from pH 2–11)
Decomposition	• By enzymes (cellulases), light, oxygen; • but more stable than other cellulose ethers
Film formation	High
Surface activity	High, good foam generator
Emulsion stabilization	Support of emulsifiers
Gelation	No gel formation
Gel strength and gel stability	No gelation in pure water
Gel transparency	—
Tendency for synaeresis	No
Impact of electrolytes (cations +, 2+, 3+)	Salts and sugar reduce the flocculation temperature
Reaction with Ca^{2+} ions	Low
Protein activity	Low
Crystallization control	High
Synergistic effects with other hydrocolloids	Viscosity win with the anionic cellulose ether Na-CMC in the pH range of 4–9 and NaCl contents of <0.5%
Other synergistic effects	Sugar and salts (NaCl) reduce the flocculation temperature
Negative interactions	—
Dosage level in foods	Low for foams: 0.1–0.5%, high for films: 10%

Table 1.14 Characteristics of carboxymethylcellulose (CMC).

	Na-CMC – cellulose gum
E-No.	E 466
Origin	Wood pulp or cotton linters
Chemical composition	Linear molecule of ß-D-glucose with negatively charged carboxymethyl substituents (COO^-)
Nutritional value (in 100 g) – metabolism	0 kJ (0 kcal); no resorption
Fibre content	Minimum 82% soluble dietary fibre
Toxicology	GRAS status, no ADI value defined, but laxative effects at 12–15 g day^{-1}
Solubility at low temperature (H_2O)	High, 100%
Appearance of an aqueous solution	Water-clear, transparent

Table 1.14 (*Continued*)

	Na-CMC – cellulose gum
Viscosity of solution in water	Low to high (30–60 000 mPa s at 2%)
Impact of heat on viscosity in water (pH 7)	Constant viscosity decrease (reversible heat-thinning)
Viscosity development in water at pH 7 ($T = 0$–$100\,°C$ (32–$212\,°F$))	100% cold water solubility; reversible viscosity decrease upon heating; re-thickening upon cooling
Shear stability	Shear-thinning with re-thickening after a rest time, pseudoplastic flow behaviour of aqueous solutions, low-substituted types are thixotropic
Thickening effect	Low to high
pH stability	Medium (pH 3–10)
Decomposition	• Insolubility at pH < 3; • degradation by enzymes (cellulases); • by cooking in acidic systems (heat + pH ≤ 3.5)
Film formation	High
Surface activity	Low, no foam generation
Emulsion stabilization	Support of emulsifiers
Gelation	Only with di- and trivalent ions
Gel strength and gel stability	No gelation in pure water; spreadable gels with sufficient amount of ions (2+, 3+)
Gel transparency	Opaque gels
Tendency for synaeresis	No
Impact of electrolytes (cations +, 2+, 3+)	High impact of monovalent ions on viscosity; precipitation or gelation with di- and trivalent cations
Reaction with Ca^{2+} ions	High (gelation or precipitation)
Protein activity	High, acts as protective colloid; solubilizing effect on protein around IEP; heat protection at acidic pH
Crystallization control	High
Synergistic effects with other hydrocolloids	Viscosity win with guar and non-ionic cellulose ethers; improved gel quality with starches, κ-carrageenan
Other synergistic effects	Crosslinking of CMC with positively charged molecules (chitosan, poly-L-lysine); crosslinking of CMC with low DS and high viscosity by acids (croscarmellose)
Negative interactions	Precipitation possible with di- and trivalent cations
Dosage level in foods	Low (0.05–0.5%)

1.7 Commercial Development – Global Market

The global market for all food hydrocolloids is estimated to be a volume of about 2100 MT (US$5.5 billion) for 2014. This market value is calculated on data from the manufacturers. Taking into account that several service companies and

Table 1.15 Global market data for food hydrocolloids 2002–2006.

Hydrocolloid	2002		2004		2006			
	Volume (Mio. $)	Market Share (%)	Volume (Mio. $)	Market Share (%)	Volume (1000 MT)	Market Share (%)	Price ($ kg^{-1})	Volume (Mio. $)
Gelatine	747	25	1095	30	164	9.6	5.68	932
Starches	743	25	880	24	1270	74.3	0.81	1029
Carrageenan	302	10	305	8	39	2.3	8.36	326
Pectin	301	10	323	9	35	2.0	10.23	358
Xanthan	221	7	253	7	41	2.4	5.68	233
Agar	142	5	134	4	10	0.6	18.59	186
LBG	122	4	106	3	9	0.5	17.93	161
Gum Arabic	101	3	234	6	36	2.1	5.72	206
Alginates	92	3	94	3	12	0.7	7.48	90
Guar	53	2	74	2	45	2.6	1.39	63
CMC	66	2	85	2	29	1.7	4.38	127
MC/HPMC	49	2	45	1	5	0.3	9.46	47
MCC	45	1	50	1	9	0.5	5.96	54
Others	18	1	30	1	6	0.4	6.60	40
Sum	3000	100	3708	100	1710	100	—	3850
Growth[a)]	—	—	**+23.6%**	—	—	—	—	**+3.8%**

Source: data provided from IMR (Dennis Seisun).

a) Growth is calculated on last number for the same category (e.g. '2006 volume $' versus '2004 volume $').

Table 1.16 Global market data for food hydrocolloids 2009–2014.

Hydrocolloid	2009				Forecast 2014			
	Volume (1000 MT)	Market Share (%)	Price ($ kg^{-1})	Volume (Mio. $)	Volume (1000 MT)	Market Share (%)	Price[a] ($ kg^{-1})	Volume (Mio. $)
Gelatine	176	9.4	4.85	853	193	9.2	4.83	932
Starches	1373	73.5	0.99	1360	1531	72.8	0.99	1516
Carrageenan	44	2.4	10.98	483	52	2.5	10.83	563
Pectin	44	2.4	11.25	495	57	2.7	11.30	644
Xanthan	50	2.7	6.52	326	64	3.0	6.47	414
Agar	11	0.6	18.45	203	12	0.6	19.33	232
LBG	10	0.5	10.10	101	11	0.5	10.64	117
Gum Arabic	40	2.1	4.05	162	46	2.2	4.04	186
Alginates	14	0.7	19.50	273	16	0.8	19.56	313
Guar	51	2.7	1.65	84	58	2.8	1.66	96
CMC	32	1.7	4.91	157	36	1.7	5.00	180
MC/HPMC	7	0.4	11.00	77	9	0.4	11.56	104
MCC	11	0.6	7.64	84	13	0.6	7.92	103
Others	6	0.3	8.17	49	7	0.3	8.14	57
Sum	1867	100	—	4707	2104	—	—	5457
Growth[b]	**+9.2%**	—	—	**+22.3%**	**+12.7%**	—	—	**+15.9%**

Source: data provided from IMR (Dennis Seisun).

a) Forecast made with constant price.

b) Growth is calculated on last number for the same category (e.g. '2014 volume $' versus '2009 volume $').

Table 1.17 Price categories for food hydrocolloids.

Price for 1 kg of hydrocolloid (US$)				
<1	**1–3**	**3–6**	**6–8**	**8–10**
Starches	Guar	• Gum arabic • CMC • gelatine • xanthan	MCC	Alginate salts
10–12	**12–15**	**15–20**	**20–30**	**>30**
• Carrageenan (SR) • MC/HPMC • locust bean gum	• Carrageenan (Ref) • HM pectin	LM pectin	Propylene glycol alginate (PGA)	Gellan gum

distribution channels are also in involved, the business is estimated to be 20–30% higher (~US$6.6–7.1 billion per year). There is still an annual growth of physical volume of 3–5%.

Starches are the major contributors to the hydrocolloid portfolio with a market share of 70–75% of physical volume and 30% of economical volume, followed by gelatine with a market share of 9–10% of physical volume and 18–25% of economical volume. These values indicate that starches and gelatine make up more than 80% of the total hydrocolloid volume and about 50% of the total business volume.

Prices are very different over the whole range of hydrocolloids. Pure gellan gum is one of the most expensive hydrocolloids at ~US$40 kg^{-1}. Native starches are at the other end of the scale (~US$0.40 kg^{-1}). Tables 1.15 and 1.16 list global market data for the 14 most-used food hydrocolloids. Price segments can be found in Table 1.17.

To give a rough indication of price per volume, the food hydrocolloids are grouped into rough prices segments. Prices are controlled by availability of raw material, demand of other (non-food) industries, processing capabilities, and logistic conditions. Changing climate conditions or new breeding measures in the case of using remnants can dramatically reduce crops. Wars or the detection of chemical contaminants may dramatically decrease the availability of raw material. Thus, food hydrocolloids are subject to more or less strong fluctuations.

References

1 Scherz, H. (1996) *Hydrocolloids: Stabilizers, Thickening and Gelling Agents in Food Products*, Food Chemistry/Food Quality, vol. 2, Food Chemical Society GDCh, Behr's Verlag GmbH, Hamburg.

2 Imeson, A. (2010) *Food Stabilizers, Thickeners and Gelling Agents*, Wiley-Blackwell, Oxford.

3 Stephen, A.M., Philips, G.O., and Williams, P.A. (eds) (2006) *Polysaccharides and Their Applications*, 2nd edn, Taylor & Francis Group, Boca Raton, FL, pp. 147–179.

2
Rheology of Food Hydrocolloids

2.1
Introduction to Rheology, Rheometry, and Visco-Elasticity

In aqueous solutions, hydrocolloids cause an increase in viscosity after hydration. Several functions and properties result from this characteristic, in particular their stabilizing functions in multiphase disperse systems such as suspensions (solid–liquid), emulsions (liquid–liquid), and foams (gaseous–liquid). Additionally, there is a close relation between rheological and sensory properties of a food product. Therefore, the measurement of rheological properties (rheometry) is of very high importance in the context of characterization and application of hydrocolloids [1].

Rheology is the science of the deformation and flow of substances. This science is a branch of physics (and physical chemistry), as the most important parameters come from mechanics: forces, deflections, and velocities. The term is derived from the Greek '*rhein*' meaning flow, stream. But rheological studies do not only include the flow behaviour of fluids but also the deformation behaviour of solids. The relation therein is that deformations caused by shear forces result in the flowing of many substances. All forms of rheological shear behaviour that are scientifically describable take place between two extremes, the flow of ideal-viscous liquids and the deformation of ideal-elastic solids. All real substances have a viscous component and an elastic component. Their behaviour can be described as visco-elastic. Table 2.1 presents an overview of different types of rheological behaviour [2].

A drop of water exhibits ideal-viscous flow behaviour. Visco-elasticity is presented by dough. A steel ball shows ideal-elastic rheological behaviour.

Rheometry is defined as the measuring technique used to capture rheological data. It includes the measuring systems, equipment, and methods of evaluation.

2.2
Definitions

Viscosity

Viscosity is defined as the resistance to flow, and so it is a measure of the toughness and the apparent internal friction of a liquid or gas when applying external

Cellulose and Cellulose Derivatives in the Food Industry: Fundamentals and Applications, First Edition.
Tanja Wüstenberg.

Table 2.1 Overview on different types of rheological behaviour [2].

	Liquids		Solids	
Behaviour:	Ideal-viscous flow	Visco-elastic flow	Visco-elastic deformation	Ideal-elastic deformation
Law/model of:	Law of Newton	Model of Maxwell	Model of Kelvin/ Voigt	Law of Hooke
Measurement:	Flow curves, Viscosity curves	Creep, relaxation and oscillation experiments		

forces. Isaac Newton (1643–1727) conducted the first studies on this material property. When a liquid between two plates, separated by distance (h) and with individual area (A), is sheared with the force (F) it starts to flow, and perpendicular to the direction of movement a velocity gradient is produced.

The dynamic viscosity (η) indicates the relationship between the applied force necessary for a liquid to overcome a defined distance:

$$\text{Dynamic viscosity } (\eta) = \frac{\text{shear stress } (\tau)}{\text{shear rate } (\dot{\gamma})}$$

The shear stress (τ) is the quotient of force (F) and area (A).

The shear rate ($\dot{\gamma}$) is the quotient of velocity (v) and plate distance (h). Synonyms are shear gradient, velocity gradient, strain rate, or rate of deformation.

The kinematic viscosity (ν) is defined as the quotient of dynamic viscosity and density of the solution:

$$\text{Kinematic viscosity } (\nu) = \frac{\text{viscosity } (\eta)}{\text{density } (\rho)}$$

The value of viscosity is not a material constant of each specific substance.

Polymers such as hydrocolloids and in particular cellulose derivatives may take various forms because of the rotatability of their covalent bonds and have a certain degree of mobility. If you apply a defined shear stress on a polymer solution and the enclosed molecule bundles, the macromolecules arrange themselves in the direction of the resulting flow. As a result, the viscosity decreases. The dynamic viscosity is temperature-dependent, because the solvent (usually water) behaves thermodynamically differently at different temperatures. To ensure comparability of viscosity data, the measurements must therefore always be carried out in the same solvent, at constant temperature, and after adequate hydration time.

The term 'apparent viscosity' is used when the ratio between shear stress and shear rate changes during application of different shear forces. It is the approach to indicate the difference to the constant viscosity of an ideal-viscous liquid.

Shear Modulus

For ideal-elastic solids, the ratio between shear stress (τ) and deformation (γ) is a material constant, when the measurement is performed at constant temperature in the reversible-elastic deformation range, the so-called linear-elastic range. This material-specific value is called the shear modulus (G) (also known as modulus of elasticity in shear, or rigidity modulus):

$$\text{Shear modulus}\,(G) = \frac{\text{shear stress}\,(\tau)}{\text{deformation}\,(\gamma)}$$

The Poisson's ratio (μ) describes the ratio of transverse strain to elongation ($0 < \mu < 0.5$). Brittle products have a μ-value of 0. Highly flexible, elastic materials and visco-elastic liquids have $\mu = 0.5$. The tensile modulus or modulus of elasticity (E-modulus) can be calculated from the shear modulus (G) using the following formula:

$$\text{Tensile modulus}\,(E) = 2G\,(1 + \mu)$$

For brittle materials: $E = 2G$ is valid.

For elastic materials, polymers behaving like visco-elastic liquids, and other incompressible isotropic products: $E = 3G$ is valid.

Calculation of the E-modulus from the G-modulus is to be considered an estimation only. The μ-values can vary within substance classes and at different temperatures.

Sol–Gel Characterization

Visco-elastic flow and deformation behaviour can be determined by oscillation experiments. Simple shear measurements have the disadvantage of irreversibly destroying gels. Additionally, the change of structure during heating or cooling can be of interest. With a dynamic-mechanical analysis (DMA), all kinds of visco-elastic substances are measurable – from low-viscosity fluids to polymer solution or melts, dispersions such as suspensions, emulsions and foams, pastes, gels, elastomers, and rigid solid bodies.

In oscillation experiments, there are unsteady shear conditions. The complex shear modulus (G^*) is to be interpreted as the stiffness of the substance, which is the resistance against deformation, during changing shear conditions ($G^* = \tau(t)/\gamma(t)$) – in contrast to the shear modulus (G) at constant shear conditions (constant deformation or constant shear stress).

The storage modulus (G') is a measure of the energy of deformation stored in the sample material during shearing. This energy is completely available after relief of the pressure and it is the driver for the re-deformation. G' represents the elastic property of the sample.

The loss modulus (G'') is a measure of the energy consumed by the sample material during shearing. The energy is 'used' to change the sample structure or is transferred to the environment, for example when the sample starts to flow partially or fully. It comes to flowing when relative movements between molecules, particles, and bigger aggregates occur, which causes friction and frictional heat. Thereby, energy dissipates, heating the sample and the atmosphere. Substances losing energy during shearing exhibit an irreversible deformation property; they change their form in a stress–relief cycle. G'' represents the viscous property of the sample material.

The storage and the loss modulus are related by the loss or damping factor tan δ (tan $\delta = G''/G'$), which expresses the relationship of viscous component to elastic component in a visco-elastic deformation ($0 < \tan\delta < \infty$). For ideal-elastic behaviour, tan $\delta = 0$, because G' completely dominates G''. For ideal-viscous behaviour, tan $\delta = \infty$, because G'' completely dominates G'. When the viscous and the elastic component are in balance, tan $\delta = 1$.

The achievement of tan $\delta = 1$ is an important evaluation criterion for gel formation and hardening processes. Then the so-called sol–gel-transition point is reached.

It is valid:

- in a liquid state (solution state): tan $\delta > 1$ (because $G'' > G'$);
- in a state of gelation (solid state): tan $\delta < 1$ (because $G' > G''$);
- at the sol–gel transition point: tan $\delta = 1$ (because $G' = G''$).

2.3 Basic Data

Table 2.2 gives the shear rates for some selected processes to give an impression of the intensity of material treatment. Viscosities of liquid and solid regulated products are listed in Table 2.3. Values of the shear modulus and tensile modulus of selected products can be found in Table 2.4

2.4 Different Types of Flow Behaviour

Depending on the relationship of shear stress and shear rate, there are different types of flow behaviour, which are explained briefly here. Additionally, these rheological properties are illustrated below in Figures 2.1–2.4. A summary is made in Table 2.5 at the end of this section.

Ideal-viscous flow behaviour, so-called Newtonian flow, is described formally by Newton's law ($\tau = \eta\dot{\gamma}$). The viscosity of a Newtonian fluid is independent of intensity and duration of a shear treatment. The relation between shear stress and shear rate is linear, passing through the origin (zero-point). Thus, the coefficient of viscosity is constant and time-independent. Examples are water, solvents, mineral oil (without polymeric additives), silicone oil, and blood plasma.

In a non-Newtonian fluid, the relation between shear stress and shear rate is different and can even be time dependent. Therefore, a constant coefficient of viscosity cannot be defined.

Non-Newtonian flow behaviour is divided into:

Table 2.2 Shear rates – typical for some materials and processes [2].

Process	Range of typical shear rates (s^{-1})	Application
Sedimentation of particles	≤0.001–0.01	Dispersion paints, fruit juice
Sedimentation of fine powders in suspensions	10^{-6}–10^{-4}	Medicines, printing inks
Level control due to surface tension	0.01–0.1	Coatings, printing colours, lacquer
Dripping, sagging of coatings (by gravity)	0.01–1	Chocolate coating, dispersion paints, plaster
Dip coating	1–100	Confectionary, French fries, bakery products, dipping lacquer
Kneading (for mixing)	1–100	Dough, elastomer mixer
Chewing and swallowing	10–100	Food (gummy bears, yogurt, cheese)
Spreading	10–1000	Butter fresh cheese
Extrusion	10–1000	Polymer melts, dough, tooth paste
Mixing and stirring	10–10 000	Processing of emulsions and semi-firm products
Tube and capillary flow	10–10 000	Pumping of juice, blood flow
Spraying and brushing	1000–10 000	Spray-drying, spraying lacquers, nasal spray aerosols, adhesives
Wet milling	1000–100 000	Pigment pastes for printing ink
Creaming	1000–100 000	Application of creams and lotions to the skin

- time-independent flow behaviour;
- time-dependent flow behaviour.

There are three time-independent flow characteristics:

1) plastic flow behaviour with a yield point;
2) pseudoplastic flow behaviour (shear-thinning);
3) dilatant flow behaviour (shear-thickening).

Plastic Fluids Plastic fluids start to flow only after an internal resistance has been overcome and the shear stress (τ) exceeds a so-called yield point. Beyond this yield point, Newtonian as well as non-Newtonian flow properties can be exhibited.

Bingham fluids are ideal-plastic substances and a special type of plastic fluids. They have a linear shear stress–shear strain relationship and require a finite yield stress before they begin to flow (the plot of shear stress against shear rate does not pass through the origin/zero-point). The surface of a Bingham plastic has peaks when it is still. By contrast, Newtonian fluids have flat featureless surfaces when still. Regulated product examples are mayonnaise, chocolate, mustard, and toothpaste.

Table 2.3 Viscosities of different food and cosmetic products[a)] (at 20 °C (68 °F)).

Product	Viscosity (mPa s)	Product	Viscosity (mPa s)
Gases/air	0.01–0.02	Sugar solution (70 °Brix)	400
Air	0.018	Liquid egg (40 °C (104 °F))	150
Water at 0 °C (32 °F)	1.79	Yogurt (40 °C (104 °F))	150
Water at 20 °C (68 °F)	1.00	Rape oil	160
Water at 50 °C (122 °F)	0.547	Chocolate sauce (50 °C (122 °F))	280
Water at 100 °C (212 °F)	0.282	Vegetable soup	450
Wine	2	Fruit mash	600
Milk	2	Whey	800–1600
Low-viscosity HPMC (hydroxypropyl-methylcellulose) (2%)	3–5		
Liqueur	10–100	Pudding (40 °C (104 °F))	1000
Cream with 30–50% fat	15–115	Apple puree	1500
Larch gum (40% solution)	23	Baby food at 40 °C (104 °F)	1500
Linseed oil (40 °C (104 °F))	30	Salad dressing	1300–2600
Olive oil at 40 °C (104 °F)	40	Glycerine at 20 °C (68 °F)	1480
Olive oil at 20 °C (68 °F)	100	Glycerine at 10 °C (50 °F)	4500
Palm oil (40 °C (104 °F))	45	Glycerine at 0 °C (32 °F)	12 000
Fruit juice (undiluted)	50	Mayonnaise	2000
Fruit juice (concentrate)	2500	Honey at 40 °C (104 °F)	2000
Cocoa butter (60 °C (140 °F))	50	Honey at 20 °C (68 °F)	About 10 000
Cocoa mass	4000	Shampoo	3000
Coconut oil	60	Hand lotion	8000
Soybean oil	60	Jam	8500
Condensed milk (40 °C (104 °F))	80	Processed cheese (60 °C (140 °F))	30 000
Condensed milk with sugar	6100	Butter (40 °C (104 °F))	30 000
Gravy (80 °C (176 °F))	110	High-viscosity cellulose gum (2%)	60 000
Sugar solution (65 °Brix)	120	Tooth paste (40 °C (104 °F))	70 000

a) Viscosities measured at 20 °C (68 °F) (if not marked customized).

Table 2.4 Values for shear and tensile modulus (*G*- and *E*-modulus) of selected products (at 20 °C (68 °F)) [2] (1 Pa = 1 N m^{-2}).

Product	*G*-modulus (G^*)	Poisson's ratio μ [a)]	*E*-modulus
Very soft gel structure (e.g. spray coatings, salad dressings)	5–10 Pa		
Soft gel structure (e.g. brush paint)	10–50 Pa		
Visco-elastic gels (typical dispersions, lotions, creams, ointments, food pastes, cosmetics)	50–5000 Pa (mostly 100–500 Pa)		
Pudding (with 5/7.5/10/15% starch)	0.1/0.5/1/5 kPa		
Adhesives before hardening:			
Soft pasty structure	0.1–10 kPa		
Strong structure (e.g. sealants)	50–500 kPa		
Gummy bears (confectionary)	10–500 kPa		
Cheese spreads	1 kPa		
Soft cheese	10 kPa		
Sliceable cheese	0.1 MPa		
Hard cheese (dried)	0.5 MPa		
Extra hard cheese	1 MPa		
Butter at $T = +10\,°C$ (50 °F)	2 MPa		
Butter at $T = +23\,°C$ (73 °F)	50 kPa		
Soft natural rubber	0.03–0.3 MPa	0.49	0.1–1 MPa
Eraser	1 MPa		
Technical elastomers	0.3–30 MPa	0.40–0.45	1–100 MPa
Hard rubber (car tire)	10–100 MPa	0.35–0.40	30–300 MPa
Thermoplastic polymers (not crosslinked)	0.1–2 GPa	0.30–0.35	1–4 GPa
Pure resins	1–2 GPa	0.40	3–5 GPa
Wood (axial)			4–18 GPa
Wood (radial)			0.3–0.6 GPa
Ice (at $T = -4\,°C$ (25 °F))	3.7 GPa	0.33	9.9 GPa
Bones			18–21 GPa
Steel	80 GPa	0.28	210 GPa
Diamond			1200 GPa

a) The higher the Poisson's ratio (μ) the more stretchy, elastic, flexible is a material ($\mu = 0.5$). The lower the μ-value, the more brittle is a material during breaking ($\mu = 0$).

Pseudoplastic Flow Behaviour (Shear-Thinning) Most hydrocolloids exhibit pseudoplasticity, being hydrated in aqueous solution. These fluids flow easier when shear forces are applied. The apparent viscosity is reduced by increasing shear rates. The shear-thinning effect of pseudoplastic solutions caused by shear force impact is completely reversible. The measured values are reproducible and independent of the time point of the measurement. Ketchup and syrups are pseudoplastic food products.

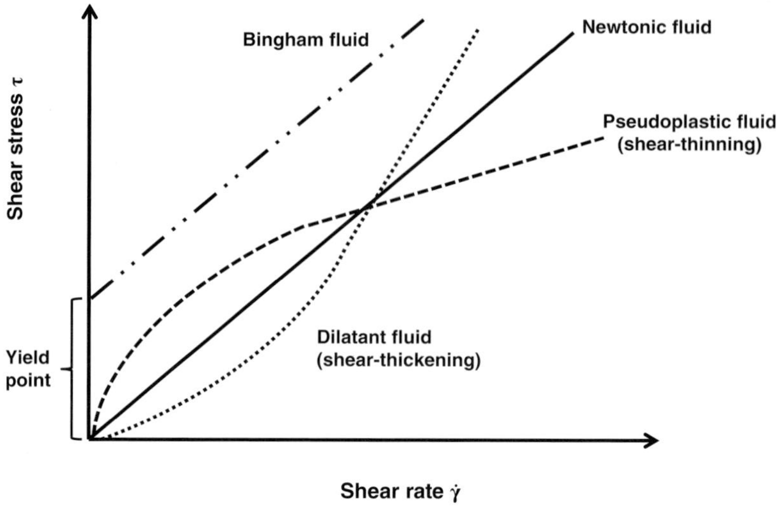

Figure 2.1 Different types of flow behaviour – relationship between shear stress and shear rate.

Dilatant Flow Behaviour (Shear-Thickening) In contrast to pseudoplastic fluids, dilatant solutions exhibit an increase in apparent viscosity when the shear rate is increased. Owing to shear forces, a fully reversible shear-thickening occurs. The measurement of viscosity is time-independent and can be repeated, giving the same results. Suspensions of corn starch or sand in water show this type of rheological behaviour.

Figures 2.1 and 2.2 demonstrate the different types of time-independent flow characteristics.

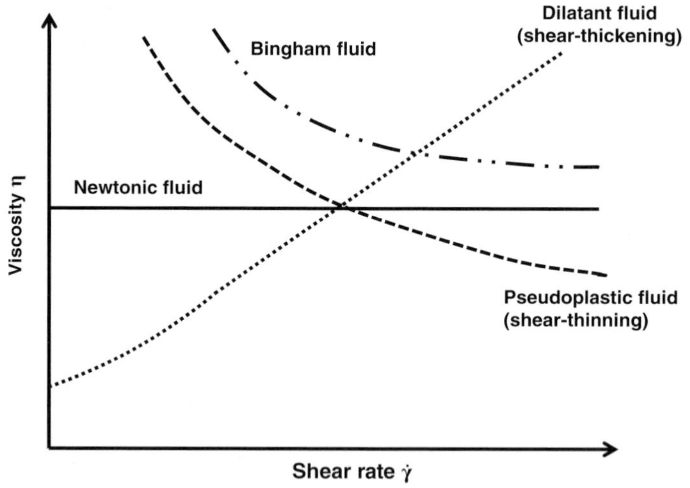

Figure 2.2 Different types of flow behaviour – relationship between shear rate and viscosity.

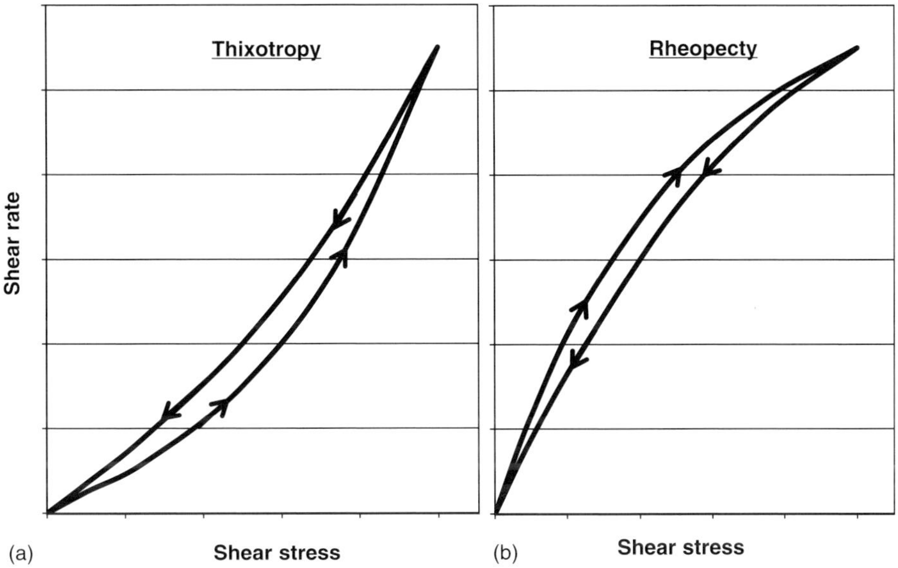

Figure 2.3 Time-dependent flow properties – thixotropy (a); rheopecty (b).

For the time-dependant flow characteristic, there is a distinction between:

- thixotropic flow behaviour,
- rheopectic flow behaviour,
- Kelvin–Voigt material.

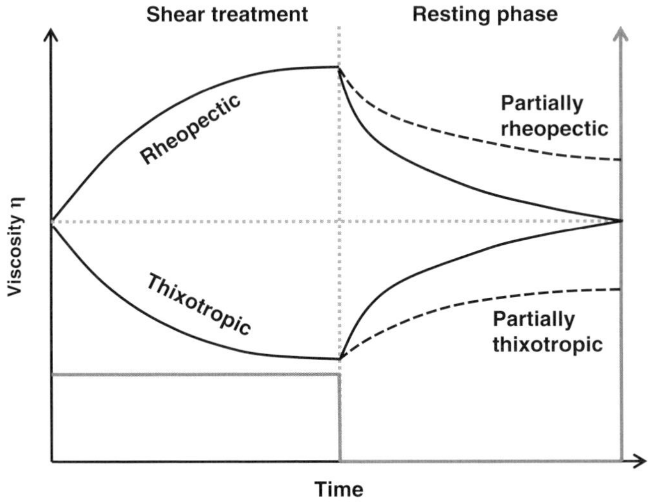

Figure 2.4 Viscosity development of thixotropic and rheopectic fluids.

Thixotropic Flow Behaviour During application of shear stress, thixotropic fluids exhibit a viscosity reduction that is fully reversible after interruption of the shear force and a specific resting time. In contrast to pseudoplastic solutions, this flow characteristic is dependent on time. At a continuous increase and subsequent decrease of shear rate, a hysteresis is observed that can be considered as a 'memory effect' of the system of the previous shear treatment. The corresponding value of shear stress (x-axis) for a given shear rate (y-axis) is higher during increasing shear rate than during its subsequent decrease. Many food products and hydrocolloid solutions have a thixotropic flow. Examples are stirred yogurt, hydrogenated castor oil, and various colloidal solutions of hydrated hydrocolloids such as xanthan gum, gelatine, pectin, methylcellulose, and cellulose gum.

Rheopectic Flow Behaviour Fluids that require a gradually increasing shear stress to maintain a constant shear rate are referred to as rheopectic. The apparent viscosity is a function of time and increases with duration of shear stress. As opposed to thixotropic flow, rheopectic fluids display an increasing viscosity when shear stress is applied. As for thixotropic systems, the shear-thickening effect of rheopectic systems is time-dependent and differs from dilatant fluids. A hysteresis is measured – the corresponding value of shear stress (x-axis) for a given shear rate (y-axis) is higher during decreasing shear rate than during its previous increase. Printing inks and gypsum paste exhibit rheopectic flow characteristics.

The difference between thixotropic and rheopectic flow behaviour is presented in Figure 2.3. The time-dependent viscosity curves are visualized in Figure 2.4.

Table 2.5 gives an overview of typical flow characteristics.

2.5
Structures of Polymers with Shear-Thinning Flow Behaviour

For shear-thinning substances, the viscosity depends on the intensity of shear treatment (shear rate or shear stress). With increasing shear force, the flow curve exhibits a decreasing slope and, accordingly, viscosity decreases. Examples are solutions and melts of polymers, organic binders such as methylcellulose, several lacquer paints, paste, and many shampoos. Shear-thinning flow characteristics can be a function of time or be time-independent (Section 2.3).

At rest, each macromolecule assumes the state offering the lowest energy consumption. Therefore, without external treatment, it occurs as a three-dimensional bundle. Viewed from outside, the outline of the coil is roughly the shape of a sphere at rest. Here, such a macromolecule is most often entangled with neighbouring molecules. During the shearing process, the molecules align themselves in the direction of shear force as well as in the direction of the resulting shear-gradient and start to flow. They unravel and

Table 2.5 Different types of flow behaviour.

Property	Flow behaviour	Description	Examples
Time-independent	Newtonian liquid	Viscosity is constant, independent of shear stress and shear rate	Water, solvents, mineral oil (without polymeric additives), silicone oil, blood plasma
	Plastic, ideal-plastic = Bingham	Substances flow after overcoming of a yield point (Newtonian or non-Newtonian flow property is exhibited beyond the yield point)	Toothpaste, lipstick, ketchup, mayonnaise, chocolate melt, mustard, clay suspensions, drilling mud, margarine, yogurt
	Pseudoplastic (shear-thinning)	Apparent viscosity decreases with increased shear stress	Ketchup, whipped cream, syrups, blood, latex paint, nail polish, hydrocolloid solutions
	Dilatant (shear-thickening)	Apparent viscosity increases with increased shear stress	Dispersions with high concentrations of solids or gel-like stiff particles such as aqueous suspensions of corn starch, sand in water, ceramic suspensions, natural rubber, dental filling composites
Time-dependant	Thixotropic	Apparent viscosity decreases with duration of shear stress	Yogurt, solutions of xanthan and cellulose gum, gelatine and pectin gels, hydrogenated castor oil, many paints, many colloidal suspension
	Rheopectic	Apparent viscosity increases with duration of shear stress	Printing ink, gypsum paste, hardening of adhesives, gel formation
	Kelvin–Voigt material	Parallel effects of elastic and viscous properties	Whipping cream, some lubricants

the purely mechanical entanglement dissolves partially, with a reduction in their flow resistance. In low-concentration polymer solutions, it may even lead to an almost complete disentanglement process. The highly-orientated molecules then move individually – with only very few points of contact – and almost independently [2].

Another reason for the flexibility of values for solution viscosity is found in the molecular structure of the polymers. The molecules have a certain degree of mobility because they can take various forms due to the rotatability of their covalent bonds.

The highest flexibility is exhibited by linear unbranched substances such as cellulose and low-substituted cellulose derivatives. A high degree of branching indeed leads to easier hydration and lower viscosities due to a lower effective volume, but also to reduced molecular flexibility. An example is acacia gum or gum Arabic.

Under very extreme shear conditions, there is the danger of degradation of macromolecules. Polymer chains can be cut or torn and the original chemical structure is destroyed. Rheological measurements should not lead to degradation of the molecules. A simple method to check if molecular destruction occurs and distorts the measured viscosity value is to perform a final measurement at very low shear rate and to compare this value with data from the first measurement. If the final value is significantly lower, molecular decomposition has taken place.

Besides pure solutions of polymers and hydrocolloids, other complex systems also exhibit shear-thinning flow characteristics. Generally, shear forces reduce the effectiveness of interaction forces and, thus, flow resistance decreases.

In suspensions, particles without strong interaction forces have no specific orientation and occur in a highly disordered state at rest. At sufficiently high shear forces, the suspended particles of rod or platelet shape arrange themselves in the direction of flow, which leads to a parallel arrangement and high order. The result is a reduction of viscosity.

In emulsions, the dispersed droplets are spherical. Shear stress causes their deformation to ellipsoids, which leads to a viscosity decrease.

In a solution, agglomerates can occur. Small primary particles aggregate to form bigger secondary particles with inclusion of part of the dispersion liquid (continuous phase), which is thereby immobilized. By shear stress, the agglomerates are destroyed and disintegrate back into the primary particles. The trapped liquid is released and can flow freely, resulting in a lower viscosity than before shearing.

2.6 Causes of Shear-Thickening of Products

Shear-thickening materials exhibit a shear-dependent viscosity that can be a function of time or time-independent (Section 2.3). The flow curve indicates an increasing slope of the curve with increasing shear force and, accordingly, the viscosity increases. Examples are dispersions with high concentrations of solids or gel-like stiff particles such as suspensions of food starch, ceramic suspensions, natural rubber, and dental filling composites.

In concentrated solutions of uncrosslinked polymers, shear-thickening can result from mechanical entanglements, in particular when the molecules are highly branched and thus relatively rigid. The higher the shear stress, the more the molecules hinder themselves from sliding to each other.

When shearing highly filled suspensions, increasing shear rate causes the indicidual particles to touch each other more often and to jam into each other. Softer, gel-like particles can be compressed. Thereby, the flow resistance increases. The shape of the particles has a decisive role. Shearing, and in particular the resulting shear gradient, leads to a rotation of the particles in the direction of shear. Rod-shaped particles or fibres also show rotational movements. Cubic particles require a higher volume during rotation than at rest; thus there is a lower free volume available between the particles for the dispersion liquid. Spherical particles occupy the same position at rest and in rotation at shear. These types of dispersions have a lower tendency to exhibit shear-thickening. The flowability of a substance can be improved by increasing the free volume between the particles. This is possible by modification of the particle shape or addition of additional dispersion liquid.

In shearing experiments with emulsions, at increasing shear rates an increasing viscosity curve is observed occasionally that is often interpreted as a shear-thickening flow property. However, in most cases, the increase is caused by a reduction of the average droplet diameter and droplet fragmentation. The high shear forces result in a constantly running dispersion process. Thereby, the volume-specific surface (the ratio of droplet surface to droplet volume) is enlarged. This causes an increasing number of interactions between the individual droplets leading to higher flow resistancy. The 'creaming or whitening effect' of cosmetic products such as creams, ointment, and lotions is an example of a continued dispersion when used on skin. Often, an uncomfortable sticky impression, sometimes with stringing, results [2].

2.7 Factors that Influence Rheological Behaviour

The rheological behaviour of a food hydrocolloid with respect to its colloidal solution depends on several internal and external factors.

The most important internal parameters are:

- *chemical structure*: molecular chain length and distribution, degree of branching, functional groups, monomeric building blocks, glycosidic bonds, presence of reactive groups interacting with other substances in solution;
- *physical nature*: swelling, dispersing, dissolution and gelling behaviour, particle size distribution, shape of particles, specific surface;
- concentration of hydrocolloid;
- *preparation and pretreatment of the hydrocolloid solution*: method of hydration, stirring and shear processes, time of hydration, temperature, order of addition;
- solvent.

Further important factors of the material system are:

- presence and concentration of:
  - solid particles in suspension,
  - hydrated polymers in solution,
  - surface-active substances in a dispersion;
- presence and concentration of other active substances with strong hydration affinity such as sugar;
- presence, type, and concentration of electrolytes (in food, in particular Ca^{2+} ions);
- complex formation with other polymers (e.g. polysaccharides, proteins);
- degradation processes by microorganisms and enzymes;
- ambient pressure;
- pH value;
- magnetic and electric field strength;
- effects of oxygen and light;
- measuring system and method of measurement and evaluation.

The essential external parameters are:

- type of treatment (force of deformation, shear rate, shear stress, etc.);
- intensity of treatment (application of low or high shear);
- duration of treatment (times of shear application and rest times);
- temperature during shear application (consideration of internal friction).

2.8 Viscosity Measurement of Thickening Hydrocolloid Solutions

Generally, industrially used hydrocolloids are specified by their viscosity value (η) in water. Viscosity in aqueous solution is the key parameter of quality. Mostly, a viscosity range is set by the manufacturer with a corresponding concentration at defined temperature and measurement conditions (equipment, speed).

For example, cellulose derivatives such as methylcellulose or HPMC are specified as aqueous solutions of 1 or 2% concentration at 20–25 °C (68–77 °F) to deliver a viscosity range of '*xy* mPa s or cP ± 20%' at defined shear conditions, measured usually with a Brookfield spindle or an Ubbelohde capillary viscosimeter (e.g. an HPMC-4000 grade has a specified viscosity range of 3200–4800 mPa s (cP)).

The individual hydrocolloid producer has an exactly defined procedure of how to prepare the solution and carry out the measurement to be able to achieve reproducible results.

Basic requirements to measure reproducible values [1]:

- The hydrocolloid solution must be produced following defined and timed conditions of incorporation technique, mechanical treatment, hydration time, and temperature before the measurement procedure begins.

- The concentration of the hydrocolloids must be precisely adjusted considering the accompanying substances, in particular water, when weighing before solubilization (weight based on dry matter).
- For preparation of the aqueous solution, water of defined quality (water hardness, electrical conductivity, pH value) must be used.
- Time factor: The development of viscous-thickening or gelling effects requires a certain period of time. Before the measurement, a defined and substance-specific waiting period must be maintained.
- The measurement must always be performed under the same conditions (temperature of solution, comparable shear stress or shear rates for non-Newtonian fluids) and should not cause irreversible changes of the hydrocolloid measurement solution.
- For meaningful characterization of hydrocolloid solutions (usually non-Newtonian flow properties), a viscosity value resulting from a single-point measurement is not sufficient; instead, a flow curve needs to be determined by multi-point measurement and application of different shear rates.
- To measure absolute values of viscosity, the solutions are processed with laminar flow and a defined, consistent shear rate. Absolute viscosities are always given in relation to their shear rate, an exception is Newtonian fluids.
- Relative values are often measured in industrial practice due to their easier handling. But the resulting values are physically inappropriate and often have a random character. They may only be used for comparison purposes within a class of substance – providing the measurement equipment and working conditions are exactly the same.

2.9 Characterization of Gels

Structural Characteristics of Hydrocolloid Gels

In general, gel-forming substances exhibit some special characteristics in which they differ from the colloidal solutions of purely thickening hydrocolloids [1]:

- Gels are mixed systems, being made of at least two components, with a high content of liquid, usually water. The liquid component is incorporated into a network of solid material, the so-called gel matrix, and thus is immobilized.
- Gels exhibit limited shape stability and elasticity. Deformation caused by external forces is reversible in the elasticity range.
- Gels have the capability to absorb large volumes of liquid with a simultaneous increase in volume (exception silica gel) or to release liquid (deswelling, synaeresis).

The ability of specific hydrocolloids to form gels depends on the length of the thread-shaped, linearly arranged molecules and is also linked to special structural conditions. Active or activated groups (e.g. $-OH$ or $-COOH$) must be present in a certain arrangement to enable crosslinking between adjacent molecules. The content of side branching must be low, because such side chains would sterically interfere with the intermolecular networking. A distinction is made between secondary valence gels, where molecules are linked by hydrogen bonds (e.g. highly esterified pectin), and primary valence gels with heteropolar linkage, frequently through calcium ions (e.g. low-esterified pectin, alginate gels). Hydrocolloid gels often have crosslinking points of different nature, which make the structural properties difficult to identify and interpret [1].

Suitable Equipment for Measuring Gel Strength

Numerous methods are available for measuring gel strength. The choice of method depends on the kind of material to be evaluated, the purpose of the test, and on tradition. The methods can be categorized in two sections: the first is to measure gel strength within the elastic limits of the gel. The alternative is to measure breaking strength. This is a fundamental difference and the results represent different properties of the gel so that a comparison is not easy.

Bloom Gelometer Bloom strength is essentially the rigidity of a gelatine gel formed under standard conditions. The Bloom Gelometer, developed and patented by the American O. T. Bloom in 1925, is used to measure the strength or compressibility of gels and certain structured foods. The procedure permits the instrument to measure the weight or force required to lower a plunger a predetermined distance into a sample, or the force required to deform the sample to a predetermined degree. The sample is raised by means of an adjustable platform until sample and spring-supported plunger are in contact. The flow of a shot is terminated by means of an automatic switching arrangement, and the weight of shot, corresponding to sample strength, is determined gravimetrically. Working with the Bloom Gelometer requires experience and skill to obtain reproducible results by exact handling.

Plunger Methods In plunger methods, strain is applied to the gel by means of a plunger and corresponding values of load and deformation are measured. Several instruments use this principle. Examples are the Boucher Electronic Jelly Tester, the Stevens LFRA Texture Analyzer, and the Instron Universal testing machine. By forcing a plunger into the gel at constant speed, a strain-versus-distance curve is obtained that can be recorded. The TA-XT2 Texture Analyser from Stable Microsystems works much like the LFRA, but the data can easily be stored as a data-file on a personal computer. Lunger methods are well suited and established in the jam and jelly industry because the gels may be tested after filling directly in the containers in which they will be sold. Modifications of this method are available. A body is embedded in the gelled test sample and the force required to pull the body out again is defined as the breaking strength (method of Lüer and Lochmüller). Herbstreith, a German pectin producer, has developed the so-called Pectinometer for analysis of pectin gels.

SAG Determination The SAG determination, developed by Cox and Higby, is an approved method for pectin standardization. The gel to be tested is prepared in a glass of standardized dimensions. After setting and curing, the gel is carefully removed from the glass and allowed to stand without support. The weaker the gel, the more it sags. The size of sagging, which is gel deformation by its own weight, is measured with a precision instrument, a so-called exchange-ridgelimeter after a specified time. Results are reported as percent SAG, that is, loss of height as percent of original weight.

Rotational Viscosimeter Measurement of gel strength by means of a rotational viscosimeter is possible under the condition that the gel structure is not already destroyed at the beginning of the measurement by shear velocities being too high. It is necessary to use an instrument that is able to work at very low numbers of revolutions. The increase in shear stress with time can be observed exactly before the substance is sheared. In particular, the equipment is suitable where shear stress can be pre-set instead of shear rate. At a certain maximum value of shear stress, the elasticity range of the gel, its gel strength, is exceeded, the measuring element starts to turn and the substance will be sheared.

2.10 Viscosimeters and Rheometers

The viscosity of a fluid is measured with a special instrument, a so-called viscometer. Viscometers only measure at one flow condition. For liquids with viscosities that vary with flow conditions, a rheometer is used. In general, either the fluid remains stationary and an object moves through it or the object is stationary and the fluid moves past it. The drag caused by relative motion of the fluid and a surface is a measure of the viscosity. The flow conditions must have a sufficiently small Reynolds number for there to be in laminar flow. Viscosimeters are calibrated using water or special calibration oils.

Standard Laboratory Viscosimeters for Newtonian Fluids

Glass Capillary and U-Tube Viscosimeter (e.g. Ostwald, Ubbelohde) The principle of measuring here is a capillary tube of glass allowing a liquid to flow along a defined distance. Another version is the Ubbelohde viscometer, which consists of a U-shaped glass tube held vertically in a controlled temperature bath. In one arm of the U tube, there is a vertical section of precise narrow bore (the capillary). Above this is a bulb, along with another bulb lower down on the other arm. In use, liquid is drawn into the upper bulb by suction and then allowed to flow down through the capillary into the lower bulb. Two marks (one above and one below the upper bulb) indicate a known volume. The time taken for the level of the liquid to pass between these marks is proportional to the kinematic viscosity. The specific weight or density of the liquid influences the measured value. Most commercial units are provided with a conversion factor, or can be calibrated with a fluid of known properties. Such viscometers are further classified as direct flow or reverse flow.

Traditionally, the viscosity of HPMC solutions is measured with Ubbelohde viscosimeters. For low-viscosity grades, the resulting values are comparable with Brookfield data. High-viscosities give extremely high values with a Ubbelohde viscometer, which is caused by the tube geometry (the viscosity measured with a Brookfield viscosimeter can be a factor of 1 : 3 lower).

Falling Sphere Viscosimeter (According to Hoeppler) Stokes' law is the basis of the falling sphere viscometer, in which the more or less transparent fluid is stationary in a vertical glass tube. A sphere of known size and density is allowed to descend through the liquid. If correctly selected, it reaches terminal velocity, which can be measured by the time it takes to pass two marks on the tube. Electronic sensing can be used for opaque fluids. Knowing the terminal velocity, the size and density of the sphere, and the density of the liquid, Stokes' law can be used to calculate the viscosity of the fluid. A series of steel ball bearings of different diameter are normally used in the classic experiment to improve the accuracy of the calculation. Glycerine is used as reference fluid. Laminar flow is required (small Reynolds number). A limiting factor on the validity of results is the roughness of the sphere being used.

A modification of the straight falling sphere viscometer is a rolling ball viscometer, which times a ball rolling down a slope whilst immersed in the test fluid.

Flow Cup (e.g. Ford-Cup, DIN-Cup, Engler Viscosimeter) The measuring principle here is to fill a beaker with a liquid of defined volume and measure the flow time until the cup is empty. The flow time increases with increasing viscosity of the fluid.

Bubble Viscometer Bubble viscometers are used to quickly determine the kinematic viscosity of known liquids. The time required for an air bubble to rise is directly proportional to the viscosity of the liquid – so the faster the bubble rises, the lower the viscosity. This method is inapplicable for hydrocolloid solutions with high viscosity or surface-active substances which tend to foam.

Equipment Suitable for Non-Newtonian Fluids

The rheometers presented here allow viscosity measurement at exact defined shear rate (velocity) and shear stress. By measuring shear stress in dependence of different shear rates, flow curves can be determined. Therefore, this equipment is convenient for the characterization of non-Newtonian hydrocolloid solutions in particular.

Rotational Viscometers Rotational viscometers use the idea that the torque required to turn an object in a fluid is a function of the viscosity of that fluid.

The torque required to rotate a disk or bob in a fluid at a known speed is measured. Examples are:

- **Measuring system with coaxial cylinders:**
  A cylindrical measuring element is located in a concentrically arranged measuring cup. The liquid is located in the annular gap. One option is that the inner cylinder rotates and its speed is measured (Searle-type). An alternative option is to rotate the outer cylinder, and one measures the power transmitted to the inner cylinder (Couette-type). The rotating cup is preferred in some cases because it reduces the onset of Taylor vortices. For Searle-types of more modern design (Haake RotoVisco and Contraves Rheomat) the speed of the inner cylinder is kept constant and the torque is measured by coil springs. The shear gradient can be determined because it is proportional to the number of revolutions. The shear stress is proportional to the torque. Viscosity results from the ratio of shear gradient and shear stress.
- **Cone-plate-system:**
  'Cone and plate' viscometers use a cone of very shallow angle in bare contact with a flat round plate that rotates. The imaginary tip of the cone touches the centre of the plate. There is again proportionality between shear gradient (angular velocity) and number of revolutions and, respectively, shear stress and torque. Thus, viscosity can be calculated in a straightforward manner. Certain problems can occur due to the very small gap width. For precise instruments of this type, the gap width is about 100 μm.

Vibrational Viscometers Vibrational viscometers date back to the 1950s Bendix instrument, which is of a class that operates by measuring the damping of an oscillating electromechanical resonator immersed in a fluid whose viscosity is to be determined. The resonator generally oscillates in torsion or transversely (as a cantilever beam or tuning fork). The higher the viscosity, the larger the damping imposed on the resonator. The resonator's damping may be measured by one of several methods:

- Measuring the power input necessary to keep the oscillator vibrating at a constant amplitude: the higher the viscosity, the more power is needed to maintain the amplitude of oscillation.
- Measuring the decay time of the oscillation once the excitation is switched off: the higher the viscosity, the faster the signal decays.
- Measuring the frequency of the resonator as a function of phase angle between excitation and response waveforms: the higher the viscosity, the larger the frequency changes for a given phase change.

The vibrational instrument suffers from a lack of a defined shear field, which makes it unsuited to measuring the viscosity of a fluid whose flow behaviour is not known beforehand. Vibrating viscometers are rugged industrial systems used to measure viscosity under process conditions. The active part of the sensor is a vibrating rod. The vibration amplitude varies according to the viscosity of the fluid in which the rod is immersed. These viscosity meters are suitable for measuring clogging fluid and high-viscosity fluids, including those with fibres (up to 1000 Pa s). Currently, many industries worldwide consider these viscometers to be the most efficient system with which to measure the viscosities of a wide range of fluids. By contrast, rotational viscometers require more maintenance, are unable to measure clogging fluid, and require frequent calibration after intensive use. Vibrating viscometers have no moving parts, no weak parts, and the sensitive part is very small. Even very basic or acidic fluids can be measured by adding a protective coating or using a stainless steel sensor.

Equipment Suitable for Comparable Measurements [1]

Brookfield Today, the measurement of viscosity of hydrocolloid solutions by Brookfield instrument is international practice. In principle, it is a rotational viscosimeter with a cylindrical or disk-shaped rotating body (spindle). The rheological parameters such as beaker size or sample volume are not exactly defined. Thus, the results are only suitable for comparative measurements of known fluids. To a limited extent, non-Newtonian flow properties can be identified using different speeds and spindle sizes.

Amylograph/Brabender Viscograph In the flour and starch producing industry, in particular, numerous devices and instruments are found, operating on the principle of rotation and using different shaped rotating bodies. The measurements are based purely on determination of torque and they are not actually viscosity measurements. The measured values are random and have no definable relation to viscosity units. Frequently, these instruments are used to determine the swelling behaviour, for example with respect to dependence of temperature, and are limited to purely comparative measurements to detect quality deviations.

2.11
Relationship between Rheology and Sensory

The relationship between rheology and sensory is the subject of numerous research and application projects at present. Only a few aspects will be presented and discussed here.

The checking of consistency of liquid and semi-solid food products that takes place in the mouth is comparable to the processes in a rotational viscosimeter during viscosity measurement. For the subjective sensory assessment, part of the fluid food is transported and placed on the surface of the tongue. The tongue is lifted und brought into circular motion, whereby part of the liquid flows between tongue and palate. The tongue exerts a certain shear stress on the liquid, which is activated to flow. It is quite conceivable that the organs of touch (touch receptors) in the mouth detect the occurring shear stress as well as the flow velocity of the liquid. In the literature, velocities of 20–50 rpm are mentioned [1].

Shear rates for chewing and swallowing of solid and semi-solid products such as gummy bears, yogurt, and cheese are specified in the range 10–100 s^{-1} [2].

Often, hydrocolloids are negatively assessed because they have a mucilaginous ('slimy') perception in the mouth. It is a subjective impression mostly caused by thick-flowing substances. These highly-viscous products cover the tongue and stay there for a relatively long time because their dilution by spittle is very slow. They cause unpleasant sensations when swallowing. Between the flow behaviour of different hydrocolloid solutions (in fact, viscosity dependence on shear rate or gradient) and the intensity of the characteristic 'slimy' perception, a very narrow correlation was detected. Surprisingly, hydrocolloid solutions perceived as mucilaginous deviate only slightly from Newtonian flow behaviour. In contrast, solutions exhibiting pseudoplastic or thixotropic flow properties (shear-thinning) and gels appear least slimy [1].

A combination of specific hydrocolloids can be advantageous. For example, consumers prefer a blend of CMC and modified starch instead of the individual substances. While a cellulose gum solution gives an oily impression on the tongue, starches tend to be dry. The blend of both is appreciated in sensory terms.

It has been reported that hydrocolloid solutions of starch, guar, and CMC exhibiting a low pseudoplasticity mask tastes like bitter and sweet more than substances with increased shear-thinning effects [1].

In most cases, foods are served at about 25 °C (77 °F) lower or higher than body temperature. Consumption is more appreciated at 10 or 60 °C (50 or 140 °F), depending on the product. Not only liquid foods, but also solid or semi-solid foods are served at these temperatures. The temperature of the product is changed in the mouth after ingestion. The rheological characteristics depend strongly on the temperature, and the temperature dependence of the elastic modulus is important.

The temperature dependence of the elastic modulus of polysaccharide dispersions may be classified into four categories:

1) so-called cold-set gels like agarose, carrageenans, and gellan that form gels on cooling of the solution;
2) so-called heat-set gels like methylcellulose and HPMC, curdlan, and konjac glucomannan that form gels on heating the solution;
3) inverse re-entrant gels like a mixed solution of methylcellulose and gelatine that form a gel at higher and lower temperatures and stays in a sol state at an intermediate temperature range;
4) re-entrant gels like xyloglucan, in which some galactose residues are removed; this forms a gel at an intermediate specific temperature range and remains in a sol state at higher and lower temperatures outside of this specific range.

The sol–gel transition temperature may be controlled by changing the concentration, molecular weight, and other molecular characteristics such as degree of branching or substitution, types of cations added, presence of salts and polyelectrolytes, sugars, and other polysaccharides.

If the transition temperature occurs at a temperature range of about 37 °C (98.6 °F), it is evident that it will drastically change the mouthfeel. For example, the flavour intensity of jellies judged by panellists correlates well with the gel strength of many gelling agents. But a gelatine gel deviates remarkably from this correlation because it melts in the mouth and gives a far better flavour release than alternative gel-forming hydrocolloids [3].

Food products can be designed with a softer or stiffer texture and a desired performance while chewing or swallowing. The textures of gels are adjusted and controlled towards a more chewy character or a more fluid character by considering the temperature dependence of the elastic modulus provided by suitable hydrocolloids and blends thereof.

References

1 Scherz, H. (1996) *Hydrocolloids: Stabilizers, Thickening and Gelling Agents in Food Products*, Food Chemistry/Food Quality, vol. 2, Food Chemical Society GDCh, Behr's Verlag GmbH, Hamburg.

2 Mezger, T.G. (2011) *The Rheology Handbook*, 3rd edn, Vincentz Network GmbH, Hannover.

3 Stephen, A.M., Philips, G.O., and Williams, P.A. (eds) (2006) *Polysaccharides and Their Applications*, 2nd edn, Taylor & Francis Group, Boca Raton, FL, pp. 147–179.

3
Cellulose

E number:	E 460
EINECS:	232–674–9
CAS number:	9004–34–6

Related hydrocolloids: microcrystalline cellulose, powdered cellulose, nano(crystalline)cellulose (NCC), microfibrillated cellulose (MFC)

3.1
Introduction, History and Development

3.1.1
Introduction

Cellulose constitutes the most abundant, renewable polymer resource available today worldwide. It has been estimated that, by photosynthesis, 10^{11}–10^{12} tons are synthesized annually in a rather pure form, for example, in the seed hairs of the cotton plant, but mostly combined with lignin and other polysaccharides (so-called hemicelluloses) in the cell wall of woody plants.

Cellulose is a polymer raw material used for two general purposes. For many centuries it has served mankind as a construction material, mainly in the form of intact wood and textile fibres such as cotton or flax, or in the form of paper and board. On the other hand, cellulose is a versatile starting material for chemical conversions, aiming at the production of artificial, cellulose-based threads and films as well as various stable cellulose derivatives used in many areas of industry and domestic life. Empirical knowledge of dyeing cellulose fibres, of burning wood, of preparing charcoal, and of the biodegradation of cellulose by rotting was already acquired thousands of years ago. Cellulose occupies a unique place in the annals of polymers [1].

Cellulose and Cellulose Derivatives in the Food Industry: Fundamentals and Applications, First Edition.
Tanja Wüstenberg.

3.1.2
Historical Origin

A resistant fibrous solid substance was first mentioned in 1838, by the French chemist Anselme Payen, that arose as a residue when different plant tissues are treated with acids and ammonia and then, subsequently, an extraction is carried out with water, where alcohol and ether remain. By treating this fibrous material with concentrated sulfuric acid Payen transformed cellulose into dextrose one year later. Thereby, he found that cellulose, starch, and sugar have the same chemical composition. In cotton, the same chemist also discovered cellulose. By elemental analysis, Payen found the molecular formula $(C_6H_{10}O_5)_n$ and also discerned the isomerism to starch [2]. He called the substance remaining after treatment of wood with nitric acid and subsequent addition of sodium hydroxide 'les cellules' (cellulose) [3]. The name 'cellulose' for this plant ingredient is first found in 1839, in a report of the French Academy on the work of Payen. However, already several thousand years before the discovery of the 'sugar of the plant cell wall', cellulose was used for making clothes, as a building material, and for energy production. Typical forms were cotton and wood as well as other available plant fibres. Cellulosic materials have shaped a considerable part of human culture since the Egyptian papyri [2].

3.1.3
Industrial Development from the Beginning to Today

As raw material for chemical reactions and productions, cellulose has been used for about 150 years [2]. Cellulose as a precursor for chemical modifications was used even before its polymeric nature was recognized and well understood [1].

Milestones on this pathway were the discovery of cellulose nitrate (commonly misnamed 'nitrocellulose') by Schönbein in 1846. This was followed by the preparation of Schweizer's reagent (a cuprammonium hydroxide solution) as the first cellulose solvent, invented by Schweizer in 1857. Then the synthesis of an organo-soluble cellulose acetate was developed by Schützenberger in 1865. Partially functionalized cellulose nitrate mixed with camphor as softener was one of the first polymeric materials used as a 'plastic' and is well known under the trade name 'Celluloid' [1].

> Celluloid is manufactured by kneading low-nitrided cellulose nitrate with plasticizers. It can be easily formed at temperatures of 80 °C (176 °F), is transparent like glass, and colourable as desired. It found use as eyeglass frames, combs, and film material.

In 1870, the Hyatt Manufacturing Company developed a technical synthesis for this first thermoplastic polymer product. It was based on the production of cellulose nitrate by reaction of cellulose with nitric acid. These developments indicated that new materials can be produced on an industrial scale by chemical modification of the biopolymer cellulose [2].

Cellulose nitrates of higher N-content have been used extensively for military purposes. Today, cellulose nitrate is the only inorganic cellulose ester of commercial interest. Regenerated cellulose filaments were obtained by spinning cellulose dissolved in cuprammonium hydroxide in an aqueous bath. By far the largest part of cellulose-based artificial fibres have been manufactured for about the last century by the so-called viscose process, invented in 1892 by Cross. This process is still practiced today, with an output of about 3 million tons annually worldwide. It makes use of the formation of cellulose xanthogenate, which is a water-soluble, less-stable anionic ester, prepared by reaction of cellulose with aqueous sodium hydroxide and CS_2 and its decomposition by spinning in an acid bath [1].

Elucidation of the structure of cellulose and the origin of cellulose chemistry as a branch of polymer research finds its beginning in 1920 with the pioneering research of Hermann Staudinger. Haworth had previously discovered the covalent bonds within the molecule and between the glucose units. Staudinger recognized that the structure of cellulose is not a simple aggregation of D-glucose units, after he had performed experiments on acetylation and deacetylation of cellulose in the 1920s and 1930s. Instead, the glucose units are firmly strung together and form long molecular chains covalently linked.

These experiments resulted in the concept of polymer-analogous reactions. According to this concept, functional groups of macromolecules – in the case of cellulose predominantly hydroxyl groups – can undergo the same kind of reactions as the corresponding low-molecular weight compounds. Further, it was observed that the supramolecular structure of the polymer may play an important role in determining the rate and final degree of conversion, as well as the distribution of the functional groups, something that has been well recognized for cellulose [1].

This discovery of the polymeric state of the cellulose molecule and the resulting polymer-analogous reactions is the origin of polymer science, which Staudinger expanded to other chain molecules and for which he was awarded the Nobel Prize in Chemistry in 1953 [2].

From today's perspective cellulose, with a share of about 2 billion tonnes of annual production of biomass, is the most common organic polymer and a very important source of raw material for the increasing demand for renewable and environmentally friendly bio-compatible materials. Pulps derived from wood are still the primary source of raw materials for the technical processing of cellulose. The vast majority (>97%) of pulps produced is used so far in the manufacture of paper and paperboard. Approximately 2.5% of the cellulose pulp is implemented in the chemical industries with only 0.24% (4.8 million tons in 2003) being used for the production of regenerated cellulose fibres and films as well as for the synthesis of various esters and ethers of cellulose [2,4].

Cellulose ethers were first reported in 1905 in a publication of W. Suida. This purely scientific work was followed in 1912 almost simultaneously – but independently – by the first patents for their production.

There were different strategies and approaches, which are given below.

The patents of the colour factories of Bayer and of H. Dreyfus sought highly substituted organo-soluble cellulose ethers that could serve as a substitute for flammable celluloid. In contrast, the patent of L. Lilienfeld describes low-substituted water-soluble products that have become, owing to their broad utility in many areas, more important over the years [5].

Lilienfeld had already written down the reaction of cellulose with dimethyl sulfate to give water-soluble cellulose derivatives in 1912. It then took until 1927 before, on the basis of work by O. Ernst and K. Sponsel, the first industrial production of methylcellulose (MC) was taken up in Germany, where cellulose was reacted with gaseous methyl chloride in the presence of aqueous sodium hydroxide solution [6].

Technical production processes for cellulose ethers with, at first, only small alkyl groups (methylcellulose, ethylcellulose) were developed by IG. Farbenindustrie in 1919–1926. Water-soluble carboxyalkyl celluloses were discovered by E. Jansen in 1918; the first water soluble hydroxyalkyl cellulose was discovered in 1920 [5].

Carboxymethylcellulose (CMC) was introduced in Germany shortly after World War I as a gelatine substitute. A larger production volume, on an industrial scale, was delayed due to high production costs and technical problems. The discovery, in 1935, that the polymer CMC improves the quality of synthetic detergents and cleaning agents facilitated the large-scale production. A little later, the economic importance of CMC increased strongly upon its use in oil drilling fluids (as borehole flushing additive) [7].

With the expansion to hydroxyethyl cellulose (HEC) and finally hydroxypropylcellulose (HPC) the range of types of water-soluble cellulose ethers was virtually brought to its present size [6]. The oil-soluble cellulose ether ethyl cellulose (EC) has been produced industrially since 1930. Since 1950, the production and application of cellulose ethers have developed strongly on a global scale.

3.1.4 Current Data for Cellulose Processing

Figure 3.1 provides an overview (from 2003) for worldwide usage of cellulose from wood. Of the approximately 1700 million tons of tree wood, nearly 80% is used for furniture manufacturing and energy production. Nearly one-tenth of the amount of wood is used for industrial purposes. Only 10% is used for pulp production. The largest share in pulp consumption is taken up by the paper industry with 97.5%. The remaining 2.5% of pulp ends up in the chemical industry. This is a volume of about 5 million tons, which is used to produce various cellulose products. Of these, a small proportion of only 12% falls into the category of the so-called 'cellulose ethers', which are used mainly in the construction

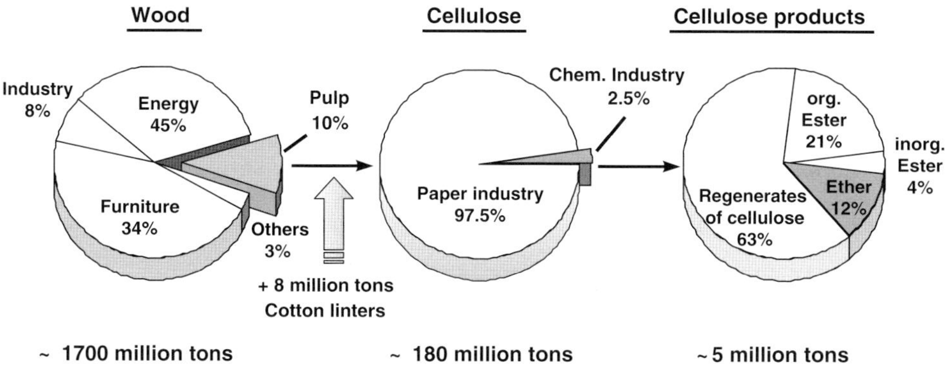

Figure 3.1 Global use of wood for cellulose ether production in 2003. Source: taken from Reference [4].

industry (90%). Regulated applications (foods, cosmetics, and pharmaceutical products) add up to less than 10% of global cellulose ether volume.

3.2 Raw Materials and Biological Origin

3.2.1 Occurrence

The main source of cellulose is the occurrence of this polysaccharide in different types of plants, often combined with other biopolymers. Of great scientific importance is access to cellulose using enzymatic and chemical methods, which have been particularly developed during the last decade [1].

3.2.2 Potential Sources for Natural Cellulosic Fibres

Cellulose is the major building block of the cell wall structure of higher plants. They use cellulose to build tensile plant tissues. Cellulose is particularly well suited to take on this task because it consists of very long linear chains of molecules that can be linked firmly together by hydrogen bonds. Since the tensile strength is such an important and difficult-to-attain property, cellulose is found in all higher plants. Because of its molecular structure, cellulose is ideal for building tensile fibre structures [8,9].

Cellulose as construction substance has a mass fraction of 40–50% of wood, 80% of flax, and 90% of cotton fibre. This makes it the most abundant organic compound on earth. Cellulose is therefore also the most abundant polysaccharide and is a constantly renewable resource. It is estimated that nature produces each year through photosynthesis a mass of 500–1000 billion tons of pure cellulose [8].

All plants produce cellulose, a natural polymer of the plant sugar 'glucose', as they mature from the 'sweet' to the 'stringy' phase. This stringy component, located in various parts of the plant, has long been exploited for textile and related uses with more than 2000 species having been spun into some kind of yarn or cord at some time or place. Only a few of these plants are presently cultivated solely for their fibres, cotton being the predominant one.

Useful fibres may be obtained from various plant parts:

- the stringy bast component of the stem – jute, flax, hemp, and so on;
- leaves – agaves, notably sisal;
- leaf petioles – Abacá (manila);
- fruit – coconut fibre (coir);
- seed – cotton, kapok [10].

Alternatively to plants, some small living organisms are able to produce cellulose as storage substance. Green algae also have cellulose in their cell walls, as do the membranes of fungi. Cyanobacteria represent a primitive source [11]. Bacterial cellulose is described in detail in the scientific literature [12]. *Acetobacter xylinum* can synthesize extracellular pellicles of cellulose from glucose. Simple marine organisms such as tunicates deposit cellulose in their cell walls [8,13].

Cellulose is a linear unbranched polysaccharide having a degree of polymerization (DP) of 1000–15 000 according to the biological origin. The molecular single block, also called a 'monomer', is β-D-glucose.

By physical methods such as intrinsic viscosity measurement, dynamic light scattering, and so on, the polymerization degrees of various cellulosic sources can be estimated [8]. Table 3.1 shows some examples.

3.2.3
Wood as Cellulose Supplier

3.2.3.1 Tree Wood and Perennial Plants

The primary occurrence of cellulose is the existing lignocellulosic material in forests, with wood as the most important source. Other alternative cellulose-containing materials include agriculture residues, water plants, grasses, and other plant substances [1]. The latter are discussed in Section 3.2.3.2.

Wood is the scaffolding and supporting tissue of perennial plants. Like all living tissues, wood is based on individual cells, whose interior cavity, the so-called lumen, is filled with an aqueous solution (serum) during the life stage. When the timber dies, the cavities originally filled with water are emptied and air takes the place of the serum. By drying, wood turns into a very solid compact material, which consists only of plant cell walls [15].

The chemical substances responsible for the development of the plant cell wall of higher plants are predominantly high-molecular-weight organic compounds (polymers). These are usually not soluble either in water or in common organic solvents (such as alcohols, acetone, and hexane).

Table 3.1 Degrees of polymerization of cellulose from various sources [8,10,14].

Source	Degree of polymerization (DP)
Wood from various species	6000–10 000
Textile flax	8000–9000
China grass	6500
Ramie	10 000
Pulp – made of spruce	3300
Pulp – made of beech wood	3050
Pulp – wood from aspen	2500
Pulp – made from fir wood	2500
Wood pulp (in general)	500–3500
Wood pulp, treated	300–1300
α-Cellulose (isolated from wood fibres)	800–1200
Cotton (raw)	10 000–15 000
Cotton linters, untreated	6500
Cotton linters, bleached	1000–5000
Cotton linters, purified	300–1500
Bacterial cellulose	2500–6000
Acetobacter cellulose	600
Rayon	300–500
Cellophane	300
Cellulose acetate	200–350
Valonia [a)]	25 000

a) The bubble alga (*Valonia ventricosa*) is a unicellular green alga that grows in tropical and subtropical oceans worldwide. With a diameter of up to 5 cm, it is one of the largest single-celled organisms of all.

In addition, there are also low-molecular-weight substances available in wood that can be dissolved in solvents and thereby be removed from the wood, so that they can be extracted. These substances are therefore referred to collectively as 'extractives'. The low-molecular-weight extractives are not discussed further at this point, since they have no meaning for cellulose production and are completely removed during the pulp cleaning steps [15].

The polymeric components form the cytoskeleton. This is the source of the 'primary fibres' for the production of paper as well as for extraction of pulp for the manufacture of more or less water-soluble cellulose derivatives. These "primary fibres" are found mainly in the herbal support materials such as the wood of the trunk, the reinforcing fibres of the stalks, and the seed hairs.

The macromolecules in wood are cellulose, polyoses (hemicelluloses), and lignin. They form the structure of the plant cell wall. Cellulose is a linear polymer, hemicelluloses are branched, and lignin has a crosslinked structure. In addition, small amounts of biopolymers occur that control the life processes of the cell. These include proteins with complicated structures and nucleic acids.

There are varieties of plants that also contain a more or less large amount of gum (exudates). An example of this is the tree *Acacia senegal*, located in the Sahel zone in Africa, from whose natural sap (syrup) the stabilizer acacia gum, also well-known as gum arabic or E 414, is obtained [15].

The actual resource for the fibre material is formed by the walls of the supporting cells. These cells are made up of a polymeric composite material containing tensile- and pressure-resistant components.

The main components are [15]:

- lignin as matrix material:
  - very strong (can accommodate compressive forces),
  - little flexibility (brittle),
  - hydrophobic, water repellent (therefore protects against water attack),
  - very slow biological degradation (protection against microbes);
- cellulose as reinforcing material:
  - high tensile strength (can accommodate tensile stresses);
- polyoses and hemicelluloses:
  - flexible adhesive (kit) between lignin and cellulose and between the individual cellulose fibrils.

The plant cell wall is a structured mixture of cellulose and accompanying macromolecules, in which these polymers are allocated and distributed in a very specific way. A special feature is that the cellulose forms microscopic fine strands (fibrils) that ensure the tensile strength of the composite material such as a textile insert in car tires or garden hose. The cellulose fibrils are packed very densely in the central cell wall layer (secondary wall). Therefore, the highest cellulose concentration is found here. The cells are encased by a layer of lignin, resulting in a particularly high concentration of lignin in the outer cell wall (of the middle lamella) [15].

Figure 3.2 shows how the components of the cell wall are distributed along a section, from the inside (lumen) to the outside (middle lamella), in terms of quantity.

The variations in the middle layer result from the packing of the cellulose fibrils [15].

While all three macromolecular substances are maintained to a large extent in the wood substrate, in the so-called pulps there are significantly less lignin and hemicelluloses while the cellulose content is greatly increased. In trees, the percentage of lignin (15–30%) and hemicelluloses (25–35%) is in sum about 50–60%. For the manufacture of cellulose derivatives, especially cellulose ethers, only pulps from pure α-cellulose are used. All accompanying polymeric substances and low-molecular-weight extractives are removed by extensive cleaning processes.

All woods show a specific fingerprint in their chemical composition. Within a particular plant species, there are still differences from plant to plant – depending on age, location, climate, environmental conditions, and so on. Even within an individual plant, there is a distinction, for example, between early-wood and late-wood, as well as compression wood and tension wood, which vary both in structure and in composition [15].

Table 3.2 gives guideline values for the composition of different woods. The pentosans are listed separately as a subgroup of the hemicelluloses, but are also

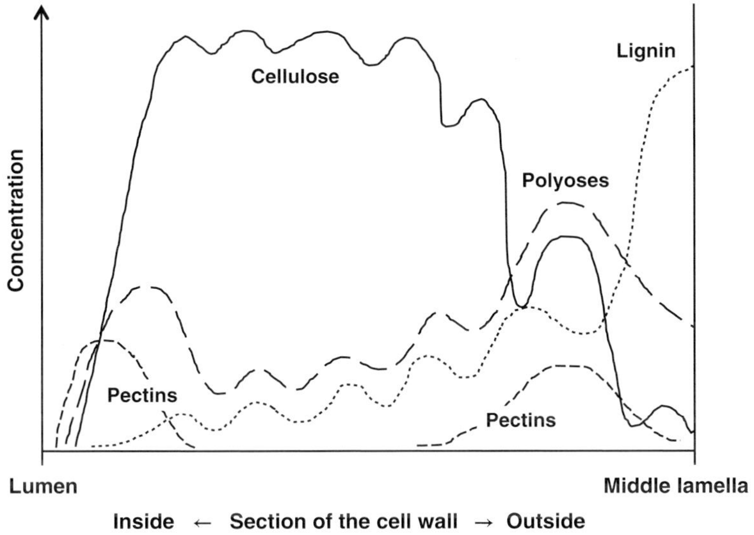

Figure 3.2 Distribution of the polymeric constituents of the cell wall of wood fibres [15]. Reprinted with kind permission of Dr. Erich Gruber.

included in the number of hemicelluloses. As 'extract', as is usual, only the organic extractives are summarized. This includes substances that are soluble in nonpolar (less polar) solvents, such as alcohols, acetone, and halogenated hydrocarbons. The main groups of organic extractives are resins, fats and oils, waxes, tannins and humic substances, terpenes, and odorants [15].

3.2.3.2 **Annual Plants**

In addition to wood, other plants are also conceivable as a source of cellulose fibres, from their seed fibres, bast, or supportive tissue fibres. The fibres from annual plants generally have lower lignin content than fibres of tree wood, so in theory they should be easier to digest chemically. However, one often finds another

Table 3.2 Chemical composition of different lumbers [15].

Source (botanical name)	Composition (%)					
	Cellulose	Hemi-cellulose	Of which pentosan	Lignin	Extract	Ash
Pine (*Pinus silvestris*)	44	26	9	28	4	0.4
Spruce (*Picea abies*)	43	27	9	29	2	0.4
Douglas fir (*Pseudosuga menziesii*)	47	22	7	30	5	0.3
Silver fir (*Abies alba*)	43	27	11	29	3	0.5
Birch (*Betula verrucosa*)	46	36	25	19	2	0.3
Beech (*Fagus sylvatica*)	45	35	17	18	4	0.3
Poplar (*Populus tremuloides*)	50	31	17	18	4	0.3

insoluble substance of incrustation in grasses, namely, silica. This is especially true for giant bamboo grass. The solid crystalline silica can be transferred into a colloidal solution only by boiling with very strong lye and be thus removed from the plant material. But through the treatment with highly concentrated alkalis, the polysaccharide cellulose and the polyoses are significantly damaged. Since the fibre cells of annual plants are also mostly slender and often have very thin walls, the pulps produced out of them show very little resistance [15].

Table 3.3 shows the fibre lengths of wood fibres in comparison with those of annual plants.

3.2.4 Bacterial Cellulose

A rather new approach to pure cellulose that is gaining increasing interest is the laboratory-scale production of the polymer by acetic acid-producing bacteria, such as *Gluconacetobacter xylinum* and *Acanthamoeba castellani.*

Table 3.3 Comparison of wood fibres with fibres of annual plants [8].

	Plant	Cellulose (%)	Lignin (%)	Fibre dimensions (mm)	
				Average length	Average width
Long fibres	**Cotton**	**85–90**	**0.7–1.6**	**25**	**0.020**
	Flax seeds	43–47	21–23	30	0.020
	Hemp	57–77	9–13	20	0.022
	Abacá[a)]	56–63	7–9	6	0.024
Medium-long fibres	**Softwood**	**40–45**	**26–34**	**4.1**	**0.025**
	Sisal	47–62	7–9	3.3	0.020
	Bamboo	44–57	21–31	2.7	0.014
	Kenaf	26–43	15–19	2.6	0.020
	Jute	45–63	21–26	2.5	0.020
	Papyrus	38–44	16–19	1.8	0.012
	Bagasse[b)]	32–37	18–26	1.7	0.020
Short fibres	**Heartwood**	**38–49**	**23–30**	**1.2**	**0.030**
	Cereal straw	31–45	16–19	1.5	0.023
	Corn stover	32–35	16–27	1.5	0.018
	Wheat straw	33–39	16–23	1.4	0.015
	Rice straw	28–36	12–16	1.4	0.008
	Esparto[c)]	33–38	17–19	1.2	0.013

a) Abaca (*Musa textilis*) also known as Manila hemp, banana fibre, or textile banana. Abaca is a native species of the genus banana in the banana family Musaceae, located in East Asia. It is used as a fibre plant and is mainly used for the production of salt water resistant ship hawser.
b) Residues from pressing sugar cane.
c) Espartogras (*Lygeum spartum*) is a grass species found in the southern Mediterranean area. The leaf fibres can be used for the production of paper (alfa paper) and ropes.

Algae form another source of cellulose, for example *Valonia ventricosa* and *Chaetamorpha melagonicum*. According to several researchers, the cellulose obtained is highly crystalline and is useful for studying polymorphs of the polymer. Cellulose of the *Valonia* type is found in fungal cell walls as well. In addition, there are several celluloses of animal origin, of which tunicin, a cell-wall component of ascidians, has been studied extensively [1].

3.2.5
Biosynthesis

The biosynthesis of cellulose is not yet completely elucidated. Moreover, contrary results have been described and discussed in many papers. During the last few years, the number of patents in this field has increased because of the interesting possibility to increase the cellulose content of plants and to construct new and more efficient plants. Because of the ability of some bacterial strains to form cellulose and because of the similarity of the biosynthetic apparatus in some aspects, much of the research has been carried out using these bacteria. Good results have been obtained with the bacterium *Gluconacetobacter xylinum* as a model organism.

The subjects of some selected papers, cited in Reference [1], on the biosynthesis and structure of cellulose are given here:

- Role of callose synthase and other (1,3)-β-glucan synthase in cellulose biosynthesis; enhancing cellulose synthesis by cellobioses.
- Effect of retardants on cellulose biosynthesis in cotton; effects on fibres and seedlings.
- Review about genes and proteins involved in cellulose synthesis in plants; role of sucrose synthase for substrate formation; orientation of the microfibril deposition; role of membrane-associated cellulase in biosynthesis process.
- Cellulose structure elucidation using atomic force microscopy.
- Estimation of the relations between cellulose I_α and I_β in wood; application of ^{13}C NMR.
- Supramolecular architecture; fibril formation, and its regulation.
- Cellulose biosynthesis as a binding factor for CO_2.
- Reviews on cellulose biosynthesis; comparison of synthesis by microorganisms and by plants.
- General reviews concerning cellulose biosynthesis by bacteria, fungi, and plants.

The subjects of some interesting patents, cited in Reference [1], on biosynthetic pathways are also given here:

- Overexpression of cellulose synthase genes for modulating expression of enzymes involved in synthesis of plant cell walls.
- Polynucleotides encoding cellulose synthase for acceleration of plant growth and up-regulation of cellulose synthase level; modifying lignin biosynthesis.

- Cellulose synthase gene from poplar; application for altering cellulose and lignin composition.
- New genes encode maize cellulose synthase polypeptides; modulation of expression of cellulose synthase in plants; production of transgenic plants expressing the new protein.
- Genes for cellulose synthase; application of these genes for improving plant stalk quality; increase of cellulose in stalks and so on.
- Transgenic plant expressing cell-wall modulating proteins as a basis for, for example, altered morphology, increased growth, modified fibre lengths, increased cellulose, and starch content.
- Isolated genes encoding polypeptides involved in cellulose biosynthesis, transgenic plants, expressed in sense or anti-sense orientation, ribozymes, co-suppression, gene-targeting molecules.

The only substrate for cellulose biosynthesis is UDP-glucose. The biosynthesis of this energy-rich compound follows the normal biosynthetic pathways in the cells, starting from glucose (Figure 3.3) [1].

The enzyme cellulose synthase accepts only UDP-glucose as a substrate. Moreover, it was noticed that by feeding modified glucoses to bacteria (*Gluconacetobacter xylinum*), as well as to plant cells or cell extracts, no significant formation of modified celluloses could be detected. Another possible source for UDP-glucose could be sucrose synthase, an enzyme associated with the plasma membrane, for example of cotton fibres. Because of its location, a direct channelling of the substrate UDP-glucose to the polymerizing enzyme is possible. However, the regulation, control, and targeting of this process is unknown in wide areas. Other possible sources for the stabilizing and transport of the substrate are annexin-like molecules, which are able to bind UDP-glucose, for example a 170-kDa polypeptide was co-purified with the cellulose synthase.

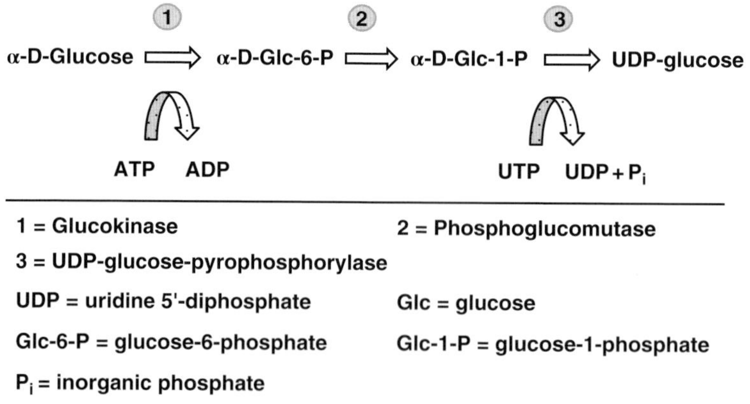

Figure 3.3 Intracellular activation of glucose as precursor for cellulose biosynthesis. Reproduced with permission from Reference [1] © 2005 Wiley-VCH.

Delmer and Brown found in the years 1997–2000 that this protein shows some homologies to the yeast β-1,3-glucan synthase. Summarizing those effects, it was stated by Brett in 2000 that UDP-glucose on the one hand or sucrose on the other, as well as further soluble intermediates from these pathways, could serve as possible precursors [1].

Different data have been described for the cellulose synthase as the active enzyme system in cellulose formation. For cellulose synthesis, different models have been developed that will not be discussed further here.

3.2.6
Biological Composition

The basic building block of cellulose is β-D-glucose ($C_6H_{12}O_6$). By condensation, that is, by the elimination of one molecule of water per two molecules of D-glucose, the molecule 'cellobiose' is formed. The cellobiose consists of two anhydroglucose units (AGUs). By condensation of many molecules of cellobiose, the cellulose molecule is formed under cleavage of further water molecules. The cellulose chains are positioned very close to each other due to their hydrophilicity. This leads to crystallization and the resultant formation of crystalline structures. The crystal structure is stabilized by hydrogen bonds along the individual chains, between adjacent chains within the cellulose lamella, and between the individual cellulose lamella.

Cellulose has a stronger tendency to crystallize than most polymers. It always occurs in partially crystalline form. The cellulose crystallites are called 'elementary fibrils'. Since these crystallites have approximately the form of long-stretched flat blocks, from whose ends disordered cellulose chains jut out like fringes, these structures are also called 'fringe crystallites' [9]. The crystallization of cellulose is discussed in detail in Section 3.4.2 ('Secondary structure').

Cellulose is formed in the plasma membrane. The individual cellulose molecules crosslink with each other to form fibrillar structures. The spatial arrangement of the cellulose fibrils is controlled by the microtubules of the cell. As smallest unit, the elementary fibrils (cellulose crystallite strands) composed of many β-($1 \rightarrow 4$) glucose chains are bundled to microfibrils, which, in turn, are assembled to macrofibrils in the cell wall. These macrofibrils are visible under a light microscope. Other components of the plant cell wall are hemicelluloses, lignins, pectins, and proteins.

The respective cellulose structures are characterized by a higher degree of order and will be surrounded and limited by substances of lower order, as the accompanying substances lignin and hemicelluloses, which means that the cellulose phases are separated by them [9].

As a naturally occurring polymer, cellulose always contains small quantities of other constituents in addition to glucose (>99%). These substances are formed during photosynthesis of the plants and are directly incorporated partially into the cellulose molecules or placed on the surfaces of the cellulose molecules. This results, for example, in lignin–cellulose complexes.

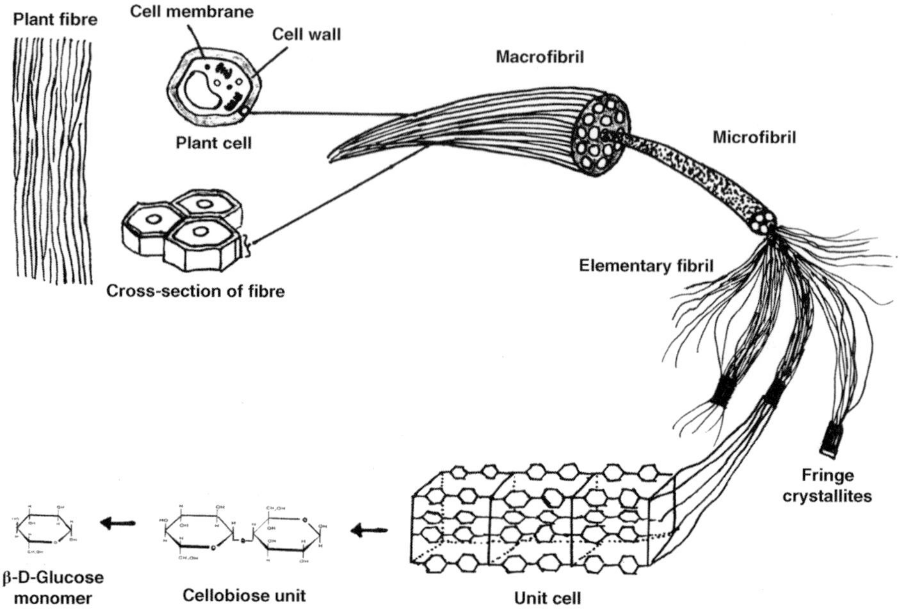

Figure 3.4 Biological origin of the cellulose in the cell wall of plants.

Most of the changes in molecular structure, however, result from secondary reactions such as hydrolytic or oxidative processes that occur while the cellulose molecules are extracted or isolated from their natural sources. Cellulose always contains carboxyl groups. In wood pulp, there is one -COOH group per 100–1000 anhydroglucose units (AGU). In cotton, one carboxyl group per 100–500 AGU can be found [10].

Figure 3.4 illustrates where cellulose is localized in a plant.

Figure 3.5 shows a cross-section through the plant cell wall.

In tree wood, cellulose is an essential part of the ingenious sophisticated fibre-reinforced composite, with long stiff cellulose chain molecules being organized in thin fibrils and thereby holding the entire plant reticulum. The hydrophobic lignin takes on the task of a wrapping material [10].

The polyoses (also known as hemicelluloses) function as a binding agent that is positioned like a flexible unit between lignin and cellulose as well as in between the individual cellulose fibrils. Figure 3.6 displays, hierarchically, how the plant cell wall is built up from the glucose to the cellulose.

Table 3.4 lists the dimensions of the structural elements of cellulose. Cellulose forms a hierarchy of structures, all of which are characterized by a longitudinal stretching.

The values presented in Table 3.4 are average numbers and can deviate for different cellulose sources. Table 3.5 shows example ranges for the sizes of microfibrils of various cellulosic materials.

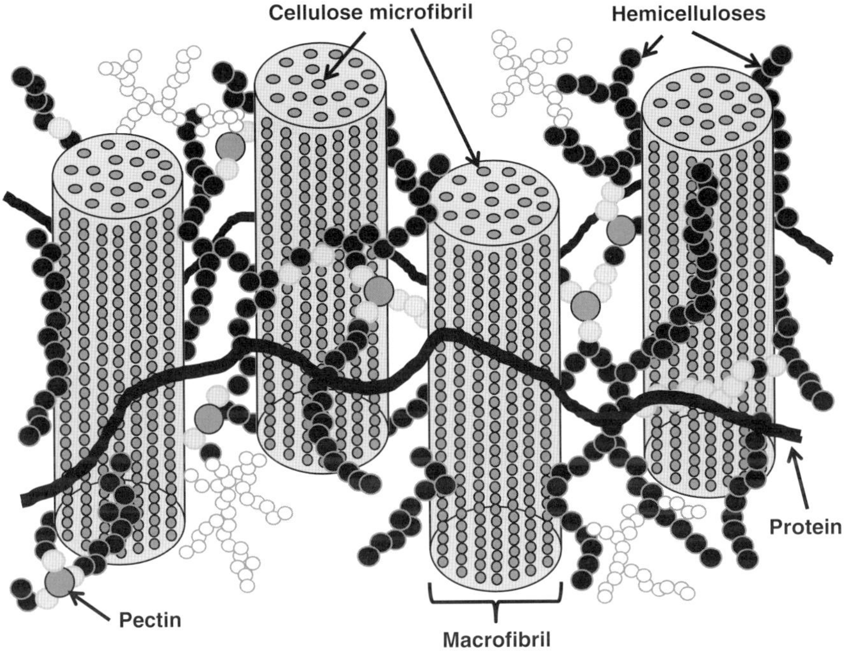

Figure 3.5 Cross-section through the cell wall of plants.

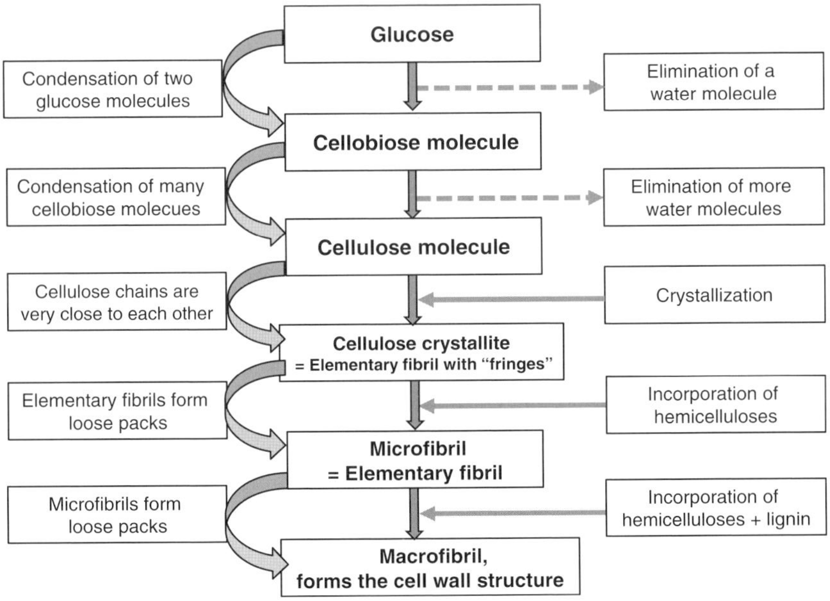

Figure 3.6 Overview of the hierarchy of structure elements from glucose to the cellulose backbone of the plant cell wall [9]. Reprinted with kind permission of Dr. Erich Gruber.

Table 3.4 Dimensions of the structural elements of cellulose [9].

Element	Width (nm)	Depth (nm)	Length (nm)
Molecule	0.82	0.39	0.52
Elementary cell	0.82	0.78	1.03
Elementary fibril	6–10	3–5	15–20
Microfibril	20–30	8–10	50–80
Macrofibril	Up to 400	Up to 100	Up to 3000

Table 3.5 Range of microfibril diameter of various cellulose samples [1].

Sample	Microfibril diameter (nm)
Bacterial cellulose	4–7
Cotton linters	7–9
Ramie	10–15
Dissolving pulp	10–30
Valonia cellulose	10–35

The dimensions of the elementary crystallites or elementary fibrils are only slightly different from native or regenerated cellulose fibres. The range varies between 24 and 40 glucose monomers. The frequently observed longer micro- or macrofibrils or their fragments are only individual elementary fibrils, lined up or placed together like in a stone wall. In native wood or cotton fibres, these elementary fibrils are arranged in various layers in the cell wall [10].

Figure 3.7 illustrates how the cellulose fibrils are positioned in cotton fibres (a) and wood fibres (b).

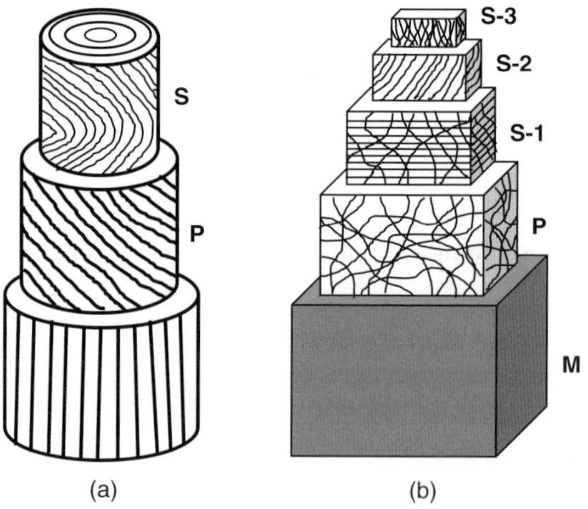

Figure 3.7 Position of the cellulose fibrils in cotton fibres (a) and wood fibres (b). Reproduced with permission from Reference [10] © 2004 Wiley-VCH.

3.2.6.1 Cotton Fibres

- In the primary cell wall (P), interlaced fibrils can be found.
- Secondary cell wall (S) – the fibrils are wound spirally around the fibre axis. At distinct distances along the fibre axis, the spiral reverses direction.

3.2.6.2 Wood Fibres

- In the middle lamella (M), lignin (casing) and hemicelluloses (binder) are concentrated.
- The primary cell wall (P) consists of unarranged positioned fibrils.
- In the secondary wall 'I' (S-1), two or more fibrillar layers cross one another and are positioned spirally along the fibre axis.
- In the secondary wall 'II' (S-2), fibrils are wound spirally around the fibre axis.
- The fibrils are tightly interlaced in the secondary wall 'III' (S-3).

Synthetic cellulose fibres, such as viscose, do not have a native morphology. Their supermolecular structure can be described as a network of elementary fibrils and their more or less random associations. This is called a 'fringe fibrillar' structure, which is shown below in Figure 3.21 [10].

3.2.7 Explanation for Industrial Derivatization

Owing to the high molecular weight and the highly-ordered structure of the β-D-glucose monomers, cellulose in its native state is not soluble in water. The individual polysaccharide chains are 'linked' relatively 'firmly' with other (adjacent) chains through hydrogen bonds (see Figures 3.16 and 3.20 below). In principle, the highly hydrophilic basic building blocks of cellulose cause its insolubility in water. Cellulose is therefore an extremely resistant fibrous substance and shows only minimal swelling in water.

With a few exceptions, food products are characterized by more or less high water content (2–99.5%). To be able to use cellulose in food applications, the molecule therefore has to be brought into a water-soluble form by means of a physicochemical modification. The resulting solubility in water covers – depending on the intensity of the physical process and the process chemicals used during the subsequent chemical reaction – the whole spectrum from light swellable (maximum 5%) to completely water soluble (100%). For example, hydrophobic substituents ($-CH_3$) may be introduced into the cellulose chains for the manufacture of methylcellulose. The methyl groups are used as placeholders, driving and keeping the various β-D-glucose strands apart. This 'artificial distancing' has the effect that water molecules can be positioned and stored in between the polysaccharide chains. It may sound paradoxical, but the insolubility of cellulose is caused by its own strong hydrophilicity and can be reduced by introducing hydrophobic, that is water-repellent, groups in the cellulose molecule, and thereby the solubility in water is developed.

The cellulose molecule has crystalline and amorphous portions (approx. 40%). In the amorphous regions, there are also bonds that can be decomposed relatively easily by acids or alkalis – the resulting product is called 'microcrystalline cellulose' [16]. This kind of cellulosic product is discussed in detail in Chapter 4. For the production of so-called 'cellulose derivatives', the cellulose is usually introduced into the industrial process in form of pulps. Pulp suppliers are the renewable resources tree wood and cotton linters. Other natural resources (marine and terrestrial plants, fruit and vegetable, bacteria) are possible, but are currently not considered for large-scale use in manufacture of food additives for economic reasons. A tree synthesizes an average of 13.7 g of cellulose per day. If all the cellulose molecules produced daily were strung together they would cover a distance of 2.62×10^{10} km. This is equivalent to about 175 times the distance between the sun and the earth [10]!

Cotton – in the form of linters – can serve as cellulose source. Since cotton linters cannot be used for textile production, they are considered as waste. The very short non-spinnable fibres result from the so-called 'third cut' after separation of the textile cotton and squeezing out the cotton oil. These fibres are made up of 98% pure cellulose and have extremely long high-quality cellulose chains. The high degrees of polymerization of cotton linters are not reached by highly refined softwood pulps. Table 3.6 lists the chemical composition of some typical cellulose-containing materials [14].

Table 3.6 Chemical composition of cellulose-containing materials [14].

Source	Composition (%)			
	Cellulose	Hemicellulose	Lignin	Extract
Hardwood	43–47	25–35	16–24	2–8
Softwood	40–44	25–29	25–31	1–5
Cotton	95	2	1	0.4
Bagasse	40	30	20	10
Coconut fibre (coir)	32–43	10–20	43–49	4
Corn cobs	45	35	15	5
Corn stalks	35	25	35	5
China grass (ramie)	76	17	1	6
Flax (unretted)	63	12	3	13
Flax (retted)	71	21	2	6
Hemp	70	22	6	2
Jute	71	14	13	2
Sisal	73	14	11	2
Kenaf	36	21	18	2
Sunn	80	10	6	3
Wheat straw	30	50	15	5

3.3 Manufacture of Pulp

3.3.1 Purification of Natural Cellulose Sources

The commercial purification of cellulose and further processing into pulp for the chemical industry is focussed on the raw materials of cotton linters, which are available in naturally highly pure form, and fast-growing wood (soft timbers). Cotton linters are used due to their extremely high cellulose contents and their unique length of the fibres. Tree wood finds use due to its availability in relative abundance and ease of harvesting. Cellulose in the natural state is difficult to purify due to its insolubility in commercial solvents [8]. As a naturally occurring material, cellulose may contain by-products leading to application problems and difficulties in chemical modification reactions. However, up-to-date cellulose isolation and purification methods yield materials of high purity and variability.

Table 3.7 gives some examples of such cellulose materials. The values of the degree of polymerization (DP) (molecular weight = DP × 162 g mol^{-1}), included in Table 3.7, manifest the huge variety in molecular weight available.

3.3.2 Cotton Linters

With the amendment of European food law in 2003 concerning the use and labelling of genetically modified ingredients and additives (EU Directives 1829/2003 and 1830/2003), the use of cellulose-based products derived from cotton linters for application in foods has been severely limited, if not made impossible.

As a result of the production routes of cotton with respect to cotton oil, evidence for GMO-freedom of linters is nearly impossible to attain or is not pursued for economic reasons. Therefore, in Europe and North America the use of cotton linters for manufacture of cellulose derivatives is limited now to

Table 3.7 Carbohydrate composition and degree of polymerization (DP) of some cellulose samples [1].

Sample	Producer	Carbohydrate composition (%)			DP
		Glucose	Mannose	Xylose	
Avicel	Fluka	100	—	—	280
Sulfate pulp V-60	Buckeye (USA)	95.3	1.6	3.1	800
Sulfate pulp A-6	Buckeye (USA)	96.0	1.8	2.2	2000
Sulfite pulp 5-V-5	Borregaard (Norway)	95.5	2.0	2.5	800
Linters	Buckeye	100	—	—	1470
Linters	Buckeye	100	—	—	2000

non-food-grade with cellulosic products being exclusively used for technical or industrial applications. The technical cellulose ether grades are handled separately in factories.

Owing to the enormous increase in prices of cotton linters, these are only used in industrial applications where very high viscosities, resulting from the high fibre lengths, are required. The fourfold increase in raw material prices for cotton linters in less than two years was created by its natural shortage. This is both the result of the constantly and strongly increasing demand volumes for cellulose derivatives and, in addition, the consequence of breeding measures. In Asia, where a large proportion of textile cotton is grown, the preference in recent years has been to plant modified varieties; these varieties have a higher percentage of pure cotton with a simultaneously very low content of linters, in order to be able to serve increased demand from the textile industry.

3.3.3 Wood

Tree wood of fast-growing timber species should – after the separation of value-adding components for the furniture and paper industry – be purified to isolate cellulose in its pure form. Cellulose is very difficult to clean in its native state as it is absolutely insoluble both in water as well as in traditional organic solvents. Alkaline solutions cause significant swelling of cellulose fibres and solubilize the hemicelluloses present. Therefore, the purification step usually consists of an alkaline digestion to remove waxes and proteins as well as hemicelluloses (25–35%) and lignin (15–30%), which account in sum for about 50–60% of the mass of wood. Pulps, entering as raw material the cellulose ether production, are often subjected to an additional extraction procedure. The aim of this process, which is also carried out under alkaline conditions, is not only the complete elimination of low-molecular-weight polysaccharides, and remaining hemicelluloses, but also to increase the content of pure α-cellulose [8]. The thus-obtained pure cellulose is finely ground and pressed into large rolls of paper. These rolls of dried pulps are then used as raw materials for the manufacturing process of cellulose derivatives, especially cellulose ethers.

The wet processing of purified wood pulp or cotton linters for the production of cellulose derivatives is another possibility. It leads to higher quality end products in terms of molecular chain length and viscosity, since the time-consuming drying process, as a quality-reducing step, is omitted.

The length of the polymer chain varies with the origin of the cellulose. Wood pulps consist of chains of 300–3300 glucose monomers. For cotton pulps, there are – depending on the type of treatment – chains of 800–10 000 anhydroglucose units (AGUs).

By selection of specific pulps and combination of corresponding pulp mixtures, the viscosity of the resulting cellulose derivative is set and adjusted. For pulp production, the following species are preferably used. The percentage given indicates the cellulose content:

- spruce (*Picea abies*) – 43%;
- pine (*Pinus silvestris*) – 44%;
- birch (*Betula verrucosa*) – 46%;
- beech (*Fagus silvatica*) – 45%;
- poplar (*Populus tremuloides*) – 50%.

3.4 Chemical Composition and Structure

3.4.1 Molecular Structure

3.4.1.1 Basic Structure

One of the certainties in cellulose science is that cellulose is an aggregate of linear polymers of D-glucopyranosyl residues in chain form, which are linked together entirely in the β-(1 → 4) configuration (see Figure 3.10). Chemical and enzymatic hydrolysis, acetolysis, methylation studies, NMR and X-ray analysis afford ample proof of this [8].

Cellulose is an isotactic β-1,4-polyacetal of cellobiose (4-*O*-β-glucopyranosyl-D-glucose). The actual base unit, the cellobiose, consists of two molecules of glucose. For this reason, cellulose can also be considered as a (syndiotactic) polyacetal of glucose.

In the cellulose chain, the cellobiose unit is repeated (see Figure 3.9), and the unit itself consists of two glucose units (see Figure 3.8), in the chair form (shown on the right-hand side), which are rotated by 180° against from each other [9].

Figures 3.8 and 3.9 show the basic building blocks of the cellulose molecule.

Consequently, one unit has the 'top' up, and the next has its 'bottom' up. Along the axis, there is the symmetry of a twofold mirror axis, as illustrated schematically in Figure 3.10.

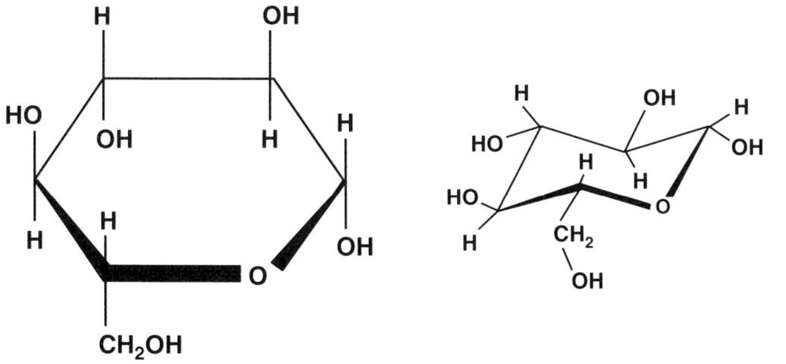

Figure 3.8 Smallest basic building block of cellulose – the β-D-glucose ring.

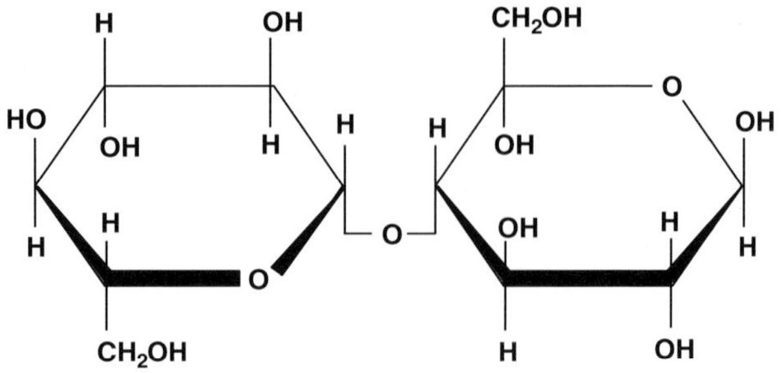

Figure 3.9 Next bigger basic building block of cellulose – the cellobiose unit.

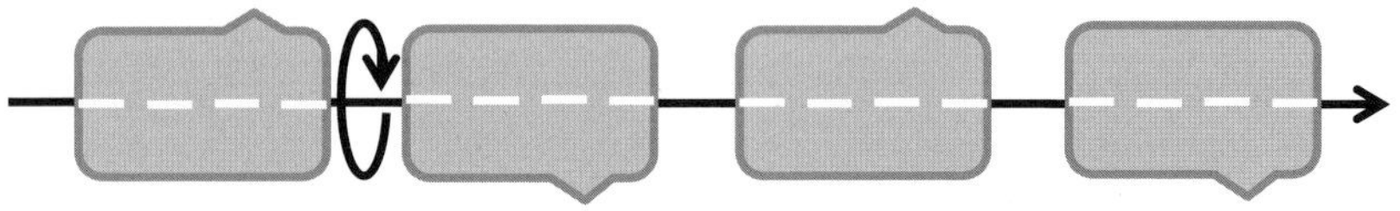

Figure 3.10 Twofold symmetry of the cellulose molecular chain [9]. Reprinted with kind permission of Dr. Erich Gruber.

The basic chemical formula of cellulose is:

$$C_{6\cdot P}H_{10(P+2)}O_{5(P+1)} \approx (C_6H_{10}O_5)_P \quad \text{or} \quad (C_6H_{10}O_5)_n$$

where P is the degree of polymerization and n is the number of glucose monomers in the cellulose chain.

The elemental composition of cellulose was discovered in 1842 by the French chemist Anselme Payen [17].

Cellulose consists of:

44.0–45.0% carbon (C),
6.0–6.5% hydrogen (H),
48.5–50.0% oxygen (O).

The molecular weight of a glucose unit is $m_0 = 162\,\mathrm{g\,mol^{-1}}$.

This results in an approximate molecular weight of the cellulose molecule of:

$$M_{\mathrm{Cell}} = m_0 P + 18 \approx 162P$$

The existence of covalent bonds within and between the glucose monomers was discovered by the British chemist and Nobel laureate Walter Norman Haworth. Final proof of the macromolecular nature of cellulose was given by the German Nobel laureate in Chemistry, Hermann Staudinger, after numerous research studies from 1920 onwards [10].

Depending on the relative position of the hydrogen atoms with respect to the plane of the pyranose ring, there are two possible conformations of the glucose monomer. In the axial conformation, the hydrogen atoms are alternately positioned above and below the plane of the ring. In the equatorial conformation, they are located in the plane of the ring. This isomerism has the lower energy content and is, as a consequence, the more stable and therefore the more likely form [17].

As analyses of glucose with X-ray crystallography methods and nuclear magnetic resonance (NMR) spectroscopy show, the glucopyranose ring takes a $^4C_1$ chair conformation. This isomeric structure is the form thermodynamically most stable. The chair conformation, in comparison to the boat or tray conformations, has a free stabilization enthalpy of 20.05 kJ mol^{-1}. In this position, the three free hydroxyl groups are placed horizontally in the ring (equatorial), while the hydrogen atoms are in a vertical position (axial). Thus, it seems very natural to assume that the same conformation of the glucose units exists in the cellulose molecule [10].

Figure 3.11 shows a section of a cellulose molecule composed of β-D-glucose monomers (glucopyranosyl residues) that are consistently associated through (1 → 4) bonds and are in the chair conformation.

3.4.1.2 Differences between Cellulose and Starch

In the plant, several structural polysaccharides are found.

The nine basic units are sugars such as [8]:

- pentoses: L-arabinofuranose;
- hexoses: D-xylopyranose, D-galactopyranose, D-glucopyranose, and D-mannopyranose;
- 6-deoxyhexoses: L-fucopyranose (6-deoxygalactose) and L-rhamnopyranose (6-deoxygalactose);
- uronic acids: D-galactopyranosyl-uronic acid and D-glucopyranosyl-uronic acid.

The plant glucans, of which both cellulose and starch are specific forms, are among the few homopolysaccharides found within the plant kingdom. Although they consist only of glucose residues, differences in their glycosidic pattern and in their secondary structure produce polysaccharides with remarkably different properties and functionalities. It is difficult to imagine, for example, two glucans with such divergent properties and functions within the plant as starch and cellulose [8].

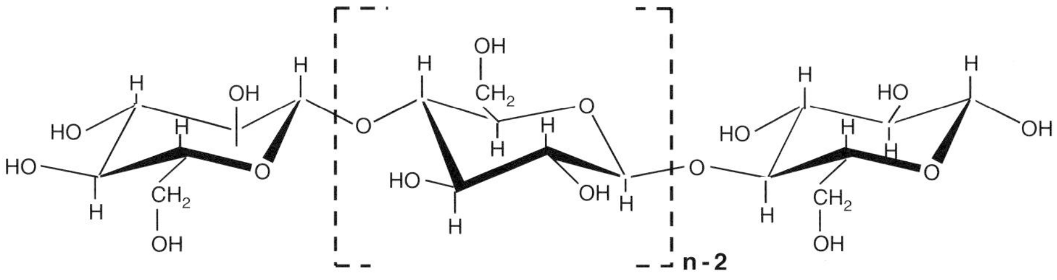

Figure 3.11 Molecule of cellulose.

The difference in the chemical structure of the basic building blocks is only small: Starch is a polymer of (1 → 4) linked α-D-glucose units. Cellulose consists of macromolecules of β D-glucose monomers with (1 → 4) linkages.

This composition affects the behaviour of the substances in the human body. Starch is fully utilized and delivers 4 kcal g^{-1} of energy. Starch digestion begins in the mouth with amylases of the spittle. Cellulose passes through the body without being degraded because of a lack of suitable enzymes (ß-amylases, cellulases).

The different characteristics of the hydrocolloids starch and cellulose, and of their numerous derivatives, found in various applications result from the secondary structure. Cellulose is a linear unbranched molecule without intermolecular atomic bonds. Its insolubility in water is caused by the fact that it has only intermolecular hydrogen bonds. In native starch, there are two basic components – amylose and amylopectin. Amylose consists of helical unbranched chains of 200–300 glucose monomers with α-(1 → 4) glycosidic bonds. The content in native starch is 10–20%. Pure amylose is water soluble. Amylopectin, the major component (80–90%), is a macromolecule of crosslinked amylose that is not soluble in water. Branched molecules of about 3000 glucose rings in α-(1 → 4) and α-(1 → 6) glycosidic linkage are found. By breeding and extraction, the ratio of amylose and amylopectin can be changed dramatically in both directions to fine tune the properties and functionalities of the starch derivatives.

Sugar Chemistry

Monosaccharides do not show all of the typical reactions of carbonylic compounds because they mostly do not occur in the open-chain form of aldehydes or ketones, but are present as intramolecular six- or five-membered hemiacetals, so-called pyranose or furanose, respectively. Through cyclization and respective ring formation, one asymmetric carbon atom on the carbon atom number 1 (C1) arises and thus forms two diastereoisomers (α and β form), which rotate the plane of polarized light to different extents. From D-glucose the 'six rings' of α-D-glucopyranose and β-D-glucopyranose are produced. In an aqueous solution, both isomers are available in gradually developing constant concentration equilibrium. Figure 3.12 shows the formation of the diastereoisomers of D-glucose and the formation of maltose and cellobiose as basic building blocks for starch and cellulose.

3.4.1.3 Structural Anomalies

As a naturally occurring polymer, cellulose always contains small amounts of other constituents in addition to glucose (>99%). These may already be partially built into or onto the cellulose molecule during biosynthesis, such as lignin–cellulose complexes. Most of the changes in molecular structure, however, result from secondary reactions, for example, hydrolysis or oxidation, during isolation from natural sources. For morphological reasons, such chemical changes occur preferably in the accessible interlinking regions between the crystallites of the elementary fibrils or their aggregations. The glucosidic links in these accessible areas, especially if oxidized sites are also present, split 1000–5000 times faster

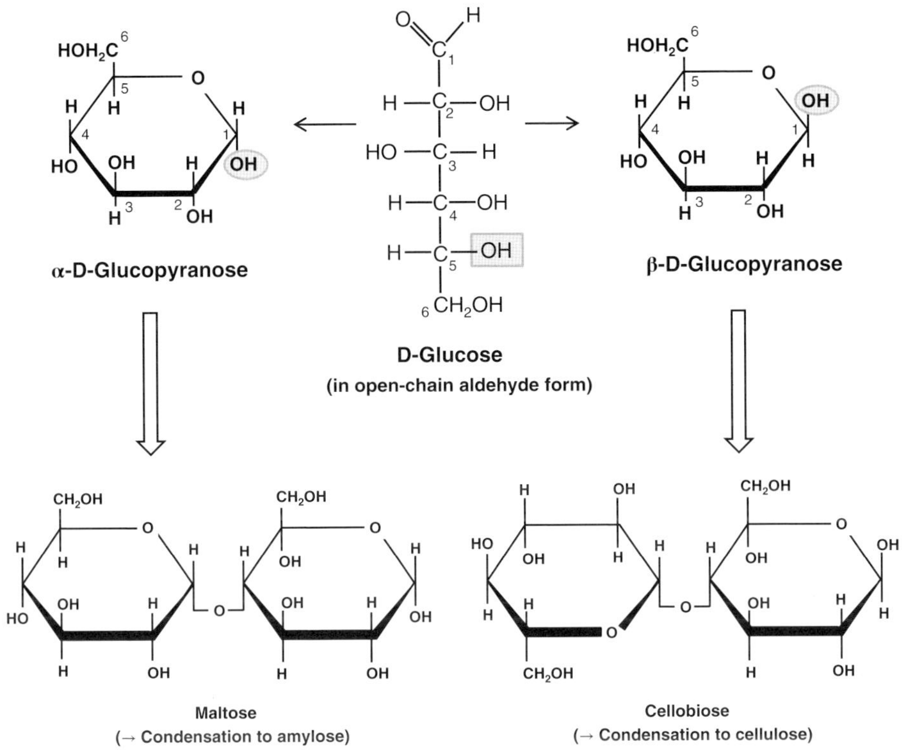

Figure 3.12 Formation of α- and β-D-glucopyranose, maltose, and cellobiose.

than glucosidic linkages inside the well-ordered crystallites. The existence of weak links, as proposed in some scientific articles, is hard to determine. In homogeneous acid hydrolysis, all glucosidic linkages split at the same rate.

Cellulose always contains carboxyl groups. In cotton, one carboxyl group per 100–500 anhydroglucose units (AGU) exists. In wood pulp, one –COOH group per 100–1000 AGU is detected [10].

3.4.1.4 Chain Length and Molecular Weight

Chain Lengths (DP/MW)

Naturally occurring cellulose has polymerization degrees of over 10 000, which correspond to a molecular weight of about 2 000 000 g mol^{-1}. Depending on the method of isolation, cellulose seem to have an average degree of polymerization of 300–3000, and thus has average molecular weights of 50 000–500 000 g mol^{-1} [17].

Definition of Viscosity

As viscosity is defined as the resistance to flow, it is a measure of the toughness and the apparent internal friction of a liquid or gas when applying external forces. Isaac Newton (1643–1727) conducted the first studies on this material property. When a liquid between two plates of area A is sheared with the force F, it starts to flow, and perpendicular to the direction of movement, a velocity gradient is produced.

The dynamic viscosity (η) indicates the relationship between the applied force necessary for a liquid to overcome a defined distance. The kinematic viscosity (ν) is defined as the quotient of η and density (ρ) of the solution.

Polymers such as cellulose and cellulose derivatives may take various forms because of the rotatability of their covalent bonds and have a certain degree of mobility. If you practice a defined shear stress on a polymer solution and the enclosed molecule bundles, the macromolecules arrange themselves in the direction of the resulting flow. As a result, the viscosity decreases.

The dynamic viscosity is temperature dependent, because the solvent (usually water) behaves thermodynamically differently at changing temperatures.

To ensure comparability of viscosity data, the measurements must therefore always be carried out in the same solvent, at constant temperature, and after adequate hydration time.

Measurement of Intrinsic Viscosity [10]

The simplest and most widely applied practical method for determination of the degree of polymerization is based on measuring the 'intrinsic viscosity', which is also known as the 'Staudinger index'. The intrinsic viscosity expresses the reduced viscosity of a solution at an infinitely small concentration. This parameter can be calculated by extrapolating the concentration of a dilute solution to the value 'zero'.

The intrinsic viscosity is defined as:

$$[\eta] = \lim_{c,G \to 0} \frac{\eta_s}{c} \qquad \eta_S = \frac{\eta - \eta_L}{\eta_L}$$

where:

$[\eta]$ is the limit value of the reduced viscosity for $c = 0$ and $G = 0$;
c = concentration of the polymer in the solution;
G = shear rate (with respect to velocity gradient);
η_S = specific viscosity.

The specific viscosity is determined from the viscosity of the solution and the viscosity of the solvent (η_L). The extrapolation to $c = 0$ can be performed graphically. Often this results in a linear relationship. The determination of $G \to 0$ can be omitted by selection of a suitable viscometer when the G value is kept small and constant.

Polymolecularity [10]

Cellulose isolated from its native sources is always polydisperse. It consists, as do all polymers, of a mixture of molecules with the same basic composition and chemical constitution but with widely differing chain length or degree of polymerization. The relative amounts of molecules of various lengths present in a given cellulose substrate can be specified by so-called differential mass distribution curves. Unlike most synthetic polymers, cellulose substrates have complicated mass distribution functions.

Knowledge of the molecular mass distribution is important for many applications. However, most methods applied in the past, such as fractionation by precipitation or selected dissolution, consume a great deal of time and are often subject to errors. In recent years, gel-permeation chromatography (GPC), especially in combination with low-angle laser light scattering, has been developed to provide a fast and highly reproducible method for this purpose.

As a convenient measure of the broadness of the mass distribution, the so-called 'non-uniformity factor' (NU) is often used. A polymer with a normal (most probable) molecular mass distribution will have a non-uniformity factor of 1. In most cases, cellulose substrates show much higher molecular non-uniformities, which indicates that there is an extremely wide distribution of molecule lengths within the cellulose mixture. Then the average molecular length of this cellulosic product is determined by the sum of high amounts of very short molecules and high amounts of very long molecules.

3.4.2 Secondary Structure

In solution, cellulose molecules exist in form of largely expanded coils. In addition to isolated and solvated molecules, cellulose and cellulose derivative solutions frequently also contain supramolecular gel particles, which are referred to as 'micelles'.

The order of the macromolecules in a cellulose fibre is not uniform throughout the whole structure. There exist regions of low order, so-called 'amorphous regions', as well as a very high crystalline order. The experimental evidence available today is adequately interpreted by a two-phase model, the fringed fibril model (Section 3.4.3.2), assuming low-order (amorphous) and high-order (crystalline) regions and neglecting the rather small amount of matter with an intermediate order [1].

In solid cellulose, high-order microcrystalline structures ('crystalline regions') alternate with those of a distinctly lower order, the amorphous zones. Cellulose is polymorph, that is, depending on the origin or the conditions during isolation or conversion, cellulose will have or can adopt various crystal lattice structures [10].

Figure 3.13 illustrates a cellulose fibre with crystalline and amorphous regions.

The proportion of crystalline sections of native cellulose is on average 45–60%. The crystalline zones are interrupted by amorphous regions where freely accessible acid- and alkali-labile bonds are located, whose hydrolysis provides the so-called 'microcrystalline cellulose'. Microcrystalline cellulose is defined as partially depolymerized cellulose, which is insoluble in water like its raw material. But at molecular weights of 30 000–50 000 $g\,mol^{-1}$, there is no visible fibrous structure [16]. In microcrystalline cellulose (MCC), the crystalline fraction is about 70% and the amorphous fraction is about 30%.

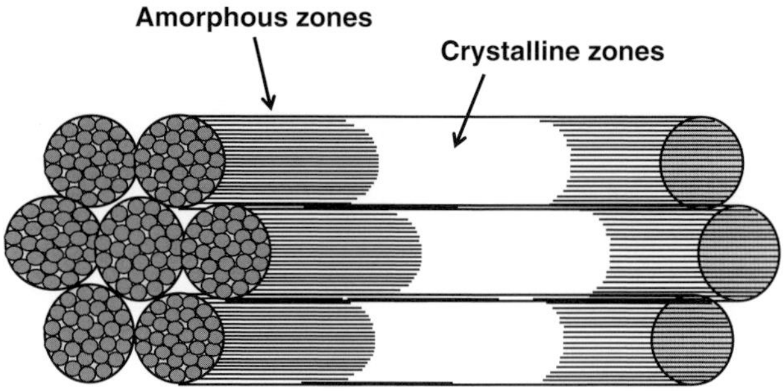

Figure 3.13 Individual cellulose fibre.

3.4.2.1 Lattice Structure and Crystallization

Cellulose, as a β-(1 → 4)-polyglucosan, forms stretched chains due to the β-linkage of the individual glucose rings. Such a cellulose chain is basically a slightly curled ribbon, which is stabilized by hydrogen bonds between the adjacent rings of the chain [9]. Figure 3.14 gives an impression of the outer shape of a portion of the chain of a cellulose molecule.

The elongated cellulose molecule can be packed very well laterally. This packing is stabilized by the formation of hydrogen bonds. The remaining free hydroxyl groups protrude left and right out of this band. They are able to develop more hydrogen bonds with neighbouring chains. In a solid cellulose substance, these chains are positioned more or less parallel to each other. In part, they form very highly-ordered (crystalline) regions [9].

In addition to the ability to participate in chemical reactions, the three free hydroxyl groups of each glucose unit of the cellulose molecule can react with each other, forming secondary valence bonds as a consequence. The strength of hydrogen bonds is about 25 kJ mol^{-1}. This is about 170 times stronger than van der Waals forces (0.15 kJ mol^{-1}), but less than one tenth of the strength of covalent O–H bonds (460 kJ mol^{-1}). The hydroxyl groups form hydrogen bonds in the cellulose molecule in two different ways. The first is by interaction between appropriately positioned hydroxyl groups in the same molecule (*intra*molecular) and the second by interaction between the hydroxyl groups in adjacent cellulose molecules (*inter*molecular). Intramolecular hydrogen bonds are formed between the hydroxyl group at C3 and the ring oxygen atom (O5) of the pyranose ring of the adjacent glucose unit in the molecular chain. The existence of another intramolecular hydrogen bond has been discussed by some researchers. It is supposed to be located between the hydrogen atom of the hydroxyl

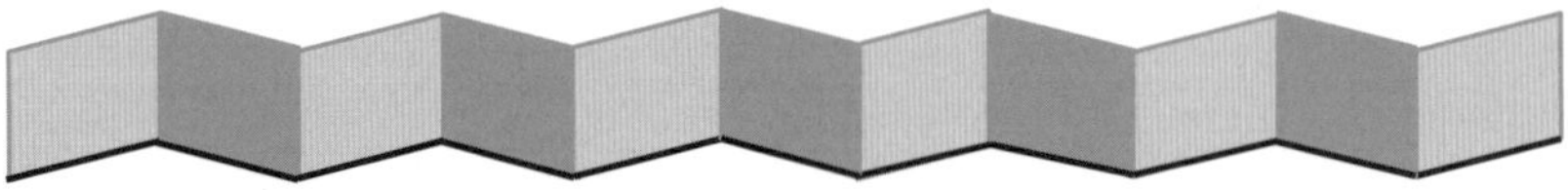

Figure 3.14 Outer shape of a segment of a cellulose molecule chain [9]. Reprinted with kind permission of Dr. Erich Gruber.

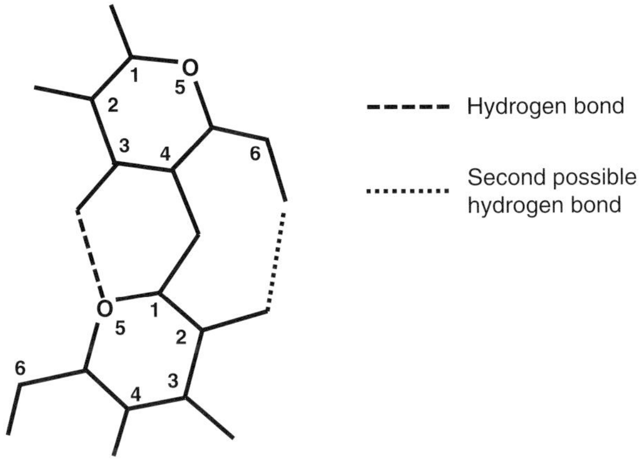

Figure 3.15 Intramolecular hydrogen bond [17].

group at C6 and the hydroxyl group at C2 of the neighbouring glucose monomers [17].

Figure 3.15 shows, schematically, the points at which a cellulose molecule can form intramolecular hydrogen bonds. Figure 3.16 shows the formation of intermolecular hydrogen bonds, whereby the cellulose chains can then 'connect' physically to organize themselves into a relatively firm structure.

Cellulose has a greater tendency to crystallize than most other polymers. It always occurs in partially crystalline form. The dominant supramolecular structure element is the cellulose crystallite [9]. Cellulose exists in several crystal modifications, differing in unit cell dimensions and, possibly, in chain polarity.

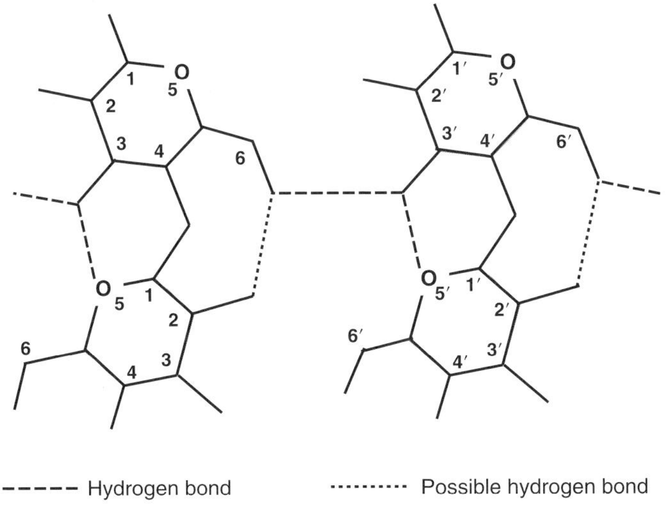

Figure 3.16 Intermolecular hydrogen bond between molecules of a lattice plane [17].

The elementary crystallites of the crystalline native cellulose have a uniform grid structure, the one so-called 'cellulose I'. For crystalline native cellulose, that is, cellulose I, Meyer, Mark, and Misch proposed a unit cell of the crystal lattice more than 70 years ago that is still applicable for practical purposes today. This model assumes a monoclinic unit cell with the space group $P2_1$ and two anti-parallel cellobiose chain segments running in opposite directions along the fibre axis. The dimensions of this cell are given in Tables 3.5 and 3.6 [1].

Figure 3.17 describes the principle and structure of cellulose crystallites as a model of the elementary cell (unit cell) according to Meyer, Mark, and Misch.

The chains, arranged in the direction of the fibre, constitute the *b*-axis of the unit cell. The chains are likely to be somewhat folded or spirally rotated so that *intra*molecular hydrogen bonds between the atoms C3 and O5 atoms as well as between the atoms C2 and C6 can be formed, which then in turn stabilize the molecular spiral. In direction of the *a*-axis, there are *inter*molecular hydrogen bondings. In the direction of the *c*-axis hydrophobic interactions occur [16].

Numerous authors have suggested that the unit cell of native cellulose may depend on the source. A doubling of the size of the unit cell has been concluded from low-temperature electron diffractograms. This eight-chain unit cell has found support but the Meyer–Misch model might also adequately represent most of the crystallographic evidence of native crystalline cellulose. Based on the more refined WAXS (wide-angle X-ray scattering) technique, from *Valonia alga* a monoclinic lattice with two parallel running chains has been proposed for cellulose and is assumed to be valid for cellulose I in general. Combined packing and X-ray intensity analysis have suggested a triclinic lattice cell with two cellulose chain segments running parallel along the fibre axis. Native cellulose has been shown to consist of two different crystal structures, cellulose I_α and I_β, by high-resolution, solid-state ^{13}C NMR spectroscopic studies. There are differences in the resonances of the C1 atoms. A singlet for cellulose I_α and a doublet for cellulose I_β appear at about 106 ppm. This rather small difference indicates a different hydrogen-bonding pattern of the glycosidic linkages. Bacterial cellulose and *Valonia* cellulose (from alga) contain a large amount of I_α modification, while in ramie, cotton, and wood cellulose the I_β phase is the dominant modification. The I_α modification has been described as a triclinic P-1 structure, with one cellulose chain per unit cell, whereas the I_β phase is assumed to be a monoclinic unit cell of the Meyer–Misch type (space group $P\text{-}2_1$ with two chains per unit cell), as concluded from electron diffraction experiments. The I_α phase is metastable and can be transformed (not completely, however) into the thermodynamically more stable I_β phase by annealing at 260–280 °C (500–536 °F) [1].

A concept for the unit cell that is fundamentally different from the Meyer–Mark–Misch model concerning the location of the cellulose molecules was developed in 1974 by Gardner and Blackwell. The difference is in the parallel arrangement of the cellulose molecules, that is, in a rectified manner [17].

Crystallization of the cellulose is always in a monoclinic crystal system, but it can show different crystal modifications. For all modifications, the grid cell is formed by four parallel chains, running in identical or opposite direction.

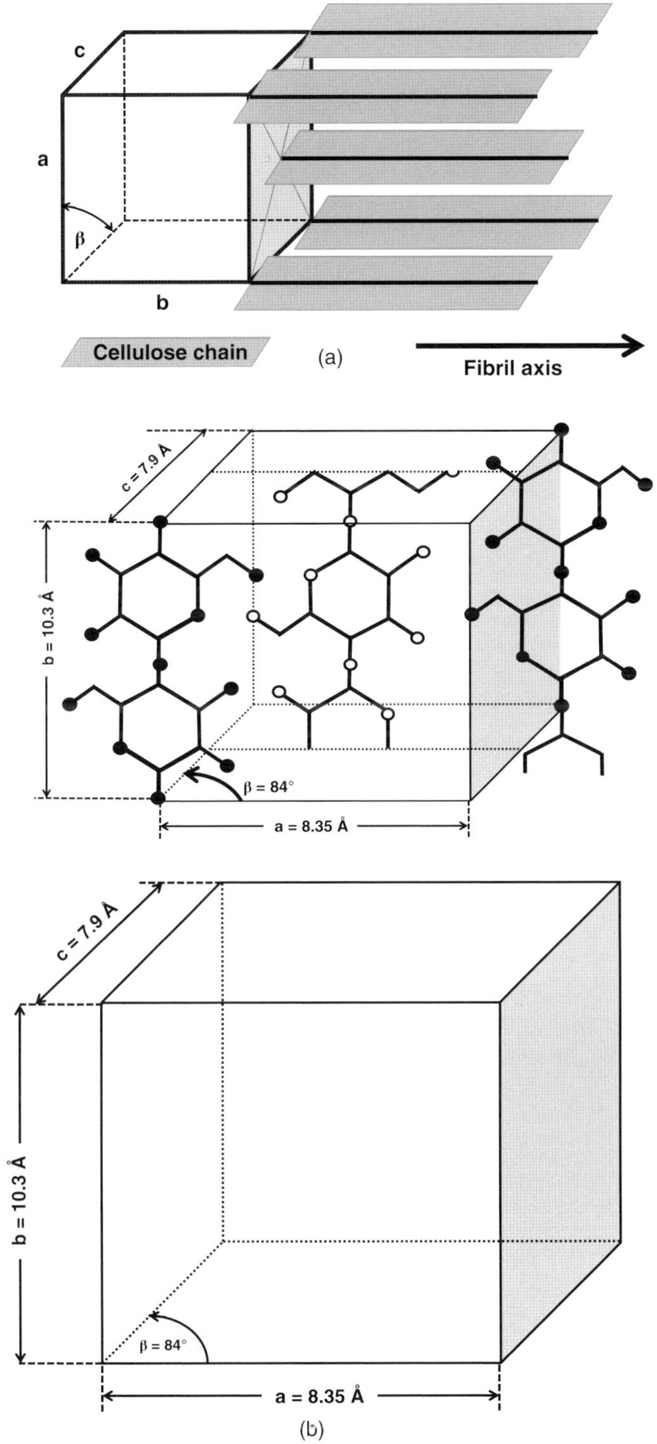

Figure 3.17 Structure of cellulose crystallites – principle (a) and model of the unit cell (b) – by Meyer–Mark–Misch [9]. (a) Reprinted with kind permission of Dr. Erich Gruber.

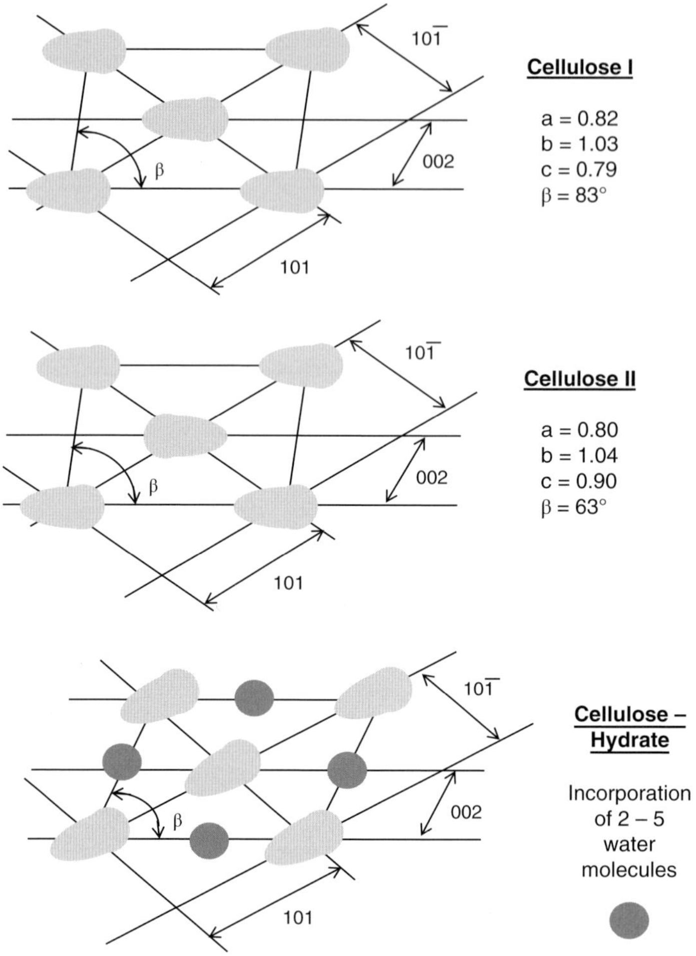

Figure 3.18 Crystal modifications of cellulose [9]. Reprinted with kind permission of Dr. Erich Gruber.

Another chain runs in the centre of the cell. The main types of crystals are 'cellulose I', and 'cellulose II', whose dimensions are shown in Figure 3.18 and Table 3.8. Furthermore, foreign molecules can be incorporated into the cellulose lattice. Thereby, the crystal planes are pushed apart and all the material swells. With the incorporation of foreign molecules, the grid cell is expanded laterally, but the longitudinal dimensions change only a little. Figure 3.18 shows a cellulose hydrate that arises after swelling in aqueous alkaline solutions by incorporation of water molecules. Various types of cellulose hydrates are possible, which differ in the number of embedded water molecules per individual grid cell. There may two to five molecules of water incorporated into the grid. In addition, other inclusion compounds are known – examples are ammonia and amines [9].

Table 3.8 Dimensions of cellulose crystals – comparison of type I and II [17].

Cellulose – I		Cellulose – II	
Side edge (Å)	Angle (°)	Side edge (Å)	Angle (°)
a-axis = 8.21	$\alpha = 90.0$	*a*-axis = 8.02	$\alpha = 90.0$
b-axis = 10.30	$\beta = 83.3$	*b*-axis = 10.36	$\beta = 62.8$
c-axis = 7.90	$\gamma = 90.0$	*c*-axis = 9.03	$\gamma = 90.0$

Native celluloses all show the so-called 'cellulose I' lattice structure. According to the Meyer–Mark–Misch model of the unit cell, each unit cell houses two countercurrently arranged cellulose molecules. The crystal lattice of 'cellulose I' is of the monoclinic sphenoidal type. The cellulose chains are in line with the *b*-axis of the unit cell (see also Figure 3.17). The cellulose molecules are aligned in the fibrillar axis and form the *b*-axis of the unit cell. The *b*-axis is 1.03 nm (10.3 Å) long, which is somewhat shorter than the extended lengths of a cellobiose unit, which suggests a slight helical twist in the cellulose chains along the *b*-axis. This twist is caused by intramolecular hydrogen bonding primarily between the hydroxyl groups on C3 of one glucose unit and the pyranose ring oxygen in the adjacent glucose unit of the same molecule. This intramolecular secondary valence bond is also responsible for the relative rigidity of the cellulose molecule. Some authors suggest a second intramolecular hydrogen bond involving the hydroxyl groups on C6 and C2 of adjacent glucose units in the same molecule (see also Figures 3.15 and 3.16). In more recent years, some researchers have suggested a unit cell for cellulose I in which the *a*- and *c*-axis of the Meyer–Mark–Misch model are doubled. However, these newer interpretations of X-ray and electron diffraction results are more or less closely related to the Meyer–Mark–Misch lattice structure. In some references, it is claimed that there is a closer agreement with the observed diffraction intensity data for a parallel arrangement of the cellulose molecules in the unit cell. However, this is still controversial. The internal cohesion of the cellulose molecules in the unit cells and crystalline domains is due to intermolecular secondary valences – partly hydrogen bonds and partly van der Waals forces. These bonds can either act between molecules situated in the same crystal lattice plane (*intra*planar bonds) or between molecules located in neighbouring lattice planes (*inter*planar bonds). The intraplanar hydrogen bonds are formed primarily between adjacent cellulose molecules in the same 002 lattice planes, giving a sheet-like structure (see also Figure 3.18). The 002 sheets are then bonded to one another by hydrogen bridges involving the hydroxyl groups on C6 and the glycosidic ring oxygen atoms of cellulose molecules favourably located in neighbouring 002 planes, or by van der Waals forces acting between neighbouring glucopyranose rings [10].

The density and bond strength of these interactions amongst the molecules and the lattice planes control to a large extent the action of swelling agents and the effect of reactants to the fibre. Involvement of the various hydroxyl groups in the intra- and intermolecular hydrogen bonds also affects the relative reactivity. From the low participation of the hydroxyl groups at carbon atom C2, the observed preferential reactivity can be understood [17].

The unit cells of the other polymorphic structures of cellulose differ basically in the lengths of their *a*- and *c*-axis and in the angle of inclination (β). The crystal type 'cellulose II' is, after cellulose-I-formation, the most important morphological structure of cellulose. The cellulose-II modification is formed as the thermodynamically most stable polymorph when cellulose fibres are treated with concentrated sodium hydroxide solution (>14%) or precipitated (regenerated) from solution [10].

The differences between the crystal modifications 'cellulose I' and 'cellulose II' are presented in Table 3.8.

Besides cellulose I, cellulose II is the most important crystalline form of cellulose from a technical and commercial point of view. There are different ways to produce cellulose II. One option is to prepare it by precipitating dissolved cellulose into an aqueous medium; this is the typical process for technical spinning of man-made cellulose fibres. It is also obtained by the so-called mercerization process, that is, by soaking cellulose in aqueous NaOH (17–20%, w/v) followed by decomposition of the intermediate by neutralization or washing out the NaOH. Mercerization is used to activate the polymer prior to the production of technical cellulose ethers.

The process of transformation of cellulose I into cellulose II is generally considered to be irreversible. Cellulose II is formed naturally by a mutant strain of *Gluconacetobacter xylinum* and occurs in the alga *Halicystis,* which were both very useful in providing an insight into the crystal structure of cellulose II [1].

In addition to the crystal types 'cellulose I' and 'cellulose II', two other polymorphic lattice structures are known, namely, 'cellulose III' and 'cellulose IV' crystal modifications [10].

The crystalline modification cellulose III is obtained by treating decomposing native cellulose with liquid ammonia (below – 30 °C (– 22 °F)) or an organic amine such as ethylene diamine, followed by washing with alcohol. Small differences in lattice dimensions exist between the two sub-modifications cellulose III_I and III_{II}. This modification has a lattice structure closely related to that of cellulose II. The cellulose-IV-modification is obtained by treating regenerated cellulose fibres in a suitable liquid at high temperature – for example, in a hot bath – and under tension to stretch the fibres. The lattice of this polymorph is closely related to that of cellulose I [1].

Some distinct differences in their infrared absorption spectra seem to indicate the existence of cellulose III and cellulose IV. Some researchers, however, doubt their actual existence [10].

Table 3.9 gives the unit cell dimensions of various cellulose allomorphs.

Figure 3.19 illustrates schematically how cellulose can be transformed into its various polymorphs.

The ability of hydroxyl groups to form secondary valence hydrogen bonds is – together with the stiff and straight chain nature of the cellulose molecule – the

Table 3.9 Dimensions of unit cells of different cellulose crystal modifications [1].

a-Axis (Å)	*b*-Axis (Å)	*c*-Axis (Å)	γ-Axis (Å)	Polymorph
7.85	8.17	10.34	96.4	Cellulose I
9.08	7.92	10.34	117.3	Cellulose II
9.90	7.74	10.30	122.0	Cellulose III
7.90	8.11	10.30	90.0	Cellulose IV

cause for the high tendency to organize into crystallites in parallel arrangement and crystallite strands (elementary fibrils), the basic elements of the supramolecular structure of cellulose fibres [10].

The dimensions of the elementary crystallites differ only slightly for native or regenerated cellulose fibres. Their length ranges between 12 and 20 nm, which corresponds to 24–40 glucose units; and they are 2.5–4.0 nm wide. The often observed larger micro- or macrofibrils (or fragments thereof) are aggregations of elementary fibrils [10].

Two questions concerning the crystal structure are still under dispute. The first question deals with the anti-parallel or parallel arrangement of the cellulose molecules in the crystal lattice as previously mentioned. The second question (still open) concerns the existence or non-existence of folded chains in the lattice. While a folded cellulose chain position in the lattice seems unlikely to most experts, the parallel molecule arrangement in the cellulose-I lattice is generally acceptable, under the condition that two cellulose-II lattice structures exist, one for heterogeneously mercerized native celluloses with parallel arrangement of the molecules and the other for regenerated cellulose substrates with anti-parallel molecule arrangement [10].

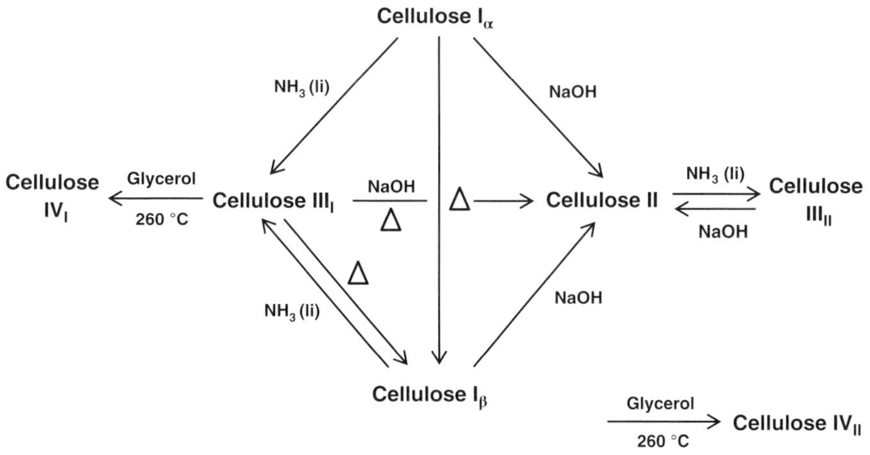

Figure 3.19 Transformation of cellulose into its various polymorphs. Reproduced with permission from Reference [1] © 2005 Wiley-VCH.

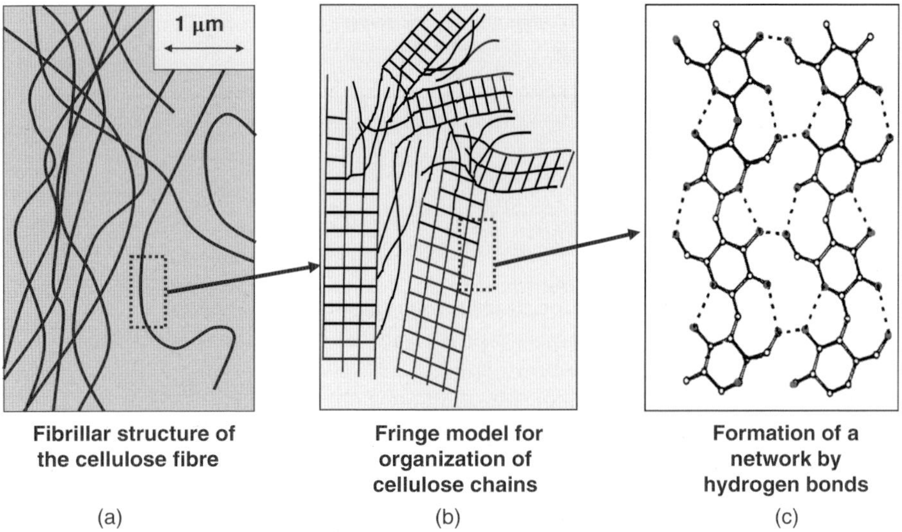

Figure 3.20 (a)–(c) Inner structure of cellulose fibres.

The construction of the inner structure of cellulose fibres is explained in the literature by different models. The best known – which are related to each other – are shown in Figure 3.20b and c.

3.4.2.2 Hydrogen Bonding

Both intra- and intermolecular hydrogen bonding occurs in cellulose. The detailed structure of this hydrogen-bond network is still a subject of discussion. The presence of intramolecular hydrogen bonds is of high relevance with regard to the single-chain conformation and stiffness. The existence of hydrogen bonds between O3–H and O–5′ (2.75 Å, by means of neighbouring AGU) of the adjacent glucopyranose unit and O2–H and O6′ (2.87 Å) in native crystalline cellulose (cellulose I) can be concluded from X-ray diffraction and NMR and IR spectroscopical data. In cellulose-II-crystallites, the hydrogen bonds are essentially the same as those proposed for cellulose I, considering the O3–H and O5′ (2.69 Å) hydrogen bond. The conformation of the C6 hydroxymethyl group differs in each chain since the chains are orientated anti-parallel in the unit cell, that is, the CH_2OH groups of the respective chains are not equivalent. Because one of the chains has one intramolecular hydrogen bond per AGU while the other chain has two, a more complex hydrogen-bonding network occurs. The intermolecular hydrogen bonding in cellulose is responsible for the sheet-like nature of the native polymer. Today, intermolecular hydrogen bonding between only the OH group at the C6′ and C3′ (″ means of the neighbouring chain) positions of cellulose molecules adjacently located in the same lattice plane (020 planes) is assumed. The intermolecular hydrogen bonding in cellulose II is significantly more complex than that of cellulose I. The anti-parallel chain model enables the formation of not only interchain but also of interplane hydrogen bonds. The most widely accepted representation of the bonding situation was been given by Kolpak and Blackwell (in 1976–1978). It should be pointed out again that the hydrogen bonding of cellulose has been interpreted in many ways [1].

3.4.2.3 Chain Stiffness of the Cellulose Molecule

Intramolecular hydrogen bonds are responsible for the relative rigidity of the cellulose molecule, and for fixing the helical arrangement of the cellulose molecules in the crystal lattice of the crystalline regions. The chain stiffness is also favoured by the β-glycosidic bond. This is reflected in the high viscosity of the solution, in the strong tendency to crystallize, and in the ability to form strands [17].

3.4.3 Supramolecular Structure of Native Cellulose Fibres

Models [17]

Besides being responsible for the chemical reactivity of cellulose, the ability of the hydroxyl groups to form secondary valence bonds with one another is also one reason for the strong tendency of cellulose to organize in parallel arrays of crystallites and crystalline bundles, the basic elements of the supramolecular structure of cellulose fibres. Individual cellulose molecules tend to arrange themselves in a highly organized way, which leads to a 'paracrystalline' state. The very regular structure of the cellulose molecule, the rigidity of the molecular chain, and the extensive hydrogen bonding capacity encourage the molecular connection and the grouping. The starting point for reflections on the structure state of cellulose fibres is the so-called 'micellar model', proposed by Naegeli and Schwendener. In this conception, the natural scaffolding substances are built from particles of the building material of colloidal size, the so-called 'micelles'. These are held together by swellable adhesive substances. Once the structure of cellulose as an oligomer of glucose with about 100 – possibly cyclically linked – segments had been recognized, a modified 'micellar model' was developed, as represented by Hengstenberg, Meyer, Mark, and Seifritz. With enforcement of the fundamental findings of Staudinger on the macromolecular nature of cellulose, the concept of the 'fringed micellar model' was drawn up. According to this structure model of the solid state of polymeric substrates, the chain-like macromolecules associate on certain segment lengths in crystalline regions. Because of their length, they are – repeatedly alternating – part of these crystalline regions and a less ordered, amorphous matrix. From light microscopic studies it is known that fibres are composed of fibrillar components. By electron microscopy, it could be detected that 'macrofibrils', observed with the light microscope, are aggregates of even smaller fibrillar elements, the so-called 'elementary fibrils'. These have in the case of cellulose fibres transverse dimensions of 2–10 nm and lengths of up to several hundred nm. According to the idea of Hess, Kratky, and Frey–Wyssling, individual cellulose molecules assemble due to their low chain flexibility on certain segment lengths and due to intermolecular interactions of the hydroxyl groups with each other, respectively, with the pyranose or glucoside oxygen atoms, in an orderly fashion to the shape of 'elementary fibrils'. Because of their length, parts of the macromolecules jut out of the ends of the individual crystallites. These extensions of the macromolecules go largely into subsequent crystallites, whereby the 'pearl ribbon' of the elementary fibril arises. Between the crystallites, the molecules go through less ordered linkage areas.

Molecular transitions from one elementary fibril to the adjacent ones can occur, to a random extent. From this idea of the construction of the elementary fibril, several models for the supramolecular and morphological structure of cellulosic fibres were created, leading finally to the assumption of the so-called 'fringe fibrillar structure', which was first mentioned by Hearle in 1964. According to this model, the structure of fibres is to be understood as a network of elementary fibrils and their more or less random associations. The molecular crossings or links from one elementary crystallite to another elementary crystallite within the elementary fibrils are responsible for their cohesion.

Fringe Fibrillar Structure

The crystal structure of the cellulose fibre is stabilized by hydrogen bonds along the molecular chains, between the individual chains within the cellulose lamella, and between the individual cellulose lamella. The crystallites have approximately the shape of elongated, flat blocks, from whose end faces disordered cellulose chains swell out as fringes. Therefore, these structures are referred to as 'fringe crystallites'. Clusters of fringe crystallites are illustrated in Figure 3.21. It is the model according to Hearle [9].

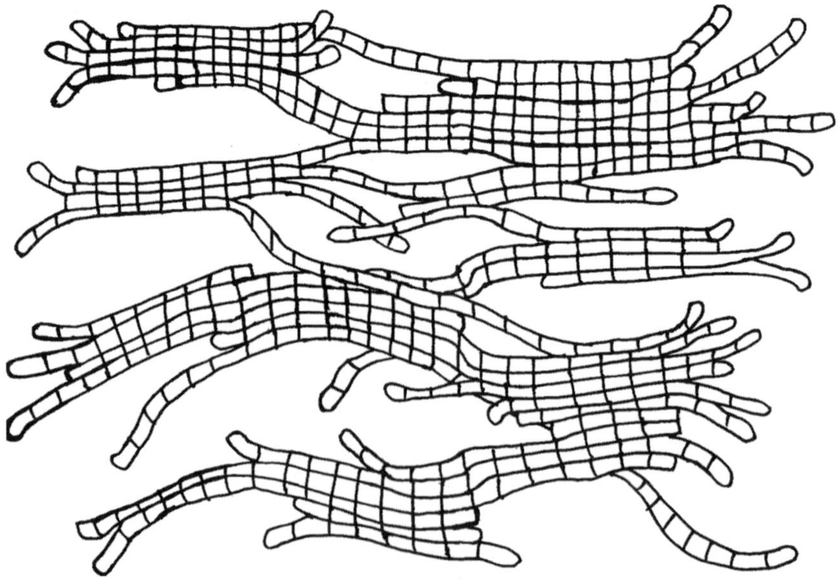

Figure 3.21 Fringe fibrillar model of the fibre structure (sketch by Hearle) [18].

These fringe crystallites are embedded in a differentiated structure hierarchy, being established through a lateral juxtaposition of the individual crystallite strands (elementary fibrils) [9].

Several elementary fibrils associate to form larger aggregations of so-called 'microfibrils'. Bundles of microfibrils group themselves to form 'macrofibrils'. The latter are the fibrils in the cell wall, which can be seen with a light microscope. The basic structural element of cellulose fibres is the so-called 'elementary fibril'. It can be seen with the electron microscope, as illustrated in Figure 3.20a. The cross-dimensions of the elementary fibrils correspond with those of the elementary crystallites. The elementary fibril is a strand of elementary crystals linked by segments of long cellulose molecules. The lateral order in the interlinking regions is distinctly less pronounced (amorphous) [10].

Synthetic fibres, such as viscose, do not have a native morphology. Their supramolecular structure can be described as a network of elementary fibrils and their more or less random associations. These fibres can also be characterized with the fringe fibrillar model (see Figure 3.20b and 3.21) [10].

3.5 Rheology

3.5.1 Dissolution Behaviour

General

Cellulose is indeed a hygroscopic material, but it can still be referred to as an inert substance, as it shows only minimal swelling in water. In addition to the water insolubility, cellulose is also insoluble in most organic solvents and dilute acids, but it swells easily. Solubilization can be achieved by concentrated acids. However, this solubility is at the cost of a considerable degradation (breakdown) of the cellulose chain through an acetalic (glycosidic) hydrolysis. Alkaline solutions lead to significant swelling of cellulose. Additionally, present hemicelluloses are dissolved.

With a 10% sodium hydroxide solution, the non-cellulosic components and short-chain cellulose molecules are dissolved in pulps. On increasing the alkali concentration to 18%, the solubility of cellulose decreases to a minimum and there are almost only non-cellulosic components – especially the polyoses – dissolved [19].

Other Suitable Solvents

Solvents such as:

- dimethylacetamide (DMA)/lithium chloride (LiCl);
- dimethyl sulfoxide/tetrabutyl ammonium fluoride;
- aqueous copper(II) tetra-ammonium hydroxide solution (Schweitzer's reagent).

are able to dissolve cellulose.

Novel Methods

A new method (patented by BASF) dissolves cellulose by a purely physical way in an ionic liquid. This liquid salt is, like sodium chloride, composed of charged particles and therefore it has a completely different solubilizing ability than water. By dissolving cellulose this way, many chemical syntheses can be performed more easily. In addition, reactions and synthetic pathways are possible that were previously unthinkable. This new approach results in a wider range of manufactured products for the construction and textile industries [20].

The extent to which this principle can also be applied in the area of regulated applications is the subject of ongoing investigations.

Characterization of Polysaccharides in Pulps

The characterization and fractionation of the polysaccharide components in pulps can be carried out with the help of concentrated sodium hydroxide solution. While polyoses (β-and γ-fraction) are soluble in an aqueous sodium hydroxide solution (17.5% concentration), α-cellulose is insoluble. The β-component can be separated from the γ-fraction by subsequent neutralization. The β-fraction precipitates, while the γ-component goes into solution [19].

3.5.2 Gelation and Behaviour with Other Ingredients

Cellulose in native form shows in water no increase in viscosity or gel formation. At all temperatures and at normal pressures cellulose is present in water as a dispersed solid only. There is no dissolution, but some limited swelling. In what follows, this behaviour is called 'inertia'. In contrast, substances that are easily dissolved with a relatively high rate of hydration are categorized as 'responsive' or 'reactive'. Cellulose is insoluble in oil and organic solvents, as well as in alcohols. Here, too, solid dispersions or slurries can only be prepared when the cellulose is first finely ground. These cellulosic dispersions must be constantly kept in motion by applying some shearing forces to prevent sedimentation of the cellulose particles. The inertia of cellulose makes it very compatible with other ingredients of polymeric nature such as proteins and soluble polysaccharides or oils. In addition, reactive substances in the formulation, such as sugars (mono-/disaccharides, dextrins), sodium chloride, and polyelectrolytes (such as phosphates, carbonates), are tolerated in large amounts without unexpected side effects by native cellulose.

3.6 Stability

3.6.1 Physical Properties

Cellulose is a relatively hygroscopic substance. Under normal atmospheric conditions (20 °C (68 °F) and 60% relative humidity) it adsorbs about 8–14% water.

Cellulose swells in water, but it is insoluble in water and dilute acids. In concentrated acids, dissolution of cellulose can be achieved – but at the cost of severe degradation of the cellulose chain. Caustic solutions cause extensive swelling and dissolution of low molecular mass portions (DP $\leq$ 200).

Cellulose is non-melting. Thermal decomposition of pure native cellulose starts at temperatures above 180 °C (356 °F) and is visible through a browning effect. The substance is completely destroyed (charring) after action of dry heat at about 250–300 °C (482–572 °F). The ignition point of cellulose is above 290 °C (554 °F).

With chlorine and zinc iodide, cellulose takes on a red-violet to blue colour. In a test with phloroglucinol (1,3,5-trihydroxybenzene, CAS number 108-73-6) in hydrochloric acid, pure cellulose must not show a red colour, because it is the detection method for lignin residues.

Additional data [10]:

- density: 1.52–1.59 g cm^{-3};
- refractive index:
  - 1.62 – parallel to the fibre axis,
  - 1.54 – perpendicular to the fibre axis;
- dielectric constant: 2.2–7.2 (at 50 Hz), highly dependent on humidity conditions;
- insulation resistance: 10^{14}–10^{17} Ω·cm, highly dependent on humidity conditions;
- electric strength: 500 kV cm^{-1};
- heat of combustion: 17.46 J g^{-1};
- heat of crystallization: 18.7–21.8 kJ mol^{-1} glucose;
- specific heat: 1.00–1.21 J g^{-1} K^{-1};
- coefficient of thermal conductivity:
  - 0.255 kJ m^{-1}·h^{-1}·K^{-1} loosely packed,
  - 0.920 kJ m^{-1}·h^{-1}·K^{-1} compressed;
- specific internal surface: 10–200 m^2 g^{-1}.

3.6.2 Stability in Food Products

At temperatures, pH values, and shear stresses normally applied during food preparation, native cellulose is highly stable. This resistance to physical influences is caused by the lack of swelling in water of untreated cellulose. Only by chemical modification of cellulose and transfer to its more or less water-soluble derivatives are the properties, and in particular its solubility, significantly changed. This results in a different stability in food and in changed process stability.

3.6.3 Chemical Reactions

General

The chemical reactions of cellulose are dictated by its polymorphic nature. The less ordered amorphous regions are significantly more reactive than the highly ordered crystalline zones. Chemical reactions always start at the disordered surfaces of the fibrils. Little or no effects are observed on the inaccessible and impenetrable crystalline structures [8].

Specific Reactivity [10]

The chemical reactivity of cellulose is largely determined by the supramolecular structure of its solid state. Most of the reactions on cellulose fibres are heterogeneous in nature. The reaction medium acts as a two-phase solid system:

- the less-ordered (amorphous) regions, which are mainly located on the surface of the elementary fibrils or their aggregations, and in the interlinking regions between the elementary crystallites in the fibrils;
- the well-ordered elementary crystallites or fused associations of the elementary fibrils.

The glycosidic links in these accessible areas, especially if oxidized sites are also present, split 1000–5000 times faster than glycosidic linkages inside the well-ordered crystallites. Any reaction will first start on the less-ordered surface of the elementary fibrils or their aggregations (topochemical reaction) and then, under favourable conditions, proceed into the interlinking regions between the elementary crystallites to penetrate from both ends into the crystallites. Therefore, as long as the reaction is limited to the accessible surface of the fibrils or fibrillar aggregations and the regions interlinking the elementary crystallites (i.e. up to degree of substitution (DS) of 1.3–1.7) there is no visible effect on the crystalline structure.

At increased degrees of substitution (to a DS of about 2.5), the X-ray diffractogram shows overlapping diffraction bands of the original cellulose-I structure and the cellulose derivative. At still higher degrees of substitution, the pure diffraction pattern of the derivative will finally result. This course of reaction implies that partially substituted cellulose derivatives are always a mixture of completely substituted cellulose, partially substituted portions (of block-polymer nature), and unsubstituted (i.e. native) cellulose.

A quasi-homogeneous reaction can be achieved when the fibre structure is loosened by swelling treatments to such an extent that all cellulose molecules can react simultaneously. A real homogeneous reaction can, however, only be achieved by bringing the cellulose into a molecularly dispersed solution.

The reactivity of cellulose substrates can be greatly enhanced by activation treatments, such as:

- swelling;
- solvent exchange;
- inclusion of structure-loosening additives;
- degradation by acids, alkalis, or enzymes;
- mechanical grinding.

All treatments aim to enlarge the area of accessible surfaces by opening fibrillar aggregations. These treatments also restore, in most cases, the loss of reactivity due to so-called 'hornification', which occurs when water is removed from cellulose by drying under severe conditions.

Swelling with water or other polar liquids is the most frequently applied activation treatment. It exclusively opens the interfibrillar interstices and swells the less-ordered surface and interlinking regions of the fibrillar elements.

For the manufacture of cellulose derivatives for technical applications, there are different effective methods using inert solvents to activate native cellulose. However, these methods cannot be applied for food-grade cellulosic products.

The solvent exchange technique is a special kind of activation from the water-swollen state. Its application allows the introduction of media that are inert in subsequent reactions and of media that are unable to swell the cellulose substrate, thus maintaining the cellulose molecules in their reactive water-swollen state.

An interesting variation of solvent exchange treatment is the so-called 'inclusion technique'. Here, inert liquids, such as cyclohexane or benzene, are introduced into the cellulose substrate by solvent exchange from the water-swollen state. During drying, they are permanently incorporated into the interfibrillar interstices or voids, thus preventing fusion of fibrils, that is, hornification. Such inclusion celluloses are very reactive.

Another very effective way of activating cellulose fibres is to enhance the accessibility of fibrillar surfaces and to open the less-ordered regions interlinking the crystallites in the fibrils by treatment with systems causing not only interfibrillar but also intracrystalline, swelling. Some inorganic acids, various salt solutions, and especially certain inorganic and organic bases achieve this at distinct concentrations. They apparently penetrate the fibre through existing capillaries and pores by opening the fibrillar interstices and entering the interlinking regions between the crystallites. From there, they enter the elementary crystallites from both ends and force them to open. At suitable concentrations and temperatures, they ultimately cause crystal lattice transfer – particularly with respect to opening the 101-plane distances (see Section 3.4.2.1, Figure 3.18).

Cellulose that is used as raw material for the manufacture of cellulose ethers for regulated applications (foods, cosmetics, and pharmaceutical products) is treated in this manner as a first processing step.

The activation of native cellulose with caustic soda is the most widely used method. It is described in more detail in Section 3.6.4 ('Possibilities for degradation').

The chemical character of cellulose is determined by the sensitivity of the β-glycosidic linkages between the glucose repeating units to hydrolytic attack and the presence of one primary and two secondary reactive hydroxyl groups in each of the glucopyranose units. These reactive hydroxyl groups are able to undergo exchange, oxidation, and substitution reactions, such as esterification and etherification.

3.6.4
Possibilities for Degradation

The native cellulose fibre can be degraded by hydrolysis either by concentrated acids or concentrated alkalis, as well as by enzymes. The selective degradation of cellulose molecules to glucose is usually carried out with concentrated acids at elevated temperatures or exposure to specific enzymes (cellulases, cellobioses).

Acids

With strong acids, cellulose can be cleaved. In this acid depolymerization, the molecular chains are each 'cut' approximately in the central region in such a manner that the glycosidic linkages between the individual anhydroglucose units (AGU) are cleaved. In homogeneous acid-driven hydrolyses, all glycosidic bonds are equally attacked. In this case, the DP is greatly reduced within a short reaction time. The acid-catalyzed hydrolysis is favoured by a low pH value. Figure 3.22 shows the principle of this degradation.

Usually, molecular decomposition has adverse consequences for the practical properties of the cellulose material, because the interior damage of the pulp fibre is not visible from outside. The tensile strength of a cellulose fibril depends on how well a force can be transmitted from one molecular chain to the adjacent chains. This force transfer is more effective the longer the adjacent chain sections are [9].

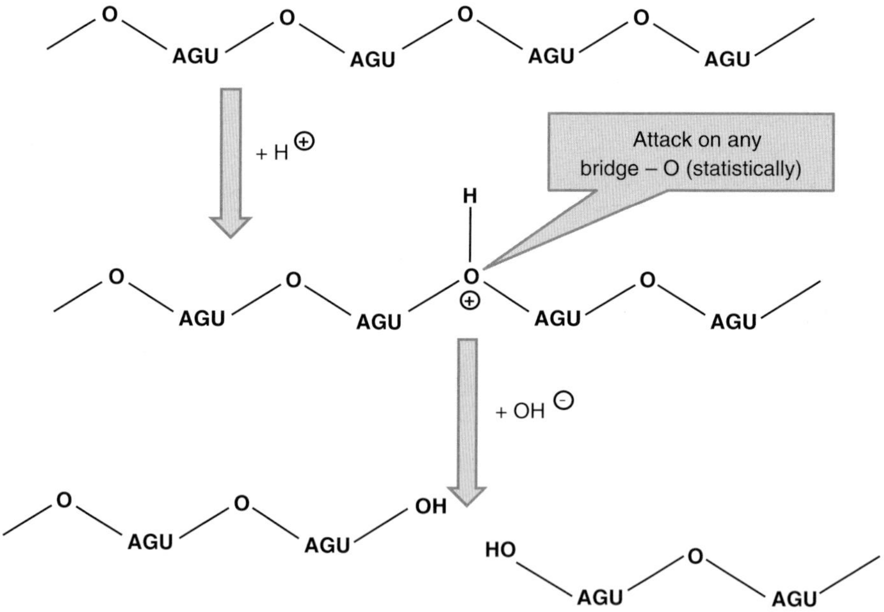

Figure 3.22 Mechanism of the acid-catalyzed hydrolysis of cellulose [9]. Reprinted with kind permission of Dr. Erich Gruber.

Alkalis

High pH values support alkaline-catalyzed hydrolysis. In an alkaline medium, degradation from the end of the cellulose chains predominates. Initially, only one glucose unit is eliminated, namely, the so-called 'reducing' end, on which there is no more additional glucose bound at the C1 atom. This splitting off produces a new end of the chain, on which a further decomposition can take place in the same way. In this manner, the chain is reduced stepwise. When the fibre is peeled off from the outside, it is called a 'peeling reaction'. The result is a slow decrease in chain length and, with time, a moderate reduction of DP [9].

Strong alkaline solutions penetrate into the cellulose fibres, which leads to a swelling of the fibres. Subsequently, the leach penetrates by capillary attraction forces into the regions between the crystalline zones. Consequently, the crystallite zones are gradually disunited and torn at the edges and thus opened, peeled, and dissolved from the outside to the inside. This process, referred to as 'mercerization' [21], is a frequently used practical method to activate the native cellulose substrate as first step during the production of cellulose ethers. This activation is generally carried out with sodium hydroxide solutions of 10–20% concentration at moderate temperatures (<20 °C (68 °F)). In this treatment, sodium-cellulose-I is formed in which the 101-plane distance is increased from 0.61 nm (6.1 Å) in native cellulose to 1.22 nm (12.2 Å). In this lattice transition, the glucopyranose rings are dislocated and aligned into 101 lattice plane. The hydroxyl groups on C2 and C6 carbon atoms are thus freely exposed and jut into the widened space between the 101 lattice planes, making them accessible for reactions (see Figure 3.18) [10]. At higher temperatures (>150 °C (302 °F)), cellulose is hydrolyzed in basic systems. In the course of this, oxidation processes can also occur [9].

Enzymes

Microbial degradation takes place by enzymatic hydrolysis of the β-(1 → 4)-glycosidic bond between the individual glucose molecules and the respective glucopyranosyl rings. For steric reasons, substituted cellulose ethers are less exposed to these enzymatic attacks, so that their relative stability is significantly higher than that of native cellulose fibres. It is further known that amorphous cellulose is more sensitive to enzymatic hydrolysis than highly crystalline cellulose. The partial degradation of native cellulose with enzyme mixtures leads to products with increased water solubility and thus improved applicability.

3.7 Analysis and Rheometry

3.7.1 Qualitative Analysis

Qualitatively, cellulose can be determined [1]:

- by X-ray spectroscopy;
- by colour reactions with KI or $ZnCl_2$;

- by reacting with a solution of iodine in sulfuric acid or phosphoric acid;
- with Congo red and others.

However, the colour reactions are not specific.

3.7.2 Quantitative Determination

For a quantitative determination, the Cross–Bevan method and the Kürschner–Hoffer method are quite appropriate. For these methods, the accessory components are removed by treatment with gaseous hydrogen chloride and subsequent heating with aqueous sodium sulfite or with nitric acid/ethanol after pretreatment with aqueous potassium hydroxide, respectively.

In the analysis of cellulose and its derivatives, instrumental methods are employed for assessing the size and chemical structure of the macromolecules within the entity of a given sample. Moreover, instrumental detection methods are also required for the monitoring of structurally relevant parameters during continuous fractionation of the polymer or chromatographic separation of its fragments.

The preferred spectroscopic methods are:

- UV/visible light
- IR
- NMR spectroscopy.

Today, predominately HPLC and gas chromatography are used to analyse the fragments obtained by acid or enzyme degradation [1].

3.7.3 Characterization of Structure of Cellulose Fibres

The following methods are used to characterize the molecular and fine structure of native and synthetic cellulose fibres [10].

1) *Degree of polymerization (DP):* the average degree of polymerization is determined by the osmotic method.
2) *Average crystallite length:* the average length of crystallite is determined by meridional X-ray low-angle scattering on slightly hydrolysed fibre samples or by measurement of the band width of the meridional 040 X-ray wide-angle reflection at half-maximum intensity.
3) *Degree of order ('crystallinity', Crl):* The degree of order – also expressed as crystallinity – is determined with a method for separating overlapping equatorial X-ray diffractions and deriving from the band width at half-maximum intensity the average cross-dimensions of the crystalline regions. Furthermore, this analysis yields information on the lattice structure, polymorphic composition, and accessibility.

4) *Degree of orientation:* the degree of orientation is determined by measuring the azimuthal intensity distribution of major equatorial X-ray diffractions arcs or by IR dichroism.

Physico-mechanical properties of cellulose fibres such as tenacities, elongations, or moduli in conditioned or wet state are determined by the following structural parameters [10]:

- average length of the fibre-forming molecules;
- average length of the elementary crystallites – with the number average 'limiting' degree of polymerization;
- degree of lateral order (crystallinity, Crl);
- degree of orientation with respect to the fibre axis;
- presence of heterogeneities (natural defects, incorporated gel or sand particles, etc.).

3.7.4 Viscosity Measurement

For the application of crystalline native cellulose in food products it is important to know that the minimal swelling ability does not allow rheological viscosity measurements of cellulose in water – as it is usual for the water-soluble cellulose derivatives. Cellulose is to be regarded in an untreated form in the respective water-based food system as a solid.

3.8 Synergies with Other Hydrocolloids

Owing to the inertia of the native cellulose and its insolubility in water, no real synergies with other hydrocolloids are observed. Cellulose occurs in the form of solid particles dispersed in the respective food system. The function is limited to that of so-called 'filler' or 'separating agent'.

3.9 Application in Food Products

Native cellulose is used as raw material for the production of partially depolymerized cellulose. The resulting products are microcrystalline cellulose (MCC) and powdered cellulose. Furthermore, highly purified cellulose in the form of pulps serves as basic material for producing functional cellulose ethers such as methylcellulose, HPMC (hydroxypropylcellulose), cellulose gum (CMC), ethylcellulose, or HPC. These partially or completely water-soluble cellulose derivatives are then incorporated as technological additives into various foods.

Untreated native cellulose is used directly as functional additive in foods only very rarely – the 'function' is limited to the role of a filling substance. One field of indirect application is in the filtration of beverages, where cellulose acts as processing aid.

3.10
Non-food Applications

Cellulose as raw material is processed in the paper industry.

In the textile and clothing industry, cellulose is used as regenerated cellulosic fibre (viscose) and linen/canvas.

A very important field of application is the construction industry, where derivatized cellulose, for example in the form of methylcellulose and MHEC (methyl hydroxyethyl cellulose), is used as an additive to improve the rheology and the water balance.

Furthermore, cellulose is the basic material for the known plastic product 'cellophane', a regenerated cellulosic film. Cellophane finds application mainly in the packaging industry. In addition, new products like transparent cigarette paper are also made with the help of cellophane.

Cellulose in form of nitrocellulose is used for the production of table-tennis balls.

The development of a process for production of bioethanol from cellulosic plant biomass for the propulsion of motor vehicles is currently being studied intensively.

3.11
Options for Derivatization of Cellulose

The possibilities by which to modify cellulose with physical and chemical methods are various (Figure 3.23). Besides the manufacture of textiles and paper, purified cellulose can be transferred into a broad range of different products by substitution of its native hydroxyl groups with chemical agents such as specific acids, chlorides, and oxides. Chemical reaction types are either esterification or etherification. The resulting products can be classified according to their solubility into three groups of:

1) pure organo-solubility;
2) mixed solubility in organic solvents as well as cold water;
3) pure solubility in water (but cold and hot water may be used as solvent) [22].

3.12
Nutritional Properties

The human body has no digestive enzymes for the degradation of cellulose. This means that the β-1,4-bonds between the single glucose monomers of the cellulose

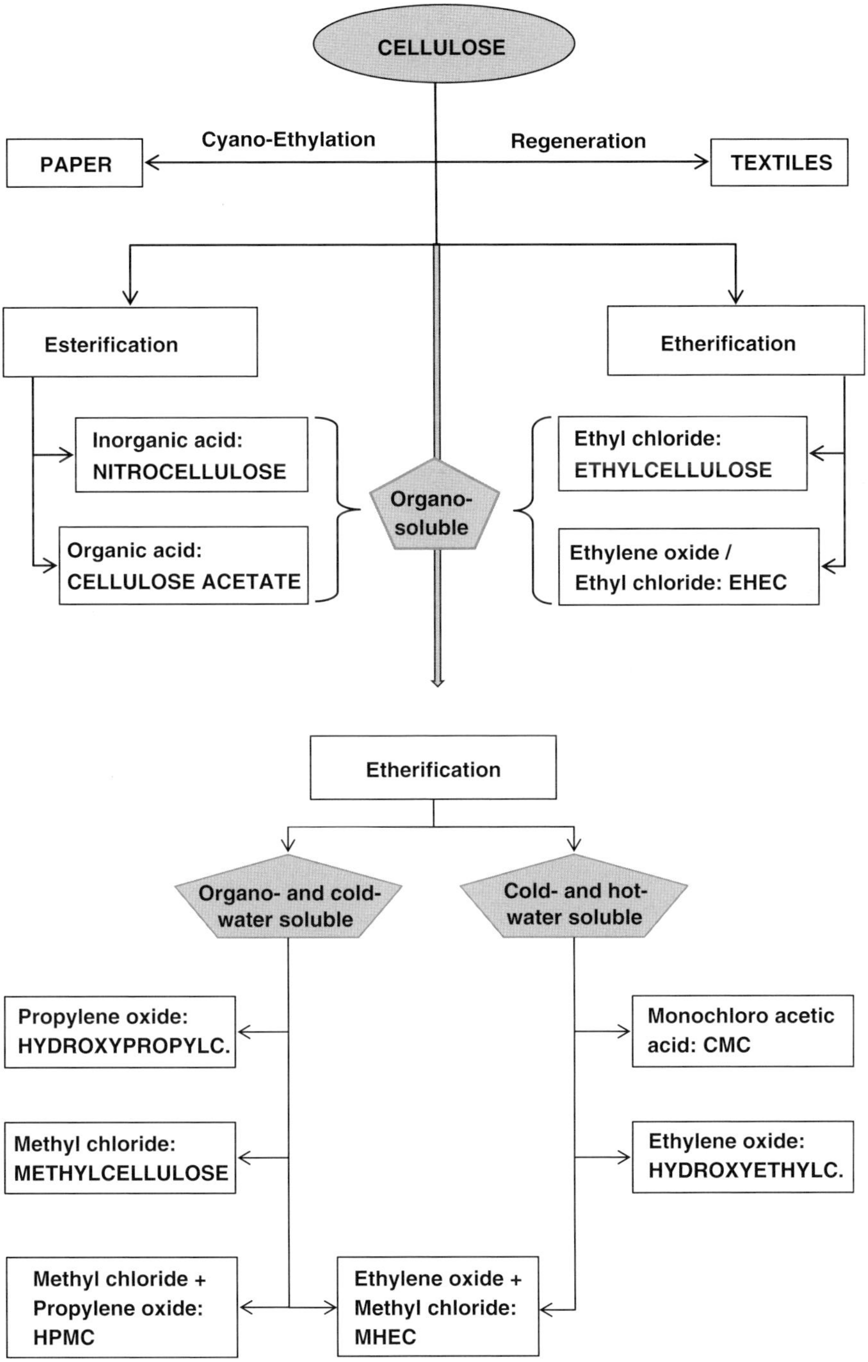

Figure 3.23 Cellulose and its derivatives (EHEC=ethyl hydroxyethyl cellulose).

chains cannot be enzymatically cleaved in the human gastrointestinal tract. For this reason, cellulose is referred to as dietary fibre. Cellulose passes through the human digestive system to the colon without molecular change. Cellulose can be partially metabolized to short-chain fatty acids by anaerobic bacteria in the colon mucosa. These fatty acids are then resorbed and metabolized by the colonic mucosa.

Numerous human and animal studies show that cellulose can be considered as nutritionally very inert ('un-reactive') and non-fermentable. Cellulose – as well as its derivatives – does not lead to the formation of gas in the human body as some other hydrocolloids and dietary fibres do.

Further studies show that the concentration of volatile fatty acids (VFAs) in faeces with cellulose-containing diets is significantly lower than in control experiments.

Therefore, cellulose increases the absorption of fatty acids. Higher levels of VFA or short-chain fatty acids (SCFAs) in the intestine are evaluated as positive for health.

Even medium-chain fatty acids (MCFAs) are more easily absorbed in the intestine, which is very helpful when a limited function of the small intestine exists [8].

Native cellulose is consumed in high concentrations when eating vegetable foods. Values for maximum daily intake amounts (ADI) are not defined.

The effect on health is generally considered as positive, since fibres such as cellulose, improve amongst other things intestinal peristalsis and absorption of many nutrients.

Cellulose and cellulose derivatives are allergen-free vegetarian or vegan additives whose harmlessness has been demonstrated in recent decades in numerous medical studies.

Toxicology The potential risk is small – cellulose is not toxic. There is only an inhalation hazard of milled cellulose in respective of finely ground pulps. Like all powdery substances they can cause irritation of the lungs.

3.13
Legislation

Cellulose as native substance is part of nearly all plants and vegetable products.

In the United States of America, powdered cellulose has the GRAS (generally recognized as safe) status and has been used as additive safely in foods for over 30 years.

In the European Union, cellulose is allowed in food products as a technological additive with unlimited concentration ('quantum satis'). A maximum dosage or ADI value is not specified. Exceptions are untreated products and foods that may not be modified by additives (e.g. butter, milk). In baby and infant foods, the use of cellulose is not allowed. As native substance, pure cellulose can be declared as 'cellulose' or with the E number 'E 460'. The description on the ingredient list can be made as 'filler', 'stabilizer', or 'separating agent'.

References

1 Klemm, D., Schmauder, H.-P., and Heinze, T. (2005) Cellulose. *Biopolymers Online*, 275–287. http://onlinelibrary.wiley.com/doi/10.1002/3527600035.bpol6010/abstract. DOI: 10.1002/3527600035.bpol6010.

2 Klemm, D., Heubein, B., Fink, H.-P., and Bohn, A. (2005) Cellulose: fascinating biopolymer and sustainable raw material. *Angew. Chem. Int. Ed.*, **44**, 3358–3393.

3 Beneke, K. (1998) *Biographies and Scientific Biographies of Colloid Scientists, Contributions to the History of Colloid Science VII*, Verlag Reinhard Knof, pp. 18–20.

4 Anon (2005) *European Coatings Journal*, **12**, 36–41.

5 Balser, K. and Iseringhausen, M. (1975) Cellulose ether, *Ullmanńs Encyclopedia of Industrial Chemistry* (eds E. Bartholomé *et al.*), vol. 9, 4th edn, Verlag Chemie, Weinheim, pp. 192–212.

6 Felcht, L-.H. (1985) *Schriftenreihe des Fonds der Chemischen Industrie*, 24, 71–76.

7 Whistler, R.L. and BeMiller, J.N. (eds) (1993) *Industrial Gums, Polysaccharides and Their Derivatives*, 3rd edn, Academic Press Inc., San Diego, chapters 18–20, pp. 475–578.

8 Stephen, A.M., Philips, G.O., and Williams, P.A. (eds) (2006) *Polysaccharides and Their Applications*, 2nd edn, Taylor & Francis Group, Boca Raton, FL, pp. 147–179.

9 Gruber, E. (2012) Fundamentals of pulp technology, Lecture notes of the course "Paper technology", Block 07 – Cellulose, Duale Hochschule Karlsruhe (from 12 September 2012).

10 Krässig, H., Schurz, J., Steadman, R.G., Schliefer, K. and Albrecht, W. (2004) Cellulose, in *Ullmann's Encyclopedia of Industrial Chemistry*, Wiley-VCH Verlag GmbH, Weinheim.

11 Nobles, D.R., Romanovicz, D.K. and Brown, R.M. Jr. (2001) Cellulose biosynthesis in the cyanobacteria with implications for the origin of vascular plant synthase. *Cellulose Community*, **8**, 161.

12 Bielecki, S., Krystynowicz, A., Turkiewicz, M. and Kalinowska, H. (2002) in *Bacterial Cellulose, Biopolymers*, vol. 5 (eds E.J. Vandamme, S. De Baerts, and A. Steinbüchel), Wiley-VCH Verlag GmbH, Weinheim, p. 37.

13 Kimura, S. and Hoh, T. (2001) Occurrence of high crystalline cellulose in the most primitive tunicate, appendicularian. *Program Technology*, **18**, 121.

14 Zugenmaier, P. (2008) *Crystalline Cellulose and Derivatives: Characterization and Structures, Springer Series in Wood Science*, Springer Verlag, Berlin, Heidelberg.

15 Gruber, E. (2009) Fundamentals of pulp technology, Lecture notes of the course "Paper technology", Block 05 – Chemistry of wood, Duale Hochschule Karlsruhe (from 25 May 2009).

16 Belitz, H.-D. and Grosch, W. (1992) *Food Chemistry*, 4th edn, Springer-Verlag, pp. 294–297.

17 Ribitsch, V. (2002) Textile fiber modification (physico-chemical characterization of viscose fibers to develop an environmentally friendly manufacturing process), Federal Ministry for Transport, Innovation and Technology, BMVIT, Reports on Energy and Environmental Research 21/2002 (in German).

18 Krässig, H.A. (1993) *Cellulose: Structure, Accessibility, and Reactivity, Polymer Monographs*, vol. 11, Gordon and Breach Science Publishers, Amsterdam. ISBN 2881247989, 9782881247989.

19 Gruber, E. (2011) Fundamentals of pulp technology, Lecture notes of the course 'Paper technology', Block 22 – Investigation and analysis of pulps 2 (Chemical Studies), Duale Hochschule Karlsruhe (from 08 March 2011).

20 Kurschat, R. (2007) Chemicals from the biorefinery. *Neue Zürcher Zeitung*, Online-Edition, 21 November 2007. http://www.nzz.ch/nachrichten/wissenschaft/chemikalien_aus_der_bioraffinerie_1.587429.html.

21 Dinard, E., Vignon, M., Chanzi, H., and Heux, L. (2002) Mercerization of primary

cell wall cellulose and its implication for the conversion of cellulose I → cellulose II. *Cellulose (Dordrecht Netherlands)*, **9**, 7.

22 Ashland, Inc. (2012) Benecel high-purity methylcellulose and hydroxypropyl methylcelllulose, physical and chemical properties (56. 001-E4 REV). From the document library at http://www.ashland.com/products/benecel-methylcellulose-mc-and-hydroxypropylmethylcellulose-HPMC.

4
Microcrystalline Cellulose

Synonyms: MCC, Cellulosegel

E number:	E 460 (i), E 460 (ii)
EINECS:	232-674-9
CAS number:	9004-34-8

Related hydrocolloids: ground cellulose, powdered cellulose, hydrocellulose

4.1
General Manufacturing Process

The raw material for the production of MCC is a selected refined alpha (α-)cellulose derived from specialty grades of wood pulp. The wood pulp is introduced into the manufacturing process in a dry compressed form as large paper rolls.

The cellulose fibres are composed of millions of microfibrils. The individual microfibril is composed of two regions – the paracrystalline region, which is an amorphous flexible mass of cellulose chains, and the crystalline region, which is composed of tight bundles of cellulose chains in a rigid linear arrangement [1].

This means that native cellulose has crystalline and amorphous chain segments. Over the entire molecule, there are acid-and alkali-sensitive bonds, which in the crystalline regions, however, are not accessible. By hydrolytic removal of these unstable bonds in the amorphous regions, microcrystalline cellulose (MCC) is produced, which is then characterized by a very high proportion of crystalline zones [2].

The proportion of crystalline sections in MCC is about 70%; the amorphous fraction is 30%. In comparison, a so-called cellulose powder as a fine-milled native cellulose fibre consists on average of 45% crystalline portions and 55% amorphous regions.

The first manufacturing step requires cooking of the purified pulp with a dilute mineral acid in water.

Cellulose and Cellulose Derivatives in the Food Industry: Fundamentals and Applications, First Edition.
Tanja Wüstenberg.

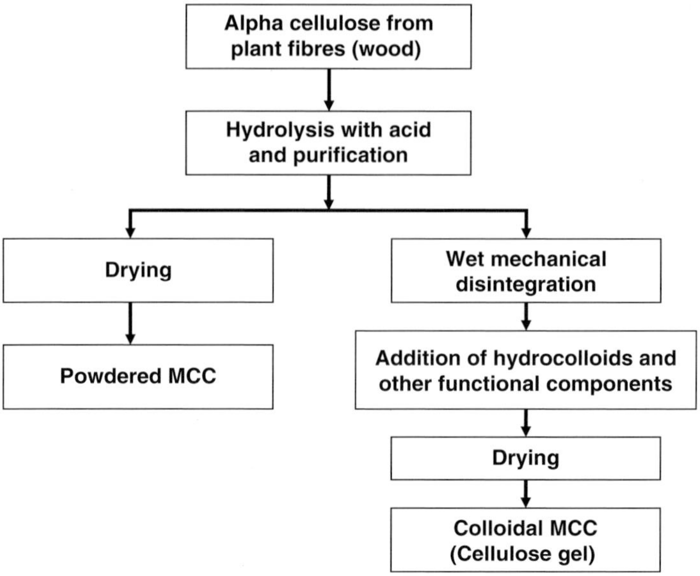

Figure 4.1 Production of microcrystalline cellulose (MCC) [1,3].

The acid preferentially attacks the less-ordered or amorphous regions of the cellulose polymer chain, thereby exposing and freeing the crystalline sites, which form cellulose crystallite aggregates. The process is carried to a point at which a LODP (level-off degree of polymerization) is attained. After completion of the hydrolysis, the cake obtained is neutralized and thoroughly washed to remove impurities [1,3].

Then the cake is processed further to either produce powdered or colloidal MCC (Sections 4.1.1 and 4.1.2). Figure 4.1 shows a flow chart of the manufacturing process.

The degree of polymerization (DP) of MCC is usually less than 400.

4.1.1
Powdered MCC

The wet-cake produced is freed of water and the dried crystalline cellulose aggregate is recovered and ground to give powdered grades of MCC [1].

Owing to the drying of the crystalline fibrils, aggregates of very porous particles are formed. This porosity is the reason for the high adsorption capacity of water or oil on the MCC particle surface. However, after transferring MCC powder to water, there is no dissolution of the MCC – a partially swollen dispersion will result. The so-called water retention capacity is low, because it is purely a surface effect.

4.1.2 Colloidal MCC

4.1.2.1 Principle of the Colloidal MCC Process

For the production of colloidal MCC products, the washed cake is then subjected to mechanical disintegration under the action of strong shear and friction forces to break up the compact structure of the cellulose fibre aggregates and to release additional microcrystals. The attrition should be sufficient to produce a mass in which not more than 1% by weight of the particles has an average length greater than 1 μm as determined by electron microscopy. When this is achieved, the material is then capable of forming a stable dispersion in aqueous and other media. After the mechanical disintegration step, the attrited microcrystals are co-processed with a hydrophilic dispersant, such as sodium carboxymethyl cellulose (CMC) [1].

This hydrophilic barrier-forming dispersing agent prevents the microcrystals from reaggregating due to hydrogen bond formation during drying. As dispersants, water soluble hydrocolloids, such as alginates, guar gum, xanthan gum, or CMC, are suitable and used in a concentration of 8–12% [3].

After co-processing with, for example, CMC, the material can be dried in several ways. Bulk-dried colloidal MCC–CMC products need homogenization to activate the microcrystals and, therefore, these are appropriate for use in frozen desserts and aerated food systems, such as whipped toppings. The spray-dried versions need shear from a high-speed mixer for effective dispersion. Another colloidal MCC product has been designed for use in dry mix, reconstituted food systems where insufficient shear is available to obtain complete dispersion of standard MCC–CMC grades. This readily dispersible stabilizing agent is produced by adding sweet whey powder to the colloidal MCC–CMC product before drying [1,4].

The latest generation of colloidal MCC is produced by co-processing the cellulose crystals with functional substances such as calcium ions and surfactants. Colloidal MCC can be dispersed in water. The MCC particles then form a three-dimensional network, which stabilizes the food product. The system is held together by weak hydrogen bonds. Dispersions of microcrystalline cellulose are not able to bind or absorb, and thereby retain, large quantities of water as other water-soluble hydrocolloids are able to. A certain amount of water is bound by the hydrophilic dispersion medium (e.g. CMC). The residual water remains free and leads quickly to synaeresis in the food product [3].

4.1.2.2 Co-processing with Other Hydrocolloids

Several types of colloidal MCC have been designed to provide unique functionalities for specific uses.

MCC + Guar Gum

The co-processing of MCC with guar gum results in powdered particles that are essentially water insoluble, shear resistant, and spherical. The processing is carried out in an aqueous medium by forming an intimate mixture of the homogeneously dispersed MCC and guar gum under controlled high shear conditions to obtain flocculated MCC–guar gum particles of the desired size. The cellulose–guar aggregates enhance the quality of fat-reduced and non-fat products by simulating the rheological, textural, and mouthfeel properties of fat [1,5,6].

MCC + Calcium Alginate

A colloidal MCC grade co-processed with an alginate salt complex has been developed to provide colloidal MCC that could be dispersed in dry mixes and milk systems. The alginate–calcium complexing reaction is exploited to provide a method for preparing the co-processed MCC–alginate blend in a form useful for producing water-dispersible MCC particles [1,7].

MCC + CMC

Several different products result from co-processing MCC with cellulose gum (Na-CMC). A critical parameter is the drying process, where the shape and the surface of the colloidal MCC are formed.

Bulk-dried colloidal MCC–CMC products need homogenization to activate the microcrystals. The spray-dried versions need shear from a high-speed mixer for effective dispersion.

Another colloidal MCC product has been designed for use in food systems where only moderate shear forces can be applied. An immediately dispersible stabilizing agent is produced by adding sweet whey powder to the colloidal MCC–CMC product before drying [1,4].

MCC–CMC for Retort Applications [1,8]

Special co-processing of MCC with CMC has been shown to be effective in providing a change in rheology under the severe heat treatment of rotary and static retort processing. This modified colloidal MCC–CMC stabilizer allows the manufacturer to meet and exceed the thermal F_0 death time requirements of retort sterilization while minimizing total process time. This is especially important in static retort processing where standard MCC–CMC grades may rapidly form a network and the product thicken, the net result of which may reduce heat penetration and, thereby, increase the process time required. This modified MCC–CMC provides a low-viscosity-suspending network that is particularly useful in retorted/canned and UHT/retort-processed beverages and foods.

MCC + Calcium Carbonate + CMC

Co-processing calcium carbonate with MCC is effective in grinding the MCC aggregates during the co-attrition process to prepare ultra-fine MCC particles and, thus, is a very effective colloidal ingredient. The product provides a low-viscosity-suspending network that is particularly suited for use in calcium-fortified, milk-based beverages [1,9].

MCC + High Methyl-Esterified Pectin

A colloidal grade of MCC co-processed with high methyl-esterified (HM) pectin has been developed to provide colloidal MCC stabilization in low-pH, UHT-processed protein-based beverage systems. The co-processed powder consists primarily of MCC particles intimately associated with HM pectin and an inorganic salt. The inorganic salt facilitates the formation of colloidal MCC during co-extrusion with HM pectin by enabling it to be subjected to a high work profile that breaks up the aggregates and releases additional colloidal MCC particles. The HM pectin functions as barrier dispersant by allowing dry, ultrafine attrited MCC particles to be formed without re-aggregation. New and/or improved colloidal cellulose properties such as low pH stability, fruit pulp suspension, protein stability, emulsion stability, bake stability, and colloidal MCC that may be dispersed in milk systems suggest new product applications. Potential areas of application include speciality low pH protein-based beverages, drinkable yogurt and cultured products, fruit sherbets, low-pH sauces, baked goods, such as fruit fillings, pie, and pastry fillings, and products labelled 'all natural' [1].

A patent pertaining to a composition and method to produce a co-dried MCC and HM pectin as a stabilizing agent in aqueous food systems is available [10].

4.1.2.3 Physical Modification with Other Functional Substances ('Alloying Concept')

Several potentially useful 'alloys' have been identified by co-processing MCC with other food-approved ingredients to provide novel functionalities for specific food uses. The wet-end processing produces physical interaction of the components and this is reinforced by the subsequent drying step.

MCC + Iota Carrageenan

The use of iota carrageenan produces a co-processed MCC–iota carrageenan powder that can be readily dispersed. It can be dry blended with other ingredients and can be dispersed in water, in the presence of these ingredients, with minimal agitation, such as spoon stirring, stirring with a wire whisk, or shaking. Other significant new properties of the co-processed MCC–iota carrageenan product are acid stability and salt stability. The powder will disperse and remain stable regardless of the acidity of the system [1,8].

MCC + Maltodextrin

Co-processed MCC–maltodextrin involves using maltodextrin alone as a barrier for the MCC particles during drying. This cost-effective MCC-based product can potentially be employed where colloidal grades of MCC are used, including frozen desserts, salad dressing, beverages, and bakery products. Other potential uses are as a bulking agent in non-aqueous food systems, for example, in peanut butter and in low-moisture food systems, as in inactive carrier or excipient for chewable tablets, and for masking the taste of drug actives such as aspirin and acetaminophen (APAP) in pharmaceutical applications [1,11].

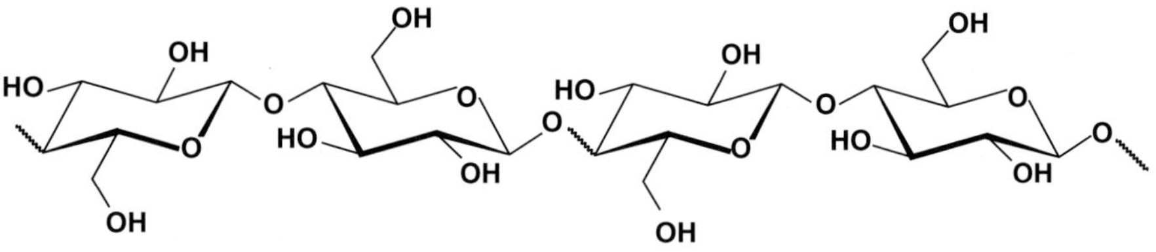

Figure 4.2 Molecule of microcrystalline cellulose (β-D-glucose polymer).

MCC + Surfactant

Colloidal MCC–surfactant compositions, composed of MCC and one or more surfactants, in which the surfactant is adsorbed onto the surface of the cellulose, have been developed to provide improved dispersibility in low-moisture or non-aqueous food applications. These composites can be used as bulking agents in low-moisture food products or in the oil phase of food systems, and are especially useful in reduced-calorie food products [1,12].

4.2
Chemistry

The chemical structure of microcrystalline cellulose corresponds to the chemical composition of the native cellulose. The molecule is a linear unbranched chain of β-D-glucose monomers that is 1 → 4-linked (Figure 4.2). The structural formula is $(C_6H_{10}O_5)_n$.

Owing to chain breakage, the molecular weight of MCC is lower than that of the raw material cellulose and ranges between 30 000 and 50 000 g mol^{-1}. The average is about 36 000 g mol^{-1} [2]. The degree of polymerization is usually less than 400. The cellulose content is at least 97%, based on the dry matter, and the product is a white or almost white, fine, odourless powder [13].

4.3
Rheology

4.3.1
Dissolution Behaviour

Like native cellulose, powdered microcrystalline cellulose is insoluble in water.

Acidic hydrolysis does not improve the solubility. The proportion of water-soluble components is less than 0.24%. MCC is also not soluble in ethanol, ether, and diluted mineral acids. In a sodium hydroxide solution, MCC may be partially dissolved [13].

Colloidal MCC grades show another hydration and respective solution behaviour than powdered MCC types. The physical properties of colloidal MCC dispersions are quite different from those of gum solutions, starch gels, and other water-soluble materials commonly used in food technology. When colloidal grades of microcrystalline cellulose are properly dispersed, the cellulose crystallites set up a three-dimensional network, with most particles less than 0.2 μm. The swelling capacity of the soluble hydrocolloid provides the dispersant function by aiding in the dispersibility of the MCC particles during reconstitution as well as in the stabilization of the resulting colloidal dispersion. It is the formation of this insoluble cellulose structural network that provides the colloidal MCC grades with their functional properties of a low-viscosity, suspending network and so on. The gel that is formed possesses the elastic properties of a solid that exhibits relatively high yield stress and a time-dependent type of flow behaviour (thixotropy). Thixotropic properties impart various desirable characteristics suitable for products such as salad dressings and mayonnaise-type applications [1].

The addition of minerals, salts, sugars, and other hydrocolloids does not improve the solubility of the microcrystalline cellulose itself. However, electrolytes are – in high quantities – able to reduce the dispersibility of the MCC in the aqueous system.

4.3.2
Gelation and Behaviour with Other Ingredients

Gelation

After an appropriate shear treatment, MCC forms a three-dimensional matrix in water, consisting of millions of insoluble microcrystals. An extremely stable thixotropic gel is formed [3].

The application of shear forces is essential for functionalization of the MCC. This gel formation by MCC is not comparable to the gelation of other cellulose-based hydrocolloids, like methylcellulose, which are 100% soluble in water.

As well as the raw material cellulose, the microcrystalline derivative shows no increase in viscosity or real gelling due to the minimal solubility in water. At any temperature and at normal pressures, MCC is – after sufficient shear treatment – a dispersed solid in water. For this reason, microcrystalline cellulose is highly compatible with other ingredients of a food formulation.

Compatibility with Proteins, Other Polysaccharides, and Fats/Oils

The compatibility of MCC with polymers such as proteins, soluble polysaccharides, and fats/oils is very good. Microcrystalline cellulose is also very tolerant to active and fast-reacting components of the recipe.

Behaviour with Sugar, Salts, and Polyelectrolytes

Sugar (such as mono- and disaccharides, dextrins), salts (such as sodium chloride or phosphates), minerals, and vitamins can be incorporated in higher levels without adverse effects in the prepared MCC dispersion. MCC acts as an inert substance – it does not react with the technologically functional components or with the substances that are enriched for nutritional reasons. Note that the

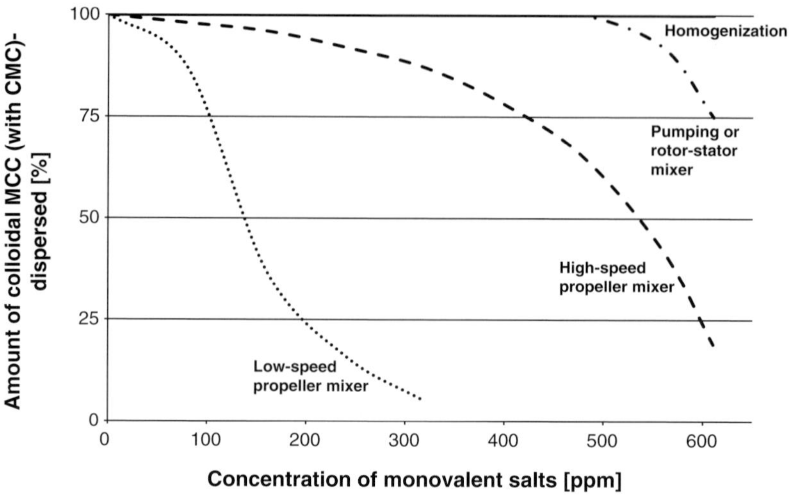

Figure 4.3 Effect of monovalent salts on the dispersion of a colloidal MCC (co-processed with CMC) [1].

presence of mono- and divalent salts can inhibit the dispersion of the MCC, so that higher shear forces are necessary to fully activate the product. Figures 4.3 and 4.4 show a MCC–CMC dispersion at varying levels of mono- and divalent salts. In general, the relationship between chemical loading and shear treatment is the following:

- Higher amounts of ions require higher shear forces.
- The effect of divalent salts is stronger than that of monovalent salts.

If the shear force is not sufficient due to the high concentration of electrolytes in the system, the MCC will not be completely dispersed and its functionality will be restricted.

Figure 4.4 illustrates the influence of different levels of divalent salts on the dispersibility of a MCC–CMC grade.

General Recommendation for Food Application

Manufacturers recommend the dispersion of MCC under shear in water in the first step and then addition of the other components of the formulation (see Section 4.6) [3].

4.3.3
Special Behaviour of Colloidal MCC with Guar Gum

The properties of MCC–guar gum aggregates are very different from those of other colloidal grades. Spherical MCC–guar aggregates, produced by controlled

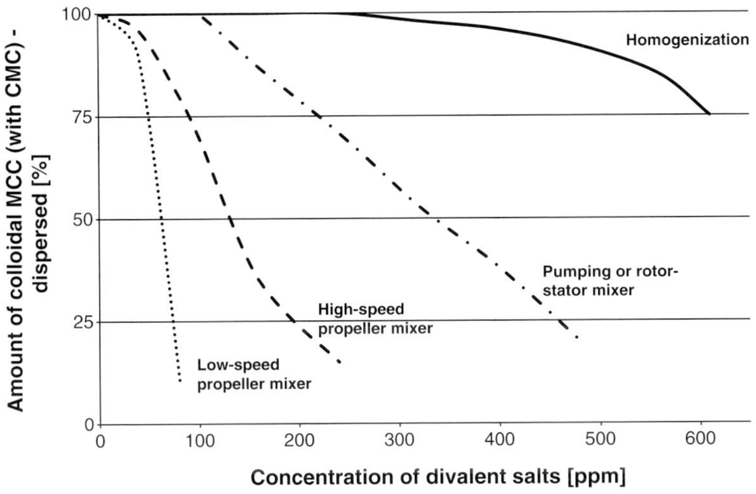

Figure 4.4 Impact of divalent salts (Ca^{2+}, Mg^{2+}) on the dispersion of a colloidal MCC (co-processed with CMC) [1].

interaction between these two polymers, have certain characteristics similar to those obtained when fats and/or oils are emulsified into in aqueous phase. It is the formation of a guar layer adsorbed onto the cellulose that gives the MCC–guar aggregates the ability to function as a fat-like substance in aqueous-based food systems. In this situation, the aggregates have smoothness and lubricity properties most closely approximating the physical and organoleptic characteristics associated with fat dispersed in water. As well as a smooth, bland mouthfeel, the MCC–guar aggregates have no detectable aftertaste or residual effect [1].

4.4 Stability

Functionality and Functional Stability

To achieve optimal functionality, the colloidal MCC must be well dispersed in water. When MCC is incorporated into an aqueous system with the corresponding shear forces, then the hydrophilic molecular portions hydrate. The microcrystals of the cellulose are released and activated – this results in a stable colloidal three-dimensional network. The highest stability of the MCC matrix is obtained when the microcrystals are activated first and all other ingredients of the formulation added thereafter. Since MCC does not bind water and thus does not use up water, the solubility of other substances is not reduced and the desired functionality of the subsequent ingredients is not adversely affected. Figure 4.5 shows colloidal MCC (as under a light microscope) – in this example, co-processed with sodium CMC –with and without sufficient dispersion by

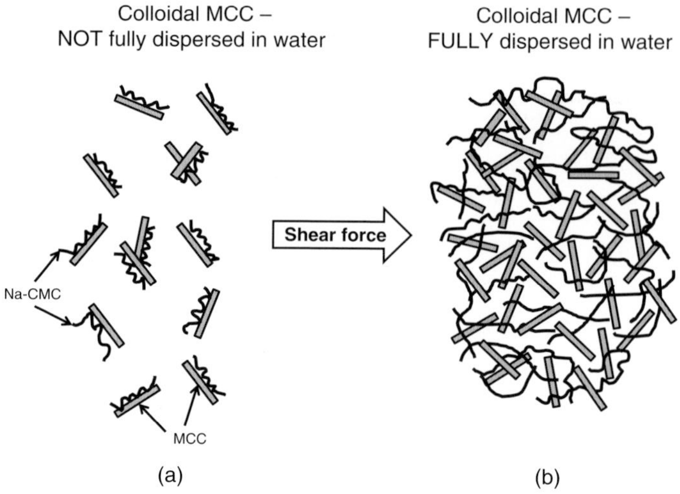

Figure 4.5 Visual impression of colloidal MCC (a) before and (b) after the activation [3].

shear forces. The network (Figure 4.5b) that develops by purely mechanical activation is three-dimensional.

Salt Stability

High amounts of electrolytes reduce the dispersion effect. Hard water, or the presence of mineral salts and divalent cations, thus requires greater shear forces for activation of the microcrystals. Complete dispersion of colloidal MCC in pure soft water will require the minimum shear time. Complex food systems based on milk or tap water with mono- and divalent ions require longer shear times and higher shear forces (Section 4.3.2, 'Gelation and behaviour with other ingredients').

Shear Stability

MCC gels react thixotropically – that is to say, the viscosity of the gel is reduced by shear forces. Following a rest period, the gel 'reforms' and the initial viscosity is recovered almost entirely. The irreversible viscosity loss after a shear treatment can be considered as low.

pH Stability

At pH values below 4.5, the addition of a protective colloid is recommended to prevent acid-induced flocculation in the system. As hydrocolloids with a protective function, xanthan gum, carboxymethyl cellulose (Na-CMC), and methyl cellulose are suitable. The level of the protective substance is approximately 10% by weight of the MCC [3].

Xanthan gum is the most effective protective polymer. However, other gums, already included in the food formulation for additional thickening, gelling, and mouthfeel effects, can also act as protective colloids [1].

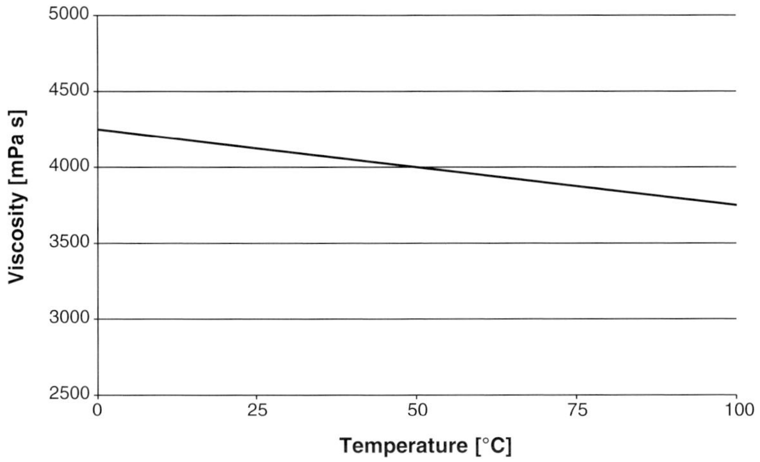

Figure 4.6 Heat stability of a MCC–CMC dispersion (2% in water) [1].

Heat Stability

Compared to other hydrocolloids, the influence of temperature on colloidal MCC dispersions is low. According to the principles of thermodynamics, distribution and dissolution processes occur faster at higher temperatures – so that during the dispersion, a slight lifting of the temperature in the area of 25–40 °C (77–104 °F) is advantageous. The thermal stability of the microcrystalline cellulose is very good over a wide temperature range. Strong increases in temperature have little impact on functionality. The viscosity loss during cooking, baking, sterilization, UHT, or microwave treatment is small. The excellent heat stability of colloidal MCC particles ensures minimal or no product breakdown at elevated temperatures. This heat stability permits great flexibility in various heat processes utilized in food manufacturing, such as batch pasteurization, UHT treatment, and hot filling. Figure 4.6 illustrates the thermal stability of a dispersion of MCC–CMC across a range of temperatures from the freezing point through chill and ambient to the boiling point. This property is essential to the utilization of MCC–CMC for ambient-stored beverages, dairy and non-dairy creams, sauces, ready-to-use bake-stable bakery fillings and fruit fillings [1].

The addition of colloidal grades of MCC to both pectin-based and starch-based bakery fillings produces a modified gel structure with improved texture, spreadability, and heat stability characteristics. The development of a distinct colloidal MCC network within the pectin gel system sustains the fibrous pectin network as a heat-stable structure [14].

Several studies have also demonstrated the value of using colloidal MCC in retortable oil-based salad dressings. Colloidal MCC has the ability to preserve textural consistency, viscosity, and emulsion stability of oil-based dressings prepared under sterilization processing conditions of 116 °C (240 °F) for 1 h [1].

Freeze–Thaw Stability

Colloidal MCC, with its tremendous surface area and capillary action, has the unique ability to manage the free water produced in frozen desserts during freeze–thaw cycles. Its unique water-absorption capability compensates for the inability or slowness of other solids in the mix to re-absorb free water upon partial melting of ice crystals during temperature fluctuations [15].

The use of colloidal MCC produces frozen dessert products with improved body and texture qualities, improved extrusion qualities, and good 'heat shock' resistance, the industry term for the cyclical temperature conditions that occur during storage and transport [1].

Emulsion Stability

The strong affinity of colloidal MCC for both oil and water results in preferential orientation of the solid particulates at the oil-in-water interface. Several basic studies, involving the use of bright field and polarized light microscopy, freeze-etch electron microscopy, and rheological measurements, have established that this colloidal network can provide a mechanical barrier of considerable strength and durability at the interface [16].

In addition, colloidal MCC thickens the water phase between the oil globules, preventing their close approach and ensuing coalescence. These attributes clearly demonstrate the potential use of colloidal MCC in stabilizing emulsion-based food systems that require higher viscosities and good shelf life stability [1].

Fat Replacement

The rheological properties of colloidal MCC products dispersed in water have been used to simulate fat in various food applications. By adding colloidal MCC into food systems, including ice cream, salad dressings, sauces, and gravies, the level of oil or fat can be effectively reduced, while preserving the physical and rheological properties [17].

Basic studies have shown that a simple emulsion containing 60% soybean oil has similar rheological properties and stability characteristics to a 20% soybean oil emulsion containing 1.0–1.5% colloidal MCC. Because of the special orientation of the colloidal MCC particulates at the oil-in-water interface, the emulsions acquire a yield value that makes them stable against 'oiling-off' or creaming [1].

Foam Stability

In aerated food systems, the insoluble colloidal MCC particulates will stabilize the water phase between air cells and provide supplementary structure to the protein film surrounding the air cells. Several application studies have demonstrated the use of colloidal MCC in improving the stand-up ability, stiffness, and foam stability of both non-dairy and dairy whipped toppings. In addition, it has been found that colloidal MCC can effectively stabilize marshmallow toppings and confectionary products and control overrun in frozen desserts [1].

4.5 Rheometry

To identify microcrystalline cellulose, and to distinguish it from other cellulose derivatives or hydrocolloids, the following methods are suitable:

1) **Solubility**
2) **IR Spectroscopy or microscopy with polarized light**
3) **Colour reaction:** 1 mg of the sample is mixed with 1 ml of phosphoric acid and heated for 30 min in a water bath. It is added to 4 ml of a 1 : 4 solution of pyrocatechol in phosphoric acid and heated for 30 min. A red colour is produced in the presence of MCC.
4) **Suspension test:** 30 g of the sample is mixed for 5 min in a high speed mixer (12 000 rpm) with 270 ml of water. The result is either a free-flowing suspension or a heavy lumpy suspension that is very viscous, hardly settles, and contains many trapped air bubbles. If a free-flowing suspension is obtained, then 100 ml is poured into a 100-ml measuring cylinder and left for 1 h. The solids settle and a supernatant liquid appears. Detection of the absence of starch is carried out by mixing 20 ml of the dispersion directly taken after the shear treatment and adding a few drops of iodine solution. There should be no purplish to blue or blue colour. Powdered cellulose is also analysed with the solubility and the suspension test method [13].

4.6 Preparation of MCC Dispersions in Food

4.6.1 Key Factors

The key to developing functional properties from the insoluble microcrystals is the creation of a three-dimensional gel network through an aqueous dispersion of the colloidal MCC. This dispersion is formed by shear force. The amount of shear required to create an effective, functional dispersion depends on the product type. Products have been developed to suit any process from high-pressure homogenization to simple dry blends. Several key factors may interfere with the dispersion and effective use of colloidal MCC [1]:

1) **Shear force:** Adequate shear must be used, that is, the shear applied during the process must be sufficient to separate and homogenize the particles of the grades of MCC selected.
2) **Order of addition:** The preferred order of addition is to add the colloidal MCC to water before the other ingredients, especially other thickeners and gelling agents, which can bind water and prevent distribution of the insoluble particles throughout the aqueous phase. In fact, as MCC does not strongly

bind water, it will not prevent the hydration of other ingredients. Starches and other hydrocolloids or gums can readily hydrate in a dispersion of MCC in water. If other polymeric thickening and gelling agents are added at the same time as MCC, the time to achieve complete dispersion is extended.

3) **Presence of ions:** Hard water and dissolved electrolytes can inhibit the dispersion of colloidal MCC. Dispersing in soft water will give complete dispersion in the minimum time. Figure 4.3 (Section 4.3.2) shows that if clean water is not used to prepare solutions, or if mixing tanks are not completely washed between different batches, rotor–stator mixers or homogenizers are needed to fully activate the MCC–CMC dispersion. The effect on activation is more pronounced for divalent salts (Figure 4.4). Products containing high levels of divalent cations, such as ice cream mixes, milk, or reconstituted milk beverages, require high-pressure homogenization to ensure that the MCC product is effectively utilized. New grades have been developed that activate more easily than traditional products. However, shear is important to activate MCC–CMC products and the dispersion cannot be over-sheared. Prolonged mixing times or very high shear pressures, such as 250 bar (4000 psi), do not affect MCC–CMC viscosity and suspending properties.
4) **pH value:** If a dispersion of MCC is acidified below pH about 4.5, a protective colloid is necessary to prevent flocculation. Most soluble gums have a protective effect, but the most effective is xanthan gum at a level of approximately 10% by weight of the MCC. However, many other hydrocolloids already included in the food formulation for additional thickening, gelling, and mouthfeel effects can also act as protective substances.

MCC–guar aggregates must be well dispersed to achieve full functionality in the aqueous phase. The shear required depends on the viscosity and composition of the aqueous phase, the concentration and type of aggregates, and the method of addition. Generally, however, it is usually independent of the temperature and the pH of the aqueous phase. As a general guideline, MCC–guar aggregates, added dry or as part of a dry blend with other ingredients, will readily disperse in water by applying relatively low shear, such as using a propeller mixer at approximately 1600 rpm for 5 min. Even with minimal shear, such as spoon stirring for dry mix applications, much of the functionality of the aggregates is obtained. Very high shear is able to disrupt the spheres and give fibrous rods – but these conditions would only occur in solutions subjected to multiple passes in a homogenizer [1].

4.6.2 Instruction and Testing

To maintain optimal functionality in a food system, microcrystalline cellulose should be dispersed in water according to the following instructions [3]:

1) addition of MCC in water while stirring;
2) addition of a protective colloid – only at pH values below 4.5;

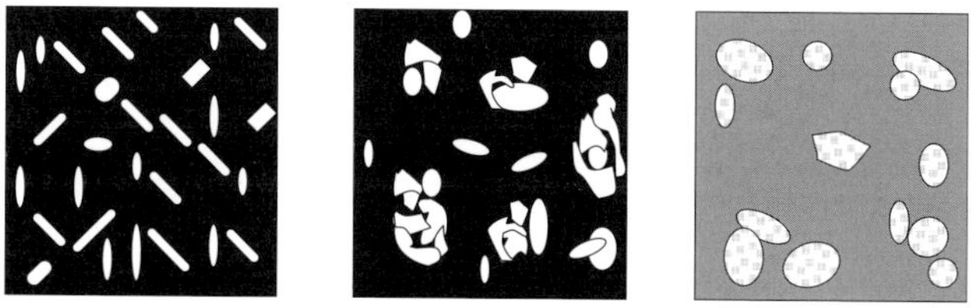

Figure 4.7 Schematic representation of the different 'physical' states of MCC dispersions [3]: (a) fully dispersed; (b) flocculated MCC (pH 3); (c) undispersed MCC.

3) shear treatment with a high speed stirrer for 5–15 min (the dispersion time depends on the type of MCC product and on the stirring equipment);
4) verification of the dispersion (polarizing microscope);
5) addition of other components of the formulation;
6) incorporation of salts and acids;
7) homogenization, if required.

Testing of the finished dispersion is carried out with polarized light at 100× magnification by light microscopy. The different physical states of MCC dispersions are illustrated schematically in Figure 4.7. After sufficient shear small rod-shaped particles are homogeneously distributed. Dispersed MCC appears as an even, crystalline pattern under polarized light (Figure 4.7a). At acidic pH values (<4.5) and without protective colloid, MCC flocculates to round larger particle aggregates that are then present in the system in an inhomogeneous distribution. Under polarized light, flocculated MCC has areas with gaps where no microcrystals are present (Figure 4.7b). Undispersed MCC powder appears in the form of irregular and very large aggregates (Figure 4.7c) [3].

4.7
Synergies with Other Hydrocolloids

The addition of small amounts of cold water-soluble hydrocolloids, such as carboxymethylcellulose, improves the hydratability and dispersibility of the MCC significantly. Additionally, cellulose gum provides protection in acidic systems and prevents flocculation of the MCC and the proteins [2].

The solubility in water is also increased with other thickening agents such as xanthan or guar gum. With starches (e.g. modified waxy maize starches), MCC shows a synergism regarding the viscosity, the shear stability and the heat resistance. The amount of starch can therefore be reduced; thereby, the sensory profile (mouthfeel, texture, taste) is also improved. Moreover, the system is more process-stable. In addition, there is less irreversible damage due to heating and freezing. Even after the application of strong shear forces, the system recovers quickly.

4.8 Functions and Properties of MCC

4.8.1 Powdered MCC

Powdered MCC is the result of drying the crystalline bundles directly after hydrolysis of the native α-cellulose. It consists of very porous particles. This porosity allows the MCC particles to absorb large amounts of water or oil onto the surface.

Powdered grades of microcrystalline cellulose are ideally suited as [3]:

- source of dietary fibre,
- non-caloric bulking agent,
- opacifiers,
- carriers,
- anti-caking agents,
- extrusion aids,
- tableting aids.

The first studies on MCC application began in 1957 with the focus on a functional stabilizer and fat replacer. The powdered MCC types were originally used as sources of fibre in low-calorie food products. Later, however, MCC found wide acceptance within the pharmaceutical industry for use in direct compression tablets.

Powered MCC grades are white, odourless, tasteless, relatively free-flowing powders that are virtually free from organic and inorganic contaminants. They are 100% insoluble in water and chemically inert, crystalline in nature, and very porous [1].

4.8.2 Specialist Powdered MCC

More recently, MCC with small particles, sized less than about 30 μm, has been used in food products as bulking agent and fat-substitute. The pure spheronized, attrited MCC has found application in very low moisture cookie fillings and coatings. The product is distinctive in that particles are produced that are substantially smooth, having a high absolute density, high loose bulk density (>0.40 $g\,cm^{-3}$), a low degree of oil absorptivity (<1.0), and substantially spherical shape, as seen at 150× magnification. This finely powder grade MCC is particularly suited for use as a bulking agent in oil-based food products or food components, which include nut butter, chocolates, ice creams, mayonnaise, lard, and cream fillings [1].

4.8.3 Colloidal MCC

Colloidal microcrystalline cellulose shows properties such as:

- establishment of a thixotropic gel–matrix system and of a 3D network;
- stabilization of two-and multiphase systems such as suspensions/dispersions, emulsions, and foams;
- replacement of fats and oils;
- improvement of texture, mouthfeel, and flavour release;
- increased heat stability;
- crystallization control during freeze–thaw-cycles.

4.9 Food Applications with Typical Formulations

Table 4.1 gives a review of suitable applications.

In non-food-applications, crosslinked or colloidal MCC is an excellent binder for tablets, granules, or any compacted material. Microcrystalline cellulose:

- is added to accelerate disintegration and dispersion;
- provides superior stability;

Table 4.1 Typical applications for MCC [3].

Application	MCC type	Concentration (%)	Functions/benefits of MCC
Mixes for power bars and candy bars	Colloidal	3.0–5.0	• Stabilizes the emulsion; • suspends the solids; • improves creaminess and pulpiness; • adds opacity
Batters and breadings	Colloidal	0.5–3.0	• Improves cling; • reduces drying time; • reduces fat absorption during frying; • reduces sogginess if finished product stored under heat lamps
Chocolate drinks	Colloidal	0.25–0.7	• Adds creaminess; • suspends the solids; • stable under high temperature processing; • adds opacity.
Confectionary	Powdered	0.5–2.5	• Controls moisture absorption; • non-nutritive bulk filler

(*continued*)

Table 4.1 (*Continued*)

Application	MCC type	Concentration (%)	Functions/benefits of MCC
Dressings	Colloidal	1.0–3.0	• Enhances the mouthfeel characteristics; • mimics the mouthfeel of oil; • stabilizes emulsions; • suspends the solids; • improves cling; • opacifier
Fillings	Colloidal	0.8–2.0	• Prevents boil-out during baking/heating (no leakage or rupture); • improves texture and flavour release
Food service	Colloidal	0.5–2.0	• Stabilizes microwave sauces; • reduces skinning on sauces held on steam table; • helps keep fried foods crisp under heat lamps; • reduces fat pick up during frying
High fibre drinks	Powdered or colloidal	0.5–1.0	• Increases dietary fibre; • adds body and creaminess; • suspends solids
Icings	Colloidal	0.2–1.0	• Controls flow and moisture migration; • imparts stability; • increases creaminess
Low-fat sour cream	Colloidal	0.35–0.5	• Imparts full-fat mouthfeel; • stabilizes the system
Puffed snacks	Powdered	2.0–5.0	• Aids in extrusion process; • increases uniformity; • promotes finer air cell structure (improves homogeneity)
Sauces	Colloidal	0.3–1.3	• Shear stability allows pumping without viscosity loss; • stabilizes emulsions; • improves cling; • adds body and creaminess; • prevents boil-out – imparts heat stability; • adds opacity
Whipped toppings	Colloidal	0.3–1.0	• Foam stabilizer
Ice cream, frozen desserts	Colloidal	0.1–1.0	• Controls the ice crystal growth: - leads to smaller ice crystals; • enhances the mouthfeel; • improves creaminess and melt-away; • replaces fat in low-fat recipes

Data taken with permission from Reference [3].

- ensures a consistent and reproducible flow rate;
- delivers improved water uptake.

Moreover MCC is used as:

- filter aid for wine, cider, and other drinks;
- release in respect of a separating agent used for chewing gum.

Below, guideline formulations for different applications are listed.

4.9.1 Baked Goods, Snacks, and Fillings

Tables 4.2–4.4 give the formulations for some baked goods, snacks, and fillings, respectively.

Table 4.2 Formulation for fat-reduced cakes – brownies (low-fat and 33% lower calorie content) [3].

Ingredient	Amount (%)
Sugar	36.70
Cake flour (wheat)	20.40
Corn syrup (42 DE)	11.60
Water	8.10
Cocoa powder, fat free	7.00
Egg white (liquid)	5.40
MCC – cellulosegel + guar	5.00
Glycerine	4.00
Salt	0.80
Instant starch	0.25
Emulsifier	0.25
Baking powder	0.25
Vanilla flavour	0.25
Sum	100.00

Data taken with permission from Reference [3].

Table 4.3 Formulation for a high protein power bar (nutritional bar/snack) [3].

Ingredient	Amount (%)
Maltitol solution	39
Protein blend[a)]	33
Glycerine	10

(continued)

Table 4.3 (*Continued*)

Ingredient	Amount (%)
Canola oil	10
MCC – cellulosegel + guar	4
Cocoa powder (12% fat)	4

Data taken with permission from Reference [3].
a) Milk protein concentrate (25.5%), whey protein isolate (25.5%), calcium caseinate (21.75%), non-fat dry milk (5.25%), soy protein isolate (22.0%).

Processing Guideline

- Dry blend all dry ingredients in a Hobart mixer (wire whisk attachment) for 5 min at speed #1;
- heat maltitol and glycerine to 85 °C (185 °F);
- utilizing a paddle attachment add the oil, followed by the heated glycerine and maltitol solution to the dry blend and mix for 5 min at speed #1;
- sheet or extrude the dough and then cut into bars;
- enrobe with chocolate coating (optional);
- seal in bags.

Table 4.4 Formulations for fillings and creams for bakery products [3].

Ingredient	Amount (%)			
	Filling, high Brix	Filling, low Brix	Pineapple filling	Cheese cake filing, low-fat
Sugar	40.0	13.0	21.20	63.1
Glucose or fructose syrup	13.6	—	—	6.0
Sugar syrup (63 DE)	—	—	47.60	—
Maltodextrin	—	—	—	5.0
Modified starch	3.0–3.5	4.0	2.00	—
Skimmed milk powder	5.0	—	—	3.0
Glycerine	—	—	—	7.0
MCC + CMC	0.8–1.0	0.8–1.2	1.0	2.0
Pectin	—	—	0.75	—
κ-Carrageen	—	—	—	0.2
Emulsifier	—	—	—	0.2
Colour and flavour	To suit	To suit	Colour 'yellow'+ pineapple flavour	Cream cheese flavour: 0.4
Sodium citrate	—	—	0.15	—
Potassium sorbate	—	—	—	0.1
Citric acid solution (1 : 1)	To desired pH	—	1.40	—
Water	To 100%	—	25.90	—
Milk, low-fat	—	to 100%	—	13.0
Sum	100.0	100.0	100.0	100.0

Data taken with permission from Reference [3].

4.9.2
Dairy Products – Ice Cream, Desserts, and Cheese Preparations

This section gives the formulations for various dairy products (Tables 4.5–4.10).

Table 4.5 Formulations for different ice cream types [3].

Ingredient	Amount (%)					
	Soft serve, full fat regular	Soft serve, fat-reduced	Soft serve, sugar-reduced	Soft serve, fat- and sugar-reduced	Ice cream, regular	Ice cream, fat-reduced
Butterfat	10.00	4.00	10.00	5.00	10.00	4.00
Milk solids, non-fat	12.00	13.00	10.00	11.00	8.25	9.00
Cane sugar	13.00	11.00	—	—	12.00	12.00
Corn syrup solids (36 DE)	—	4.00	—	—	6.00	8.00
Whey (dried)	—	—	—	—	2.75	3.00
Maltitol	—	—	8.00	8.00	—	—
Polydextrose	—	—	8.00	4.00	—	—
Sorbitol	—	—	—	4.00	—	—
Sucralose	—	—	0.012	0.01	—	—
Stabilizer blend **A** with MCC, CMC, Carrag.[a)]	0.5–0.6	0.6–0.7	0.5–0.6	0.6–0.7	—	—
Stabilizer blend **B** with MCC, CMC, Carrag.[b)]	—	—	—	—	0.45–0.55	0.55–0.65
Total solid content (approx.)	35.5	32.6	36.5	32.6	39.5	36.6

Data taken with permission from Reference [3].
a) A: monoglycerides, MCC, CMC, TSPP (tetrasodium pyrophosphate), carrageenan, NaCl.
b) B: MCC, mono- and diglycerides, CMC, maltodextrin, Polysorbate 80, carrageenan, NaCl.

Table 4.6 Formulation for gelatine-free mousse with cold processing [3].

Ingredient	Amount in final dessert (%)	Gram per 1000 ml
Sugar	11.0–13.0	140.7–177.6
Cocoa	5.0–6.0	63.9–82.0
Whipping base (53% vegetable fat, 40% glucose syrup, 5% caseinate, 2% emulsifier E 472 a)	3.6–4.4	45.4–58.2
MCC	1.4–4.8	18.5–23.8

(continued)

Table 4.6 (*Continued*)

Ingredient	Amount in final dessert (%)	Gram per 1000 ml
Instant starch	0.8–1.8	10.2–24.6
Milk	To 100%	

Data taken with permission from Reference [3].

Processing Guideline

- Mix all dry ingredients;
- pour cold milk into a bowl;
- add the dry mix and stir slowly and carefully, until the mix is dispersed;
- whip at high speed for 3 min;
- fill the mousse into a bowl or single portion containers;
- leave the mousse at 5 °C (40 °F) for one hour before serving.

Table 4.7 Formulation for pudding – fat-free and fortified with calcium [3].

Ingredient	Amount (%)
Skimmed milk	50.0
Water	32.5–33.0
Sugar	10.0
Starch	2.5
Cocoa (10–12% fat)	2.0
MCC (fortified with calcium)	1.5–2.0
κ-Carrageenan	0.5
Salt	0.3
SSL (sodium stearyl lactate)	0.2

Data taken with permission from Reference [3].

Table 4.8 Formulations for fat-reduced sour cream [3].

Ingredient	Amount (%)	
	5% Fat	12% Fat
Cream (40% fat)	4.5	23.5
Modified starch (E 1422)	3.0	1.0–3.0
MCC + CMC + carrageenan	0.5	0.35
Culture (fermentation bacteria)	0.02	0.02
Milk (3.5% fat)	To 100%	To 100%

Data taken with permission from Reference [3].

Processing Guideline

- Blend all dry ingredients;
- add the dry blend to a mix of cream and milk;
- preheat to 70–72 °C (158–162 °F);
- homogenize at 50 bar and 70 °C (158 °F);
- pasteurize – typically at 95 °C (203 °F) for 5 min;
- cool to 22 °C (72 °F), fill into cups and inoculate with culture;
- ferment for approx. 20 h at 22 °C (72 °F) to pH 4.4;
- apply thermization – heat treatment to 75 °C (167 °F);
- refrigerate.

Table 4.9 Formulation for fresh cheese preparations [3].

Ingredient	Amount per 1000-l milk
MCC + Guar gum	1 kg
Carrageenan[a)]	290 g
Non-fat dry milk solids – to disperse the MCC uniformly	3–5 kg
Milk (approx. 3.2% butterfat)	1000 l
$CaCl_2$ solution (45%)	187 g
Starter cultures (*Lactococcus lactis* subspecies *cremoris*)	Optional – 16 g
Chymosin rennet (100%), double strength	108–136 g
Salt – depending on flavour preference	12–17 kg

Data taken with permission from Reference [3].

a) Key benefits: increased yield, reduced syneresis, creamier product, allows fat reduction without affecting the texture and mouthfeel.

Table 4.10 Formulation for low-fat processed cheese preparation [3].

Ingredient	Amount (%)
Standard skim milk cheese	54.0–56.0
Non-fat dry milk solids	3.8
Sweet dairy whey powder (WPC)	3.8
Buttermilk solids	3.0
Lactic acid (10% solution) to pH 5.4–5.8	Approx. 3.0 (to adjust pH)
NaCl	2.1
Melting salts (citrates E 330–333, lactates E 325–327, phosphates E 339–341, E 450–452)	1.4
MCC + guar	1.0
Sodium citrate	0.8
Carrageenan (kappa- and iota)	0.6 + 0,3
Sorbic acid	0.1

(*continued*)

Table 4.10 (*Continued*)

Ingredient	Amount (%)
Colours (ß-carotene, annatto extract)	0.02 each
Water to achieve the final moisture[a)]	Target – approx. 58% water in finished product

Data taken with permission from Reference [3].
a) To obtain a firmer final product, the amount of carrageenan can be increased and/or the amount of water can be reduced.

Processing Guideline

- Prepare a blend of all dry ingredients;
- grind skim milk cheese to fine particles;
- add cheese to cooker and begin heating;
- add water, then add the dry blend, lactic acid, flavour and colour to the mixture;
- heat to approximately 74 °C (165 °F) for about 3 min or whatever is standard in your operation;
- fill into beakers and cool.

4.9.3 Whipping Creams – Dairy and Vegetable Fat Based

Tables 4.11–4.13 show the formulations for three types of whipping cream.

Table 4.11 Formulation for dairy whipping cream with 30–35% fat [3].

Ingredient	Amount (%)
Cream (with 40% butterfat)	75.0–87.5
Stabilizer (one combination to be chosen)[a)]:	0.3–0.5
a) MCC + CMC	0.3–0.5
b) MCC + CMC + carrageenan	0.3–0.5
c) MCC + starch + CMC	0.4–0.6
Full fat milk	To 100%

Data taken with permission from Reference [3].
a) The optimal use level of any dairy stabilizer depends on milk/cream quality in general and specifically on the amount and quality of milk proteins present in the dairy system.

Processing Guideline

→ Part 1 (activation of the stabilizer system):

- disperse the stabilizer mix in cold milk;
- homogenize at a minimum pressure of 50 bar (725 psi); the shear caused by steam injection (direct heat processing) is sufficient to adequately disperse MCC-based stabilizers;

- the MCC-milk-dispersion is mixed into the cream using a propeller mixer until uniformly incorporated.

→ Part 2 (UHT sterilization):
- sterilization can be done by direct or indirect heat treatment;
- homogenization should not exceed 40 bar (580 psi) and should happen after the heat treatment during the cooling phase at 65–75 °C (149–167 °F);
- filling temperature should be as low as possible – preferably at 8–10 °C (46–50 °F).

Table 4.12 Formulation for vegetable fat whipping cream with 15–27% fat [3].

Ingredient	Amount (%)		
	27% Fat	22% Fat	15% Fat
Fat phase:			
Hydrogenated palm kernel oil	27.00	22.00	15.00
Lactic acid ester of mono- and diglycerides of fatty acids	0.50	0.70	1.00
Tartaric ester of mono- and diglycerides of fatty acids	0.10	0.10	0.10
De-oiled soybean lecithin	0.15	0.15	0.15
Aqueous phase:			
Sucrose	11.00	11.00	11.00
MCC	0.45 or: 0.35	—	—
MCC + CMC	—	0.45	0.45
MCC + guar	—	—	0.50
Sodium caseinate	0.60–1.00	0.7–1.15	0.80–1.30
Salt	0.40	0.40	0.40
Sorbitol (70% solids in water)	0.85	0.85	0.85
β-Carotene powder (10%)	0.003	0.003	0.002
Disodium phosphate	0.015	0.015	0.015
Sodium citrate	0.01	0.012	0.015
Flavour	0.05	0.05	0.05
Water	To 100%	To 100%	To 100%

Data taken with permission from Reference [3].

Processing Guideline

- Disperse the MCC with respect to the MCC–CMC–stabilizer in water using a high shear mixer at approximately 1800 rpm for 10 min; a Silverson-type rotor–stator mixer will disperse the stabilizer in 7 min at full speed;
- add the MCC–guar–stabilizer (for the 15% fat product) and mix for approximately 5 min;
- add citrate, phosphate, and sorbitol and mix for 1 min;
- add sodium caseinate plated on sugar and mix at 1400 rpm for 4 min;
- add salt, flavours, and colours and mix for 1 min;

- heat to 70–75 °C (158–168 °F) and hold;
- dissolve emulsifiers in fat by heating to 70–75 °C (158–168 °F);
- add the fat phase to the heated aqueous phase and emulsify; homogenize one-stage at 140 bar (2000 psi) at a temperature of 70–75 °C (158–168 °F);
- rapidly cool to a temperature of less than 10 °C (50 °F);
- package and store at refrigeration temperature.

Table 4.13 Formulation for whipped topping (freeze–thaw-stable) [3].

Ingredient	Amount (%)					
	20% fat		24% Fat			
	With MSNF	Protein-less*	With MSNF	Sodium caseinate	Sodium caseinate	Soy protein
Water	59.85	64.55	58.08	61.18	60.58	61.08
Hydrogenated vegetable fat	20.00	20.00	24.00	24.00	24.00	24.00
Sucrose	8.00	8.00	7.00	7.00	7.00	7.00
Corn syrup solids (62 DE)	6.00	6.00	5.00	5.00	5.00	5.00
Non-fat dry milk (MSNF)	5.00	—	5.00	—	—	—
Sodium caseinate	—	—	—	2.00	2.50	1.00
Soy protein	—	—	—	—	—	1.00
MCC + CMC	0.60	0.50	0.40	0.30	0.40	0.40
CMC – 20 000 mPa s (2%)	0.10	0.10	0.10	0.10	0.10	0.10
Polysorbate 60	0.30	0.30	0.30	0.30	0.30	0.30
Monoglycerides (distilled)	0.15	0.15	0.12	0.12	0.12	0.12
HPC (low viscosity)	—	0.40	—	—	—	—
Sum	100	100	100	100	100	100

Data taken with permission from Reference [3].

Processing Guideline for 'Formulation Without Protein (*)'

→ Part 1 (hot water + HPC + emulsifier):

- disperse HPC into 12.55% water with 60 °C (140 °F) and agitate for 5 min; add Polysorbate;
- maintain water temperature at 60 °C (140 °F) and continue mixing for 5 min.

→ Part 2 (cold water + MCC + CMC):

- disperse MCC–CMC-blend into cold water (52%) with high speed agitation;

- carefully add CMC (slowly!) and continue agitating for 15 min (alternatively dry blend CMC in sugar before transferring into cold water).

→ Part 3 (sugar + corn syrup):
- blend (remaining) sugar and corn syrup solids and add to the batch;
- begin heating the batch.

→ Part 4 (vegetable oil + emulsifier):
- blend and melt hydrogenated oil and emulsifiers at 60 °C (140 °F);
- add the melted fat to the batch.

→ Part 5 (heat and pressure treatment):
- pasteurize batch at 71 °C (160 °F) for 30 min;
- homogenize at 138 bar (2000 psi) first stage and 35 bar (500 psi) second stage;
- cool batch rapidly to 5 °C (40 °F) and age for 24 h.

4.9.4 Confectionery

This section gives details of the formulation of different types of 'nougat' (Table 4.14) and caramels (Table 4.15).

Table 4.14 Formulation for sugar-based confections – type 'nougat' [3].

Part	Ingredient	Amount (%)	
		Nougat type 'A' (soft, aerated, chewy)	Nougat type 'B' (short texture, clean break, reduced stickiness)
1	Corn syrup (42 DE), warm −50 °C (120 °F)	10.0	10.0
	Water	8.7	8.7
	Powdered sugar	5.0	5.0
	Egg white (dried)	2.0	2.0
	MCC + CMC	0.6	—
	MCC (fine powder)	—	1.0
2	Corn syrup (42 DE)	32.5	32.5
	Sugar	31.6	31.6
	Water	8.8	8.4
	Cocoa (10–12% fat)	0.8	0.8
	Sum	100.0	100.0

Data taken with permission from Reference [3].

Processing Guideline for Nougat type – Production with Hobart Mixer

→ Part '1':
- disperse MCC–CMC-powder in water under high shear (Silverson mixer at 9000 rpm for 3 min, or Hobart mixer with whip attachment for approximately 10 min);

- dry blend powdered sugar and egg white solids in Hobart mixer;
- add MCC–CMC-dispersion to the dry blend and mix for 1 min at low speed. Allow to hydrate for 5 min;
- heat corn syrup to 50 °C (120 °F), then add to the mix;
- beat to a stiff peak at speed #3 using the whipping attachment.

→ Part '2':

- combine corn syrup, sugar, and water;
- heat to 135 °C (275 °F).

→ Combination of part '1' + '2':

- slowly add this hot solution into the Hobart mixer containing the part '1' mixture and stir with whip attachment until a uniform mix is achieved;
- add cocoa and mix at low speed for 2 min, or until uniform;
- transfer to cooling slab and cut, or extrude into desired size or shape;
- nougat can be enrobed with chocolate or other coatings, if desired.

Processing Guideline for Nougat Type B with Hobart Mixer –The Dispersion Step is Omitted

→ Part '1':

- dry blend powdered sugar and egg white solids in Hobart mixer;
- add water to the dry blend and mix for 1 min at low speed; hydrate for 5 min;
- add corn syrup and the fine powder MCC;
- then beat to a stiff peak at speed #3 using the whipping equipment;

→ all other steps are the same procedure for nougat type 'A' (see method of part '2').

Table 4.15 Formulations for caramels – fat-reduced and without fat [3].

Ingredient	Amount (%)		
	Caramels, fat-free	Caramels, 4% fat	4% Fat with chocolate
Sweetened condensed skim milk	40.40	20.20	20.20
Corn syrup (42 DE)	29.10	33.65	33.65
Sucrose	14.60	17.70	17.70
Water	9.00	19.30	16.70
Fructose (crystalline)	5.00	2.50	2.50
Chocolate liqueur	—	—	6.00
Butter oil, clarified	—	3.00	—
MCC + guar	1.30	2.80	2.40
Vanilla flavour	0.30	0.30	0.30
Lecithin	—	0.20	0.20
κ-Carrageen	0.20	0.15	0.15
Salt	0.10	0.20	0.20
Sum	100.0	100.0	100.0

Data taken with permission from Reference [3].

4.9.5
Beverages (RTD)

Dairy and Protein-Based Beverages

Tables 4.16–4.19 list the formulations of different dairy and protein-based beverages

Table 4.16 Formulations for fat-free milk with full-fat-body (pasteurized) [3].

Ingredient	Concentration (%)		
	Type A With 0.3% TiO_2 (white colour), enriched with Ca^{2+}	Type B With cocoa powder	Type C Highly enriched with Ca^{2+} (500 mg, half daily dose)
Low-fat milk (0.3% fat)	To 100	To 100	To 100
MCC + CMC + Ca^{2+}	0.300	0.250	0.300
Carrageenan	0.024	0.028	0.024
Calcium carbonate	—	—	0.080

Data taken with permission from Reference [3].

Table 4.17 Formulations for economic chocolate milk (UHT) with reduced milk content [3].

Ingredient	Concentration (%)					
	20% Milk reduction			40% Milk reduction		
	Type A	Type B	Type C	Type D	Type E	Type F
Viscosity (in mPa s resp. cP)	167	177	194	113	189	112
Cost reduction against reference with 0.02% carrageenan (%)	5	12	22	24	23	38
Skim milk (0.3% fat)	65.9	69.9	69.8	45.75	49.05	47.25
Water	20	20	20	40	40	40
Sugar	8	6	8	8.5	9.2	6
Cream (40% fat)	4	—	—	4	—	—
Skim milk powder	—	2	—	—	—	3
Cocoa powder	1.5	1.5	1.5	1.5	1.5	1.5
Inulin	—	—	—	—	—	2
MCC + CMC + Carrageenan + Ca^{2+}	0.6	0.6	0.7	0.25	0.25	0.25
Sum	100.0	100.0	100.0	100.0	100.0	100.0

Data taken with permission from Reference [3].

Table 4.18 Formulations for soy milk beverages – UHT (fortified with calcium) [3].

Ingredient	Concentration (%)				
	Type A	Type B	Type C – chocolate	Type D – chocolate	Type E – with green tea
Soy milk, fluid (7–9% TS)	90.34	90.34	91.3	90.0	95.4[b)]
Sugar	9.0	9.0	7.0	8.2	2.0
Maltodextrin (15 DE)	—	—	—	—	2.0
Tricalcium phosphate[a)]	0.31	0.31	—	—	—
Calcium carbonate[a)]	—	—	0.3	0.32	0.17
Cocoa powder	—	—	1.0	1.2	—
MCC + CMC + carrageenan	0.35	—	0.4	—	—
MCC + CMC	—	0.35	—	0.28	0.4
Flavour (green tea, honey)	—	—	—	—	0.025 + 0.005
Sum	100	100	100	100	100

Data taken with permission from Reference [3].
a) 120 mg Ca^{2+} in 100 ml soy beverage.
b) Alternative: 3.75% soy protein isolate + 1.25% soybean oil + 90.4% demineralized (or soft) water.

Processing Guideline

- Blend all dry ingredients;
- disperse in fluid soy base with medium shear for 10–15 min;
- UHT process (142 °C (288 °F) for 3–5 s);
- cool to approximately 70 °C (158 °F) and homogenize with 150 and 50 bar (alternatively upstream homogenization can be applied);
- cool to 10 °C (50 °F) and fill aseptically;

Table 4.19 Formulations for acidified protein beverages – pasteurized, based on milk or soy [3].

Ingredient	Concentration (%)		
	Soy beverage pH 4.2–4.4	Milk mix beverage with pH 4.0–4.3	Whey drink with pH 3.9–4.7
Skim milk	—	20.0	—
Apple, orange or pear concentrate (70 °Brix)	1.6–8.2	4.2	Optional
Sugar	8.0	8.0	9.0
Whey protein isolate	—	—	3.0
Skimmed milk powder	—	1.8–5.0	—
Soy protein isolate	1.5–3.0	—	—
Citric acid (1:1)	0.45	0.25–0.4	0.3
MCC + CMC[a)]	0.4–0.6	0.4	0.4
CMC – 100 to 2000 mPa s (2%)[a)]	0.2–0.4	0.3	0.3
Flavour	Vanilla: 0.13	—	To suit

Tricalcium phosphate (TCP)	0.1	—	0.3
Water	To 100%	To 100%	To 100%

Data taken with permission from Reference [3].
a) HM pectin can alternatively be used for protein stabilization.

Processing Guideline for 'Soy Beverage'

→ Phase 1 (acid + stabilizer):
- heat 50% of the water to 71 °C (160 °F);
- add juice concentrate;
- pre-blend pure CMC powder with sugar;
- disperse CMC–sugar-blend and MCC–CMC-stabilizer in the water–juice mixture at 71 °C (160 °F). Stir for 10 min;
- add citric acid and mix for 5 min.

→ Phase 2 (soy protein phase):
- add soy protein isolate to remaining water (50%);
- add TCP;
- heat to 71 °C (160 °F) for 15 min.

→ Cool both phases to approx. 43 °C (110 °F);

→ add phase 2 (soy protein) into phase 1 (juice + stabilizer):
- adjust for water loss;
- reduce foaming by adding suitable anti-foamer;
- ultra-pasteurize at 91 °C (195 °F) for 15 s;
- homogenize at 173 bar (2500 psi) in two stages (138 bar/35 bar (2000 psi/500 psi))
- cool to 10 °C (50 °F) and store under refrigeration.

Coffee and Wellness Drinks

The formulation for coffee beverages and for some nutritional drinks is shown in table 4.20

Table 4.20 Formulations for coffee and nutritional beverages – UHT treated [3].

Ingredient	Concentration (%)	
	Coffee beverage	Nutritional beverages (for adults)
Water	90.1–90.3	74.6–74.91
Corn syrup solids (24 DE)	—	8.6
Sucrose	4.5	6.6
Non-fat dry milk solids (NFDM)	3.0	3.3
Coffee, instant	1.0	—
Vitamin and mineral mix	—	0.15
Fat	Butter: 0.3	Corn oil: 2.5
Soy protein isolate	—	0.79
MCC	0.4–0.6	0.5–0.8

(continued)

Table 4.20 (*Continued*)

Ingredient	Concentration (%)	
	Coffee beverage	**Nutritional beverages (for adults)**
Salts – for stabilization	Dipotassium phosphate: 0.25	Potassium citrate monohydrate: 0,30; NaCl/KCl: 0.07 + 0.23; dipotassium phosphate: 0.20; calcium carbonate: 0.10
Flavour	Optional	Vanilla: 0.26
Emulsifier mix	0.15	—
Emulsifier	Polysorbate 80: 0.03	Soy lecithin: 0.28
Cocoa (10–12% fat)	0.07	1.2
Carrageenan	Optional[a]: 0.02	0.01–0.02
Sum	100.0	100.0

Data taken with permission from Reference [3].
a) Carrageenan is recommended in coffee beverages to enhance the creamy texture and mouthfeel.

Processing Guideline 'Coffee Beverage'

- Disperse MCC in water using moderate to high shear:
- blend remaining dry ingredients and add to the MCC dispersion while slowly heating. Mix for 10 min;
- melt butter. Add blend of emulsifiers to melted butter;
- add blend of oil and emulsifiers at 60–66 °C (140–150 °F); continue heating to 83 °C (180 °F) and hold for 10 min;
- homogenize beverage at 138 and 35 bar (first 2000 psi, second 500 psi);
- fill bottles and close them;
- retort at 122 °C (250 °F) and 2 bar (29 psi) for 20 min. Static or rotary retort can be used in formulations without carrageenan. Recipes with carrageenan as stabilizer require rotary systems to allow for proper dispersion and set of the carrageenan. Bottles should be cooled to 58 °C (135 °F) before allowing to set and cool quiescently;
- remove bottles from retort, shake several times and cool.

Processing Guideline 'Nutritional Beverage'

- Disperse MCC in water using medium shear (a propeller-type mixer) for 10 min;
- dry blend vitamins, minerals, proteins, and carrageenan; add to the MCC-dispersion and continue mixing for 30 min;
- add oil and all remaining ingredients. Mix for 5 min;
- HTST process at 80 °C (175 °F) for 3 s;
- homogenize beverage at 207 and 35 bar (first 3000 psi, second 500 psi);
- fill bottles and close them;
- retort in a rotary system to the appropriate F0 value; cool to 32 °C (90 °F).
- alternatively, static retort at 118 °C (245 °F) and 2 bar (29 psi) for 20–30 min. Upon removal from retort, shake bottles to disrupt any localized gel structure. Continue to shake several times while cooling.

Alcoholic Beverages

Tables 4.21 and 4.22 give the formulation for piña colada and an alcoholic milk-based drink, respectively

Table 4.21 Formulation for piña colada [3].

Part	Ingredient	Concentration (%)
A	Water	39.5
	Sugar	8.0
	MCC	0.5
	CMC – 10 000 mPa s (2%)	0.1
	Potassium sorbate	0.1
B	Water	33.4
	Sugar	8.0
	Pineapple concentrate	6.3
	Creamed coconut, unsweetened (98.5% solids)	4.0
	Polysorbate 60 as emulsifier	0.1
A + B	*Sum*	100

Data taken with permission from Reference [3].

Processing Guideline

→ Part 'A:'

- disperse MCC in water with high speed propeller mixer for 10 min;
- blend CMC with sugar and add to the MCC-dispersion;
- continue to agitate for 25 min while heating to 46 °C (155 °F);
- add potassium sorbate.

→ Part 'B:'

- heat water to 43 °C (110 °F) and dissolve the sugar;
- add creamed coconut, pineapple concentrate and Polysorbate 60. Maintain 43 °C (110 °F) temperature and mix for 14 min.

→ Combination of part 'A' and 'B' under medium to high agitation:

- pasteurization at 88 °C (190 °F) for 30 min;
- homogenization – in two steps at 138 bar and 35 bar (2000 psi and 500 psi);
- hot filling.

Table 4.22 Formulation for an alcoholic milk-based drink with a creamy consistency [3].

Ingredient	Concentration (%)	
Evaporated milk	35.2	*Method of preparation:*
Rum – 90° proof dark	32.5	1. disperse MCC in water with high shear;
Water	20.6	2. addition of evaporated milk; mix for approx. 3 min;
Fructose syrup (68% DM)	11.2	3. add fructose syrup and lecithin. Stir approx. 3 min;
MCC	0.3	4. pasteurization at 85 °C (185 °F) for 2 min;
Lecithin[a)]	0.2	5. cooling to 21 °C (70 °F);
Sum	100	6. addition of rum

Data taken with permission from Reference [3].
a) Lecithin is used as anti-foaming agent.

4.9.6 Soups, Sauces, Salad Dressings, Marinades, and Spreads

This section covers the formulation of various creams, sauces, and salad dressings. Formulations are presented in Tables 4.23–4.30.

Table 4.23 Formulations for cooking creams (UHT) as base for a variety of sauces [3].

Ingredient	Amount (%)			
	5% Fat	12% Fat	20% Fat	30% Fat
Cream (40% fat)	12.5	30.0	50.0	75.0%
Tapioca starch (modified)	2.0–2.5	2.0–2.5	1.0–2.0	1.0–2.0
Emulsifier E 471	0.2	0.2	0.2	0,2
MCC + CMC + carrageen[a)]	0.35	0.32	0.28	0.25
MCC + CMC + xanthan[a)]	0.7	0.65	0.6	0.55
Flavour	To suit	To suit	To suit	To suit
Skimmed milk	To 100%	To 100%	To 100%	To 100%

Data taken with permission from Reference [3].
a) The cream is either stabilized with a blend of MCC, CMC, and carrageenan or as alternative with a blend of MCC, CMC, and xanthan.

Preparation Procedure

- Blend all dry ingredients;
- add the dry blend to a mix of cream and milk;
- preheat to 75 °C (167 °F);
- homogenize at 100 bar and 20 bar (approx. 1500 psi and 250 psi);
- UHT treatment at 140 °C (284 °F) for 3–5 s;
- cool to 15 °C (59 °F);
- fill aseptically;
- the resulting cooking cream is used as a base for various sauces.

Modification to 'Wine Sauce'

- 150 ml wine is placed in a saucepan and concentrated to a final content of 100 ml;
- addition of 150 ml cooking cream to the concentrated wine;
- heating to 60 °C (140 °F).

Modification to a Meat Sauce

- 25 g butter is melted in a saucepan;
- addition of 100 g bouillon;
- this mixture is boiled for 30 s;
- vinegar (2 g) is added under stirring;
- cooking cream (100 g) is added to the sauce, mixed, and heated to 70 °C (158 °F);
- the meat sauce is thickened to the desired consistency.

Table 4.24 Formulation for white sauce with 75% fat reduction (2.2% fat) [3].

Ingredient	Amount (%)
Water	85.9
Milk powder (28% fat)	7.5
Modified starch (E 1422)	2.6
MCC + guar	2.0
MCC	0.75
Salt	0.6
Dried cream extract	0.5
Spices (onion powder, white pepper)	0.1 + 0.05
Flavour	Optional
Sum	100.0

Data taken with permission from Reference [3].

Table 4.25 Formulations for sauce 'Béarnaise' and 'Hollandaise' [3].

Ingredient	Amount (%)	
	Béarnaise	Hollandaise
Water	50.198	52.912
Vegetable oil	25.0	25.0
Vinegar (7% acid)	8.5	8.0
Egg yolk (liquid)	7.0	4.5
Sugar	2.5	2.0
Modified starch (E 1422)	2.0	2.5

(*continued*)

Table 4.25 (*Continued*)

Ingredient	Amount (%)	
	Béarnaise	Hollandaise
NaCl	1.5	2.0
MCC	1.0	1.0
Lemon juice (5% acid)	1.0	1.0
Skim milk powder	0.5	0.5
ß-Carotene (10%)	0.002	0.004
Spices	Tarragon: 0.8; optional other spices	Mustard powder: 0.5; white pepper: 0.06; garlic powder: 0.012; onion powder: 0.012
Sum	100.0	100.0

Data taken with permission from Reference [3].

Table 4.26 Formulation for tomato-based sauce [3].

Ingredient	Amount (%)
Water	76.7
Tomato concentrate (30 °Brix)	13.0
Vegetable oil	5.0
Modified starch (E 1422)	3.5
MCC	0.9
NaCl	0.9
Sum	100.0

Data taken with permission from Reference [3].

Table 4.27 Formulation for Béchamel sauce [3].

Ingredient	Amount (%)
Water	85.55
Skim milk powder	7.0
Modified starch (E 1422)	3.5
Margarine	2.0
MCC	1.0
NaCl	0.8
Nutmeg	0.1
Black pepper	0.05
Sum	100.0

Data taken with permission from Reference [3].

Table 4.28 Formulations for salad dressings – fat-reduced, low-fat, and fat-free [3].

Ingredient	Amount (%)		
	10% Fat	5% Fat	1% Fat
Corn syrup (42 DE)	12.0	12.0	12.0
Soybean oil	10.0	5.0	1.0
Vinegar	5.0	5.0	5.0
Sugar	4.0	4.0	4.0
Cultured buttermilk powder	4.0	4.0	4.0
MCC (alternative: MCC + guar or MCC + CMC)	1.3	1.7	2.0
NaCl	2.0	2.0	2.0
Maltodextrin	—	1.0	2.0
Egg yolk powder	0.5	0.5	0.5
Xanthan gum	0.4	0.4	0.4
Spices	Garlic powder: 0.18; onion powder: 0.18; mustard flour: 0.05; pepper: 0.03; parsley: 0.02		
Potassium sorbate	0.1	0.1	0.1
Flavour + colour	Optional	Optional	Optional
Water	To 100%	To 100%	To 100%

Data taken with permission from Reference [3].

Table 4.29 Formulation for instant sauce 'Hollandaise' (without fat) [3].

Base recipe	Amount (%)
(A) *Dry mix:*	
Starch	35.0
Sweet whey (dried)	20.0
MCC	11.4
Egg yolk powder	12.0
Maltodextrin	12.0
Tetrasodium pyrophosphate	4.0
Citric acid	2.0
Flavour (butter + sherry)	Each 1.1
Flavours (egg + lemon)	0.3 + 0.2
Lemon herb seasoning	0.3
Salt	0.6
Colour	Optional
(B) *Finished sauce:*	
10% dry mix + 90% water or milk	

Data taken with permission from Reference [3].

Table 4.30 Formulations for cheese sauce and cheese dip [3].

Ingredient	Amount (%)	
	With MCC only	With MCC + starch
Water	49.23	48.83
Cheddar cheese (grated)	25.0	25.0
Cheddar cheese powder	10.0	10.0
Soybean oil	10.0	10.0
Milk (3.5% fat)	2.5	2.5
Modified waxy maize starch	—	1.0
MCC	1.25	0.65
Disodium phosphate	1.0	1.0
NaCl	0.75	0.75
Emulsifier	0.15	0.15
Lactic acid	0.05	0.05
Annatto colour	0.035	0.035
Oleoresin paprika	0.035	0.035
Sum	100.0	100.0

Data taken with permission from Reference [3].

4.9.7
Meat Products

Colloidal MCC has recently been studied as a stabilizer and fat replacer in cooked meats and emulsified products, such as sausages and pâté. The addition of cellulose is able to raise product yields by retaining moisture during cooking, even at the high temperatures required for microbial safety in poultry products. The insoluble cellulose particles interrupt the protein structure in the cooked meat and avoid cold shortening, the rubbery consistency produced by excessive protein extraction by seasonings during fresh meat storage. In pâté, MCC gives a smooth consistency to the product, which spreads easily, giving a good flavour release and full-fat texture [1]. Table 4.31 gives the formulation for a fat-free burger.

4.9.8
Applications for Powdered Cellulose

The powdered grades of MCC are commonly used in food products as a high-quality, inert fibre source and a zero-calorie bulking agent. Very finely powdered MCC can be readily suspended in ready-to-drink dietetic products to provide fibre. It can also impart much of the body and opacity usually contributed by fat. It has a clean flavour release with no significant flavour contribution or flavour-masking properties. Additionally, the porous and free-flowing nature of powdered grades of MCC makes them ideal as carriers of liquid materials such as

Table 4.31 Formulation for fat-free beef burger [3].

Ingredient	Amount (%)
Beef (less than 2% fat)	53.00
Water (as ice)	38.95
Salt	1.8
Modified starch	1.5
Spices	1.4
Flavour	1.0
Sugar	0.5
Dextrose	0.5
MCC	0.5
Konjac flour	0.4
Phosphate blend	0.3
Cure salt (6.25% sodium nitrate)	0.15
Sum	100.0

Data taken with permission from Reference [3].

essential oils. This characteristic has been widely used to prevent caking in shredded cheese [1].

4.10 Non-food Applications

Colloidal or crosslinked MCC is used in tablets and granules, to accelerate their decomposition or dissolution. Moreover, the MCC acts as a binder in tablets, granules, or other compressed materials. During the manufacturing process, the flow rate of the dry substances is adjusted with MCC, so that the method is reproducible. The final product shows an increased stability and improved water absorption in the application [3].

In the metal and chemical industries MCC serves as a processing aid additive in the clarification of production wastewater where wash filter systems are used.

4.11 Nutritional Properties

The MCC component of powdered and colloidal grades is composed of insoluble fibres only and does not contribute any other material of nutritional significance [1]. This means that microcrystalline cellulose is a very good source of dietary fibre. It contributes no or only very little calories to the food system [3]. Native cellulose itself is completely indigestible in the human body. Hydrolysis of the cellulose chains to produce powdered MCC does not change the behaviour of the molecules in the intestine or add calories. Therefore, the energy content of powdered MCC is 'zero'.

Table 4.32 Quantitative nutritional analysis of MCC (colloidal or powdered) [3].

Substance (in 100 g)	Colloidal MCC[a)]	Powdered MCC[a)]
Energy content	20 kcal	0 kcal
Fat	0 g	0 g
Dietary fibre (total)	93 g	98 g
Soluble fibres	5 g	0 g
Carbohydrates/sugar	N.d.	N.d.
Protein	N.d.	N.d.
Vitamin A	N.d.	N.d.
Vitamin C	N.d.	N.d.
Sodium	934 mg	4 mg
Iron	0.5 mg	0.24 mg
Calcium	2.0 mg	0.1 mg
Ash	2 g	1 g

Data taken with permission from Reference [3].
a) N.d. = not detectable.

The breaking up of cellulose aggregates to release the microcrystals during the production of colloidal MCC has no impact on the nutritional value of the product. However, some colloidal grades provide calorific content due to the soluble hydrocolloids used during co-processing with MCC. Guar gum (1 g has 0.7 kcal) and starches (1 g delivers 4.18 kcal) are examples. Cellulose gum and carrageenan do not add any calories to colloidal MCC [3].

Table 4.32 lists typical nutritional information for MCC.

In general, the energy content of microcrystalline cellulose is low (≤20 kcal per 100 g). MCC is a non-resorbable and non-digestible fibre. At small particle sizes and small molecular sizes, it is possible that persorption occurs. Persorption is the mechanical passage through the intestine into the organism with yet undetermined significance and implications for the human body. The dietary fibre content of food is significantly increased by the use of MCC, wherein it is mainly to the fraction of insoluble fibres, since the soluble fraction is less than 5%.

MCC is virtually inert and will not interfere or interact with other nutrients added to food products for fortification, such as vitamins and minerals. In fact, MCC is an excellent suspending agent and can provide stable suspensions of even heavy particulates like ferrous fumarate or other minerals. Some newer grades are uniquely co-processed with calcium to offer enhanced calcium content in nutritionally fortified foods and beverages [3].

4.12 Legislation

In the United States of America, microcrystalline cellulose has GRAS ('generally recognized as safe') status and has been used safely in foods for over 30 years.

In Europe, MCC is listed in Annex I of the European Parliament and Council 'Directive 95/2/EC of 18 March 1995 on Food Additives other than Colours and Sweeteners'. It is approved as 'E 460 (i)' in the list of permitted emulsifiers, stabilizers, thickening, and gelling agents for use 'quantum satis', the level required to achieve a given technological benefit. Purity standards of food grade materials are listed in EC directive 2008/84/EC of 27 August 2008. MCC has been evaluated by both the EC Scientific Committee for Food (SCF) and the Joint FAO/WHO Expert Committee for Food Additives (JECFA). Both committees determined the maximum available daily intake (ADI) to be 'not specified'. The copolymers used with MCC, calcium alginate, pectin, guar gum, and CMC, are also listed in Annex I of directive 95/2/EC and are approved for use under the E numbers E 404 a, E 440, E 412, and E 466, respectively.

Recently, cellulose gum was authorized as a synonym for CMC, which aligns EU und USA ingredient declarations for this material [1].

For several years now colloidal MCC may be referred to as 'cellulose gel' in the European Union as an alternative name to 'Microcrystalline cellulose' or the E number E 460 (i). The function of the MCC on the ingredient list can be described as 'filler', 'stabilizer', 'release agent', or 'carrier'. In untreated products, and foods for babies and infants, MCC may not be used [18].

References

1 Imeson, A. (2010) *Food Stabilizers, Thickeners and Gelling Agents*, Wiley-Blackwell, Oxford.
2 Belitz, H.-D. and Grosch, W. *Textbook of Food Chemistry*, 4th edn, Springer-Verlag, page 294–297.
3 FMC Corporation (2014) (Division formerly 'FMC Biopolymer' now FMC Health and Nutrition) Cellulose Gel – product description, application guide and recipes. http://www.fmcbiopolymer.com/Food/Ingredients/CelluloseGel/Introduction.aspx (accessed on 19 March 2014).
4 McGinley, E.J. (1982) Stabilizing agent for dry mix food products, US Patent 4,311,717.
5 McGinley, E.J. and Tuason, D.C. (1993) Fat-like bulking agent for aqueous foods comprising microcrystalline cellulose and galactomannan gum, US Patent 5,192,569.
6 Tuason, D.C., McGinley, E.J., and Selinger, E. (1995) The development of Novagel^{TM} RCN, in *Food Technology International Europe*, Sterling Publications Ltd., London.
7 Tuason, D.C. and McGinley, E.J. (1994) Colloidal microcrystalline cellulose and barrier dispersant composition and method for manufacture, US Patent 5,366,742.
8 Tuason, D.C., Selinger, E., Krawczyk, G.R., Sewall, C., and Hogan, D.T. (2002) Rapidly peptizable microcrystalline cellulose-based stabilizing agents, US Patent 6,391,368.
9 Venables, A.C. and Buliga, G.S. (2000) Ultra-fine microcrystalline cellulose compositions and process, US Patent 6,037,380.
10 Krawczyk, G.R., Tuason, D.C., and Fischer, G.A. (2005) MCC/hydrocolloid stabilizers and edible compositions comprising the same, US Patent Application 20050233046.
11 Buliga, G.S., Tuason, D.C., and Venables, A.C. (2002) Texture and stabilizer composition, US Patent 6,440,474.
12 McGinley, E.J., Krawczyk, G.R., and Selinger, E. (1995) Co-processed particulate bulking and formulating aids: their composition, production and use, US Patent 5,441,753.

13 European Union (2008) Commission directive 2008/84/EC of 27 August 2008, laying down specific purity criteria on food additives other than colours and sweeteners. *Official Journal of the European Union*, **51**, L253, 94–102. Available at http://eur-lex.europa.eu/LexUriServ/LexUriServ.do?uri=OJ:L:2008:253:0001:0175:EN:PDF.

14 McCormick, R.D. (1974) Controlling heat stability for bakery fillings, toppings. *Food Product Development Magazine*, **8** (10), 24, 27.

15 Keeney, P.G. (1979) Confusion over heat shock. *Food Engineering Magazine*, **52** (6), 116–118.

16 Oza, K.P. (1988) Colloidal microcrystalline cellulose-stabilized emulsions, Ohio State University, Columbus. *Dissertation Abstracts International, B*, **49** (2), 372–373.

17 McGinley, E.J., Thomas, W.R., Champion, J., Philipps, G.O., and Williams, P.A. (1984) The use of microcrystalline cellulose in oil in water emulsions, in *Gums and Stabilizers for the Food Industry*, vol. **2** (eds G.O. Philipps, D.J. Wedlock, and P.A. Williams), Pergamon Press, Oxford, pp. 241–249.

18 For food legislation Germany see: Anon (2000) *Federal Law Gazette*, 1521–1536. Available at http://bundesrecht.juris.de/zzulv_1998/anlage_4_16.html.

5
Fundamentals of Water-Soluble Cellulose Ethers and Methylcellulose

Synonyms: MC, methyl ether of cellulose, cellulosemethylether, Methocel A

E number:	E 461
CAS number:	9004-67-5
FDA code:	21 CFR 182.1480

Related hydrocolloids: mixed ethers such as hydroxypropylmethylcellulose

5.1
Manufacturing Process of Cellulose Ethers

5.1.1
General Principles

Cellulose as a Raw Material for Etherifications

Cellulose as a polyvalent alcohol (with three hydroxyl groups per anhydroglucose unit) is basically accessible to the same etherification reactions as the lower alcohols.

The hydroxyl groups of cellulose, however, form inter- and intramolecular hydrogen bonds, which impede direct etherification. Before the ether synthesis begins, therefore, the H-bridges need to be dissolved by appropriate measures such as alkalization. The production of cellulose ethers generally starts with the preparation of the so-called 'Alkali cellulose'.

Preparation of Alkali Cellulose

Usually, caustic soda is used as reagent to generate the reactive alkali- or soda cellulose. Typical concentrations are between 18% and 50%. This alkalinization is performed on a large scale for all finished products (food, non-food). The aim of this treatment is, in each case, expansion of the lattice structure of the cellulose. Figure 5.1 describes schematically how the caustic soda penetrates between the associated cellulose chains, then breaks the hydrogen bonds between the adjacent OH groups, and thus loosens the rigid structure.

Cellulose and Cellulose Derivatives in the Food Industry: Fundamentals and Applications, First Edition.
Tanja Wüstenberg.

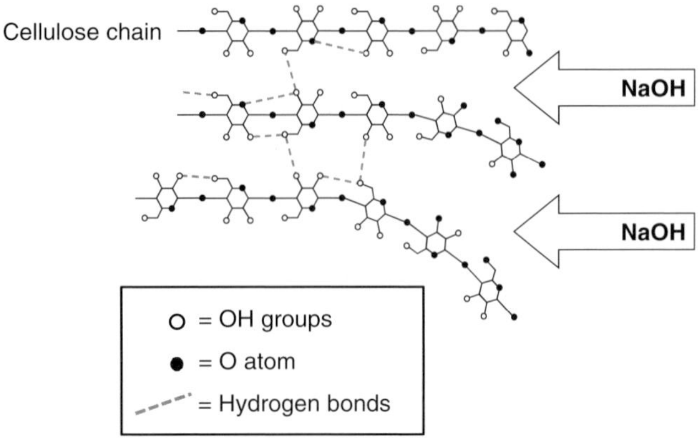

Figure 5.1 Alkalization of the cellulose – preparation for etherification.

The swelling state – and thus the lattice expansion – is dependent on the ratio of cellulose, sodium hydroxide, and water at a specified temperature. This ratio is customized individually to the final product. The alkaline pretreatment is crucial for the number and distribution of substituents introduced and thus for the properties of the resulting cellulose ethers. Technically, the alkalization of cellulose is performed with the diving press method, the mash alkalizing process, and the slurry process. It is advantageous to implement mechanical disruption of the pulp before treatment with NaOH. Thereby, the amount of alkali employed can be reduced. Sodium hydroxide impedes the access of hydrophobic etherifying agents to the OH groups of the cellulose, as a layer enveloping, and even partially penetrating into, the crystalline regions.

Maturation of Alkali Cellulose

The desired viscosity of the final product is achieved by so-called 'maturation' of the alkali cellulose, which is a controlled alkaline oxidation. Parameters for this degradation are time and duration, temperature, and content of oxygen or air. Maturation can be carried out in a shorter time at elevated temperature. The higher the alkali content of alkali cellulose, the shorter is the time for the desired viscosity reduction. Oxygen carriers (special metal salts) can catalytically accelerate the hydrolysis. If high viscosities of cellulose ethers are desired or the pulp already has a low degree of polymerization, maturation can be eliminated completely.

Synthesis of the Cellulose Ethers [1]

Until now, the basic chemical reactions used industrially for etherification of cellulose are confined largely to the ether synthesis by Williamson (a) and base-catalysed oxyalkylation (b) (Figure 5.2).

The Williamson synthesis (Figure 5.2a) is the most common method for etherification of cellulose. In this case, the alkali cellulose is reacted with alkyl halides. In the reaction, a cellulose ether and a salt are formed.

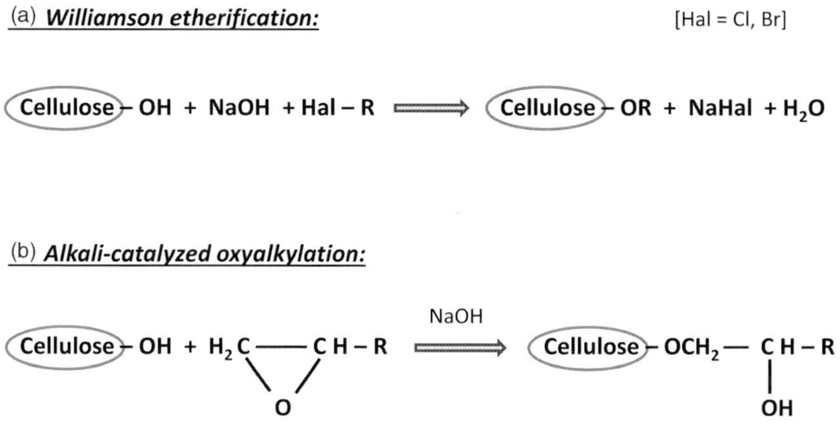

Figure 5.2 Basic reactions for synthesis of cellulose ethers [1].

For the addition of epoxides to the OH groups of cellulose, the cellulose has to be 'opened' also with alkalis. Oxyalkylation (Figure 5.2b) runs at higher pressures. The caustic soda only serves to expand the lattice of cellulose and to reactivate the etherification agent (ring opening of the epoxides). Since it is not consumed, smaller quantities of sodium hydroxide solution are needed compared to the Williamson ether synthesis.

The variety of the spectrum of cellulose ether products can be justified by the fact that the raw material cellulose is a multifunctional molecule. The structure of cellulose displays is known to be a polydisperse linear molecule of anhydroglucose rings linked by acetal bonds. The cellulose molecule contains three reactive hydroxyl functions per anhydroglucose monomer. The primary OH group is located at the C6-position. Two secondary hydroxyl groups are located at C2 and C3. These hydroxyl functions are sites for the alkylation reaction, from which emerges the cellulose ethers. Other chemical reactions, such as esterification (reaction of cellulose with organic and inorganic acids), also occur on the OH groups of atoms C2, C3, and C6. An example of a technical cellulose ester is cellulose nitrate, which is also known as nitrocellulose or guncotton. During etherification, substituents with high molecular weight prefer the primary hydroxyl group on carbon atom number 6 (C6) of the anhydroglucose unit (AGU). To a lesser extent, the hydroxyl groups on C2 or C3 of this unit, but never both at the same time, are etherified. Substituents having lower molecular weight are distributed in approximately equal proportions to the hydroxyl groups C2 and C6. Generally, the distribution of substituents depends on their size, reactivity and polarity, and on the reaction conditions. Figure 5.3 illustrates the location of the reactive OH groups in the individual anhydroglucose unit of a cellulose molecule.

Uniformity of substitution, which is particularly important for the solubility characteristics, is not only controlled by the course of reaction but primarily by the production of uniform alkali cellulose. Notably, its swelling state has a key role.

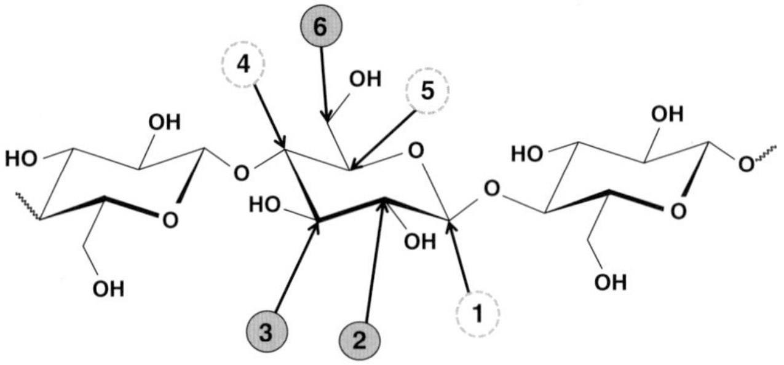

Figure 5.3 Positioning of the free hydroxyl groups in the glucose monomer (AGU) [2].

Today's technically performed processes of etherification obviously do not run in a homogeneous intermediate phase, but are heterogeneous. The ether syntheses lead – in addition to the desired end products (e.g. monoethers such as methylcellulose or mixed ethers such as HPMC) – to several by-products. Therefore, a multistep cleaning follows each reaction to remove these unwanted materials [2].

Purification of the Produced Cellulose Ether

The undesired by-products formed depend on the type of ether. Substances such as alcohols, alcoholates, or alkoxides, or other ethers, can be formed. Or in the case of CMC, the by-products can be glycolic acid, sodium glycolate, and sodium chloride (common salt). The separation and removal of these substances is required to meet the strict criteria of purity for use in regulated applications (food, cosmetics, pharmaceuticals). The type of cleaning process applied is dependent on the physical and chemical properties of the target product. Cellulose ethers such as methylcellulose and HPMC, which are insoluble in hot water, are washed several times with hot water and then go through some filtration steps. Products such as sodium carboxymethylcellulose, which are soluble in both cold and hot water, require washing in a solvent system containing aqueous ethanol or acetone. After cleaning, the still moist products are dried and brought to the desired particle size, using appropriate methods (grinding, screening, sifting). In the final step, the specification parameters are analysed and checked for compliance with the applicable legal requirements. This is followed by transfer of the products into suitable packaging materials [3].

5.1.2
Production of the Monoether Methylcellulose

For the production of methylcellulose, there are three commercial processes.

The *discontinuous gas circulation process* (patent of IG Farbenindustrie, 1925) consists of the following steps:

- The starting material used for etherification is alkali cellulose, produced by appropriate alkalization and maturation (Section 5.1.1).

- The alkali cellulose is introduced into the pressure reactor equipped with a stirrer.
- After evacuation, for every 100 kg of cellulose some 80 kg of methylene chloride is introduced, which has been previously pre-heated to about 60 °C (140 °F).
- The reaction starts immediately and provides a pressure of 4–5 bar (58–73 psi).
- The gas is withdrawn from the top of the reactor and passes through a cooler (condenser), whereby methanol and water are separated.
- After leaving the condenser, the methylene chloride is again heated to reaction temperature, and after the addition of further quantities of methylene chloride from the supply tank it is pressed into the reactor by a pump.
- Under continuous circulation of the gas, the temperature is increased to 90–110 °C (194–230 °F) to accelerate the reaction.
- This cycle is continued until the alkali of the alkali cellulose is nearly completely consumed (about 6–8 h).
- The excess etherifying agent is withdrawn and recovered by condensation.
- After finishing the reaction and cooling of the reaction mass in the gas stream, the methylcellulose formed is transferred from the reaction vessel into a mixing container, which is supplied with softened water at 95 °C (203 °F). Here, the salt formed during reaction is removed and the remaining alkali is neutralized.
- The reaction mass is treated in a centrifuge and washed with boiling water and steam.
- The pulp contains about 50% water. It is compressed in a screw press and then crushed in a hammer mill while still wet. Thereafter, the product is dried in a drum dryer.
- A grinding step can be carried out before the final filling.

In the modern version of this method, a kneader is used as reactor. Its axis and kneading blades are hollow, so that the gaseous etherifying agent can be inserted into the centre of the cellulosic mass. Figure 5.4 illustrates the manufacturing route.

The *etherification method in liquid phase* (Patent of Bayer, 1912) works with liquid methyl chloride. At reactions temperatures of 60–70 °C (140–158 °F), pressures of 20 bar (290 psi) arise. For better penetration of the alkali cellulose with methyl chloride, the etherifying agent is used in extreme excess (5 : 1). Owing to high resulting pressures and acid cleavage of the excess methyl chloride, the process leads to considerable equipment difficulties [2].

A third way is the *continuous process* (Kalle AG, 1965), in which the alkali cellulose is produced as a mash. The activated cellulose mass is then pressed with the aid of a double screw extruder through a horizontal continuous reactor. The reaction product is washed under pressure. Methylene chloride and dimethyl ether as by-product are pumped in a circuit [2].

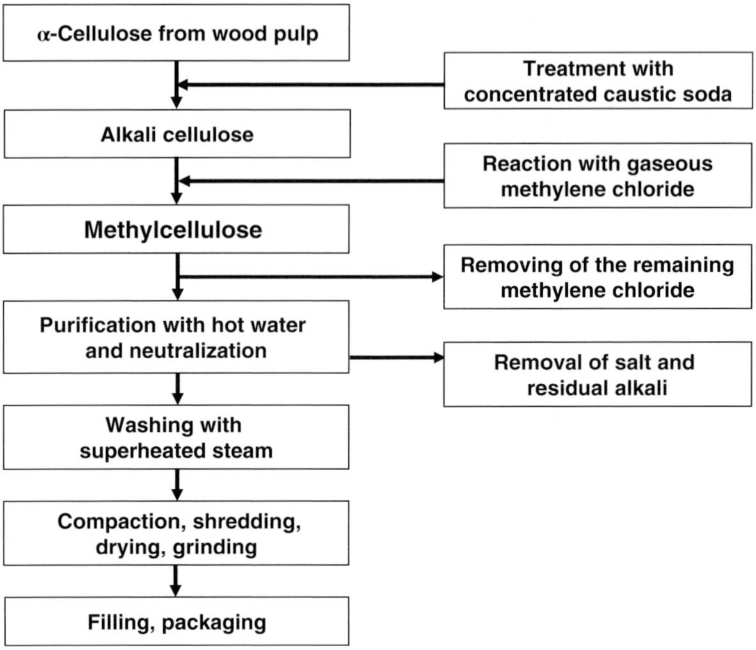

Figure 5.4 Production of food-grade methylcellulose (MC).

5.2 Chemistry

The base of the methylcellulose molecule is formed by the polysaccharide cellulose and its $1 \rightarrow 4$–linked β-D-glucose units. On average, two of the three free hydroxyl groups are replaced by methyl groups (Figure 5.5). The DS (degree of substitution) of this molecule shown is exactly 1.75.

The general chemical formula for the methyl monoethers is: $C_6H_7O_2(OR^1)(OR^2)(OR^3)$, wherein R^1, R^2, and R^3 can be:

- H
- CH_3
- CH_2CH_3.

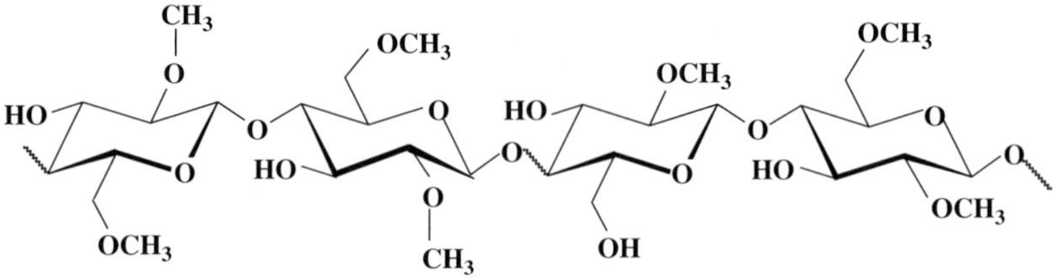

Figure 5.5 Methylcellulose in chair form notation.

The DS of water-soluble methylcellulose types is between 1.3 and 2.6 methyl groups per anhydroglucose unit. The gravimetric content of methoxy groups ($-OCH_3$) is 25–33%. According to the EU criteria of purity, a hydroxyethoxyl amount ($-OCH_2CH_2OH$) of up to 5% is allowed. The molecular weight is 20 000–380 000 $g\,mol^{-1}$, which corresponds to a degree of polymerization of 100–2000 glucose monomers. The molecular weight of a methylated anhydroglucose unit is, for example, 179 (with DS = 1.2) and 193 (with DS = 2.2) [2,4,5].

5.3
Rheology

5.3.1
General Swelling and Dissolution Behaviour of Cellulose Ethers

5.3.1.1 Factors of Influence

Etherification causes transfer of the water-insoluble cellulose into a swellable to completely water-soluble derivative. For the properties of the cellulose ethers – especially their solubility – the following factors are relevant [2]:

- **Type and size of the substituents**
- **Uniformity of substitution:**
  We distinguish between a uniform and even, known as a statistical distribution, and an irregular and uneven block-wise distribution (Figure 5.6). The more evenly distributed the substituents are, the lower the degree of substitution for which a specific solubility of the cellulose ether is achieved.

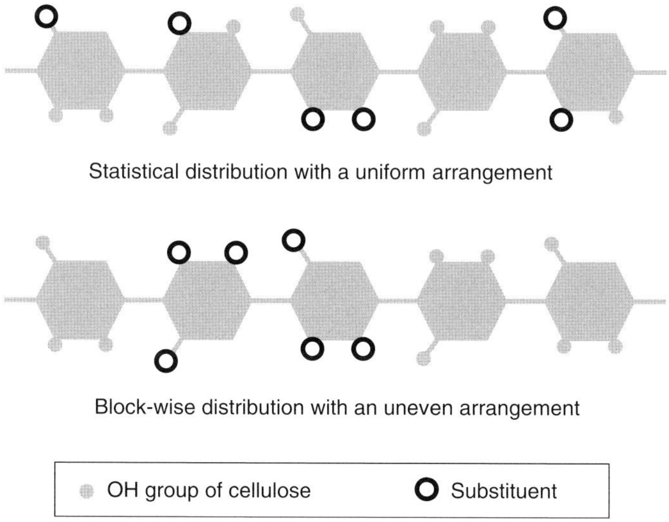

Figure 5.6 Possibilities for the distribution of substituents.

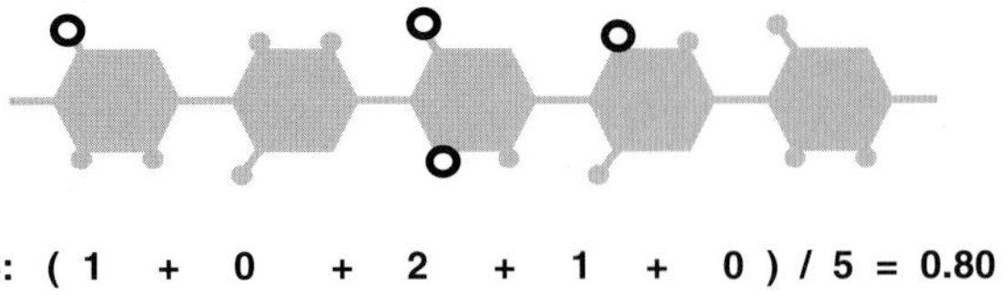

Figure 5.7 Graphical presentation of the degree of substitution (DS).

- **Degree of substitution (DS):**
  The degree of etherification or average degree of substitution is defined as the number of etherified hydroxy groups per anhydroglucose unit (AGU):

  $$\mathrm{DS} = [\mathbf{mol\,“substituent”/mol\,“AGU”}]$$

  Since each individual glucose monomer contains three substitutable OH groups, the DS is between 'zero' (= native cellulose) and 'three' (full substitution). Figure 5.7 describes how to calculate the degree of substitution. The presented cellulose ether molecule has a degree of substitution (DS) of 0.8.
- **Molar substitution (MS):**
  Molar substitution is the average number of moles of substituted groups per glucose monomer. The degree of substitution alone does not always give complete information about the overall level of substitution because, in some ether types, already substituted groups are open to be connected with further foreign groups through ether bridges. Figure 5.8 illustrates the molar substitution by means of a model cellulose ether with a degree of polymerization of five glucose units.

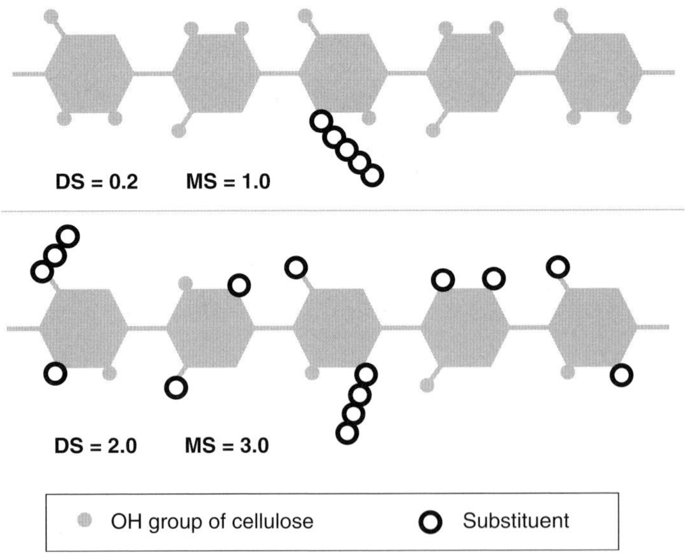

Figure 5.8 Difference between degree of substitution (DS) and molar substitution (MS).

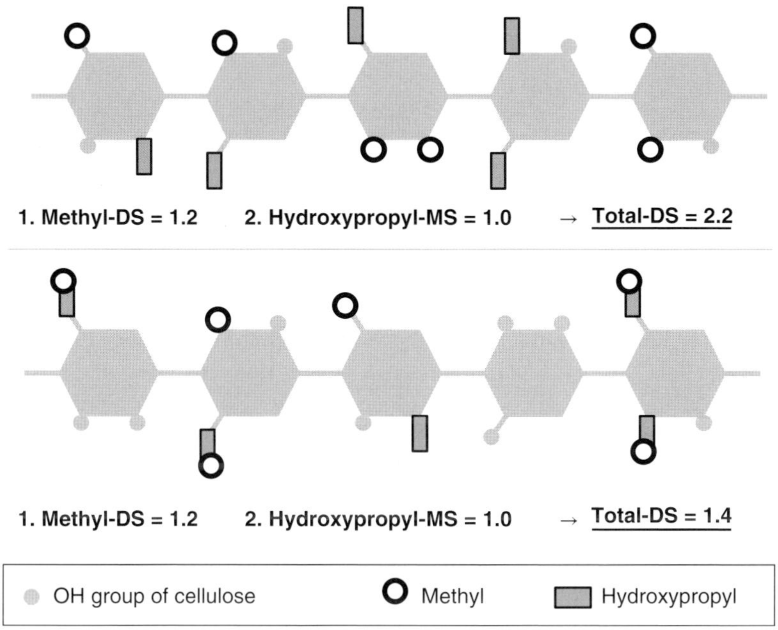

Figure 5.9 Influence of molar substitution on the total DS.

Figure 5.9 gives an example of how the degree of substitution of cellulose ether (as total DS) can vary because of the molar substitution. The number of introduced groups ('methyl' and 'hydroxypropyl)' remains constant. In each cellulose molecule, six methyl groups and five hydroxypropyl groups are 'incorporated'. The total DS can adopt values from 1.2 to 2.2.

- **Degree of polymerization (DP, molecular weight):**
  Long-chain cellulose derivatives dissolve more slowly because the viscosity development inhibits hydration. In Figure 5.10 the degree of polymerization is plotted. The presented cellulose ether has a DP of 5.
- **Particle size distribution or granulometry:**
  The finer the particles are, the faster the hydration process will be. (Note: cold-water-soluble cellulose ethers tend to form lumps!)

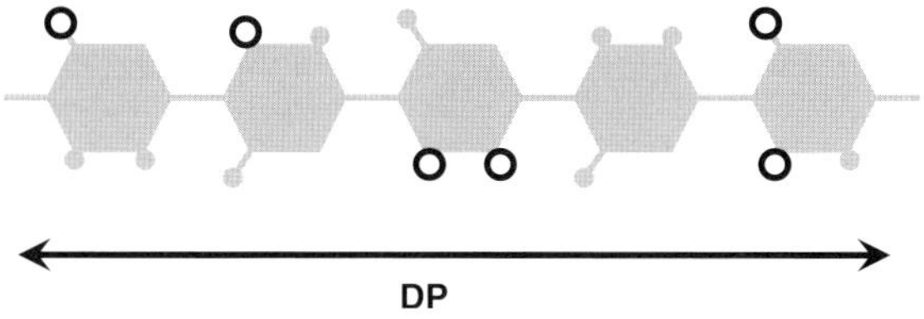

Figure 5.10 Degree of polymerization – a cellulose ether with a DP of 5.

5.3.1.2 Stages of Swelling to Dissolution

Depending on the nature of the substituent and level of substitution, cellulose ethers go through all stages of solubility.

For *hydrophilic* substituents, the scale begins with swelling and then goes with increasing degree of substitution (DS) through an alkali-soluble stage and achieves the stage of water solubility at relatively low DS (about 0.5), which is maintained up to full substitution (DS = 3). An example is Na-CMC.

While the scale of *hydrophobic* substituent starts in the same way, with the stage of swelling, it goes through a level of alkali-solubility and achieves solubility in water at a degree of substitution of more than one (DS > 1.0). With a further increase in DS, the solubility in water is reduced and solubility in organic liquids arises. Finally, at very high degrees of substitution, cellulose ethers with solely hydrophobic substituents are no longer water-soluble; they are only soluble in organic solvents.

For example, a methylcellulose having a DS of about 1.3–2.6 is soluble in water. But a methylcellulose with a DS of about 2.5–3 is only soluble in pure organic liquids or polar–nonpolar solvent mixtures.

The dissolution behaviour – in particular, the dependence of water solubility on the degree of substitution – can be explained when the molecular structure of the cellulose chains and their supramolecular linkages in a partially crystalline lattice within the fibre structure is considered. By etherification, even with hydrophobic substituents, the cellulose lattice is expanded. In this process, the hydrogen bonds are cleaved and the hydrophilic hydroxyl groups of the anhydroglucose units are accessible for hydration. With increasing substitution, the number of OH groups and thus the solubility in water decreases, while the organophilicity increases.

Another crucial factor is the distribution of substituents along the cellulose chain. A very homogeneous distribution results in a better solubility of the product as an uneven, albeit higher substitution.

Figure 5.11 shows, schematically, the stages passed through by different water-soluble cellulose ethers during swelling and dissolving:

1) The unswollen fibre of a cellulose ether shows no difference compared to a non-swollen native cellulose fibre.
2) On contact with water, the outer secondary wall (S1) is burst by the swelling middle layer (S2).
3) Through the swelling pressure of the middle layer (S2), the outer layer (S1) is loosened and at resistant regions pushed together to form narrow parallel windings (cuffs). Thereby, the characteristic spherical swelling (ball form) with the string-of-pearls like structure of the fibre is formed.
4) During further dissolution of the outer layer, spiral bands are developed that separate slowly. Moreover, there is an expansion on the cuffs. The spherical swelling shifts into a can-like swelling form.
5) The outer layer dissolves slowly.

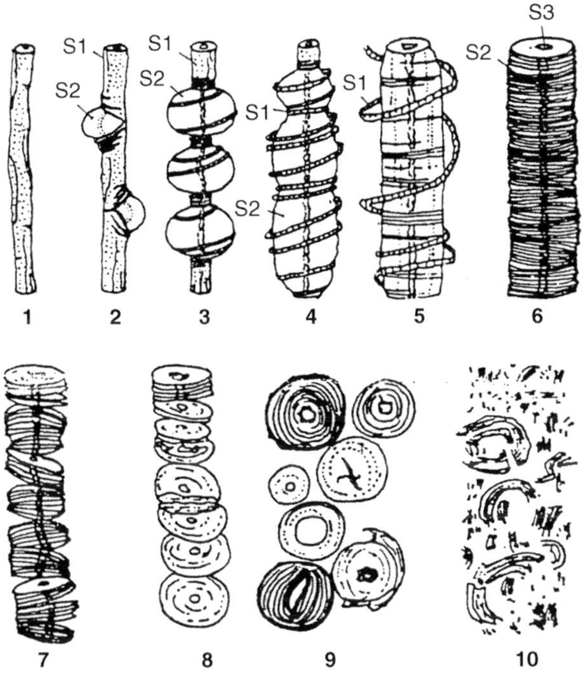

Figure 5.11 Typical stages of the swelling and dissolution process of cellulose (see text for details) [2,6].

6) The exposed middle layer reveals cross structures. The inner layer (S3) widens rapidly.
7) With formation of disc packages, transverse decomposition of the fibre begins.
8) The disc packages fall apart like money rolls in individual slices.
9) The discs themselves show concentric rings that detach from the outer layers up to the core.
10) Further breakdown into small fragments, which become visible under a microscope while dissolution progresses.

Most cellulose ethers go through these stages very quickly. A small part swells only slowly and stops at some stage without reaching full solubility. With good stirring, the final state is reached already after 90–180 min and remains constant over days.

For methylated cellulose derivatives, lower temperatures (below 20 °C (68 °F)) favour the rate of the dissolution process because at elevated temperatures the surfaces of the fibres start to swell to give a gel and thus the path for the solvent into the interior of the fibre is blocked [2,6].

5.3.2 Dissolution Behaviour of the Monoether Methylcellulose

5.3.2.1 Suitable Solvents

Methylcellulose with a medium to high degree of substitution (DS 1.3–2.6) is very soluble in cold water at 0–15 °C (32–59 °F). In this case, a transparent colloidal solution is formed. The solution viscosity is in a range between very low and high – depending on the concentration of MC and its degree of polymerization (molecular chain length). Owing to the high rate of hydration, powdery methylcellulose tends to clump when incorporated purely in pure water as solvent. Section 5.3.4 describes methods by which MC powder can be integrated lump-free into various food systems.

If the solution shows a cloudy appearance, methylcellulose is not completely hydrated. Cooling to about 5 °C (41 °F) can improve the solubility and thus the transparency.

Another reason for the opaque appearance of the solution can be an 'overdose' of the substance – there is not enough water available in the system to optimally and completely dissolve the MC. As consequence, the efficiency and functionality is reduced.

Methylcellulose (of DS 1.3–2.6) is insoluble in:

- hot water at any temperature from 40–60 °C (104–140 °F) and higher;
- ethanol (100%) – MC is only soluble in mixtures of water–EtOH;
- other ethers and chloroform;
- fats, oils, and medium-chain triglycerides (MCT).

The monoether can be dissolved in glacial acetic acid (concentrated acetic acid). Highly substituted methylcelluloses with degrees of substitution of 2.5 and above are soluble in organic solvents (pure alcohols or polar–nonpolar mixtures). Low-substituted MC types are soluble in 4% sodium hydroxide solution if their DS is between 0.4 and 0.6. With degrees of substitution below 0.4, solubility can only be achieved during freezing of dilute sodium hydroxide solution (4–8%).

The addition of sugars and salts has only a negative impact on the solubility of methylcellulose; at very high concentrations of non-polymeric ingredients there is no longer enough free water available for its hydration. Sugar concentrations above 32% are critical for methylcellulose.

At salt contents of 10–15%, the solubility is limited – depending on the temperature of the medium. The lower the water temperature, the better the hydration will be. Smooth gelation occurs at salt concentrations below 10%.

Minerals can be tolerated in large amounts.

Methylcellulose is very 'responsive' and reacts rapidly, that is, the rate of water absorption is high. Slow-acting polymers such as proteins and other hydrocolloids may be limited in their solubility and functionality when they are not able to absorb water quickly enough in sufficient volume. This physical problem can be resolved by using a modified process technology.

5.3.2.2 Physicochemical Data

Here, some physical data and properties of powdered methylcellulose and solutions prepared therefrom are listed.

To define the demarcation of monoether methylcellulose from mixed-methylated cellulose ethers, the product data of hydroxypropyl methylcellulose are used in part as HPMC shows the closest resemblance [7].

Physicochemical Data of Pure Methylcellulose (Powder)

- Appearance:
  white to off-white or cream coloured fine to coarse powder;
- molecular weight according to EU Directive on the criteria of purity:
  - 20 000–380 000 (MC),
  - 13 000–200 000 (HPMC);
- biological oxygen demand BOD_5: 0–0.1 mg O_2 per mg product;
- bulk density:
  - powder: 0.2–0.5 g cm^{-3} (12–32 lb ft^{-3})
  - granular: 0.3–0.6 g cm^{-3} (19–38 lb ft^{-3});
- decomposition temperatures:
  - browning at 200 °C (392 °F);
  - charring at >220 °C (428 °F).

Properties of Methylcellulose Solutions

- Viscosity: 5–50 000 mPa s (2% at 20 °C (68 °F) and 20 rpm Brookfield RV);
- temperature of gelation (at 2% of the viscosity level 4000 mPa s):
  - for methylcellulose: 35–55 °C (95–131 °F),
  - for HPMC: 55–90 °C (131–194 °F);
- surface tension (at 20 °C (68 °F)) of 0.1% solutions:
  - 53–59 mN m^{-1} (dyne cm^{-1}) for methylcellulose;
  - for comparison:
    - 45–55 mN m^{-1} (dyne cm^{-1}) for HPMC,
    - 72 mN m^{-1} (dyne cm^{-}) for tap water;
- interfacial tension (at 25 °C (77 °F)) of 0.1% solutions versus paraffin oil:
  - 18–24 mN m^{-1} (dyne cm^{-1}) for methylcellulose;
  - for comparison:
    - 17–29 mN m^{-1} (dyne cm^{-1}) for HPMC,
    - 45 mN m^{-1} (dyne cm^{-1}) for tap water;
- pH value (1% solution at 20 °C (68 °F)): 5.5–8.0;
- pH stability: 2–11;
- freezing point of 2% solutions 0 °C (32 °F);
- density of 2% solutions (at 20 °C (68 °F)): 1.0068 g ml^{-1};
- refractive index (2% at 20 °C (68 °F)): 1.336;
- specific heat energy: 0.93 ± 0.05 cal g^{-1} from 20 to 90 °C (68 to 194 °F).

Table 5.1 Viscosity as function of MW and DP [8].

Viscosity of solution at 2% and 20 °C (68 °F), BF (mPa s)	Intrinsic viscosity (dl g^{-1})	Average molecular weight (kDa)	Average degree of polymerization (DP)
5	1.2	10	53
10	1.4	13	70
40	2.1	20	110
100	2.7	26	140
400	3.9	41	220
1500	5.7	63	340
4000	7.5	86	460
8000	9.3	110	580
15 000	11.0	120	650
19 000	12.0	140	750
40 000	15.0	180	950
75 000	18.4	220	1160
100 000	21.0	246	1280

5.3.2.3 Molecular Weight and Viscosity

When methylcellulose is dissolved in water, or in any other suitable solvent, the viscosity of the solvent increases. This thickening effect is dependent on the degree of polymerization (DP), which corresponds to the length of the cellulose chain and is therefore coupled to the molecular weight (MW). The higher the degree of polymerization (and the resulting molecular weight), the higher is the viscosity of the solution at a constant concentration. Table 5.1 gives an overview of the viscosity as function of molecular weight and degree of polymerization.

5.3.2.4 Rheological Profile – Factors Influencing the Cold Viscosity

General

In the preparation of aqueous solutions of the cellulose ethers methylcellulose and HPMC, several factors influence the rheology and thus control the viscosity development and thickening effect. The dissolution behaviour of the non-ionic cellulose derivatives methylcellulose and HPMC differs significantly in some aspects from that of the anionic carboxymethylcellulose (CMC). Differences are always detected and measured when the properties are largely controlled by the substituted groups and the effect and importance of the cellulose chain is lessened.

Owing to the neutrality of MC and HPMC molecules, resulting from uncharged methyl and hydroxypropyl groups, the influence of pH value is considered to be low. In addition, the presence of monovalent salts and polyelectrolytes has no, or only minor, impact on the concentrations normally used. However, the temperature-dependent rheology is very complex for methylated cellulose derivatives. When methylcellulose or HPMC are dissolved in pure water, the clear smooth solutions obtained are pseudoplastic and non-thixotropic. The pseudoplastic behaviour of methylcellulose is a function of molecular weight, wherein molecular weight and pseudoplasticity have a proportional correlation (see Figure 5.14) [3].

The type and level of substitution do not directly affect the rheological properties of MC and HPMC solutions at 'cold' temperatures below the gelation point. However, the degree of substitution has, through its contribution to molecular weight, an indirect influence on the flow behaviour.

In this section, the effect of substance concentration, morphology/particle size, shear force, and water with respect to solvent temperature are discussed in more detail.

Relationship between Concentration and Viscosity

Viscosity is a function of concentration. The higher the applied quantity of hydrocolloid or cellulose derivative, the bigger is the thickening performance in an aqueous system.

The relationship between concentration and viscosity of the resulting solution does not follow logarithmic laws, but it is a function of the eight-root. As a rule of thumb, doubling the quantity of MC will increase viscosity by a factor of about 6–10.

Figure 5.12 illustrates the correlation between concentration and thickening effect for MC types having different chain lengths. The number next to each curve represents the viscosity level of methylcellulose (measured as a 2% solution). The measurements were carried out at room temperature (20 °C (68 °F)).

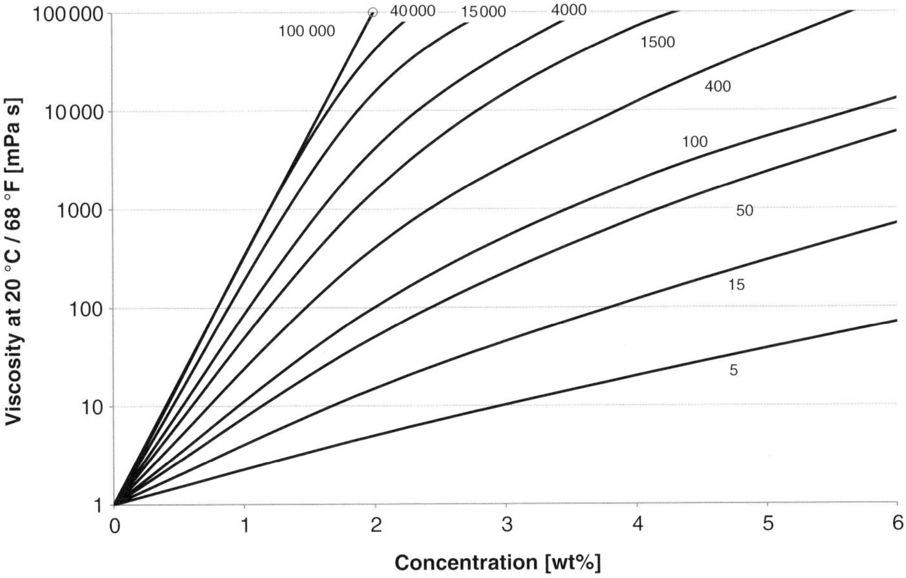

Figure 5.12 Relationship between concentration and viscosity for methylcellulose (see text for details).

These curves can be directly transferred to HPMC and Na-CMC. The length of the cellulose chain is the parameter that significantly controls the cold viscosity.

Influence of Morphology

The term morphology describes the physical form and size of cellulose derivatives.

Cellulose-based hydrocolloids are morphologically very different to starches. Starch particles are usually much smaller and – depending on the applied drying method – of smooth-round or jagged-rough and squarer shape. Particles of etherified celluloses (but also powdered cellulose and MCC) further have an elongated and three-dimensional (3D) fibre shape with wave-like surface structures and they are several times larger than starch particles.

The morphology of a cellulose ether has a crucial role in its functionality in applications.

In addition to chain length and level of substitution, the morphology controls, as a parameter of 'particle size distribution', the time point of hydration, and therefore the extent of dissolution and solubility, and ultimately the course of viscosity development – under the assumption that there is enough free water of corresponding temperature present in the system.

Figure 5.13 indicates how the morphology and particle size of a cellulose ether determine its viscosity development. The developing viscosity is represented as torque, which is the internal force and resistance of the solution. A powder and granules of the same viscosity level of a methylcellulose were measured. Lump formation, which usually occurs when adding powder fractions undiluted to water, is prevented by prior wetting of the particles with small amounts of propylene glycol.

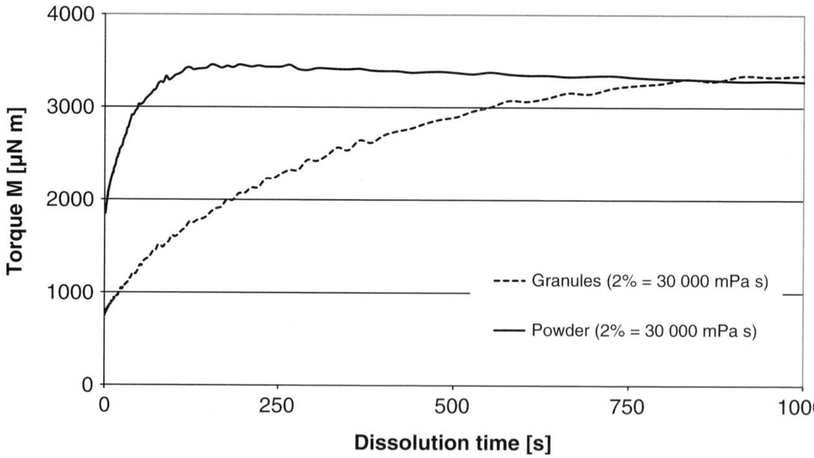

Figure 5.13 Relationship between morphology resp. particle size and viscosity.

The level of substitution has no impact at cold temperatures (10 °C (50 °F)) – the curves would run for HPMC or CMC at the same viscosity level with the same slope.

In a food application, the development of viscosity shifts mostly to longer dissolution times, because the amount of water present is lower, and the cellulose derivatives compete with other substances for available free water.

The same curve occurs when a product is stirred during hydration at different speeds.

The solid line in Figure 5.13 (powder) shows the dissolution and viscosity development of a powder at high stirring speed. The dotted line (granular) would represent the same powder at a low shear rate.

Note that methylated cellulose ethers (MC, HPMC), owing to their surface activity, tend to form a foam and, consequently, fine air bubbles are incorporated into the solutions. In particular, high-viscosity types 'degas' due to their strong thickening power only very slowly, so that moderate stirring speeds are recommended to prevent foam formation from the outset.

Effect of Shear Forces

Below the specific gelation temperature, the rheology of aqueous methylcellulose solutions is controlled by applied shear force. Depending on concentration and molecular weight or degree of polymerization, there are specific viscosity values at corresponding shear rates. The effect of warm temperatures and the presence of other dissolved substances are neglected here. All data refer to pure solutions of methylcellulose in water. The rheology of aqueous solutions of MC or HPMC is neither affected by substitution type nor by level of substitution (DS) at temperatures below the gelation temperature. Consequently, viscosity values can also be transferred to the analogue iso-viscosity types (comparable molecular weight and comparable chain length) of HPMC, HPC, and Na-CMC as long as concentration and temperature remain constant. At most shear rates, solutions of methylcellulose exhibit pseudoplastic flow behaviour (thixotropy). The shear-dependent viscosity reduction is fully reversible at moderate temperatures ($T < T_{\text{Gelation}}$) and neutral pH values. The quantity of viscosity reduction, the so-called shear-thinning, increases with increasing molecular weight (and DP) and, respectively, with increasing concentration.

At low shear rates MC solutions show nearly Newtonian flow characteristics. The shear stress at which a solution has this behaviour increases with decreasing concentration and with shorter cellulose chains. After a rest period without input of mechanical power, the initial viscosity returns completely. Upon heating, methylcellulose solutions form a 3D gel structure and as result of this gelling an extremely thixotropic flow is observed. At very high quantities of use, short-chain MC or HPMC molecules may show thixotropic flow behaviour even below their specific gelation temperatures. Figure 5.14 visualizes how the viscosity of methylcellulose solutions with different degrees of polymerization or molecular weight varies with shear treatment. All MC types

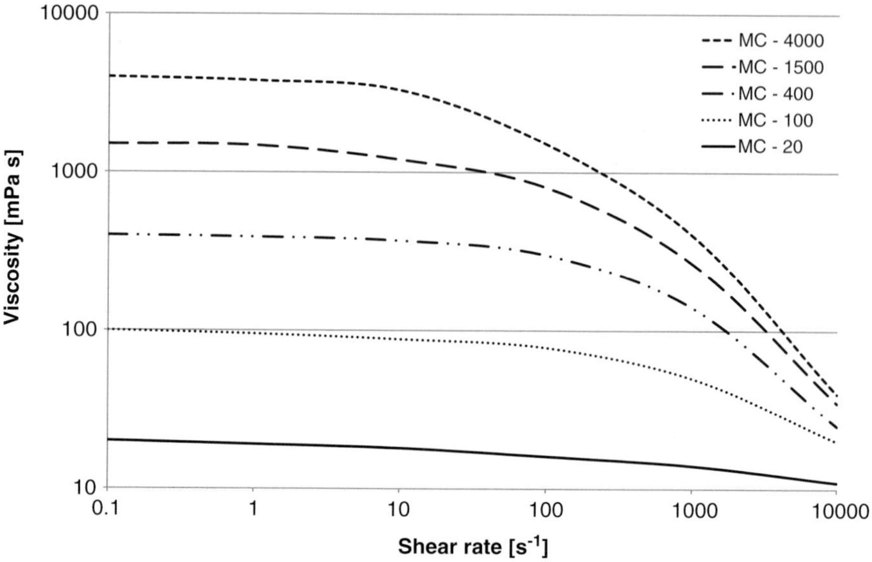

Figure 5.14 Effect of shear forces on viscosity of various methylcelluloses (2%).

are used in the same concentration (2%). Measurements were carried out at room temperature (20 °C (68 °F)).

Figure 5.15 illustrates how shear forces act on methylcellulose solutions of different concentrations. The lower the hydrocolloid concentration and initial viscosity, the lower is the impact of shear forces.

Effect of Dissolution Temperature

For methylated cellulose ethers (MC, HPMC), the dissolution temperature is of central importance. Above the specific gelation temperature, these hydrocolloids are 100% insoluble in water, they act as solid particles – in hot (boiling) water there is only a dispersion of MC and HPMC. Particles must be kept in suspension by shear forces. Without agitation, the solids sink to the bottom of the vessel and form sediments.

There is no gelation because the necessary prior hydration of methylcellulose or HPMC is missing. Dissolution begins with solubility occurring when the temperature of the solvent (water) is below the critical gelation temperature, and the specific hydration temperature is achieved by cooling. Viscosity development is therefore dependent on level of substitution and on type of substitution, that is, the content of methyl and hydroxypropyl groups. Section 5.3.4 describes the hydration process. The specific dissolution temperatures are listed in Table 5.7.

Figure 5.16 illustrates the influence of dissolution temperature on viscosity development. Obviously, a solids dispersion of HPMC (type E) requires significantly lower cooling energy for a higher cooling temperature than does a slurry of methylcellulose (type B). This HPMC starts dissolving at 95 °C (203 °F); its full

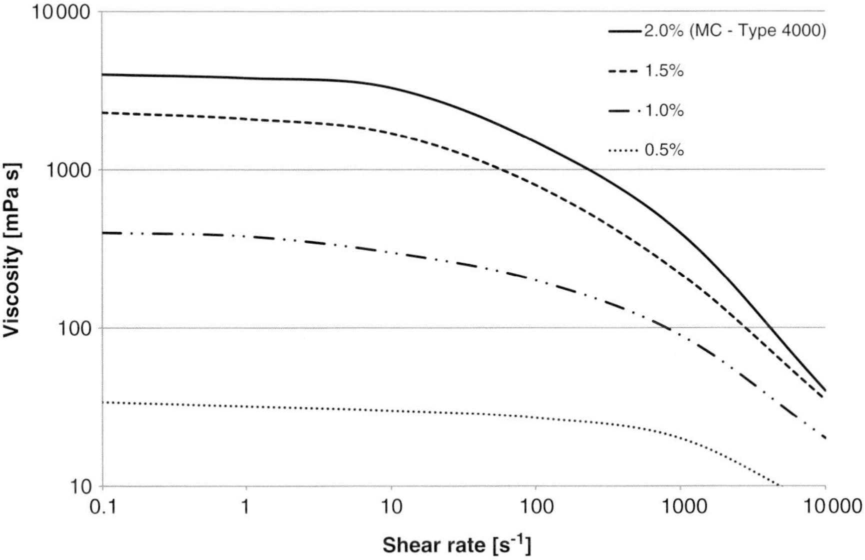

Figure 5.15 Combined effect of shear force and concentration at MC.

viscosity is achieved at about 50 °C (122 °F). MC will remain as a solid – undissolved – form until 20 °C (68 °F). Dispersions of HPMC types C and D would be located approximately in the central area between the two curves, and would also require less cooling energy than methylcellulose.

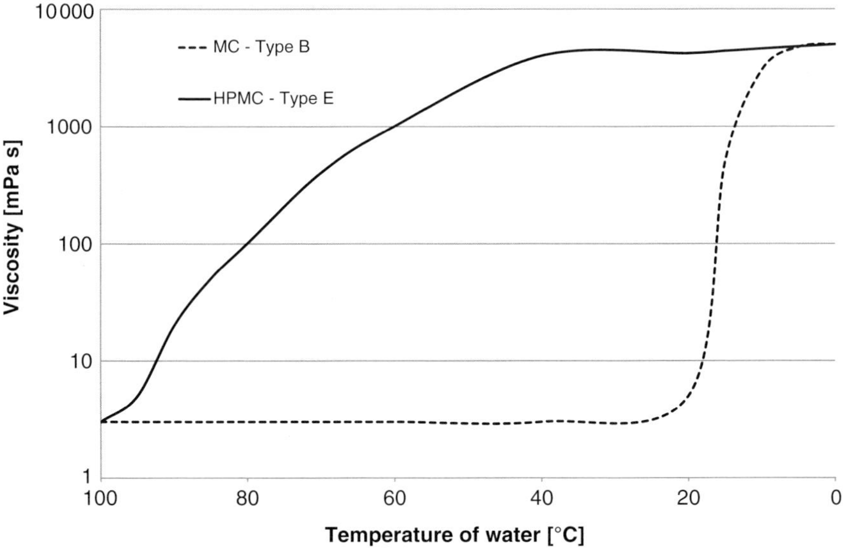

Figure 5.16 Influence of dissolution temperature on viscosity development.

5.3.2.5 Surface Activity of Aqueous Solutions

Methylcelluloses and hydroxypropylmethylcelluloses (HPMCs) differ from other hydrocolloids and cellulose ethers in that they are surface active. MC and HPMC reduce the surface tension of aqueous solutions and also reduce the interfacial tension between two immiscible phases. Thus, these hydrocolloids support the emulsification of two- and multi-phase systems. The lower the surface tension of a solution, the higher is the surface activity. For all types of MC and HPMC, the surface tension of their solutions decreases with increasing temperature. A minimum in surface tension is measured between 30 and 50 °C (86–122 °F). The minimum is specific for a particular MC or HPMC with a defined type and level of substitution for a defined molecular weight and concentration.

The relationships depicted in Figure 5.17 are believed to be a consequence of the unusual thermal dependence of methylcellulose interfacial hydration. Methylcelluloses exhibit inverse solubility characteristics characterized by an exothermic heat of dilution. When heated, aqueous solutions undergo gradual viscosity reduction followed by the development of a gel with a high yield value. If the surface tension of a methylcellulose or HPMC solution is determined as a function of temperature, the observed value using a 'Du Noüy' tensiometer shows a gradual reduction in magnitude followed by a significant increase. The curve labelled *cumulative surface force* in Figure 5.17 typifies the behaviour observed. This cumulative surface force is the result of two

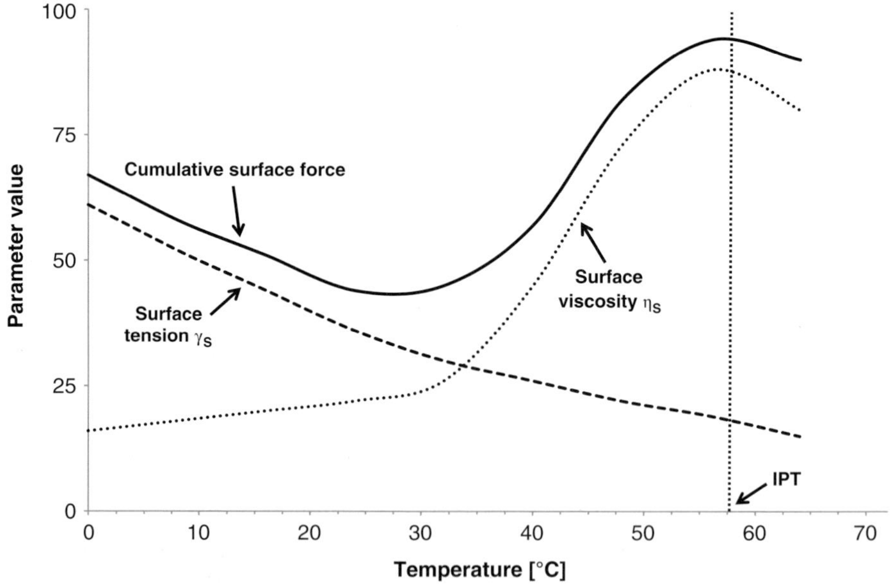

Figure 5.17 Surface tension of methylcellulose solutions as measured with a Du Noüy tensiometer [8].

primary components, the true *surface tension* of the solution (γ_S) and the surface thermal gelation contribution, which can also be labelled as *surface viscosity* (η_S). Hydrocolloid solutions, including MC and HPMC solutions, normally exhibit gradual decreases in surface tension, that is, increased surface activity, with increasing solution temperature. However, as the temperature of a methylcellulose solution is increased, polymer dehydration ensues. At the air–water interface, re-entry of dehydrated polymer into the bulk phase is inhibited. The onset of partial dehydration occurs at temperatures as low as 5–10 °C (41–50 °F) for methylcellulose; for most commercial HPMC products it is not observed until the temperature is above 25 °C (77 °F). The intrinsic surface activity of methylcellulose (or HPMC) coupled with interfacial dehydration causes the polymer concentration at the air–water interface to increase at the expense of polymer dissolved in the bulk phase.

It appears that this interfacial sorption phenomenon is not limited to the classic monolayer, but rather continues monotonically with time or increasing temperature. With sufficient surface film ageing, the interfacial film polymer concentration becomes exceedingly high (>5 wt%) even though the original bulk solution might have been very dilute (e.g. below 0.1 wt%). The onset of surface gelation occurs when a sufficient amount of polymer accumulates at the surface to cause development of mechanical structure features observed for bulk-phase thermal gelation. As with bulk-phase thermal gelation, the interfacial phenomenon is probably the result of hydrophobic associations between methoxyl substituents. These surface films have only recently been studied, but they are known to possess unusually high surface viscosities with characteristic elastic moduli.

Interfacial phenomena become predominant above room temperature prior to reaching the incipient precipitation temperature (IPT). In fact, even at MC solution temperatures as low as 30 °C (86 °F), the interfacial sol–gel transition proceeds to the extent that film development can be observed within minutes.

The overall effect of the various surface forces is to provide the observed cumulative surface force curve. Evidently, one might erroneously conclude from 'Du Noüy' data that the surface tension of these solutions is increasing when, in fact, it is declining. The development of mechanical film properties apparently proceeds until perturbed at a sufficiently high temperature by the formation of a suspended, poorly hydrated precipitate or by the process of bulk thermal gelation. This unique combination of surface properties promotes the formation of emulsions and foams having unusual stability at elevated temperatures [8].

Tables 5.2 and 5.3 give values for the surface and interfacial tensions of various methylated cellulose derivatives in accordance with the DS and molar substitution (MS). Table 5.2 shows how the surface tension of MC and HPMC solutions with different concentrations varies when the chemical composition – expressed as methoxyl content (DS) and hydroxypropoxyl content (MS) – is considered.

In summary, the surface tension decreases with increasing hydrocolloid concentration and with increasing hydroxypropyl content, which is to be regarded

Table 5.2 Surface activity of MC and HPMC solutions [8].

Substitution of cellulose		Surface tension γ_S (dyne cm^{-2}) at concentration (%)			
DS methoxyl	**MS hydroxypropyl**	**0.01%**	**0.10%**	**0.25%**	**0.50%**
1.86	—	56.4	54.0	—	—
1.76	0.14	53.4	52.0	—	—
1.53	0.26	52.2	49.7	48.3	46.8
1.84	0.24	48.5	47.4	46.8	46.4
1.38	0.75	47.4	45.4	43.9	—

Table 5.3 Interfacial tension of MC and HPMC solutions versus paraffin oil [8].

Substitution of cellulose		Interfacial tension (IFT) (dyne cm^{-2})
DS – Methoxyl	**MS – hydroxypropyl**	**At 0.1% and 25 °C (77 °F)**
1.85	—	19–23
1.75	0.15	19–23
1.85	0.25	18–19
1.50	0.25	26–28
1.40	0.75	26–30

as a characteristic parameter. This leads to the conclusion that HPMC types with high HP content have the greatest surface activity and are therefore particularly suitable for the formation of foams. The lowest activity is measured for high methylated methylcelluloses and HPMC types; the latter should have the lowest possible HP content.

Table 5.3 gives values for the interfacial tension (IFT) of various MC and HPMC solutions relative to paraffin oil. Here, too, the influence of chemical composition (Me and HP content) is used as parameter.

The interfacial tension of a 0.1% methylcellulose solution is between 19 and 23 dyne cm^{-2}; the commercially available HPMC products have values between 18 and 30 dyne cm^{-2}. Increasing hydroxypropyl content leads to a significant increase in interfacial tension. The sum of substituted groups ('Me–DS' + 'HP–MS') cannot be used as an indicator of the expected surface or interfacial tension, since a high methoxyl content counteracts this effect.

5.3.2.6 Mixtures and Blending for Viscosity Adjustment

Blending of different viscosity grades of methylated cellulose ethers to set an intermediate viscosity is possible in principle but, from a technological perspective, is a critical step, representing a significant risk for the production and product safety. If a viscosity level is to be prepared by blending, the starting materials should be as similar as possible – have a comparable substitution, and not drift

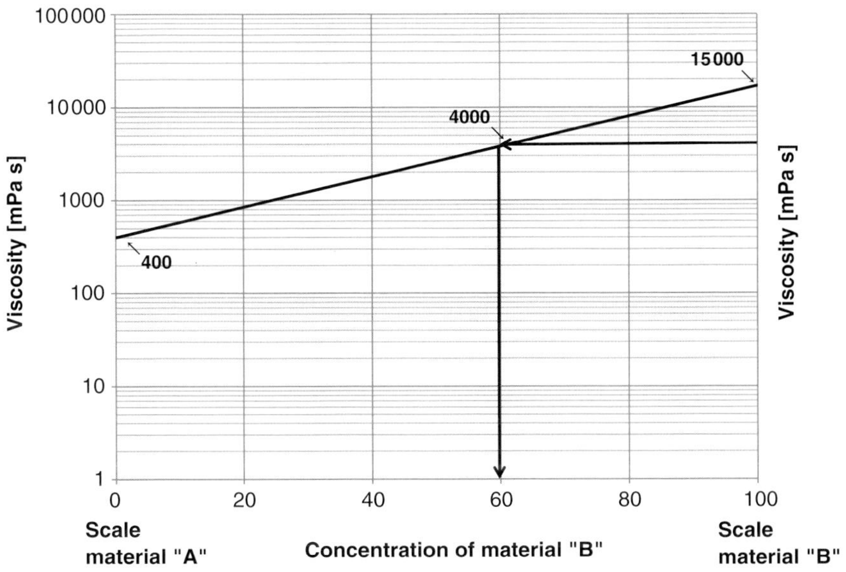

Figure 5.18 Blending chart with example.

too far apart even in terms of chain length. Technologically, however, an iso-viscous mixture can nevertheless lead to surprises regarding functionality compared to a pure (unmixed) product.

The relationship between viscosity and concentration is not linear, but exponential. Therefore, the viscosity of a mixture cannot simply be calculated as the arithmetic mean.

The illustrated composition on the blending chart given in Figure 5.18 is used as an example:

From starting material 'B' (2% = 15 000 mPa s) and starting material 'A' (2% = 400 mPa s) a target product 'C' with a viscosity of 4000 mPa s (at 2%) is to be produced by mixing. Mathematically and graphically this arises in a mixture of 60% material 'B' and 40% material 'A'.

The mathematical relationship can be expressed by the following formula. It is based on the Arrhenius equation and provides a correlation between viscosity and polymer concentration:

$$V_S{}^{1/8} = nV_1{}^{1/8} + (1 - n)V_2{}^{1/8}$$

V_S = viscosity sought;
n = wt% of the first component (1) in the blend;
V_1 = viscosity of the first component (1) in the blend
V_2 = viscosity of the second component (2) in the blend

The Blending Chart is used in the Following Way

The points corresponding to the viscosities of the two starting materials are selected on the two vertical scales (A and B) and connected with a line. The point corresponding to the desired final viscosity is then located on one of the vertical scales and a horizontal line drawn from it to the first line drawn. A vertical line drawn from this intersection to the bottom scale (x-axis) will reveal the percent of the right-hand (Scale B) material needed to make up the blend.

This mixing chart may be also used for other water-soluble cellulose ethers such as HPMC, HPC, and CMC. For cellulose-based products with very limited solubility in water, which only swell partially – for example microcrystalline cellulose (MCC), ethylcellulose (EC), or croscarmellose – the mixing chart has no relevance. It provides no meaningful results, since the degree of solubility cannot be estimated theoretically.

5.3.2.7 Available Solid Forms of MC

Methylcellulose gum products are available in powder and granular form from different manufacturers. An indication of particle size is:

- powder: on 0.125 mm (ASTM no. 120): max. 8%;
- powder: on 0.8 mm (ASTM no. 20): max. 1%;
- granular: on 0.2 mm (ASTM no. 70): 55–90%.

The primary benefit of powder products is rapid dissolution. Granular products have reduced dusting tendency and are more easily dispersed. Both powder and granular products may be treated with dispersing agents to make dissolution easier, but these products cannot be used in foods or in contact with foods [7,8].

5.3.3 Gelation of Methylcellulose

5.3.3.1 Basics and Relationships

It is characteristic of methylcellulose and hydroxypropylmethylcellulose (HPMC, MHPC) that the viscosity of their solutions initially decreases with increasing temperature, and then at a temperature dependent on substituent and degree of substitution a fully reversible gelation occurs [9].

To form a typical gel, a concentration of at least 1.5 wt% of cellulose ether is necessary. The lower the degree of polymerization or the shorter the cellulose chain, the higher is the required quantity of MC or HPMC. For example, 6% of methylcellulose (type B) of viscosity level 15 mPa s (measured at 2%) is necessary to bring this product to form a cut-resistant gel. By heating pure solutions with lower concentrations of this low-viscosity MC type, only a turbid solution with slightly increased viscosity is produced (compared to the viscosity of the initial solution).

MC and HPMC thus behave inversely to the classic gel formers gelatine and κ-carrageenans. At room temperature, there is a viscous solution that gels

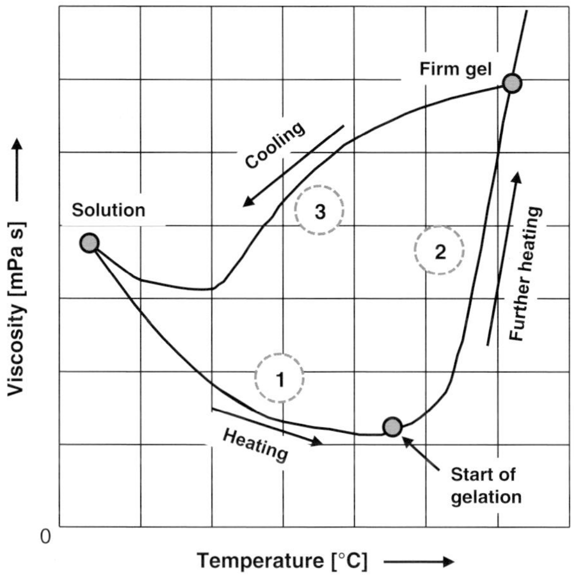

Figure 5.19 Temperature-dependent viscosity and gelation of methylcellulose [8].

('flocculates') upon heating and then, upon cooling, reverts 100% reversibly back into the viscous state. Figure 5.19 shows the behaviour of methylated cellulose ethers (methylcellulose, hydroxypropylmethylcellulose) during heating and cooling.

During heat treatment, there are three phases, which are briefly described here:

Phase 1:
At low temperatures, methylcellulose is present fully hydrated. The solution is absolutely clear and transparent when there is sufficient water for solubilization of MC. With increasing temperature, the polymer chains lose some of their physically bound water and viscosity drops. The decrease in resistance is measurable with viscometers, but is usually not visible with the naked eye.

Phase 2:
Upon further heating, dehydration of the polymer chains leads to intensive polymer–polymer interactions. The solution begins to gel at temperatures between 38 and 70 °C (100 and 158 °F). The mechanism of gelation is described below in more detail.

Phase 3:
Methylated cellulose ethers form gels when heated and go back into solution when the resulting gels are cooled. Above the specific gelation point, methylcellulose and HPMC are insoluble. If the gels are stirred during their development, a flocculation effect is observed. Then, white string-like particles and flakes 'swim' in the solution. Flocculation is – like gel formation – a completely reversible process. Upon cooling, the strings and flakes dissolve again and become invisible. The initial viscosity value is returning.

The curve in Figure 5.19 shows that there is a hysteresis between gelation and the so-called 'melting back' of the gel. It means that the melting of a gel back into solution takes place at lower temperatures than the development of this gel. For example, the gel of a 2% solution of methylcellulose having a viscosity of 4000 mPa s is formed at 50–55 °C (122–131 °F) and it melts back at about 25–30 °C (77–86 °F). Below 20 °C (68 °F), a viscous solution is available again, as it was before heating.

These gels of methylcellulose are technologically important since they are 100% reversible and can be made as often as desired and melted back again.

The gelation temperature is controlled and influenced by the following factors:

- type and level of substitution:
  - content of methyl groups (Me) in the molecule,
  - content or presence of hydroxypropyl groups (HP),
  - ratio of 'Me' to 'HP' groups;
- concentration of cellulose ether in solution;
- total solid content in the system;
- type of solvent;
- presence of additives that increase or decrease the gelation point.

Hydrophobic methyl groups are responsible for gelation. Therefore, the higher the quantity of Me groups in a methylcellulose or HPMC, the lower is the gelation temperature, and the stronger is the resulting gel. Or, vice versa, the higher the hydroxypropyl substitution, the softer is the gel, and the higher is the temperature at which gelation occurs. Hydroxyalkyl substituents stabilize the hydration shell around the macromolecule, and thereby raise the gelling temperature. By changing the ratio of 'methyl groups : hydroxypropyl groups', the gelation temperature of HPMC types can be varied within wide limits [9].

Table 5.4 provides an overview of commercially available methylated cellulose ether types with their chemical composition.

Figure 5.20 illustrates the five different substitution types of methylcellulose and HPMC commercially available.

Table 5.5 presents the influence of type and level of substitution on the temperature of gelation and the properties of the resulting gels.

Figure 5.21 describes the correlation between hydroxypropyl content (HP-MS) and gel strength. Investigations were carried out with solutions of methylcellulose and HPMC at 2% concentration. The quantity of methoxyl groups, measured as degree of substitution (Me-DS), ranges from 1.6 to 1.9. Solution viscosities are between 500 and 10 000 mPa s. As already shown in Table 5.5, it is clear that gel strength is a function of HP content – with reversed proportional behaviour. The higher the hydroxypropyl proportion in the molecule, the lower is the gel strength, that is, the softer the resulting gel will be.

Table 5.4 Overview of commercially available methylcellulose and HPMC grades [3].

Cellulose ether[a),b)]	Methyl content (%)	HP content (%)	Degree of substitution 'methylation' (DS)	Degree of substitution 'hydroxy-propylation' (MS)
Methylcellulose (A)	31–33	—	>1.8	—
Methylcellulose (B)	28–31	—	1.6–1.8	—
HPMC – type C	28–30	8–12	1.63–1.85	0.1–0.3
HPMC – type D	28–30	4–8	1.0–1.8	0.1–0.2
HPMC – type E	20–24	6–10	1.1–1.4	0.1–0.3

a) EU Directive for methylcellulose: 25–33% Me, max. 5% hydroxyethoxyl.
b) EU Directive for HPMC: 19–30% Me, 3–12% HP.

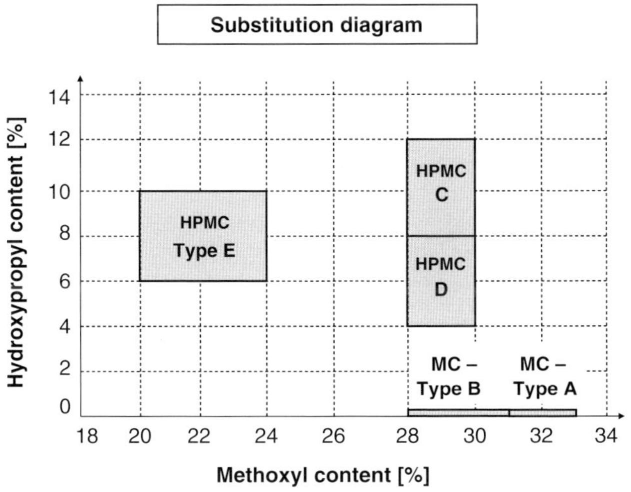

Figure 5.20 Different substitution types of MC and HPMC.

The concentration of cellulose ether and its viscosity (degree of polymerization) also have an effect on gelling characteristics. The higher the concentration and the viscosity or the chain length of the polymer, the lower is the resulting gelation temperature.

Example

Methylcellulose having a viscosity of 4000 mPa s (at 2%) gels at 50 °C (122 °F).

Increasing the concentration to 3% lowers the gelling temperature to 47 °C (116.6 °F).

On dilution of a solution to 0.5% methylcellulose, the setting temperature rises to 65 °C (149 °F).

Table 5.5 Relationship between Me-/HP-content, gelation temperature, and strength.

Cellulose ether[a), b)]	Methyl content (%)	HP content (%)	Temperature, when gelation starts (at 2% conc.) [°C (°F)]	Temperature of melt-back [°C (°F)]	Strength of gel
Methylcellulose (A)	31–33	—	38–44 °C (100–111 °F)	15 °C (59 °F)	Very firm
Methylcellulose (B)	28–31	—	50–55 (122–131)	25 (77)	Firm
HPMC – type C	28–30	8–12	58–64 (136–147)	35 (95)	Semi-firm
HPMC – type D	28–30	4–8	62–68 (144–154)	35 (95)	Semi-firm
HPMC – type E	20–24	6–10	70–90 (158–194)	50 (122)	Soft

a) EU Directive for methylcellulose: 25–33% Me, max. 5% hydroxyethoxyl.
b) EU Directive for HPMC: 19–30% Me, 3–12% HP.

Figure 5.22 illustrates the relationship between concentration of cellulose ether and level of gelation or flocculation temperature. The inclination of the curve is dependent on viscosity level and content of methyl groups.

A decisive role is taken by the total solids content of the food product. The presence of sugars, proteins, and other polymers in the formulation can lower the gelation temperature of this example-methylcellulose so far that a gel is formed with a concentration of only 0.5% MC already at 40 °C (104 °F). A high concentration of salt (NaCl, phosphates, carbonates, other polyelectrolytes) in the system can also affect gelation. Owing to the reduced availability of water, the gelation temperature of methylated cellulose ethers (MC, HPMC) decreases.

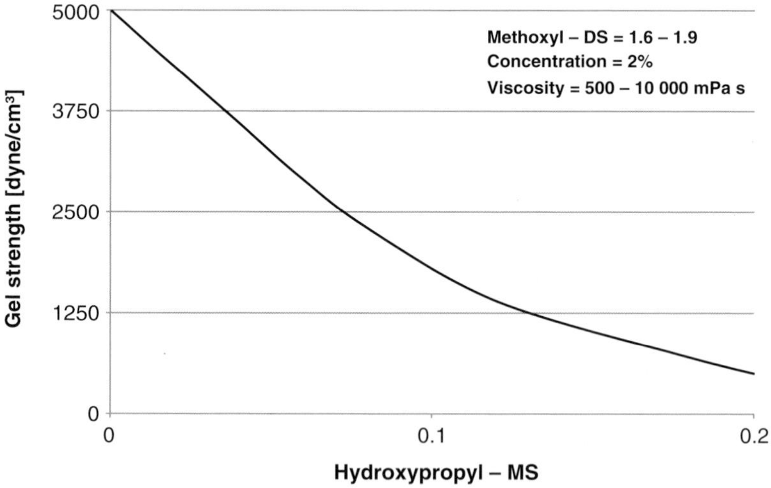

Figure 5.21 Gel strength as function of hydroxypropyl-MS for methylcellulose/HPMC [8].

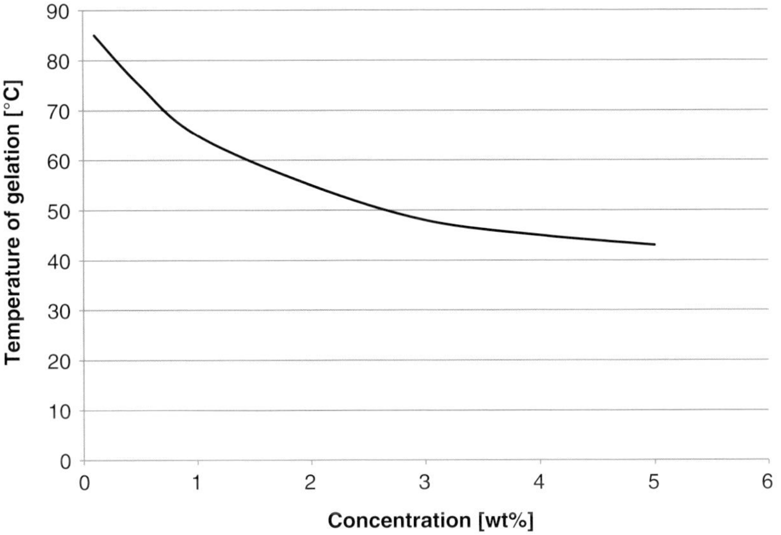

Figure 5.22 Concentration-dependent gelation temperature of methylcellulose [7]. Reprinted with kind permission of Ashland Inc.

For example, a 2% HPMC solution, which gels/flocculates in pure water at 65 °C (149 °F), will form a gel already at room temperature on addition of 4% sodium carbonate (Na_2CO_3) or sodium phosphate (Na_3PO_4). The influence of different additives on the temperature of gel formation is shown in Table 5.6. Gelation temperature is determined using an amylograph.

Table 5.6 Effect of additives on gelation temperature of MC and HPMC solutions (2%).

		Gelation temperature [°C (°F)]		
Additive	**Concentration (%)**	**Methylcellulose 1500 mPa s – type B**	**HPMC 50 mPa s – type D**	**HPMC 4000 mPa s – type E**
No addition	—	54 (129)	64 (147)	86 (187)
$MgCl_2$	5	43 (109)	54 (129)	68 (154)
Na_2CO_3	5	Salting out	Salting out	Salting out
Na_3PO_4	2	33 (91)	42 (108)	53 (127)
Sucrose	5	52 (126)	67 (153)	84 (183)
Sucrose	20	45 (113)	60 (140)	63 (145)
Sorbitol	20	30 (86)	45 (113)	50 (122)
Glycerol	20	35 (95)	60 (140)	65–70 (149–158)
Ethanol	20	>75 (167)	>75 (167)	>75 (167)
Propylene glycol	20	60 (140)	>80 (176)	>80 (176)

The investigations carried out in Table 5.6 allow some conclusions:

- Most polyelectrolytes as well as sugar and glycerol reduce the temperature of gelation, because they have a higher affinity for the freely available water and dehydrate the cellulose ethers as a result.
- The reduction of gelation temperature is a function of ion type, which is provided by the additive used. Generally, the size of the ion charge correlates to the reduction of temperature of gelation.
- Some liquids such as ethanol and propylene glycol cause an increase in temperature of gel formation. Although these solvents have a stronger affinity to the medium water than the cellulosic polymers, they increase their solubility and thus the temperature of gelation. The efficiency of gel temperature-raising additives differs – depending on the type of substitution and polymer length of the methylated cellulose ether. Section 5.3.3.2 takes a closer look at this relationship.

In addition, the gelation temperature is influenced by two other factors:

1) a fast rate of heating raises the temperature of gelation;
2) an increasing shear rate also results in a gelling at higher temperatures.

In addition to temperature-induced gelation in solutions, methylcellulose and HPMCs with short chain lengths may also indicate a gelation in the interfaces between two phases as a result of their surface activity. 'Small' molecules of cellulose ethers migrate into the air–water interface, where they can form a gel even at room temperature due to their high relative concentration (due to the small layer thickness) at low absolute concentration (from 0.01 wt% on). The gel adds greatly to the long-term stabilization of this highly sensitive system.

5.3.3.2 Influence of Solvent on Gelation

Solvents such as ethanol and propylene glycol raise the temperature of gelation of methylcelluloses and HPMCs. The reason for this is that these fluids improve the solubility of substituted cellulose molecules, and thus move the tendency to flocculation to higher temperatures. The extent of the effect of the solvent on gelation depends on the chemical structure of the solvent itself, the quantity used, and the type and level of substitution of the respective cellulose ether.

For example, the addition of 20% propylene glycol has the following effects on gelation of a 2% solution of MC or HPMC (see also Table 5.6 and Figure 5.23):

- methylcellulose – type A or B: increase in gelation temperature by about 6 °C (plus 11 °F);
- HPMC – type D: increase in gelling point by 16 °C (plus 29 °F);
- HPMC – type E: reduction of gelation temperature by at least about 5 °C (minus 9 °F).

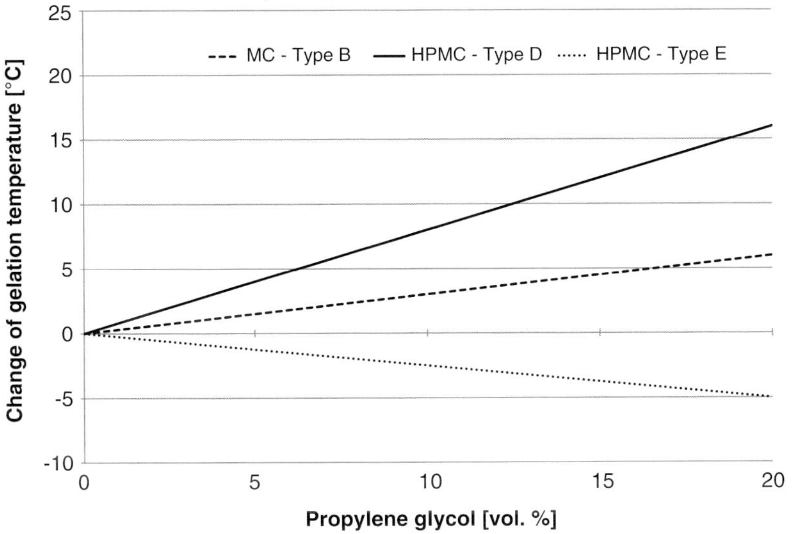

Figure 5.23 Effect of propylene glycol on gelation temperature of MC and HPMC (2%).

Figures 5.23 and 5.24 present the relationship between the amount of solvent present and its effect on temperature of gel formation for selected MC and HPMC types.

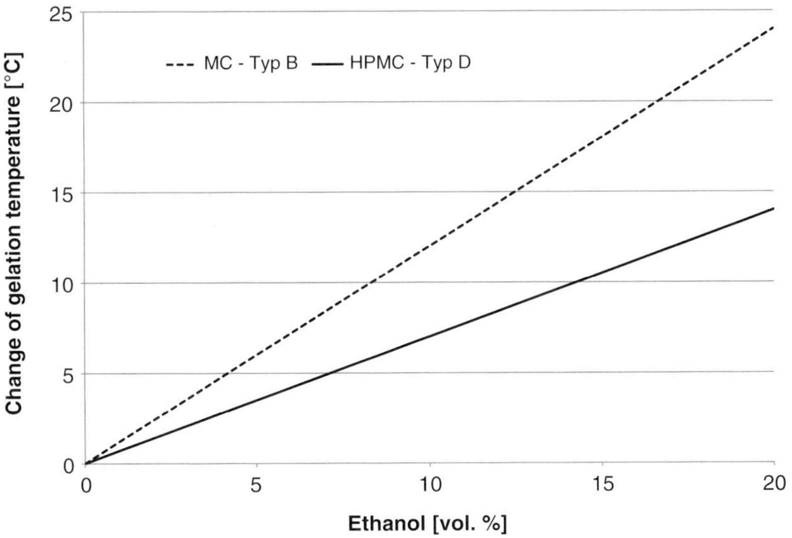

Figure 5.24 Effect of ethanol on gelation temperature of MC and HPMC (2% solution).

Ethanol has the effect of gel-point increase especially on hydrophobic methyl groups, as is clearly visible in Figure 5.24. Methylcellulose forms a gel in mixed solvent systems having an ethanol content of 20 wt% at significantly higher temperatures than in pure water. An effect of ethanol on hydrophilic HPMC also exists, but is much smaller than on MC.

5.3.3.3 Mechanism and Cause of Heat-Induced Gelation

In colloid chemistry, the terms 'sol' and 'gel' as opponents or antagonists are closely related. A 'sol' is defined as a molecularly-disperse hydrated substance in a solvent (e.g. a homogeneous polymer solution). It leads to formation of a so-called 'gel' when partial phase separation takes place in this fluid after the bundle of polymers has associated with each other without the solvent being completely excluded. A gel can thus be understood as a swollen network in which polymer chains are associated with each other to form a 3D network by means of physical adhesion forces and/or chemical crosslinking. These physicochemical linkages mean that gels exhibit the characteristics of elastic solids [10,11].

In 1935, E. Heymann first described the formation of such gels upon heating of aqueous methylcellulose solutions. On heating, there is a reversible transformation, the result of which is clouding. As cause of the temperature-induced gelation, Heymann assumed the dehydration of hydrated methylcellulose chains [12].

In recent decades, numerous investigations have been carried out on the mechanism of gel formation. To this day, it still has not been sufficiently and finally clarified as to how the 3D network is formed during thermal treatment of methylcellulose solutions. Listed below are the most famous theories and models used to explain the relationships.

The theory postulated by Sakar is based on the idea that water molecules surround the hydrophobic methoxyl substituents of methylcellulose at low temperatures, and form so-called 'cage structures'. Water solubility of these polysaccharides results from the shielding of the hydrophobic CH_3 groups. By increasing the temperature, these structures are increasingly deformed until they even break. Thus, the hydrophobic regions of these cellulose ethers are opened and released, and an association of the polymer chains is possible [13].

Through warming, the vibrational and rotational energies of the water molecules increase to such an extent that these forces exceed the strength of the hydrogen bonds. The latter provide – in the 'cold state' – the orientation of the dipolar water molecules around the polymer chain like a protective shield. The energized water molecules decouple from the fragile protective cover made out of orientated water around the chain. The dehydrated hydrophobic polymer segments begin to associate with each other. With further increasing temperature and/or with increasing duration of heating, the number of hydrophobic interactions increases. This results in a gel becoming more and more firm, because the number of crosslinking points increases proportionally and the network becomes evermore dense [3].

Figure 5.25 illustrates the theoretical profile of thermal gelation of methylcellulose (model by Sakar).

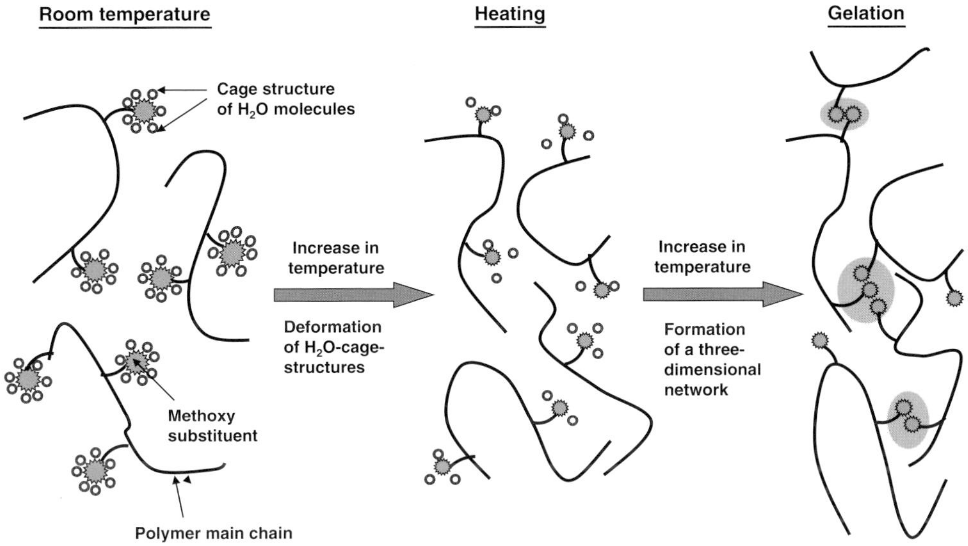

Figure 5.25 Mechanism of gelation of methylated cellulose ethers [10]. Reprinted with kind permission of Dr. Matthias Knarr.

E.R. Morris published a mechanism for gelation of methylcelluloses in which existing crystalline regions of the polysaccharide chains combine to form aggregates at low temperatures. On heating, these superstructures increasingly dissolve and disappear. Finally, gelation is initiated by hydrophobic interactions [14].

An opposite approach is followed by T.P. Lodge, who does not interpret gelation of methylcellulose as a uniform process, but finds a relation to the phenomenon of liquid–liquid phase separation. Accordingly, hydrophobic interactions lead to the formation of aggregates in solutions. These aggregates are very different in shape and size. These concentration inhomogeneities become optically visible in solution during the turbidity observed during gelation [15].

The research work of J. Desbrières and coworkers supports this conception. They describe the state of dissolution of methylcellulose at low temperatures as a molecular dispersion and free of aggregates. On heating, hydrophobic interactions between the individual methyl groups could increasingly be detected by fluorescence spectroscopic methods. When there are highly substituted regions in the cellulose ether, the increase in temperature leads to formation of a clear gel. In addition, in the areas of high concentrations, pseudo-crystalline regions were analysed that have a similar structure to that of trimethylcellulose. With increasing temperatures, concentration fluctuations caused by phase separation lead to the formation of compact aggregates. The size of these aggregated units is limited by gelling, since mobility of the polymer chains is reduced. According to Desbrières, the existence of trisubstituted areas is therefore the basic requirement for gelation of methylcelluloses [16].

5.3.3.4 Non-thermal Gelation

In addition to thermally-induced gelation methylcellulose and HPMC also gel upon addition of sufficient quantity of coagulative cosolute (substances promoting flocculation). The derivatized cellulose molecules are summarized below under the heading 'polymer'. Substances belonging to this category of 'flocculation accelerators' are salts such as phosphates, sulfates, and carbonates. The action of these salts tends to strip water molecules away from the polymer via disruption of the hydrogen-bonding forces. This is analogous to the effect of thermal energy as a hydrogen bond disruptor.

The net result is that certain polymer segments have an insufficient attraction to the electrolyte solvent – concentration differences and inhomogeneities of solvent in the system occur, partially dehydrating the chain and allowing the formation of hydrophobic interactions and regions whereby several methyl groups can 'agglomerate'. The mechanism of salt-induced gelation has clear parallels to the thermal gelation. Several salts have been found to be most effective for lowering the gelation temperature of methylated cellulose ethers. Trisodium polyphosphate and sodium sulfate have this effect, as well as sodium carbonate and magnesium chloride.

Methylcellulose is more sensitive in this respect than are the various HPMC types. In MC, the percentage reduction of flocculation temperature is higher because the content of methyl groups is higher and no hydrophilic HP groups are present that could provide for attenuation or dilution. Firm gels can be formed at room temperature by the addition of 3% trisodium polyphosphate to a 2% solution of MC. Table 5.6 (Section 5.3.3.1) gives an overview of the effect of additives on gelation temperature of MC and HPMC solutions. The use of salts to form gels with non-ionic cellulose ethers is presently being investigated. The opportunities offered by gelation at 'cold' and room temperatures for formulators of food applications are obvious [3].

A magnetic resonance probe has been developed for mapping of the temperature profiles in gels [17].

5.3.3.5 Difference between Methylcellulose and HPMC

The incorporation of hydroxypropyl groups into a molecule of methylcellulose always reduces the strength of the resulting gel, and causes an increase in the critical gelation temperature. Hydroxypropyl substituents (HP) are significantly more hydrophilic than the hydrophobic methyl groups – so they are also able to better retain the water of hydration when exposed to heating. By stronger (but still purely physical) binding of water through HP groups, the temperature required to disperse the substituted groups of an HPMC has to be higher than for a pure methylcellulose.

Furthermore, the associated balance between water and HP-substituents allows the formation of a more hydrophilic and more watery gel than would be possible with a methylcellulose [3].

Separation of water – which means visible formation of a 'water's edge' – is often observed when heating pure MC gels in a microwave oven.

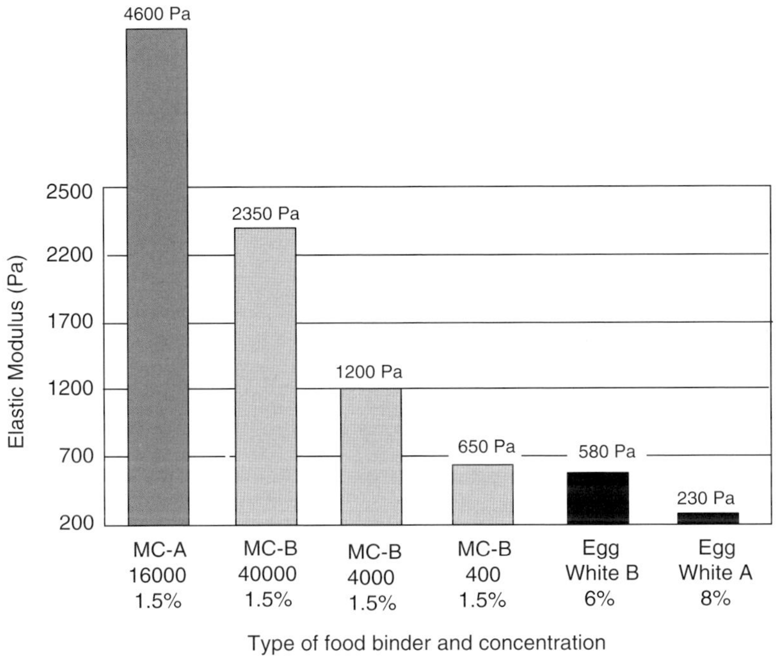

Figure 5.26 Mechanism of gelation of methylated cellulose ethers.

5.3.3.6 Gel Strength of Different Food Binders

To give an indication of the gel strength of methylcellulose, different MC types can be compared with each other and two types of egg white (Figure 5.26). The elastic modulus is measured of the MC hot gels at 1.5% concentration. For egg white, higher amounts of 6–8% are necessary to obtain reasonable data. The highest gelation power and therefore strongest gel is achieved by MC type A. The gel strength also depends on the polymer lengths. The same concentration of a long-chain MC (e.g. MC-B 40 000) results in a stronger gel than a solution of a short-chain MC (e.g. MC-B 400). The gelling power of egg white is much lower. In food systems, methylcellulose can be used in relatively low concentrations (<1%) to achieve a strong binding. This property is used to create vegetarian products such as soy patties, vegetable sticks, or meatless burgers.

5.3.4 Dispersion and Hydration

The extremely hydrophilic character of the methylated cellulose ethers MC and HPMC combined with their surface activity can lead to problems when pure powders are to be incorporated in pure water. When there is sufficiently rapid dispersion, lumps are formed. Attempts to destroy the aggregates by means of mechanical forces with a high-speed mixer end in the formation of more or less stable foam.

After adequate dispersion of the powder in water, proper hydration and dissolution of MC or HPMC must be ensured in the second step. Only completely dissolved cellulose ether can cause the desired functionality in the final application. The main mistake occurring during the first application trials was to increase the dosage when the achieved effect is considered to be too low. The reason for failure is mostly not an insufficient quantity of MC or HPMC but, instead, limited functionality due to insufficient hydration, so that only 20–30% of the additive or even less is fully operational.

The following methods are recommended to ensure optimal dispersion and hydration of methylcellulose or HPMC powder in food and thus to achieve maximum functionality.

Methods for Incorporation of Methylcellulose or HPMC Powders in Food Systems

1) **Preparing a dry mixture:**
Methylcellulose or HPMC is mixed homogeneously with other dry non-polymeric substances such as sugars, salts, spices, pigments, or cold-water-insoluble components of the formulation such as flours and cook-up starches. This homogeneous dry blend is then transferred into the water phase. 'Dilution' of powders of cellulose ethers by means of other dry ingredients separates the individual particles from each other, and they will not stick together during swelling in water. The mixing ratio of 'dry substances : methylcellulose' should ideally be at '7–10 : 1', but at least at '3 : 1'. Note that the temperature of the aqueous phase must be below the hydration temperature of the respective cellulose ether (Table 5.7).
2) **Dispersion in anhydrous liquids:**
Methylcellulose and HPMC can be dispersed in non-aqueous low-viscosity media such as vegetable oil, glycerol, propylene glycol, or polyethylene glycol (PEG). Another possibility is prewetting with alcohol (ethanol, isopropanol) or the inclusion in highly concentrated sugar syrups (at least 50% sugar). A suitable mixing ratio is 5–8 parts fluid and one part of MC

Table 5.7 Relationship between type and level of substitution and hydration temperature.

Cellulose ether[a),b)]	Methyl content (Me) (%)	Hydroxypropyl content (HP) (%)	Hydration temperature	
			(°C)	(°F)
Methylcellulose (A)	31–33	—	<10	<50
Methylcellulose (B)	28–31	—	13	55.4
HPMC – type C	28–30	8–12	25	77
HPMC – type D	28–30	4–8	25	77
HPMC – type E	20–24	6–10	30	86

a) EU Directive on methylcellulose: 25–33% Me, maximum 5% hydroxyethoxyl.
b) EU Directive on HPMC: 19–30% Me, 3–12% HP.

or HPMC. In these water-free systems, methylcellulose does not dissolve – there exists a dispersion of solids in which the singulated particles of cellulose ether are coated with a water-repellent substance. In a second step, this slurry is introduced in an aqueous phase that is appropriately temperature controlled, and by stirring the MC particles hydrate slowly. This dissolution process takes longer – in particular when using oils as fluid, since the oil layer needs to be 'washed away' first before hydration begins.

3) **Hot–cold technique:**
The technological basis for this method is the insolubility of methylcelluloses and HPMC types in hot water. In formulations containing high quantities of added water, about one-third of the water is heated. The water temperature must be above the specific gelation temperature (see Table 5.5). The cellulose ether powder is dispersed in hot water only, but it does not dissolve. The remaining cold water is then added to the hot slurry. Water can be used also as ice. By slow stirring, a transparent solution with a suitable viscosity results after a short time. If the amount of water cannot be split, the product must be taken to a storage place. Use of a vessel with a double jacket cooling system is another option. It is important that before use again the solution is thoroughly mixed, so that resultant sediments and concentration inhomogeneities can be compensated.

A properly dissolved methylcellulose or HPMC is completely transparent and clear, like water. An opaque appearance indicates a lack of hydration, the causes of which are usually an incorrect temperature control or a lack of water as solvent. After proper dispersion, it is to be taken into account that the water temperature is below the specific hydration temperature.

If the use of suitably cold water is not possible, then methylcellulose (or HPMC) is present in a not-completely hydrated and only partially dissolved form. In this state, its functionality in the application is limited. This problem can be resolved by cold storing of semi-finished food, and stirring them up once again before further processing, so that the now dissolved hydrocolloid may be homogeneously and continuously distributed in the system.

Once dissolved, cellulose ethers can be incorporated warm into hot food products – above the gelation temperature. MC or HPMC are then, however, only distributed 'spot'-like in the system and are not very functional. To achieve the optimal effect of the substances, the food must be stirred again after cooling to below the temperature of gelation. Final mixing (before shaping or filling) ensures the formation of a continuous phase of dissolved MC or HPMC with excellent functionality.

Table 5.7 indicates the effect of substitution type and level of substitution on hydration temperature of methylated cellulose ethers. These levels of substitution have their origin in the pharmaceuticals arena and are deduced from the US pharmacopoeia. Sometimes, food-grade MC and HPMC have broader specifications in terms of viscosity and, mostly, they are cheaper due to less analytical effort.

In the area of food-approved MC and HPMC products, various other substitutions are possible. At present, these modification forms are not manufactured industrially on a larger scale, so that commercial use is not possible.

There are two 'rules':

- The higher the content of methyl groups, the lower is the temperature of hydration.
- The total amount of substituted groups (sum of 'Me + HP') affects the hydration temperature inversely proportionally. This means that highly substituted cellulose ethers require rather cold temperatures for solubilization.

5.3.5
Technique of Delayed Hydration

Using methylated cellulose derivatives such as pure MC or HPMC, there is the possibility to apply the 'Technique of delayed hydration' because of their reversible thermal gelation. Analogous to the 'hot–cold technique' (see point '3' in Section 5.3.4.1) this method is based on the fact that methylcellulose (or HPMC) can be incorporated into hot aqueous systems without being hydrated and without gelling.

This is achieved when the temperature of the liquid phase is higher than the specific gelation temperature of methylcellulose (or HPMC). Owing to the hot temperature of the system, the cellulose ether is not hydrated and not dissolved. The hydrocolloid is only dispersed in the liquid, but it does not thicken at this time point. After the heating process, the product is cooled. Below the specific gelation temperature, methylcellulose or HPMC dissolve slowly and develop the corresponding viscosity.

This technique of delayed hydration offers some advantages:

1) **Viscosity of the product is lower than in conventional processes**
   a) The result is a lower energy demand by an optimized thermal treatment and simplified transportation;
   b) the product is protected because:
      i) less shear force must be applied; this allows, for example, the transport and pumping of products at lower pressures;
      ii) heat transfer in low-viscosity liquid media is easier and improves the microbiological safety;
   c) less wear and therefore reduced maintenance costs;
   d) reduced cleaning costs through minimized adhesion of product on inner surfaces of vessels, piping, and heat exchangers.
2) **Enhanced stability of oil-in-water emulsions:**
   High viscosities and small droplet diameters lead to stable OW emulsions. However, a high viscosity of the crude emulsion during the homogenization process leads to a large droplet diameter, which allows the droplets to easily coalesce and creaming. Most gelling and

thickening agents behave at hot temperatures like their names indicate and strongly thicken the aqueous phase. By use of MC (or HPMC) instead, or alternatively partial replacement of these traditional thickeners by MC/HPMC, low-viscosity systems can also be created at high temperatures, which are homogenized subsequently. The resulting oil droplets have the desired small particle size. When the emulsion cools, methylcellulose hydrates and thickens and causes the final viscosity increase. In addition, MC or HPMC molecules, being surrounded by a hydration shell, place themselves around the individual oil droplets. This forms a barrier to prevent them coalescing and increases emulsion stability further. A prerequisite for successful application of this principle is that the homogenization is carried out at a temperature above the specific gelation temperature of methylcellulose (or HPMC).

5.3.6 Functional Properties Resulting from Methylation

To predict the effect of hydroxypropylated or pure methylcellulose in food formulations it is important to know the technological properties of these substances in their native and therefore unmixed form.

Methylcellulose and HPMC show the following characteristics, functions, and benefits:

- 100% cold water solubility;
- insoluble in hot water – possibility of delayed hydration;
- increasing viscosity, thickening of the aqueous phase;
- wide range of viscosity is available (2% from 5–250 000 mPa s) for rheology control, texturizing (mouthfeel) and stabilization of foods;
- binder/adhesive;
- reversible thermal gelation – soft to firm gels that melt back on cooling;
- good water retention results in extended shelf-life and allows dry, crispy dough pockets with juicy fillings;
- control of water balance during freezing, slow growth of ice crystals;
- shape retention of food products during heating, stabilization of matrix systems over a large temperature range from −20 to +200 °C (−4 to +392 °F);
- effectiveness without increasing the viscosity, thus high compatibility with other thickening hydrocolloids;
- resistance to amylases;
- surface activity and emulsifying properties – stabilization of two- and multiphase mixtures, foam generation;
- high resistance to salts and acids (to pH 1);
- film formation and barrier function – reduced fat absorption during heating in oil;
- suspending of ingredients – reduced sedimentation and phase separation in liquid systems with particles;

- lubricant and slip aid, reduced adherence and sticking on surfaces of production equipment and walls of containers;
- substitution of animal or allergenic proteins (e.g. egg, gluten) with maintenance of constant process stability and comparable texture of the final product;
- replacement of fats while maintaining the full and creamy mouthfeel;
- high efficiency – use in low concentrations (0.015–0.8%).

5.3.7
Behaviour of Methylcellulose with Other Ingredients

Methylcellulose is a non-ionic polymer that when dissolved has a high compatibility with other ingredients. Owing to the hydrophilic nature of the glucosidic chain and the hydrophobic character of the substituted methyl groups, a heterogeneous molecule results that preferentially places itself at interfaces of multiphase systems. Technologically, both pure methylcellulose as well as HPMC are emulsifying substances used for stabilization of different mixtures like simple OW emulsions, multiple emulsions (type WOW), and oil–water–air systems (such as foams).

We note that methylcellulose cannot completely substitute classical emulsifiers (such as mono-/diglycerides, various esters, lecithin, caseins) in low-viscosity products, but only support them.

In high-viscosity foods, the limited emulsifying capacity of methylcellulose is compensated through the high dynamic viscosity (Stokes' law) so that the movement of particles can be sufficiently reduced, and the emulsion remains stable.

Negative interactions between methylcellulose and other components of the formulation are almost exclusively caused by the strong hydrophilicity and the associated high water absorption rate and thus related to the large water binding and retention capacity of these cellulose derivatives.

To ensure smooth application of methylcellulose, HPMC, and CMC, different strategies can be followed:

1) **Increasing the amount of water:**
   Taking into account the water activity of the system (a_w value), 5–10% additional water may be introduced in the formulation.
2) **Stepwise processing:**
   In a first step, slow acting substances such as proteins are added and 'activated' (by mechanical disruption, diffusion of salts, etc.). When the 'lazy' proteins and ingredients are – after first addition of water – present in a swollen state optimally, methylcellulose is added with a second quantity of water.
3) **Change of the molecular weight:**
   If the amount of water cannot be increased, a cellulose ether having a shorter chain length, and thus lower viscosity level, should be used instead.

4) **Use of another substitution type and/or level of substitution:**
   High sugar concentrations (>30%) are critical for pure methylcellulose.
   Hydroxypropylmethylcellulose accepts higher sugar quantities (up to 50%). The highest compatibility with sugar is detected for HPMC types with a low content of methyl groups, that is, with low total DS.
5) **Use of a different type of cellulose ether:**
   If the salt tolerance of carboxymethylcellulose (CMC) is not sufficient, it can be replaced with methylcellulose or HPMC.
   CMC can be used in high-Brix systems with up to 65% sugar. However, due to its liquefaction during heating ('heat-thinning effect of CMC') it is only rarely suitable as replacement for thermal-gelling hydrocolloids such as MC or HPMC.

General

The following rules of thumb are valid for the compatibility of MC and HPMC with other substances (sugar, salt, phosphate, etc.) in aqueous solutions:

- the lower the molecular weight of a cellulose derivative, the better is its solubility.
  Short-chain polymers usually have a lower water demand when hydrating and can therefore accept higher quantities of other dissolved substances. In addition, short-chain cellulose ethers do not impede their own hydration, as do mid- and long-chain grades through their fast development of viscosity.
- the lower the content of methyl groups, the higher the tolerance;
- the compatibility with sugar is greater than that with charged substances (polyelectrolytes);
- monovalent anions and cations are better accepted than multivalent ions;
- the lower the temperature, the higher the rate of hydration.

Table 5.8 gives an overview of compatibility of methylated cellulose derivatives with other substances.

Table 5.8 Tolerance of methylcellulose and HPMC with other additives.

	Amount of additive (in grams) that is tolerated by 100 cm^3 (cc) of a 2% solution of the cellulose ether.					
Additive	**Methylcellulose 15 mPa s type B**	**Methylcellulose 4000 mPa s type B**	**HPMC 50 mPa s type D**	**HPMC 4000 mPa s type D**	**HPMC 100 mPa s type E**	**HPMC 4000 mPa s type E**
NaCl	12	8	18	12	20	13
$MgCl_2$	12	9	36	26	42	39
Na_2CO_3	5	4	6	5	5	5
Na_3PO_4	3.0	2.7	3.9	3.7	4.9	4.4
Saccharose	100	67	123	81	162	116

Sugar

Compatibility of MC and HPMC with sugars such as glucose, fructose, and sucrose is restricted. The reason for this is the limited availability of free water for hydration.

The lowest tolerance is found for long-chain methylcelluloses. Owing to their higher hydrophilicity, HPMC types can tolerate larger quantities of sugar than MC. The greatest tolerance is measured for HPMC types with a low content of methyl groups (such as HPMC – type E). At sugar concentrations above 30%, there is no longer enough free water available to hydrate pure medium- and high-viscosity methylcelluloses. If thermal gelation is required for product quality or smooth processing, a short-chain MC has to be used with minimal thickening effect. Another option is the application of a more sugar-tolerant HPMC, which accepts up to 45% sugar. Sugar leads to a disproportionate increase in viscosity (by a factor of 2–5) when heating solutions of MC or HPMC.

Salts (NaCl Salt, Phosphates)

Salts affect the availability of water and thus compete like glucose with methylcellulose and HPMC for the amount of water present. This can:

- delay hydration and viscosity development;
- increase viscosity;
- lead to precipitation of MC or HPMC out of solution;
- lower the gelation temperature.

Compared with other hydrocolloids, methylcellulose and HPMC types are relatively salt-tolerant and salt-stable in the presence of sodium chloride because of their neutral molecule characters. As with sugar, HPMCs are more tolerant to salts in general than methylcelluloses. This is due to the more hydrophilic character of hydroxypropylmethylcelluloses. HPMC type E has, with the smallest quantity of methyl groups, the greatest tolerance in the presence of salts.

Hydration and solubility of cellulose ethers is – depending on substitution type and on molecular weight – limited to salt concentrations of about 7.5–15%. Hydration is additionally temperature-dependent and best at low temperatures of 0–5 °C (32–41 °F).

Complete gelation is only possible at salinities below 5–15%.

Reversible salting-out effects are observed at 8–20% salt content (NaCl), depending on the type of methylated cellulose ether.

Sensitivity of these cellulose ethers in the presence of phosphates is significantly higher (see Table 5.8). However, the phosphate concentrations approved and applied in food products are not critical for methylcelluloses and HPMC.

In brine applications, this salting-out-effect is used advantageously to suppress and delay viscosity development.

Pigments/Fillers

Water-insoluble substances such as pigments and fillers generally do not affect the hydration and functionality of methylated cellulose ethers. Solutions of methylcellulose or HPMC are often used as a dispersion medium for the 'distribution' of these substances.

Acids/Alkalis

As negatively charged groups, which could be protonated through addition of acid, are missing in molecules of MC and HPMC, these substances do not react noticeably until a pH of 2 when pH-lowering agents are added. In addition, alkaline liquid shows no effect up to a pH of 12 because the methyl and hydroxypropyl groups are already present in the neutral and, thus, more stable form. At pH values lower than 2 and more than 12, acid- or base-catalysed depolymerization of the cellulose chain may occur. But such extreme systems are not suitable for direct human consumption and have to be 'diluted' first to be food-grade.

Minerals

The tolerance of methylcellulose and HPMC to minerals is high. Mono-, di-, and trivalent metal ions do not lead to precipitation of these cellulose derivatives. The neutral (non-ionic) character of the methyl and HP groups ensures that the molecules are not complexed in the presence of other charged substances and therefore cannot be easily precipitated. It is possible that methylated cellulose ethers are salted out of solution by polyelectrolytes when the concentration of multivalent ions and other dissolved substances (such as sugars) exceeds a specific limit. The salting-out effect can be explained by competition of the cellulose ether with other substances for the freely available water, which can lead to dehydration of already dissolved MC/HPMC or a visible reduction in their hydration. Highly concentrated mineral salts affect gelling of methylated cellulose ethers, so that gelation can occur already room temperature. The course of gelation temperature is a function of type of ion. The higher the ionic charge, the greater is the reduction of gelation temperature (see Table 5.6). However, as these highly loaded systems cannot be consumed for health and sensory reasons without dilution, the use of MC or HPMC is not affected. Hydroxypropylmethylcellulose tolerates higher salt levels than pure methylcellulose. The difference between the different substitution types of HPMC is low (see Table 5.8).

Proteins (Except for Gelatine)

Methylcellulose exhibits a high compatibility with proteins. The high 'rate of reaction' of a cellulose ether – in comparison to the low-speed proteins – can usually be adjusted simply by modifying the production technology. Protein-containing emulsions of both each texture (liquid, medium-viscosity, or firm) and any processing- and storage-temperature are thickened with methylcellulose and HPMC and are stabilized. Examples are milk shakes, ice cream, sauces, egg-containing foods, cooked sausages, and vegetarian soya products. Stabilization of low-viscosity neutral milk drinks (with a viscosity below 10 mPa s) with MC or HPMC is only recommended with restrictions. It may lead to flocculation and phase separation of the caseins present. Dehydration of casein micelles has been discussed as cause of this effect. Decomposition (dismantling) of the protective hydration shell is followed by agglomeration of milk proteins, which ultimately leads to destabilization of the system. The natural synergy of κ-carrageenan with α-S_2- and κ-caseins and the resulting very low dosage (0.02%) seems to make the use of other hydrocolloids in dairy products with pH ~7 pointless anyway.

Gelatine

Methylcellulose does not lead to a significant increase in gel strength of the resulting gelatine gel. An addition of 1% of a medium-viscosity methylcellulose (type B with 2% = 4000 mPa s) has no significant effect on measured gel strength of a gelatine gel of 3% concentration. High-viscosity methylcellulose with a higher methyl group content (e.g. MC – type A with 2% = 16 000 mPa s) increases gel strength by at least 20%, wherein the dry matter content rises by about 33%. The transparent gel obtained after cooling a pure gelatine solution becomes opaque, dull, and grainy by addition of methylcellulose. The mixture of gelatine (0.8%) with medium- and high-viscosity methylcellulose (1%) leads to an improved whipping volume (overrun increased up to 300%) with extended shelf life and stability for cold-manufactured foams.

Modified Starches

Low-viscosity methylcellulose improves the performance of modified starches (E 1422) during cooking processes and the sensory properties of the final product (see Section 5.6 on 'synergistic effects'). While starches deliver the required cold viscosity, methylcellulose ensures thickening at higher temperatures from about 60 °C (140 °F) without forming a gel. Positive interactions were observed with tapioca and waxy maize starches. Combination of these antagonistically acting hydrocolloids results in a viscosity profit, so that both substances can be used with 30–50% reduced concentration. The resulting product – for example, a sauce – has a stable viscosity in each temperature range and is no longer subject to stronger starch-induced fluctuations. In addition, the boiling process is controlled and, thereby, product safety is improved. It results in fewer starch-related process problems such as burn marks on pipe walls, vessels, containers, and other equipment. Cleaning is simplified because cellulose-based films are easy to detach from steel surfaces. An additional positive effect is that methylcellulose compensates during freeze–thaw processes for the weaknesses of starches, which cause syneresis, by catching released water. Furthermore, the sensory properties of this starch–MC combination are rated positively. Recommended dosages for pourable sauces are 1.5–3% of modified starch and 0.3–0.6% of low-viscosity MC (of viscosity level 2% = 15 mPa s).

Carrageenans

General observations on the nature of gels by the addition of methylated cellulose ethers (with 1%) to κ-carrageenan (at 1%) are listed below. Gel strength is measured with a Texture Analyser at 4 °C (39 °F):

- Addition of a cellulose derivative leads to improved flexibility and stability of a gel of κ-carrageenan. The mixed gel is less brittle and fragile.
- The gel strength of the hydrocolloid mixture increases with increasing viscosity of MC or HPMC.
- A difference between methylcellulose and HPMC of the same initial viscosity cannot be determined.
- Lowering the pH value results in higher gel strength.
- Owing to the molecular flexibility, particularly low-viscosity HPMC integrates itself very well into the network of the carrageenan and increases the gel strength disproportionately. An acidic pH has an additional positive effect on the strength and stability of the resulting gel.

Other Polymers/Hydrocolloids

In the literature and in practice, no negative interactions of methylcellulose or HPMC with other polymers are known so far. At present, there are intensive investigations to find positive interactions and synergistic effects. Through the existence of low-viscosity methylcellulose and HPMC types, generally high stability, and absence of genuine intolerances, the creative combination of hydrocolloids with MC and/or HPMC to develop innovative formulations and processes has few technological limits.

Preservatives

It is known from practical experiments that the combination of methylcellulose with salts of benzoic acid used as preservatives can lead to negative interactions. In particular in beverages, this reaction may cause an undesired flocculation and phase separation.

Fats/Oils

General trends in the addition of oils to methylcellulose solutions are:

- reduced gel strength,
- reduced foam stability with whipped MC–oil mixtures,
- reduced overrun,
- increased viscosity.

A deviation from these generally valid tendencies is observed in soybean oil, for instance:

- soybean oil increases the gel strength of MC gels;
- soybean oil leads to an improved overrun with methylcellulose.

Short-chain types of hydrophobic methylcellulose (15 mPa s at 2%) can be used successfully in systems with high fat and low water contents, because they only require a low amount of water for hydration.

5.3.8
Methylcellulose and HPMC for Fat Reduction in Coatings

Through their film-forming and gelling properties, Methylcellulose and HPMC influence the 'fat budget' especially of fried foods. With a specific surface treatment, the oil absorption of products, heated in fat, can be controlled and/or reduced. During and after the frying process the gel generated by MC or HPMC acts as oil barrier and keeps the moisture in the crust and the oil out of the crust. The reason for this lies in the closure of pores on the product surface by the thermal gel. Thereby, the oil cannot enter the product and be absorbed by the core.

Principle of Fat Absorption during Frying [18]

How does frying increase the fat content of products?

There are two main mechanisms that exist during and after the frying process:

1) **Moisture loss – during heating/frying:**
   - During frying water in the crust (or batter or dough layer) evaporates through the pores.
   - Oil penetrates the crust where water has evaporated, resulting in a fat-concentrated crust.
   - Methylcellulose and HPMC form reversible thermal gels on the product surface and in the dough that hold moisture in and reduce oil and water migration.
   - The cellulose derivatives allow water to remain in the crust and block the penetration of oil, resulting in fried food with reduced overall fat content.
2) **Condensation – during cooling:**
   - When food is removed from the fryer or oil bath, steam in the pores begins to cool and condense. On cooling, a pressure (hypotension) is created that pulls the remaining oil droplets from the product surface into the pores. Up to 80% of final content may be absorbed this way [19].
   - The MC and HPMC gels function as barriers to prevent the steam escaping. The hydrocolloids help provide balanced pressure conditions.

Figure 5.27 demonstrates how methylated cellulose derivatives (MC, HPMC) hold moisture in the crust/batter/coating layer, so that oil cannot be absorbed from food when the product leaves the oil bath and cools. Figure 5.27a (left-hand side) demonstrates what happens to a standard product without using MC or HPMC when heated in an oil bath. Water evaporates abruptly through the pores in the crust/batter and oil penetrates into

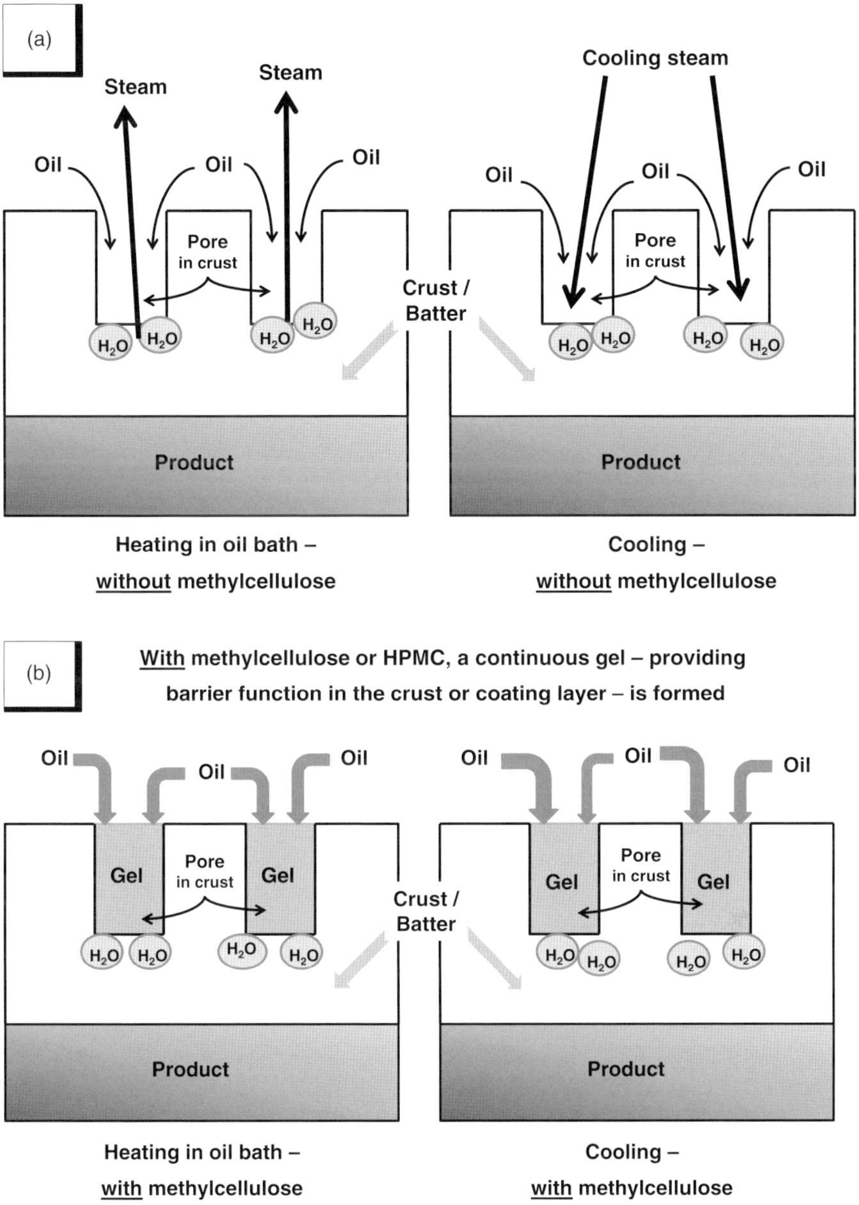

Figure 5.27 Principle of reduced fat absorption by coating with MC/HPMC [18]. Reprinted with kind permission of Ashland Inc.

the product through the open pores. Figure 5.27a (right-hand side) represents a standard product during the cooling phase. The cooling steam creates a pressure, respectively hypotension or suction, that pulls any oil through the pores into the product.

If a batter or breading for a frying product contains sufficient quantity of dissolved methylcellulose or HPMC, then a continuous gel is formed within the coating system. Figure 5.27b (left-hand side) shows a crust with methylcellulose/HPMC, which form a thermal gel upon heating and thus closes the pores present on the product surface, so that water cannot evaporate out of the product and oil cannot enter the core. This gel is also stable during cooling to low temperatures and can thus prevent oil being 'sucked into' the product due to changing pressure conditions. This process is shown on the right-hand side of Figure 5.27b.

Coating with Films of MC

Figure 5.28 indicates how fat uptake of fried vegetable burgers and potato chips or strips or French fries can be influenced by spraying or dipping the products into a MC solution.

By applying a very thin layer of methylcellulose (as a 2–3% solution) on the product surface, the oil absorption can be reduced by 30–35%. For example, a coating with a 2% MC solution lowers the oil absorption from 8.3 to 5.5 g of oil per 100-gram burger, which is 33% less fat. A moderate drying with infrared light before frying increases the effect. In industrial production, the dissolved

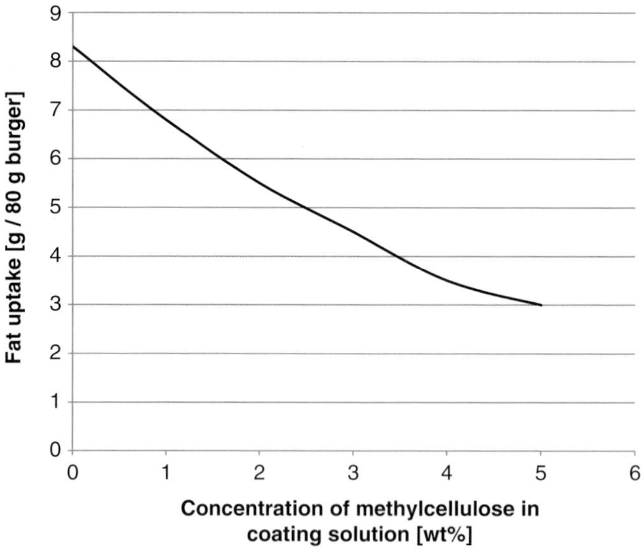

Figure 5.28 Fat uptake after application of different MC concentrations.

methylcellulose is fixed on the product surfaces by the heat of the oil during a pre-frying process.

The effect of methylcellulose increases as its concentration on the outer surface of the food rises. Therefore, the best effect is achieved by coating the frying product with a pure and as high as possible concentrated solution of methylcellulose. Unsurprisingly, methylcelluloses with low DP and low-viscosity are the first choice for this kind of application. Short-chain HPMC types are a less effective alternative.

Batter Systems with Medium-Viscosity MC and HPMC

Cellulose derivatives can also be used as a component of breadings. Another factor influencing the efficiency of oil absorption in fried foods is the temperature of the breading and the temperature of breading preparation. Only complete hydration of methylcellulose or HPMC ensures optimal functionality. The temperature of the liquid batter must be necessarily below 15 °C (59 °F) when using pure methylcellulose (see Table 5.7). If cold processing is not possible, HPMC should be incorporated as an alternative. Here, HPMC grades with higher methyl contents have the greatest impact on oil absorption – fat uptake may be reduced by up to about 30%.

The emulsifying properties of methylated cellulose ethers are a further interaction with fats and oils. These relationships are described in detail in the next section.

5.3.9 Emulsifying Properties of Methylcellulose

Methylated cellulose derivatives (MC, HPMC) stabilize emulsions by reducing the surface and interfacial tensions of the phases involved. In addition, the hydrocolloids increase the viscosity of the aqueous phase.

The HLB value, proposed in 1954 by W.C. Griffin for the hydrophilic–lipophilic balance (HLB), is between 10 and 12 for methylcellulose and HPMC. The lower the value, the more lipophilic the substance is. W/O emulsifiers have values between 3 and 8. For O/W emulsifiers, HLB values are between 8 and 18.

Notably, neither methylcelluloses nor HPMC grades can fully replace conventional emulsifiers such as mono- and diglycerides, lecithin, or caseins in low-viscosity systems. The emulsifying properties of the hydrophilic or lipophilic groups are usually temporary. After a certain time, simple oil–water mixtures separate again into their individual phases.

Permanent stabilization can only be achieved by a combination of methyl groups present and an adequate chain length of the cellulose backbone. Long cellulose chains cause a strong thickening effect of the aqueous phase. Through this significant viscosity increase, the multiphase mixture is permanently stabilized.

Two examples illustrate this:

Recipe A	Recipe B
24–49% Sun flower oil	24–49% Sun flower oil
49–74% Water (cold)	49–74% Water (cold, 10 °C (50 °F))
2% HPMC (with 2% = 100 mPa s)	2% Methylcellulose (with 2% = 40 000 mPa s)

Both emulsions are prepared using the same method:

1) dispersing MC or HPMC in oil;
2) slow addition of cold water by the use of high shear forces;
3) storage for 7 days.

Result: The low-viscosity emulsion 'A' separates into two phases after only a few hours of lifetime. The highly viscous emulsion 'B' is stable for months.

Figure 5.29 shows the operation of a low-viscosity methylcellulose, or HPMC, with migrating properties.

Importantly, the emulsifying effect of a low-viscosity cellulose derivative alone is not sufficient to successfully stabilize an emulsion over a long time period (see recipe 'A'). The addition of a thickening hydrocolloid or of a classic emulsifier is mandatory. Even the shortest cellulose ethers have a relatively high molecular length and are therefore – compared with the much more 'compact' molecular structures of classical emulsifiers – significantly slower and less flexible in the occupation of interfaces. This also explains the relatively large droplet size of the oil droplets in emulsions stabilized with MC or HPMC. Figure 5.30 shows how a high-viscosity methylcellulose acts in emulsions.

In this case (as shown in recipe B), the methylcellulose (or HPMC) alone is sufficient to make the high-viscosity emulsion storage-stable by continuously

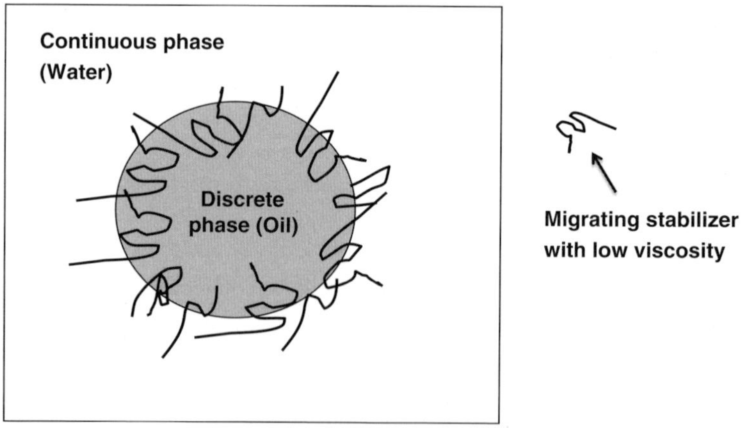

Figure 5.29 Short-chain migrating methyl cellulose/HPMC as emulsifier.

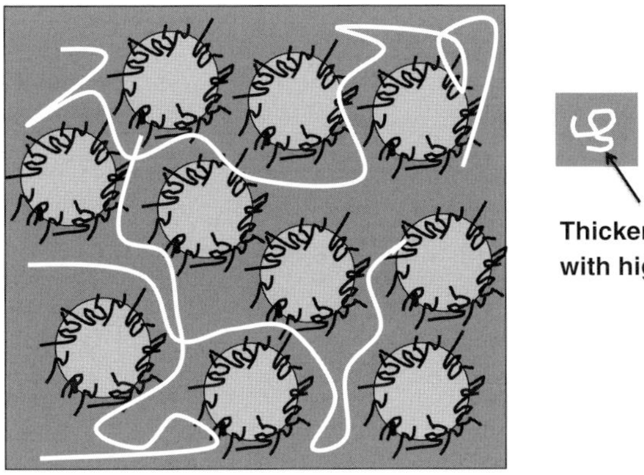

Figure 5.30 Long-chain methylcellulose as emulsifier with thickening properties.

ensuring thickening of the water phase. These types of emulsions are also very freeze–thaw stable and very heat stable.

The use of MC or HPMC in crude emulsions can also lead to benefits during the homogenization process and positively influence the quality of emulsions. These relationships are discussed in detail in Section 5.3.5 on the technique of delayed hydration.

5.4 Stability

5.4.1 Storage Stability of Cellulose Ethers

The following data are generally valid for methylated cellulose derivatives:

Physical appearance:

- white to slightly yellowish powder, odourless and largely tasteless;
- bulk density/weight: 0.25–0.70 $g\,cm^{-3}$;
- specific (relative) density: 1.39;
- browning temperature: 190–200 °C (374–392 °F);
- decomposition temperature: 225–230 °C (437–446 °F).

Cellulose derivatives are neutral powder products with high light-fastness. In solid form they can be stored indefinitely if they are kept as hygroscopic substances in closed containers.

During heating of moist cellulose ethers, keratinization is observed (from about 95 °C (203 °F)). By further heating, intensive browning occurs at about

190–200 °C (374–392 °F). If a further temperature increase is carried out, decomposition (charring) of the cellulosic product follows.

Cellulose derivatives are, as organic polymers, combustible under certain conditions (temperature, oxygen availability); a fire can be extinguished using conventional extinguishing agents.

During storage and use, the dust content of the air must always be taken into account to avoid reaching a critical level for dust explosion. If cellulose ethers are present as very fine dusty powders, an explosive powder–air mixture can occur under the action of oxygen. For example, a critical level for a methylcellulose or HPMC with a particle size of 74 μm and smaller (finer than 200 mesh) is achieved at a concentration of 28 $g\,cm^{-3}$.

Cellulose derivatives should not be stored in the immediate vicinity of peroxides and other oxidizing agents.

5.4.2 Microbiological Stability

As opposed to starches, cellulose-based polymers provide a poor culture medium and breeding ground for mould and bacteria. These microorganisms can only develop if they are able to form cellulases. The term 'cellulase' refers to an enzyme complex, which consists of three different types of enzymes.

The so-called endocellulases break the connections within the cellulose molecule to dissolve the crystalline structure, and thus produce shorter chains.

The exocellulases separate from units consisting of two glucose molecules from the ends of the small chains, which generated by the endocellulases. In this case, the disaccharide cellobiose is produced.

Finally, *cellobiases* or *β-glucosidases* cleave the β-glycosidic bond between the two glucose molecules in the cellobiose units.

In the presence of cellulases, there is an attack on the unsubstituted glucose molecules of the chain and the mechanism of enzymatic cleavage is started. This microbially-induced hydrolysis leads to a strong decrease in viscosity and can result in complete loss of the thickening effect. The more highly the cellulose derivatives are substituted, the more stable they are, therefore, against intramolecular degradation by enzymatic hydrolysis. With an average degree of substitution of 2 (DS = 2.0), water-soluble cellulose ethers are considered to be largely resistant to the attacks of bacteria and fungi.

In the preparation of aqueous solutions, which are to be stored for a long time (more than one week at room temperature), the addition of a preservative is suggested.

Basically, cellulose derivatives and also native celluloses are biologically (enzymatically) degradable. The desired enzymatic hydrolysis of the completely insoluble microcrystalline cellulose is a complex process that must be externally controlled to run successfully to glucose as final product.

5.4.3
Process Stability after Hydration

Films

Films of methylcellulose and HPMC are formed by moderate drying below the specific gelation temperature of their highly concentrated solutions. These unplasticized films are transparent, clear, rough, and relatively brittle, and so they are only partially elastic. However, the elasticity of films of MC and HPMC is higher than that of CMC films. The elastic properties can be further improved by addition of plasticizers, so that very flexible and stable films result. Table 5.9 summarizes physical data for pure MC films.

Suitable plasticizer compositions include water and polyhydric alcohols such as propylene glycol, poly(ethylene glycol), and glycerol. The use of thermomechanical drawing techniques results in films with increased tensile strength and modulus of elasticity. These films are stable in normal room air and dissolve completely in cold water after some time. They are used to provide a temporary barrier to aqueous systems. In hot water, with the action of superheated steam or at very humid warm air, gel formation is observed. The gels are not 100% impermeable to water vapour as mistakenly stated in the literature. But, these films do form a permanent and effective barrier to fats

Table 5.9 Properties of unplasticized methylcellulose films [8].

Property	Typical data	
Specific gravity	1.39	1.29
Area factor (inch^2 per lb per mil)	24 000	25 860
Moisture-vapour transmission rate (at 38 °C (100 °F) and 90–100% relative humidity)	67.5 g/100 inch^2/ 24 h/mm	65 g/100 inch^2/ 24 h/mm
Oxygen transmission rate (at 24 °C (75 °F))	25 cc/100 inch^2/ 24 h/mm	70 cc/100 inch^2/ 24 h/mm
Tensile strength (lb inch^{-2}) (at 24 °C (75 °F) and 50% relative humidity)	8500–11 400	8500–8850
Elongation (at 24 °C (75 °F) and 50% relative humidity)	10–15%	5–10%
Stability to ultraviolet radiation (500 h, Fadeometer exposure)	Excellent	Excellent
Resistance to oils and most solvents	Excellent	Excellent
Light transmission (2-mm film) (%):		
– 400 nm	55	82
– 290 nm	49	34
– 210 nm	26	6
Refractive index, $n_D^{20°}$	1.49	1.46
Softening point [°C (°F)]	—	240 (464)
Melting point [°C (°F)]	290–305 (554–581)	260 (500)
Charring temperature [°C (°F)]	290–305 (554–581)	270 (518)

and oils. Stability upon exposure to ultraviolet radiation is high. HPMC films containing no propylene gylcol and as much as 50% titanium dioxide form continuous films for pharmaceutical tablet coating and solid food supplement products. Inclusion of pigments to a level as high as 40% has been reported to have little effect on tensile strength but increases brittleness and reduces film toughness [8].

Influence of Temperature

Hydrated methylcellulose and HPMC is stable in the temperature range −50 °C to +135 °C (−58 to 275 °F). At pasteurization temperatures ($T < 100$ °C (212 °F), at atmospheric pressure), these cellulose derivatives show high durability also at low pH values. During sterilization (121 °C (250 °F), 2 bar) of very acidic foods (pH ≤2.5), a slight (and irreversible) reduction in viscosity is observed, which can be compensated by a higher dosage of 10–20%. The influence of heating rate and shear rate is to be mentioned. Rapid temperature increases and high shear rates lead to an increase in gelation temperature. At freezing temperatures below 0 °C (32 °F), solutions of methylcellulose and HPMC are not subject to phase separation such as that of most other hydrocolloids. Solubilization of these cellulose ethers is improved when the temperatures fall upon cooling, which is shown through an increasing clarity of solutions until the freezing point is achieved.

Based on measurements of heat of fusion, it can be shown that methylcellulose and HPMC retain, even at very cold temperatures, their affinity for water and are constantly surrounded by a hydration shell. Part of the water is still 'bound' and remains in a deeply supercooled state; it will not be frozen. The amount of heat that is normally released during freezing is reduced by the amount of this extremely cold, so-called 'supercooled', water. This effect is shown in Figure 5.31. The dotted line indicates the quantity of heat theoretically expected to be released, based on the concentration of methylcellulose or HPMC. The solid line shows the heat energy that is actually released. The measured heat of fusion is significantly lower than the theoretical value, because the system still contains unfrozen ('supercooled') water, which lowers the total energy content.

The rate of formation of ice crystals is controlled and retarded by methylcellulose and HPMC. Only the smallest crystals are formed. The 'capacity' of the cellulose ethers is not reduced or used up. The freeze–thaw–freeze process can be repeated any number of times without significantly increasing the size of ice crystals and without negatively affecting the quality of frozen products.

Effect of pH Value

Because of the neutrality of methylcellulose and HPMC molecules – the substituted groups are not charged – they do not respond significantly to the addition of substances that alter the pH over the wide pH range 2–12. Extremely high quantities of acid or alkalis lead to hydrolysis of the polymers, which is clearly visible in a strong decrease in viscosity.

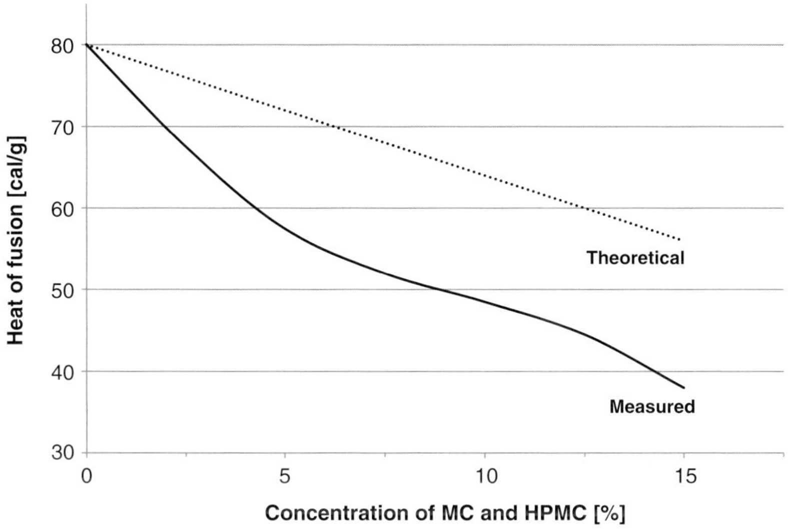

Figure 5.31 Heat of fusion as a function of the hydrocolloid concentration.

Surface Activity

A decisive difference between methylated cellulose ethers (MC, HPMC) and other hydrocolloids is their surface-active properties. Strikingly, when measuring solutions of different gelling and thickening agent, both methylcellulose and HPMC have a lowering effect on surface tension. The values are shown in Table 5.10. Pure water has a value of $72\,\mathrm{mN\,m^{-1}}$ and is taken as zero-reference value. The highest surface activity is detected for solutions of methylcellulose and HPMC. Within the group of methylated cellulose ethers, highly substituted HPMC types (high HP content with medium methyl content) have the greatest effect.

The lowering of surface tension by methylated and hydroxypropylated cellulose derivatives is useful for the preparation and long-term stabilization of foams. The molecules place themselves on the interfaces of the two-phase 'air–water' mixture. Short-chain polymers can even diffuse into the extremely thin

Table 5.10 Dynamic surface tension of hydrocolloid solutions.

Hydrocolloid	Surface tension γ_s (mN m^{-1})
Water	72
Xanthan	69
CMC	68
Sodium alginate	62
Propylene glycol alginate (PGA)	58
Methylcellulose (MC)	53–59
HPMC	43–55

boundary layer. There, due to the high molecular concentration, temperature-independent gelation takes place that causes the high stability. The absolute concentration of cellulose ethers is usually very low – as little as 0.01 wt% can be sufficient to 'trigger' this effect.

With MC and HPMC, whipping products of every recipe and consistency can be made storable. The replacement of protein and casein as traditional foam generators opens up new possibilities for development of vegetarian and vegan foods. Examples are egg-free meringues – plain – chocolate, nuts, or coconut milk. Another option is vegetarian 'cream'.

Notably, the formation of foam during manufacturing processes can be a disruptive disadvantage. High-speed stirring or other intensive shearing activities for preparation of MC and HPMC solutions are not recommended. Application of moderate shear forces with a subsequent rest period is the recommended processing way to prevent undesired foam formation.

Effect of Shear Forces – Rheological Behaviour

The rheology of aqueous MC and HPMC solutions is influenced by several factors:

- molecular weight or chain length (DP),
- concentration,
- temperature,
- presence of other dissolved substances.

Below their specific gelation temperature, solutions have pseudoplastic flow behaviour at almost all shear rates. Pseudoplasticity increases with increasing molecular weight and/or increasing concentration. At very low shear rates, solutions flow in Newtonian fashion rather than pseudoplastic. This is best seen with short-chain methylcelluloses or HPMC types in very low dosage. At temperatures below gel formation, rheology is controlled neither by type of substitution nor by degree of substitution. All methylcellulose and HPMC types exhibit a comparable viscosity profile at cold temperatures ($T < T_{\mathrm{Gelation}}$) when applying constant concentrations and shear rates.

Upon heating the solutions, a 3D gel structure is formed and the flow behaviour becomes strongly thixotropic. If very high quantities of low-viscosity methylated cellulose derivatives are dissolved in water, their solutions can also react thixotropically below the gelation temperature.

In summary it can be stated that:

- mechanical forces (like stirring, kneading, homogenizing) affect the viscosity;
- viscosity is reduced due to shear (shear thinning);
- this effect is completely reversible in the pH range 2–12 for HPMC and methylcellulose;
- after a rest phase, viscosity returns to its initial value with a delay (thixotropy).

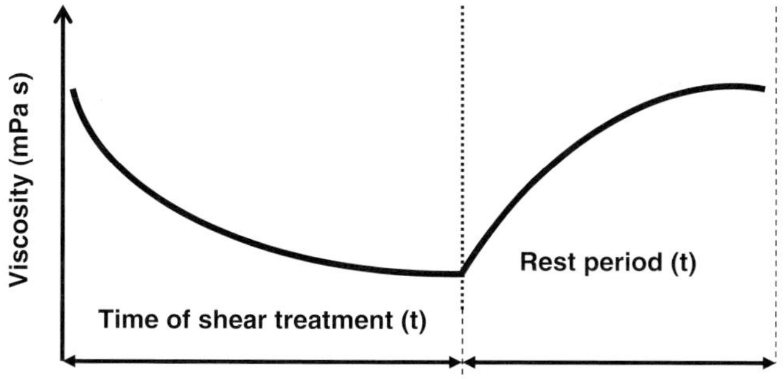

Figure 5.32 Change of viscosity as a function of time after a shear stress.

Figure 5.32 illustrates schematically the effect of shearing.

5.5 Rheometry

5.5.1 Analytical Procedures to Define Substitution of MC

Analysis of the methoxy, hydroxypropoxy, and hydroxyethoxy contents of methylcelluloses may be performed via hydroiodic acid cleavage of the ether function with conversion into the corresponding alkyl iodides. The procedure described in ASTM method D-3876[1)] for the analysis of methoxy and hydroxypropoxy substitution is based on a gas chromatographic modification of the classical Zeisel distillation procedure. The sealed tube method is advantageous in that it allows simultaneous determination and differentiation of several substituents. Hydroxyethoxy substitution may be analysed by using the method of Lemieux and Purves. Other procedures based on infrared spectroscopy and nuclear magnetic resonance (NMR) spectroscopy not only allow quantitative determination of substitution but may also supply additional information regarding the distribution of substituents among the three reactive sites on the D-glucopyranosyl units. Analytical procedures using thin-layer chromatography, high-performance liquid chromatography (HPLC), and gas chromatography (GC) deliver results for the quantitative determination of substituent distribution in cellulose ethers [8].

1) ASTM D3876–96(2007) = Standard Test Method for Methoxyl and Hydroxypropyl Substitution in Cellulose Ether Products by Gas Chromatography.

5.5.2
Determination of the Molecular Weight

Conventional methods, including light scattering, sedimentation, and osmotic pressure, have been applied to the determination of molecular weight and molecular weight distribution of methylcelluloses.

More recently, gel permeation chromatography procedures have been developed for molecular weight and distribution analyses of HPMC and MC [8].

5.5.3
Methods of Identification and Quantitative Determination

Methods of identification and quantitative determination of methylcellulose derivatives in solution include enzymic methods, colour reactions, and precipitation reactions. These procedures allow quantitative determination of cellulose ethers in solution at concentrations as low as 1–10 ppm [8].

Flocculation or precipitation of methylcellulose by tannins in neutral to weakly acidic solution can be used to detect the methylcellulose analytically.

For simple differentiation of methylcelluloses from other cellulose derivatives or other hydrocolloids their solution behaviour and gelling properties can be used (see Sections 5.3.2 and 5.3.3).

5.6
Synergies with Other Hydrocolloids

Low-viscosity methylcelluloses (15–400 mPa s at 2%) show positive interactions with modified starches (E 1422, acetylated distarch adipates) from tapioca and waxy maize in developing hot viscosity in food. However, this viscosity profit only plays a minor role and is in most cases not the only reason for combining both stabilizers. The crucial facts are controlled behaviour during heating with less boiling and reduced splashing and an improved sensory profile. The dusty dry mouthfeel of the starches is compensated by the 'oily', or sometimes also perceived as sticky to slimy, cellulose ether.

Cooking starch and methylcellulose act in an antagonistic way, as described below.

The starch provides the desired viscosity already at lower temperatures and methylcellulose thickens – depending on concentration, total dry matter content, and amount of available water – at higher temperatures, from about 60 °C (140 °F); this, however, because of the low MC concentration, occurs without the formation of a gel structure. The temperature-dependent viscosity curve shows for the starch–MC composition no further extreme fluctuations, which is the case when using starch as single thickener. Additionally, methylcellulose increases the freeze–thaw stability of a food product. When starch granules release water during thawing (syneresis), the water is caught and

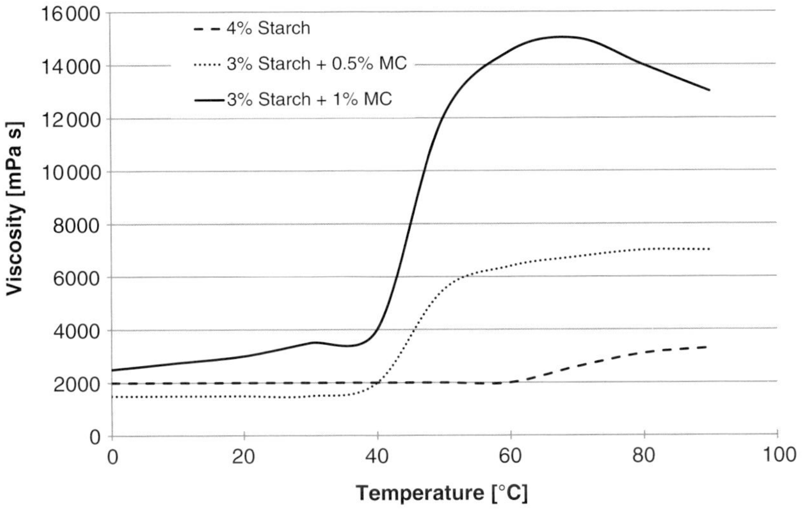

Figure 5.33 Interaction of low-viscosity methylcellulose with modified starch (E 1422).

bound immediately by MC. The tendency for synaeresis is reduced accordingly. By combination of these hydrocolloids, the dosage of each can be reduced by 30–50% compared to traditional recipes. For thin-flowing sauces, the recommendation is 0.3–0.6% of a low-viscosity methylcellulose and 1.5–3.0% of a modified starch. Figure 5.33 schematically illustrates the synergism between starch and MC.

An exemplary formulation in which this MC–starch synergism is applied is given in Section 5.7.5.2. Further intensive research is currently carried out to find more synergistic effects.

5.7 Food Applications with Typical Formulations

There are many possible applications in foods resulting from the functional properties of methylated cellulose ethers described in Section 5.3.6.

The choice of substitution type and amount as well as the viscosity level depend on:

- formulation – and presence of other substances that increase the viscosity or require water;
- method of manufacture;
- temperature control;
- desired texture and mouthfeel;
- and the desired consumption temperature.

The methylcellulose and HPMC grades, listed in subsequent tables, can be replaced with each other when necessary. For this reason, these two cellulose derivatives are summarized below.

Explanation of the Descriptions and Abbreviations in the Tables

- **Viscosity value:**
  Measured as a 2% solution of cellulose ether in water at 20 °C (68 °F) with a Brookfield spindle viscometer, the viscosity correlates with the molecular weight and, respectively, the chain length of the cellulose backbone:
  - low = 6–100 mPa s (2%),
  - medium = 400–4000 mPa s (2%),
  - high = 15 000–250 000 mPa s (2%);
- **gel strength (refers to** Table 5.5**):**
  - very firm (e.g. 'MC – type A');
  - firm (e.g. 'MC – type B');
  - semi-firm (e.g.: 'HPMC – type C' or 'HPMC – type D');
  - soft (e.g. 'HPMC – type E').
- **RTU/RTD** = 'ready-to-use' or 'ready-to-drink'

5.7.1
Bakery Products

5.7.1.1 Functions and Overview

Methylcellulose and HPMC fulfil multiple functions in baked goods:

- increase the baking volume;
- shape retention upon heating in an oven, microwave, or oil bath;
- improved appearance of the crumb (more elasticity, homogeneous pore structure);
- good water retention, reduced drying, longer lasting freshness and shelf life;
- reduced toughness after heating in the microwave;
- excellent barrier properties in frozen layer cakes, fillings, and coatings;
- inhibition of crystal growth during freezing processes, which results in smaller ice crystals and leads to less destruction of the dough matrix (resulting in significantly improved quality of frozen products);
- reduced staling during storage – by film formation around the starch grains, starch retrogradation is retarded;
- improved dough structure and texture while maintaining the short bite, no increased toughness, no rubbery chew characteristics;
- replacement of gluten – HPMC takes over the function of gluten with respect to adhesive proteins in gluten-free products;

Table 5.11 Application of methylcellulose and HPMC in bakery products.

Application	Product	Viscosity	Gel strength	Concentration (%)
Bread and bread rolls, buns	MC or	Medium	Firm	0.2–0.8
	HPMC	Medium	Soft	
Pizza dough plates	HPMC	Low	Semi-firm	0.2–0.4
Laminated dough	HPMC	High	Soft	2.0–3.0
Pan cakes/waffles	HPMC	Low	Semi-firm	0.2–0.4
Doughnuts	MC or	Medium	Firm	0.3–0.5
	HPMC	Low	Semi-firm	
Muffins	MC or	High	Very firm	0.1–0.5
	HPMC	Low	Soft–semi-firm	
Brownies	HPMC	Medium	Soft	0.2–0.4
Sweet rolls	HPMC	Medium	Soft	0.2–0.4
Ice cream cones	HPMC	Medium	Soft	0.2–0.4
Biscuits	HPMC	High	Soft	0.2–0.4
Cakes	HPMC	Low–medium	Soft–semi-firm	0.2–0.8
Cookies	MC or	Low	Firm	0.3–0.5
	HPMC	Low	Soft–semi-firm	
Tortilla chips (crispy)	MC	Low	Firm	0.3–0.5
Soft tortillas (foldable)	HPMC	Medium	Soft	0.1–0.2
Baked goods – gluten-free	HPMC	Medium–high	Soft	0.5–1.5

- reduced fat uptake in products that are heated in oil;
- minimize the stickiness on surfaces – which leads to decreased adhesion of the dough and thus to more rapid cleaning of equipment;
- easy detachment of moulded products from the mould;
- increased elasticity, strength, and flexibility – resulting in improved properties at subsequent processing steps, such as curl, wrinkle, and so on (less breakage during production, transport, and storage);
- lightweight fluffy structures of pancakes and waffles with good dimensional stability;
- increased throughput in production;
- lubricant in extrusion processes;
- reduced cleaning input (less time, possibility of water as cleaning agent).

Table 5.11 presents an overview of baked products in which methylated cellulose ethers are used.

5.7.1.2 **Typical Formulations**

Some typical formulations for bakery products are shown in Tables 5.12–5.16.

Table 5.12 Formulation for fat-free pita bread for sandwich pockets.

Ingredient	Amount (%)
Wheat flour	53.40
Water	43.80
Dry yeast	1.05
Salt	0.60
Shortening	0.50
Monoglycerides as emulsifier	0.30
Methylcellulose – type B with 4000 mPa s	0.35
(alternative: HPMC – type E with 4000 mPa s)	
Sum	100.0

Table 5.13 Formulation for frozen bread rolls for the microwave.

Ingredient	Amount (%)
Water to moisten the dry mix	42.7
Flour (wheat-rye mix or corn flour)	31.9
Maltodextrin	7.5
Skimmed milk powder	5.7
Sugar	5.4
Powdered egg	3.7
Baking soda (sodium-free)	1.3
Shortening with emulsifier	1.1
Lecithin	0.2
Salt	0.1
Methylcellulose – type B with 4000 mPa s	0.4
Sum	100.0

Table 5.14 Formulation for tortilla chips (crispy).

Ingredient	Amount (%)
Water	54.6
Corn flour for tortilla chips	45.0
Methylcellulose – type B with 15 mPa s	0.4
Sum	100.0

Table 5.15 Formulation for a batter for small cakes – replacement of hard fat.

Ingredient	Amount (%)
Wheat flour – type 450	31.9
Sugar	26.7
Cold water of 0–10 °C (32–50 °F)	22.1
Whole egg, liquid	10.0
Vegetable oil	4.7
Skimmed milk powder	2.5
Baking soda	1.3
Emulsifier E471	0.4
Salt	0.2
Vanilla extract	0.4
Methylcellulose – type A with 50 000 mPa s	0.4
Sum	100.0

Processing guideline:

- Flour, sugar, salt, baking powder, and vanilla extract are mixed together to prepare a dry blend;
- dispersing of methylcellulose and emulsifying agent (E 471) in oil;
- dissolving of the milk powder in cold water;
- production of the matrix emulsion as hard-fat substitute by transfer of the water phase into the oil phase under high shear forces;
- addition of the emulsion to the flour mixture while stirring slowly until a dough is formed;
- addition of the eggs and then stirring at medium speed;
- 45 g of dough is set into a paper form and baked for about 15 min in a preheated oven at 170 °C (338 °F).

Table 5.16 Formulation for doughnuts (yeast dough).

Ingredient	Amount (%)
Cold water of 0–10 °C (32–50 °F)	32.75
Wheat flour + soy flour	50.0 + 3.0
Sugar	6.0
Vegetable fat	3.0
Skimmed milk powder	2.0
Dry yeast	1.6
Egg yolk powder	0.8
Salt	0.5
Methylcellulose – type B with 4000 mPa s	0.35
Sum	100.0

5.7.2 Fillings

5.7.2.1 Functions and Overview

Methylcellulose and HPMC have the following functions in fillings:

- thermal gelation reduces the boil-out while heating and baking;
- shape retention upon heating in the oven, oil bath, or microwave (baking stability);
- reduced migration of free water, less soaking of the dough layers, increased crispiness;
- predictable product consistency at hot and cold temperatures, even during freeze–thaw cycles;
- high stability in acid systems and at low a_W values;
- good compatibility with high fat contents due to the lipophilic character.

Table 5.17 gives an overview of typical uses of methylated cellulose derivatives in fillings.

5.7.2.2 Typical Formulations

Tables 5.18–5.20 show some typical formulations for fillings.

5.7.3 Glazes, Predusts, Coatings, and Batters

5.7.3.1 Functions and Overview

Methylated cellulose derivatives are used in glazes, coatings, and batters for various reasons and functions:

Table 5.17 Use of MC and HPMC in fillings.

Application	Product	Viscosity	Gel strength	Conc. (%)
Fruit fillings (high Brix)	HPMC	Medium–high	Soft	0.2–0.4
Fruit fillings (without additional	MC or	Low	Firm	0.5–0.8
thickening)	HPMC	Low	Semi-firm	
Creamy fillings	MC or	Medium	Firm	0.2–0.4
	HPMC	High	Soft	
Cheese fillings	MC or	Medium–high	Firm	0.2–0.8
	HPMC	Medium	Soft	
Savoury fillings	MC or	Low–medium	Firm	0.2–0.5
	HPMC	Medium	Soft	

Table 5.18 Formulations for a bake-stable fruit preparation.

Ingredient	Amount (%)		
	Variant 1	Variant 2	Variant 3
Frozen fruits, diced/mashed	71.3	50.0	50.0
Sugar	19.1	24.0	9.2
Water (10 °C (50 °F))	6.1	19.5	Juice: 40.0
Methylcellulose – type B with 15 mPa s	MC or HPMC: 0.8	0.8	—
HPMC – type D with 4000 mPa s		—	—
Methylcellulose – type B with 40 000 mPa s	—	—	0.4
CMC 30 000	—	—	0.4
Modified starch (E 1422)	2.5	4.7	—
Citric acid (as solution 1:1)	0.1	1.0	—
Potassium sorbate	0.1	—	—
Flavours and colours	Optional	Optional	—
Sum	100.0	100.0	100.0

Table 5.19 Formulations for sweet bake-stable fillings based on milk based – with starch [20].

Ingredient	Amount (%)		
	Milk cream filling	Cheese cake filling	Cheese cake (6% fat)
Water	73.85	—	60.0
Fresh cheese	—	62.0	—
Fat substitute, instantized (modified corn starch)	5	—	11.7
Milk powder (MP)/fresh cheese powder (FCP)	MP (skimmed): 4.0 + MP (full fat): 3.5	—	FCP: 12.0
Flavour, natural	Optional	—	0.2
Granulated sugar	13.0	19.0	9.0
Fructose (crystalline)	—	—	3.0
Whole egg	—	Liquid: 10.0	Dried: 0.5
Egg white powder	—	—	1.0
Milk, fat-reduced (1.5% fat)	—	5.2	—
Flour or starch	—	Flour: 3.0	Starch: 1.5
Salt	—	—	0.1
Vanilla	Optional	0.4	0.4
Disodium phosphate	0.15	—	—
Methylcellulose – type B with 4000 mPa s	0.5	0.4	0.6
Sum	100.0	100.0	100.0

Table 5.20 Formulations for heat-stable fillings with cheese.

Ingredient	Amount (%)	
	Cheese filling for pockets	Ricotta cheese filling
Water	71.25	—
Cheese powder (full fat)	14.00	Parmesan: 2.25
Ricotta cheese 'light' (reduced-fat)	—	70.00
Ricotta cheese, full fat	—	25.00
Milk, low fat	10.00	—
Starch	3.00	—
Sugar syrup	1.00	—
Mustard powder, white pepper, paprika	Each 0.10	—
Garlic powder, onion powder	Each 0.10	Each 0.25
Parsley	—	0.50
Salt	—	1.00
Methylcellulose – type B with 400 mPa s	0.25	—
Methylcellulose – type B with 40 000 mPa s	—	0.75
Sum	100.0	100.0

- thermal gelation reduces the upheaval of breading, no blistering;
- increased durability of products and frying oils;
- replacement of vegetable and animal protein and gelatine;
- improved barrier properties;
- fast dissolution and hydration properties in dry mixtures;
- good adhesion to meat surfaces and plant substrates;
- excellent freeze–thaw stability prevents bursting and breaking of batters and coatings;
- increased homogeneity of coatings;
- edible films of MC reduce the moisture loss at increased juiciness of products and decrease the oil migration between different components;
- increase in yield due to reduced water (vapour) loss.

Table 5.21 shows which methylcellulose and HPMC grades are used in coatings.

5.7.3.2 Typical Formulations

Typical formulations are presented in Tables 5.22 and 5.23.

5.7.4 Reformed and Extruded Products

5.7.4.1 Functions and Overview

Methylcellulose and HPMC have different functions in reformed products and extruded foods:

Table 5.21 Application of methylcellulose in coatings.

Application	Product	Viscosity	Gel strength	Conc. (%)
Dry glazes	MC or	Low	Firm	0.75–3.2
	HPMC	High	Soft	
Batter mixes (dry blends for meat, poultry and seafood)	MC or	Medium	Firm	1.0–2.5 (in dry mix)
	HPMC	Low	Semi-firm	
	or			
	HPMC	High	Soft	
Edible films	HPMC	Low	Semi-firm	5.0–10.0 (before drying)
French fries – dipping solution	MC	Low	Firm	2–3
Liquid glazes	MC or	Medium	Firm	0.3–1.0
	HPMC	Medium–high	Soft–semi-firm	
Moisture barrier (dry powder)	HPMC	Low	Soft	1.0–5.0
Wet batters for animal products	MC or	Medium	Firm	0.2–0.5
	HPMC	Low	Semi-firm	
Egg-free batters	HPMC	Low	Soft–semi-firm	0.3–0.8
Adhesive for seasonings	HPMC	Low	Semi-firm	0.5–6.0

Table 5.22 Formulation for coating solutions of French fries and bakery products.

Ingredient	Amount (%)
Water	97–98
Taste-giving ingredients such as spices, herbs, seeds, and so on	Optional
Methylcellulose – type B with 15 mPa s or HPMC – type C with 15 mPa s	2–3
Sum	100

Processing guideline:

The MC or HPMC solution is sprayed onto the surface of the foods. Alternatively, the products go through an immersion bath. The wafer-thin layer is fixed on the product outer surface with warm air or hot oil (during pre-frying).

- increase of yield and volume due to reduced water (vapour) loss in all thermal processes;
- retention of shape and structure during manufacturing processes, during heating and cooking, during transport and storage, in all freeze–thaw processes, and in final preparation before serving;
- connection and gluing of particulate components of the formulation, stable consistency;
- replacement of other 'glues' and binders such as whole egg, egg white, milk proteins, gluten – vegetarian and vegan formulations are possible;

Table 5.23 Formulations for a dry mix for breading [21].

Ingredient	Amount (g)
Wheat flour	77.3
Maize or wheat starch	11.9
Salt	6.4
Baking soda	1.3
Raising agent ($Na_2H_2P_2O_7$ or $NaHCO_3$)	1.8
Xanthan	0.05
Methylcellulose – type B with 1500 mPa s	1.25
Sum	100.0

Data taken with permission from Reference [21].

Processing guideline:

1. The dry mixture is packed in 150 g portions;
2. a packet of batter mix (150 g) is mixed with 230 g of cold water (0–10 °C (32–50 °F));
3. the liquid batter is used to coat chicken, fish, or vegetable pieces;
4. excess batter is allowed to drain on a sieve tray;
5. this is followed by fixation in hot oil at 175 °C (347 °F).

- lubricant in extrusion processes, reduced adhesion to equipment surfaces, easy cleaning, increased throughput;
- good mouthfeel – not a dry and dusty impression as occurs with starches – and improved juiciness;
- reduced oil absorption when heated in hot oil – healthier cooking.

Table 5.24 indicates how methylcellulose and HPMC are used in reformed and extruded products.

5.7.4.2 Typical Formulations

Tables 5.25–5.30 show typical formulations of various reformed and extruded products.

5.7.5 Soups and Sauces, Salad Dressings, and Marinades

5.7.5.1 Functions and Overview

Methylcellulose and HPMC perform various functions in liquid systems:

- thermal stability, good adhesion to substrates, no flowing away during heating/cooking/grilling, suitable for UHT processes;
- reduction of undesired cooking volume (less gas bubbles) and boiling over;

Table 5.24 Application of methylated cellulose ethers in reformed and extruded products.

Application	Product	Viscosity	Gel strength	Conc. (%)
Vegetarian sausages and burger, soy products	MC	Medium–high	Firm–very firm	0.7–2.0
Meat croquettes	MC	Medium	Firm–very firm	0.25–0.35
Potato products	MC or HPMC	Medium Low–medium	Firm Soft–semi-firm	0.2–0.6
Seafood	MC	Medium–high	Firm–very firm	0.4–1.0
Onion rings	MC	Medium	Firm	0.8–1.2
Pasta	MC	Medium	Firm	0.5–0.75

Table 5.25 Formulations for vegetarian patties, burgers, and 'meatballs' based on soy protein [7,20].

Ingredient	Amount (%)		
	Variant 1	Variant 2	Variant 3
Water of 0–10 °C (32–50 °F)	60.0	55.0	64.0
Texturized soy protein concentrate (TSPC)	21.0	20.0	19.0
Thickener	Maize starch: 2.0; potato starch: 2.0	—	Gluten: 4.0
Breadcrumbs	—	3.8	—
Vegetarian broth (powder)	—	3.0	—
Spices and herbs	Onion, dried: 1.5; seasoning and flavour: 0.5	Onion, dried: 0.7; pepper, black: 0.3; cumin: 0.6; parsley: 0.5; soy sauce: 3.0	Spice mix: 2.0; beef flavour, natural: 1.8; soy sauce: 1.5
Hard cheese, grated	—	5.0	—
Soy protein isolate	—	2.0	3.0
Vegetable oil	10.0	5.0	3.0
Salt	1.0	—	0.5
Dextrose	—	—	0.5
CMC – 30 000 mPa s	—	0.3	—
Methylcellulose – type B with 4000 mPa s	—	0.8	—
Methylcellulose – type A with 16 000 mPa s	2.0	—	0.7
Sum	100.0	100.0	100.0

Data taken with permission from References [7,20].

Processing guideline for variant 2:

1. Soy protein textured (TSPC) is mixed with about ⅓ of the water. The rehydrated soy protein is stored cold for some time to allow it to swell sufficiently;
2. methylcellulose, CMC, breadcrumbs, broth (dry), and spices are premixed dry;
3. incorporation of the dry mixture (of step 2) into the swollen soy protein;
4. addition of cheese, soy protein isolated, soy sauce, and water to the mass of step (3). Mix well, and leave to stand for about 5–10 min;
5. addition of oil, and then mix for 1–2 min;
6. form little dumplings or balls (about 20 g) and rods (40 g);
7. bake at 160–180 °C (320–356 °F) for 7–10 min;
8. cold storage or freezing;
9. directly before consumption, bake at 160–170 °C (320–338 °F) for about 10 min.

Table 5.26 Formulation for bake-stable cheese.

Ingredient	Amount (%)
Grated cheese (Edam, Gouda, Maasdam, Cheddar, etc.)	75.0
6% Methylcellulose solution of MC – type B with 40 000 mPa s (concentration in the finished reformed cheese = 1.5%)	25.0
Sum	100.0

Table 5.27 Formulation for meatless sausages based on fish protein.

Ingredient	Amount (%)
Fish	94.0
Cold water at 0–5 °C (32–41 °F)	3.2
Oil	2.3
Methylcellulose – type A with 50 000 mPa s	0.5
Sum	100.0

Processing guideline:

1. Dispersion of the methylcellulose in oil;
2. adding cold water to the MC–oil dispersion and preparation of an emulsion with shear forces;
3. addition of spices and MC–O/W emulsion to the minced fish (frozen) and mix at low speed until a homogeneous mass is formed;
4. filling of the fish mass in casings;
5. heat treatment for protein fixation and preservation;
6. cold storage.

Table 5.28 Formulation for 'potato rosti' or hash browns.

Ingredient	Amount (%)
Raw potatoes, grated	99.42
Salt	0.25
Nutmeg, black pepper, and so on	To taste
Methylcellulose – type A with 16 000 mPa s	0.33
Sum	100.0

Table 5.29 Formulations for extruded onion rings [7].

Ingredient	Amount (%)		
	Variant 1		Variant 2
Water	73.0–73.9		—
Onion, dried	12.0		—
Onions, strips and chopped	—		65.0 + 32.2
Wheat flour	8.0		—
Corn starch, native	4.0		—
Sodium alginate (+ 5% $CaCl_2$ solution as dipping bath to gel the hydrocolloid)	—		1.0
Salt	1.0		0.5
Sodium phosphate	-		0.5
Methylcellulose – type B with 4000 mPa s	1.7 or:	—	—
Methylcellulose – type B with 40 000 mPa s	—or:	0.8	0.8
HPMC – type E with 100 000 mPa s	0.3		—
Sum	100.0		100.0

Data taken with permission from Reference [7].

Table 5.30 Formulations for extruded shrimp and seafood [7].

Ingredient	Amount (%)	
	Variant 1	Variant 2
Water	30.0	30.0
Soy protein	10.0	10.0
Shrimp	54.5	32.5
Fish	—	22.0
Salt	0.5	0.5
Bread crumbs	2.0	2.0
Potato starch, native	1.5	1.5
Flavour	Optional	Optional
Methylcellulose – type B with 4000 mPa s	1.5	1.5
Sum	100.0	100.0

Data taken with permission from Reference [7].

Table 5.31 Using MC and HPMC in emulsions and other liquid systems.

Application	Product	Viscosity	Gel strength	Conc. (%)
Cheese sauces	MC	Low–high	Firm	0.3–0.5
Condiments	MC or	Medium	Firm	0.2–0.5
	HPMC	Low–high	Soft	
Creamy sauces	MC or	Low–medium	Very firm/firm	0.2–0.4
	HPMC	Low–medium	Soft–semi-firm	
Gravies – dry mixes	HPMC	Medium–high	Soft	2.0–5.0
Gravies, liquid (RTU, ready to use)	MC	Low	Firm	0.5–0.8
Red sauce based on	MC or	Low–high	Firm	0.1–0.5
tomato	HPMC	Low–medium	Soft	
Soups	MC	Low–medium	Firm	0.2–0.4
Salad dressings –	MC or	Medium	Firm	1.0–3.0
dry mixes	HPMC	Medium	Semi-firm	
Pourable salad	MC or	Medium	Firm	0.25–0.35
dressings	HPMC	Low–medium	Soft–semi-firm	
Spoonable dressings and dips (high-viscosity)	HPMC	Medium–high	Soft–semi-firm	0.2–0.5
Marinades (liquid)	MC	Low–medium	Firm	0.2–0.4
Dry marinades	HPMC	High	Soft	2.0–5.0

- 100% cold water solubility, control of viscosity, and consistency at a low dose;
- possibility of delayed hydration allows low viscosities during manufacturing – resulting in high process flexibility;
- synergy with modified starches in the creation of hot viscosity; very compatible with other thickeners;
- development of reduced calorie products – replacement of fat/oil while maintaining the creamy mouthfeel of comparable full-fat recipes;
- emulsifying properties prevent the separation of oil during storage;
- high stability in acid systems;
- very good freeze–thaw stability.

Table 5.31 gives an overview of the possible applications of methylated cellulose derivatives in soups, sauces, salad dressings, and marinades.

5.7.5.2 Starch–Methylcellulose Synergism in Sauces

In subsequent recipes (Tables 5.32 and 5.33) the texture and functionality are controlled by the synergistic effect between starch and low-viscosity MC.

5.7.5.3 Traditional Soups, Sauces, Toppings

Some typical formulations are shown in Tables 5.34 and 5.35.

Table 5.32 Formulation for cheese sauce.

Ingredient	Amount (g)
Water	77.5
Cheese powder	11.0
Powdered milk (fat-free)	9.0
Starch	1.7
Methylcellulose – type B with 400 mPa s	0.3
Mustard powder (senfpulver)	0.1
Garlic powder	0.1
Onion powder	0.1
White pepper	0.1
Paprika	0.1
Sum	100.0

Processing guideline:

1. Homogeneous mixing of dry ingredients;
2. provide the water and then stir the dry mixture into the liquid phase;
3. heating at medium temperature for 5 min;
4. cooling the sauce for 30 min.

Table 5.33 Formulation for white sauce.

Ingredient	Amount (g)
Skimmed milk	86.43
Melted butter	8.35
Wheat flour	2.15
Starch	2.00
Salt	0.74
Methylcellulose – type B with 15 mPa s	0.30
White pepper	0.03
Sum	100.0

Processing guideline:

1. Homogeneous mixing of dry ingredients;
2. slow addition of melted butter to the dry mixture;
3. add the milk to the mix of step 2;
4. heat to boiling with constant stirring;
5. the best effect is achieved when the sauce is stored cold and reheated.

Table 5.34 Formulations for pizza sauce with increased bake stability and reduced tomato content.

Ingredient	Amount (g)		
	Standard, with increased baking stability	**Bake-stable, 30% reduction in tomato**	**Bake-stable, 40% reduction in tomato**
Water	41.7–41.9	56.0	60.7
Tomato concentrate (three times conc.)	47.5	33.2	28.5
Vegetable oil	2.7	2.7	2.7
Salt	1.4	1.4	1.4
Sugar syrup	1.4	1.4	1.4
Sugar	1.3	1.3	1.3
Dried onion	1.3	1.3	1.3
Starch	1.0	1.0	1.0
Garlic powder	0.5	0.5	0.5
Pepper	0.4	0.4	0.4
Oregano	0.3	0.3	0.3
Methylcellulose – type B with 40 000 mPa s	MC: 0.3	—	—
Methylcellulose – type B with 4000 mPa s	or	0.5	0.5
HPMC – type D with 4000 mPa s	HPMC: 0.5	—	—
Sum	100.0	100.0	100.0

Table 5.35 Basic formulation for cream soups.

Ingredient	Amount (g)
Water	61.4
Milk	30.1
Vegetable oil	4.0
Modified starch (E 1422)	2.0
Chicken broth	1.5
Salt	0.5
Pepper	0.1
Methylcellulose – type B with 15 mPa s	0.4
Sum	100.0

5.7.5.4 Emulsions

Tables 5.36–5.38 show some typical formulations that involve emulsions.

Table 5.36 Formulation for a matrix emulsion as a substitute of hard fats.

Ingredient	Amount (%)
Vegetable oil	49.0
Cold water from 0–10 °C (32–50 °F)	49.0
Methylcellulose – type A with 50 000 mPa s	2.0
Sum	100.0

Table 5.37 Formulations for mayonnaise-type salad dressings with reduced fat content.

Ingredient	Amount (%)	
	30% Fat, freeze–thaw stable, cold process	10% Fat, cooking process
Water	39.9	62.2
Sunflower oil	30.0	10.0
Vinegar (5%)	13.0	14.0
Lactic acid (10%)	4.0	—
Lemon juice	—	1.0
Sugar	4.0	4.0
Maltodextrin (DE 20)	4.0	—
Fructose syrup	—	2.0
Starch (E 1442)	—	2.5
Salt	2.0	2.0
Mustard	2.0	1.3
CMC – 30 000 mPa s (2%)	0.8	0.5
Methylcellulose – type A with 50 000 mPa s	0.3	0.5
Sum	100.0	100.0

Table 5.38 Formulations for liquid marinades.

Ingredient	Amount (%)	
	Mustard marinade	Steak marinade
Water	54.14	54.14
Dijon mustard	26.70	—
Steak sauce	—	26.70
Sugar syrup with high fructose content	6.70	6.70
Vegetable oil	4.80	4.80
Vinegar	2.50	2.50
Salt	2.00	2.00
Instant starch	1.00	1.00
Xanthan	0.18	0.18
Onion powder	0.13	0.13
Garlic powder	0.10	0.10

(*continued*)

Table 5.38 (*Continued*)

Ingredient	Amount (%)	
	Mustard marinade	**Steak marinade**
Potassium sorbate + sodium benzoate	Each 0.70	Each 0.70
Methylcellulose – type B with 4000 mPa s	0.35	0.35
Sum	100.0	100.0

Table 5.39 Formulations for gummy bears with reduced gelatine content.

Ingredient	Amount (%)	
	25% Gelatine-reduction	**50% Gelatine-reduction**
Water out of the gelatine mass	16.8	16.8
Water out of the sugar solution	7.3	8.6
Total water content	**24.1**	**25.4**
Sucrose	28.7	29.1
Sugar syrup (43 DE)	37.2	37.8
Gelatine (250 Bloom, 8 mesh)	7.3	4.82
Citric acid	1.2	1.2
Colouring	0.06	0.06
Flavour 1 + flavour 2	0.21 + 0.06	0.21 + 0.06
Starch	0.82	0.82
Methylcellulose – type A with 50 000 mPa s	0.35	0.53
Sum	100.0	100.0

5.7.6
Confectionery and Desserts

The formulations of several confectionary products and desserts are given in Tables 5.39–5.42.

5.7.7
Meat and Fish Products

See Tables 5.43–5.47 for formulations of meat and fish products.

5.7.8
Pet Food and Animal Feed

Methylcellulose and HPMC are used to perform various tasks in pet food and animal feed:

- improved stability of emulsified products;

Table 5.40 Formulations for chocolate desserts.

Ingredient	Amount (%)	
	Standard	Reduced fat
Milk (3.5% fat)	55.0	—
Skimmed milk (0.3% fat)	—	55.0
Fructose syrup	20.0	20.0
Water	18.9	18.9
Starch (E 1442)	2.0	1.8
Cocoa powder, de-oiled	3.0	3.0
Salt	0.4	0.4
Vanilla	0.2	0.2
Sodium stearoyl lactylate (SSL)	0.1	0.1
Methylcellulose – type A with 16 000 mPa s	MC or HPMC: 0.3	MC or HPMC: 0.4
HPMC – type E with 4000 mPa s		
HPMC – type E with 100 000 mPa s	0.1	0.2
Sum	100.0	100.0

Table 5.41 Formulation for 'Crème brûlée' – standard.

Ingredient	Amount (%)
Cream (30–40% fat content)	66.0
Sugar	16.0
Whole egg, liquid	13.2
Egg yolks, liquid	4.0
Vanilla extract	0.5
Methylcellulose – type A with 700 mPa s	0.3
Sum	100.0

Table 5.42 Formulation for 'Crème brûlée' – low-fat (maximum 2% total fat content).[a)]

Ingredient	Amount (%)
Water	57.6
Milk powder (skimmed, max. 1% fat content)	15.0
Vanilla liqueur (30% alcohol by volume)	10.0
Sugar	9.0
Egg yolk, liquid	6.0
Cinnamon	0.2
Vanilla extract	0.2
Methylcellulose – type A with 700 mPa s	2.0
Sum	100.0

a) For the topping, additionally brown sugar (5%) and cinnamon (0.2%) are mixed and caramelized.

Table 5.43 Formulation for low-fat chicken patties.

Ingredient	Amount (%)
Chicken breast	85.0
Water (cold)	13.0
Phosphate + salt	0.2 + 0.5
Chicken broth (powder)	0.1
Methylcellulose – type A with 50 000 mPa s	1.2
Sum	100.0

Table 5.44 Formulation for a meatloaf.

Ingredient	Amount (%)	Amount (g)
Minced beef	63.9	383.4
Minced pork	20.0	120.0
Cold water with 0–5 °C (32–41 °F)	6.0	36.0
Breadcrumbs	4.5	27.0
Onions, chopped	2.4	14.4
Green pepper + black pepper	1.6 + 0.3	9.6 + 1.8
Salt	1.0	6.0
Methylcellulose – type A with 700 mPa s or type B with 400–4000 mPa s	0.3	1.8
Sum	100.0	600.0

- suspension of solid particles, reduced sedimentation and phase separation;
- possibility of delayed hydration allows processing at low viscosities;
- providing filling viscosity;
- dust control in the production of pellets;
- enhanced durability of pellets, less abrasion/breakage;
- increased lubricity ('lubricant');
- positive influence on chewing attributes.

Table 5.45 Formulation for a salmon mousse.

Ingredient	Amount (%)
Salmon, cooked and shredded	79.6
Cold water with 0–10 °C (32–50 °F)	12.5
Mayonnaise	7.5
Spices	Optional
Methylcellulose – type A with 16 000 mPa s (alternatively, HPMC – type E with 4000 mPa s)	0.4
Sum	100.0

Table 5.46 Formulation for fish balls stuffed with sauce, bake-stable.

	Ingredient	Amount (%)
(A) **Fish matrix**	Water	24.5
	Frozen Alaska pollock fillet	72.2
	Salt	0.8
	Instant potato starch	2.0
	Methylcellulose – type A with 50 000 mPa s	0.5
(B) **Lemon butter sauce filling**	Water	72.2
	Spices (lemon-butter-flavour)	4.5
	Salt	0.7
	Butter, liquid	20.0
	Instant potato starch	1.5
	Xanthan	0.2
	Methylcellulose – type A with 50 000 mPa s	0.3
	HPMC – type E with 100 000 mPa s	0.5
(C) **Final product**	Filling resp. sauce (frozen)	5–10 g
	Fish dough	40–50 g

Table 5.47 Formulations for burger from fish fillet and calamari.

Ingredient	Amount (%)	
	Fish fillet burger	Calamari burger
Cod fillet	95.2	—
Calamari pieces (thawed)	—	45.1
Calamari crushed	—	45.1
Potato flakes	3.0	8.0
Salt	1.0	1.0
Methylcellulose – type A with 16 000 mPa s	0.8	0.8
Sum	100.0	100.0

Table 5.48 indicates the use of methylcellulose and HPMC grades in various animal feed products.

5.8 Non-food Applications

The technical possibilities of pure methylcelluloses are very diverse. The property profiles of MC and hydroxypropylated methylcelluloses (HPMC) are very similar – so that in many end-products, replacement of MC with HPMC – and

Table 5.48 Recommendations for methylated cellulose ethers in animal feed.

Application	Product	Viscosity	Gel strength	Concentration (%)
Pellets	MC or	Low	Strong	2.0–3.0
	HPMC	Low	Semi-firm	
Semi-moist products	HPMC	Medium–high	Low–semi-firm	0.1–0.3
Canned products	MC or	Medium	Strong	0.2–0.5
	HPMC	Low–high	Low–semi-firm	

vice versa – can be realized without major changes in formulation and process [8]:

1) **Adhesives:**
   - **Wallpaper glue:**
     Methylcellulose gums alone or in combination with native and modified starches are employed as wallpaper adhesives. MC is more resistant to microbial growth than are starches and is often used as heavy-duty wallpaper adhesive with papers that have low water permeability, such as vinyl-coated papers. It is claimed that inclusion of ethylene–vinyl acetate copolymer emulsions, poly(ethylene glycol), or quartz provides improved strippability to the pasted paper.
   - **Electronics:**
     Methylcellulose may be used as adhesive in electronic components. For example, MC is used with zinc oxides in the construction of low-loss condensers.
   - **Generally:**
     Methylcellulose gums may be used in adhesive for bonding of leather, textiles, and paper. Combined with antimicrobial, fungicidal, and anticorrosion agents, they provide effective adhesives for porous substrates at solids levels of about 3%. MC grades are also used for fixing in the restoration of sensitive artefacts. Methylcellulose is used to thicken various adhesives based on aqueous polymer emulsions of poly(vinyl acetate), acrylate ester copolymers, and ethylene–vinyl acetate copolymers. Other adhesive applications include library paste, book binding, cigar overwraps, and pencil lead binders.
2) **Agricultural applications:**
   Agricultural applications of methylcelluloses are centred primarily around the capability of the polymer to act as a binder, film former, and release agent.
   - **Fertilizers:**
     The binding properties of methylcelluloses allow their use as a granulating agent for fertilizer compositions and as a medium to bind fertilizer to seeds. The use of MC grades to bind fertilizer to seeds is particularly beneficial in areas where climatic conditions are adverse. In

addition, the use of such coatings on fine seed can result in a larger coated seed that is easier to hand plant.

- **Pesticides:**
  Free-flowing pesticide aggregates that are iso-dimensional and readily dispersed in water are prepared using methylcellulose gums as a binder.

  Herbicidal, insecticidal, and fungicidal products are possible. In addition, controlled-release compositions can be obtained by employing MC grades as granulating or film-forming agents. Methylcelluloses may be used as binders, or 'seed stickers', to prevent dusting of bioactive substances during the seed coating process. Typical use levels based on the weight of seeds are 25–50% of a low-molecular-weight MC grade. Methylated cellulose ethers are also used to stabilize flowable pesticide formulations to provide compositions that are readily dispersible in water. In aqueous herbicide and pesticide sprays, MC gums, at levels as low as 0.015%, help to wet the surface of the foliage and to adhere the pesticide to the plant, When incorporated into agricultural dusts (6–10%), the hydrocolloids promote adhesion of the dust to plant foliage.
- **Animal feed:**
  methylated cellulose ethers serve as coating and binding agent for feed pellets, flakes, and extruded products for animal nutrition.
- **Miscellaneous:**
  methylcelluloses have been used as a soil stabilizer and as coating for egg to prolong freshness on storage.

3) **Ceramics:**
   The use in ceramics has been a particularly active growth area for MC and HPMC. Development of new ceramic formulations and the demand for close-tolerance, extruded, and pressed ceramic fabrications have resulted in industrial demand for highly effective processing aids. Both the water-holding properties and the thermal gelation of methylcelluloses improve processing of ceramics by virtue of inhibition of binder migration while achieving desired plasticity and excellent green strength. Studies of the effects of MC grades on the structure and rheology of concentrated mineral suspensions have led to improved insight into the effects of these binders on uniformity of dispersion of the mineral particles and the presence of hydrated macromolecular adsorption layers at the surface of particles. Effective use levels of methylcellulose gums as binders and extrusion aids range from as low as 0.1% to as high as 10% based on solids. In addition to controlling dimensional stability and green strength of ceramic extrusions, MC products help control the density and porosity of the resulting ceramic. The hydrocolloids are used in refractories and in the preparation of shaped metal articles formed by sintering, wherein the gum reduces the fragility of shaped articles prior to sintering.

Methylcellulose is also used as a binder and an adhesive for the application of glazes to ceramics.

4) **Paints and protective coatings:**
   - **Paints:**
     The use of methylcelluloses, particularly hydroxyalkyl-modified methylcellulose gums, as thickeners for latex paints has been a major application. In recent years, competitive pressure from hydroxyethylcellulose has resulted in erosion of the market share of MC grades in latex paint systems, but MC use continues to supply a large market. Levels in latex paints range from as low as about 360 g per 100 l (3 lb per 100 gallon) of paint for high-molecular-weight cellulosics with 2% aqueous solution viscosities of around 75 000 mPa s to about 720 g per 100 l (6 lb per 100 gal) for lower-molecular-weight cellulose ethers. Evaluation of the effect of molecular weight on paint properties indicated that high-shear rheology, spatter resistance, and foaming tendency are significantly affected by molecular weight. Low-shear rheology, scrub resistance, and colour acceptance are, though, not affected. Cellulose ethers used to thicken latex paints may be exposed to cellulase enzymes or to residual polymerization catalysts, both of which may effect molecular weight reduction and loss of thickening efficiency. Cellulose derivatives that have a high DS, such as HPMC products, show greater resistance to viscosity degradation in latex paint systems than cellulosics with a lower total DS.
   - **Paint stripper:**
     Methylcellulose derivatives with organic solvent solubility are used in paint stripper systems to provide viscosity so that the stripping solvent does not drain from vertical surfaces and to prevent rapid evaporation from the thickened system by promoting surface-film formation. Many paint stripper systems contain volatile solvents such as methylene chloride, methanol, toluene, and acetone. Control of volatility by thickeners and waxes is important for improved performance of the stripper composition. Both scrape-off and flush-off paint stripper formulations use methylcellulose gums at levels of approximately 1.44 kg per 100 l (12 lb per 100 gal) of stripper. A surfactant is included in flush-off formulations to promote water flushing of the stripping mixture.

5) **Construction:**
   The largest single industrial use for methylcellulose and its modifications is in construction applications. Particularly important uses include formulation of tape joint compounds for gypsum board panelling, gypsum spray plaster formulations, ceramic tile adhesive, grout, and mortar applications. These gums are valuable additives in Portland cement mortars, cement plasters, stucco, and crack filler compositions, as rheology control and water retention aids. Recent new applications include use in cement board extrusion, low-density foamed cement construction compositions,

and cement-based coatings on foamed polystyrene in insulating wall structures.

- **Gypsum plasters:**
  The use of methylcellulose and HPMC in gypsum spray plaster compositions is probably the largest application for this gum family in Europe. Concentrations of 0.2–1.0% based on gypsum improve pumpability, trowelability, and water retention of the gypsum plaster. Rapid drying of the plaster is prevented so that satisfactory curing, bonding, and finishing are achieved.
- **Tile adhesive:**
  Another important field of application is tile adhesives. The quality of the water balance and the processing properties are significantly improved by addition of MC, HPMC, or HEMC. The rheological profile then allows a flexible, non-sticky application and homogeneous distribution of the joint compound with optimum consistency, increased yield, and high stability.
- **Modified cement:**
  The use of methylcelluloses in cement formulations leads to cement mortars, stuccos, and plasters with improved workability, curing, and water retention.

  Normal use levels for HPMC or MC are in the 0.03–0.2% range based on solids in the modified cement formulation. Methylcellulose reduces shrinkage and enhances mortar strength. The consequences of addition of plasticizers, air-entraining aids, and other additives on compressive and tensile strength cannot be accurately predicted; they should be determined experimentally.
- **Lightweight components made of cement:**
  Lightweight, or air-bubbled, cement is desired for various applications, including non-load-bearing cement walls and cement roofing. The surface-active properties of methylcelluloses and HPMC at a concentration of 0.5–1.0% based on dry ingredients in the cement formulation and the use of additional surfactants allow the formation of foamed cement systems with adequate foam stability, resulting in a lightweight cured concrete system.
- **Other applications:**
  Portland-cement-based adhesives and grouts for stone and ceramic tile may also employ methylated cellulose ethers as workability and water-retention aids. Small amounts of 0.02–2% based on cement enhance workability, open time, water retention, and bonding. Methylcelluloses (0.05–0.8% based on cement) have found application as workability improvers and water retention aids in exterior plaster and insulating wall systems.

6) **Detergents:**
   Non-ionic methylcellulose gums such as MC and HPMC have been used in laundry detergents as anti-redeposition aids. Synthetic fabrics,

notably polyester and polyester blends, and restrictions on use of phosphate builders required new formulations employing highly substituted modified methylated cellulose derivatives (0.5–1.5% based on the detergent formulation). In addition to the efficacy of these gums as anti-redeposition agents, they have also been recognized as effective substances in imparting a 'soil shield' to polyester textiles. A high degree of substitution of the cellulose ether is particularly important to achieve this effect. One proposed explanation is that a reversible thermal coagulation of the highly substituted cellulosic onto the textile results in an irreversible adsorption on the textile. Subsequent coverage by a layer of hydratable methylcellulose decreases soil redeposition and promotes detachment of oils. A highly substituted methylcellulose is more hydrophobic and is therefore expected to associate closely with the hydrophobic polyester. Hydroxybutylmethylcellulose, a product that is no longer commercially available, has been reported to be most effective as an anti-redeposition agent and soil-shield for polyester.

7) **Paper:**
   Methylcellulose gums find application in speciality papers, including uses as a microencapsulating agent for pressure-sensitive copying paper and as a coating for paper used as a separator in dry cells. They find additional uses in paper products as greaseproof coatings, adhesive coatings, sizing, and release coatings. Because of their approval as direct food additives, premium-grade methylcellulose products may be employed in paper used in contact with food and other paper-based food packaging materials.
8) **Textiles:**
   Methylcelluloses are used:
   - as thickeners in printing paste formulations;
   - as coatings for sheets used in transfer printing on textiles;
   - as dilute MC solutions to treat cotton yarn to reduce linting and friction in the knitting process;
   - as thickeners for latexes in textile coating applications.
9) **Suspension polymerization:**
   Methylcellulose gums are used to stabilize and adjust droplet sizes of monomer suspensions in water so that a uniform particle size polymer is obtained when the monomer is polymerized. This use is particularly important in the polymerization of vinyl chloride. In addition, these cellulosic stabilizers are also reported to be of value in the suspension polymerization of acrylonitrile, vinylidene chloride, and styrene. Typical use levels range from about 0.05 to 0.1% based on the weight of monomer.
10) **Cell culture/virology:**
   Methylcellulose is used as a basic material for cell cultures to study the reproduction of viruses.
11) **Cosmetics and personal care:**
   In particular, hydroxypropylated methylcelluloses (HPMCs) are commonly used in shampoo and hair conditioning compositions; 0.2–0.4% of

the hydrocolloid provides body and thickening to such formulations and functions as an emulsion stabilizer. The high compatibility of the non-ionic gum with surfactants and salts contributes to its popularity in this application. Its solubility in alcohol–water mixtures is also valuable in preparing a compatible composition that does not separate upon storage. The unique combination of thermal gelation and surface-active properties gives exceptional foam stabilization. Other possible applications include:

- toothpaste and tooth cleaning formulations;
- lip gloss compounds;
- hand lotions;
- antiperspirants;
- cosmetic creams;
- gelled alcoholic systems for cosmetics.

Normal levels range from 1 to 3% of the formulation. A long-lasting insect repellent includes a methylcellulose gum. Upon application, a protective film is obtained that prevents rapid loss of the active ingredient.

12) **Pharmaceutical applications:**

Methylated cellulose derivates (MC and HPMC) are used:

- as binding and thickening agents as well as granulating agent for powder products;
- for the coating of tablets;
- as bulk laxatives;
- for stabilization of emulsions and multiphase mixtures, such as wound dressings and ointments;
- as thickening agents for drugs for eye diseases (ophthalmics);
- as matrix in the preparation of controlled-release drug systems;
- as replacement of gelatine for production of hard-shell medicinal capsules;
- to cause immediate and permanent immobilization of human spermatozoa (*in vitro*).

For many years, methylcelluloses and hydroxypropylmethylcelluloses have been used as binders, thickeners, stabilizers, and coatings of diverse pharmaceutical products. A highly promising new growth area is the use of MC or HPMC to control drug delivery rates in sustained-release pharmaceutical formulations.

The main application of methylcellulose and HPMC can be described as follows:

- **Tablet film coatings:**

Because methylcelluloses are water soluble and may have some degree of organic solubility (depending on the substitution type and DS), they are used as aqueous-based film coatings, as coatings delivered from organic solvents, and for coating from solvent–water mixtures. Typical formulations employ lower molecular weight grades of cellulose ethers, normally with solution viscosities of from 5 to 50 mPa s at

2% concentration. The final coating weight is usually 1–3% of the tablet. The coating formulation also contains coloured pigments or TiO_2 and plasticizers. The effects of molecular weight and plasticizers on mechanical properties and permeability of methylcellulose films have been examined.

- **Controlled and delayed release of active ingredients:**
  Medium- and high-viscosity HPMC and MC types are used to create a matrix in the tablet, which then effects controlled time-delayed release of the active ingredients. The cellulose derivative is slowly hydrated by a swelling process, and thus the active ingredient is moved very slowly and controlled from the tablet into solution. The active substance is homogeneously distributed and thereby gradually released. In addition to the molecular weight and corresponding resulting viscosity, the particle size of the MC or HPMC has a decisive effect on the delay of release. Drug systems can be tablets, impregnated webs or films, and mixed polymer film coating systems. The administration of controlled-release drugs includes oral, buccal, sublingual, transdermal, and suppository methods.
- **Ophthalmological applications:**
  Methylcelluloses are used as thickeners and stabilizers for ophthalmic solutions and ointments for the treatment of eye diseases. Other ophthalmic applications are:
  - liquids for wetting contact lenses for cleaning and storage;
  - artificial tears;
  - over-the-counter eye drops.

  The reduction in surface tension due to mild surfactancy, and the film-forming property of methylated cellulose derivatives, improves wetting and spread of the solution over the eye. Methylcellulose may be valuable in the treatment of chemical burns of the eye.

13) **Special effects:**
    The slimy, sticky appearance of a suitable solution of methylcellulose in water, with its toxicologically harmless non-allergenic property, means that MC is frequently used for special effects in photography and television, when mud or viscous substances (e.g. artificial blood or oil) are required.
14) **Other miscellaneous applications:**
    Methylcellulose and HPMC are used:
    - as thickeners in re-usable gel pads for cooling or warming body parts after sports injuries and inflammation;
    - as thickeners and binders in the production of reconstituted tobacco films;
    - as gelling agents in alkaline batteries;
    - as binders for graphite in lead and coloured pencils;
    - as drilling fluids for oil drilling;
    - to isolate rubbers from their aqueous dispersions;

- in oil-well stimulation;
- to control soil erosion;
- as mould release agents for cement products and ceramic moulded articles.

5.9 Nutritional Properties

Quantitative Analysis

Table 5.49 lists typical nutritional information for methylcellulose (MC). This analysis is also valid for hydroxypropylmethylcellulose (HPMC).

Toxicity

Numerous feeding studies have been conducted to verify the safety of methylcellulose products. No significant absorption of MC from the gastrointestinal tract following single and multiple oral doses to rats was observed. Methylcellulose, like other hydrocolloids, acts as a bulk laxative by retaining water in the digestive tract. For this reason, ingestion of large quantities can lead to retention of high levels of water in the stool, which is manifested as diarrhoea [8].

Methylcellulose is – just like native cellulose itself and all other cellulose derivatives synthesized from it – not degradable and absorbable by the human digestive system.

The substance passes through the body without molecular changes. The practical energy content is $0\,\text{kcal}\,\text{g}^{-1}$. The theoretical energy value can be higher –

Table 5.49 Quantitative nutritional analysis of methylcellulose and HPMC.

Ingredient (in 100 g)	Methylcellulose resp. HPMC
Energy content (measured)	0 kcal
Water	5.0 g
Ash (including salts and iron)	1.5 g
Fat (total)	0 g
• Saturated fats;	0 g
• trans fats	0 g
Protein/amino acids	0 g
Carbohydrates/sugar	0 g
Dietary fibres (total)	93.5 g
Soluble fibres	93.5 g
Sodium	390 mg
Chloride	610 mg
Iron	15 mg
Calcium	0 g
Phosphorus	0 g
All vitamins	0 g
All mineral salts	0 g

depending on the valid food law with its classification of the concerned hydrocolloid. No ADI have been defined. At doses higher than 12–15 g, laxative effects are possible. With normal nutrition and with the consumption of standard food, this quantity of methylcellulose cannot be achieved because, in general, the MC dosage is low (0.05–0.5%).

Higher-viscosity methylcellulose is used for the treatment of intestinal diseases because it binds high amounts of water and, if present, water-dissolved unhealthy substances, and thus can 'channel' these harmful materials out of the body.

5.10 Legislation

European Union (EU)

Celluloses and cellulose ethers fall under regulations stemming from EEC directive 74/329/EEC and its amendments – covering emulsifiers, stabilizers, thickeners, and gelling agents [3].

In the European Union, methylcellulose is approved as food additive with the E number 'E 461'. In the directive, MC is classified under miscellaneous additives in Annex I as 'additive being generally approved for foods with the exception of some special foods'. Annex 4 regulates the exceptions like untreated food, and those food products that may not be modified by additives (e.g. water, butter, honey, etc.). Methylcellulose may be used as a technological additive to achieve a particular effect with unlimited dosing ('quantum satis'). The function of methylcellulose may be described as 'stabilizer', 'emulsifier', 'thickener', or 'coating composition' [22].

The labelling of methylcellulose in Europe may be done with the E number, with the chemical name, or a combination of both. The terms 'modified cellulose' or 'modified vegetable gum' – common in the USA – are not sufficient, although some cellulosics manufacturers use them as product names on their specifications.

United States of America

The food-grade methylated cellulose ethers MC and HPMC meet the specifications of the Food Chemical Codex (FCC). Authorized synonyms of methylcellulose for labelling are 'MC' and 'modified cellulose'. Another common name of methylcellulose used by US food producers is 'modified vegetable gum'.

Food additives are regulated in the Unites States by the Food and Drug Administration (FDA) and the US department of Agriculture (USDA). Methylcellulose as chemical additive is generally recognized as safe (GRAS). The specific approval is listed in the US Code of Federal Regulations (CFRs) Title 21, MC under 21 CFR 182.1480. There are other specific approvals for cellulose derivatives such as those that define their use in 'Adhesives and Coatings for Food Use' (21 CFR 175.300) or specific food products such as 'Artificially

Sweetened Fruit Jellies' (21 CFR 150.141). The USDA has regulatory authority over meat products in the USA. The approvals for cellulose ethers in meats are different to those for other food products. According to 9 CFR 318.7, MC is approved as an extender or stabilizer in meat and vegetable patties. For poultry products, MC is approved as extenders and stabilizers according to 9 CFR 381.147 [3].

Canada, Australia, and Japan

In Canada, methylcellulose is listed in the 'Food and Drugs Act and Regulations' as food additive that may be used as emulsifying, gelling, stabilizing, and thickening agent.

In Australia, MC has been approved for inclusion in the 'Food Standards Regulations' under Section A 10, group 1, and referencing modifying agents.

In Japan, the methylated monoether has approval for food use. The materials must meet specifications for the Japanese Pharmacopeia or the Japanese Food Codex [3].

Other Countries and Areas

In other countries and areas, methylcellulose needs to undergo local approval before being applied. As the product has been intensely examined in Europe and the United States of America, the successful EU and US FDA approvals will be helpful for implementation in other markets. In Russia, manufacturers currently apply for the admission of methylcellulose.

References

1 Felcht, L-.H. (1985) *Schriftenreihe des Fonds der Chemischen Industrie*, 24, 71–76.
2 Balser, K. and Iseringhausen, M. (1975) Cellulose ether, *Ullmanńs Encyclopedia of Industrial Chemistry* (eds E. Bartholomé *et al.*), vol. 9, 4th edn, Verlag Chemie, Weinheim, pp. 192–212.
3 Stephen, A.M., Philips, G.O., and Williams, P.A. (eds) (2006) *Polysaccharides and Their Applications*, 2nd edn, Taylor & Francis Group, Boca Raton, FL, pp. 147–179.
4 Scherz, H. (1996) *Hydrocolloids: Stabilizers, Thickening and Gelling Agents in Food Products*, Food Chemistry/Food Quality, vol. 2, Food Chemical Society GDCh, Behr's Verlag GmbH, Hamburg.
5 European Union (2008) Commission directive 2008/84/EC of 27 August 2008, laying down specific purity criteria on food additives other than colours and sweeteners. *Official Journal of the European Union*, **51**, L253, 94–102. Available at http://eur-lex.europa.eu/LexUriServ/LexUriServ.do?uri=OJ:L:2008:253:0001:0175:EN:PDF.
6 Grosse, L. and Klaus, W. (1972) The analysis of water-soluble cellulose ether. *Fresenius Journal of Analytical Chemistry.*, **259** (3), 195–203.
7 Ashland, Inc. (2012) Benecel high-purity methylcellulose and hydroxypropyl methylcelllulose, physical and chemical properties (56.001-E4 REV). From the document library at http://www.ashland.com/products/benecel-methylcellulose-mc-and-hydroxypropylmethylcellulose-HPMC.
8 Whistler, R.L. and BeMiller, J.N. (eds) (1993) *Industrial Gums, Polysaccharides and Their Derivatives*, 3rd edn, Academic

Press Inc., San Diego, chapters 18–20, pp. 475–578.

9 Belitz, H.-D., Grosch, W., and Schieberle, P. (2009) *Food Chemistry*, 4th edn, Springer-Verlag, pp. 294–297.

10 Knarr, M. (2003) Rheological investigation and modeling of the sol-gel characteristics of methyl cellulose and k-carrageenan. PhD thesis, University of Hamburg.

11 Böhm, N. (1998) Isolation, characterization and investigation of the Gelling of (1.3) (1.4)-ß-glucans. PhD thesis, University of Hamburg.

12 Heymann, E. (1935) Sol-gel transformations. The inverse sol-gel transformation of methylcellulose in water. *Transactions of the Faraday Society*, **31**, 846–864.

13 Sakar, N. (1995) Kinetics of thermal gelation of methyl cellulose and hydroxypropyl methyl cellulose in aqueous solutions. *Carbohydrate Polymers*, **26**, 195–203.

14 Haque, A. and Morris, E.R. (1993) Thermogelation of methylcellulose. Part I: molecular structures and processes. *Carbohydrate Polymers*, **22**, 161–173.

15 Kobayashi, K., Huang, C.-I., and Lodge, T. P. (1999) Thermo-reversible gelation of aqueous methyl solutions. *Macromolecules*, **32**, S.7070–7077.

16 Hirrien, M., Chevillard, C., Desbrières, J., Axelos, M.A.V.. and Rinaudo, M. (1998) Thermogelation of methylcellulose: new evidence for understanding the gelation mechanism. *Polymer*, **39**, 6251–6259.

17 Sun, X., Litchfield, J.B., and Schmidt, S.J. (1993) Temperature mapping in a model food gel using magnetic resonance imaging. *Journal Food Science*, **58** (1), 168.

18 Ashland Inc. (2012) Benecel^{TM} methylcellulose and hydroxypropyl methylcelllulose in fried coated food applications (VC-721). From the document library at http://www.ashland.com/products/benecel-methylcellulose-mc-and-hydroxypropylmethylcellulose-HPMC.

19 Ufheil, G. and Escher, F. (1996) Dynamics of oil uptake during deep-fat-frying of potato slices. *Lebensmittel-Wissenschaft Technologie*, **29** (7), 640–644.

20 Murray, J.C.F. (2000) Cellulosics, in *Handbook of Hydrocolloids* (eds G.O. Phillips and P.A. Williams), Woodhead Publishing Limited, Cambridge, pp. 219–230.

21 Ashland Inc. (2012) Batter coating for frying, dry mix with Benecel^{TM} methylcellulose and Supercol^{TM} XG xanthan gum (formula 01–1048A). Formulation from the document library at http://www.ashland.com/products/benecel-methylcellulose-mc-and-hydroxypropylmethylcellulose-HPMC.

22 For food legislation Germany see: Anon (2000) *Federal Law Gazette*, 1521–1536. Available at http://bundesrecht.juris.de/zzulv_1998/anlage_4_16.html.

6
Ethylcellulose

Synonyms: EC, cellulose ethyl ether, ethyl ether of cellulose, Ethocel

E number:	E 462
CAS number:	9004-57-3
FDA code:	21 CFR 172.868

Related hydrocolloids: none (EC is very specific) ethylcellulose, food applications, manufacture, non-food applications, rheology, stability

Note

Owing to the particular solubility and application of ethylcellulose, other materials and substances are also identified that may not be used for the manufacturing of food products. Because, nevertheless, the applications are relevant, like packaging materials and safekeeping containers, which come into direct contact with food, the present chapter is extended to cover this 'technical' area. For every special-use case the food-juridical situation of the desired additives must be checked.

In addition, the special qualities of ethylcellulose and its diversity can be demonstrated by excursions into other fields.

6.1
Manufacturing

Low-substituted water-dissolvable ethylcellulose is not produced for commercial purposes on account of less expensive technological alternatives. Highly etherified and thus water-insoluble types are produced in significant volumes primarily in the USA, by two different procedures. The basic material is pulp from wood with α-cellulose contents of at least 86%.

The batch process for the production of ethylcellulose was developed before 1928 by I.G. Farbenindustrie and is still running to this day virtually unchanged.

Cellulose and Cellulose Derivatives in the Food Industry: Fundamentals and Applications, First Edition.
Tanja Wüstenberg.

Long grain milled cellulose is homogeneously mixed in a kneader with 50% sodium hydroxide for 3–4 h at 20–30 °C (68–86 °F). The viscosity is adjusted by the dosage of the amount of air. The alkali cellulose is added with five times the amount of sodium hydroxide to ethyl chloride and is initially charged in a coated stirred autoclave. After keeping the autoclave at 90 °C (194 °F) for 4 h, the exothermic etherification reaction leads to a rise in temperature to 110 °C (230 °F). Next, the reaction mixture is maintained for another 7–8 h at 110 °C (230 °F) and then the internal temperature is increased to 130–140 °C (266–284 °F). During the 10–20 h continuous reaction only about ¼ of the ethyl chloride content reacts with the cellulose, the remaining agent is converted into the by-products ethanol and diethyl ether. During the reaction, the pressure initially increases to 20–25 bar (290–363 psi) and decreases with decreasing content of ethyl chloride to 12 bar (174 psi) towards the end of the reaction. In the next step, the solvents are removed and distilled.

The product EC (ethylcellulose) is precipitated in raw, salty form, then neutralized with acetic acid, washed, centrifuged, dried, and ground. The yield of ethylcellulose is about 125–130 kg from 100 kg of cellulose.

Subsequently, a semi-continuous process has been developed: First, the cellulosic web runs continuously through a bath of aqueous sodium hydroxide solution (50–76%) at 55–130 °C (131–166 °F). Then the cellulose is ripened for a few seconds at a slightly lower temperature, and finally cooled. The viscosity is adjusted by the oxidation. The alkali cellulose contains about 4 moles of NaOH and 3 moles of water per mole of anhydroglucose. It is then reacted with ethyl chloride until an average degree of substitution of 1.7–2.0 is reached. To complete the reaction, solid sodium hydroxide is added. The presence of aromatic hydrocarbons reduces the pressure to 9–12 bar (130–174 psi). About half of the applied ethyl chloride reacts with the cellulose. The product is then precipitated, washed, and dried [1].

6.2 Chemistry

Ethylcellulose is a polymer of β-D-glucose monomers with 1 → 4 linkages, in which the free hydroxyl groups are substituted with ethyl groups almost completely. The degree of substitution is about 2.5. Figure 6.1 shows a portion of the molecule of ethylcellulose.

Ethylcellulose is a polymer of substituted anhydroglucose units with the general chemical formula [2]:

$$[C_6H_7O_2(OR^1)(OR^2)]$$

where R^1 and R^2 can be H or CH_2CH_3.

The content of ethoxyl ($-OCH_2CH_3$) is at least 44% and not more than 52% (based on dry weight). This corresponds to a maximum of 2.8 ethoxyl groups per

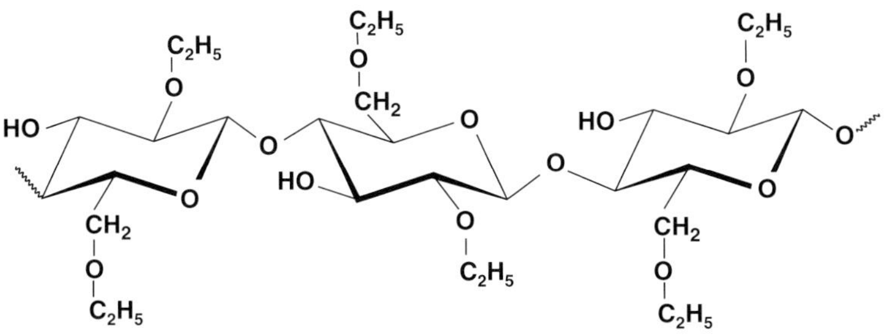

Figure 6.1 Molecule of ethylcellulose.

anhydroglucose unit. Theoretically, complete substitution of all available free OH groups of the native cellulose molecule would be possible, thus ensuring a triethyl ether. Ethylcellulose with a DS (degree of substitution) of 3.0 and an ethoxyl content of 54.88%, however, has no significant commercial importance, since it lacks strength and flexibility. This product is not thermoplastic and has a very low compatibility with other substances, and only very limited solubility [3].

Commercially relevant EC products with food approval and EC grades that come into contact with food have degrees of substitution from 2.25 to 2.6 ethyl groups per anhydro-glucose monomer and ethoxyl contents of 44–50%. The low ash content (up 0.4%) allows a virtually residue-free combustion and, thus, the use in highly sensitive technical applications (see Sections 6.7 and 6.8) [2]. Ethylcellulose is a white coarse powder with a bulk density of ~0.4 g cm^{-3} (400 g l^{-1} (2.6–2.8 lb gal^{-1})).

6.3 Rheology

6.3.1 Physicochemical Parameters

Ethylcellulose shows the following product parameters and physicochemical properties. The values were determined for an EC – type 'B' [3]:

- appearance: white to off-white or slightly grey coarse powder;
- available degree of substitution (DS):
  - low (type A): 45.0–47.2% ethoxyl, corresponds to DS = 2.22–2.41,
  - medium (type B): 48.0–49.5% ethoxyl, corresponds to DS = 2.46–2.58,
  - medium (type C): 49.6–51.5% ethoxyl, corresponds to DS = 2.58–2.73,
  - high (type D): 50.5–52.5% ethoxyl, corresponds to DS = 2.65–2.81;
- viscosities (measured as 5% solution in '80 : 20' toluene–ethanol):
  - low viscosities: 4, 7, 10, 14, and 22 mPa s (cP),
  - medium viscosities: 50, 100, 200, and 300 mPa s (cP);

- specification of food-grade EC types with the highest approval class (Section 6.10), which are allowed for direct use in food products:
  - 48.0–49.5% ethoxyl; DS = 2.46–2.58,
  - viscosity range = 7–100 mPa s (cP) as 5% solution in 80 : 20 toluene–EtOH;
- bulking density (as granular powder): 2.6–2.8 lb gal^{-1} (~310–340 g l^{-1}),
- specific gravity: 1.14 N m^{-3};
- specific volume (in solution): 23.9 $inch^3$ lb^{-1} (~3 cm^3 kg^{-1});
- bulking value (in solution): 0.099–0.104 lb gal^{-1} (~12 g l^{-1});
- colour (Hazen value of the solution): 2–5;
- discolouration by sunlight: very slight;
- softening point: 152–162 °C (306–324 °F);
- light transmission – practically complete: 310–400 nm;
- light transmission – better than 50% complete: 280–310 nm;
- moisture absorption (by film in 24 h at 80% RH): 2%;
- refractive index (cast film): 1.47;
- tensile strength (3 mm film, dry): 6800–10 500 lb $inch^{-2}$ (~98–152 g cm^{-3});
- tensile strength – wet (% of dry strength): 80–85;
- water vapour transmission in 24 h (3-mm film): 890 g m^{-2};
- flexibility as folding endurance (3-mm film, double folds): 160–2000;
- hardness index of a 3-mm film (Sward): 52–61;
- elongation at rupture (3-mm film, conditioned at 25 °C (77 °F) and 50% RH): 7–30%;
- odour of flakes/granules: slight;
- taste: none;
- electrical properties:
  - dielectric constant (at 25 °C): 2.8–3.9 mC (3.0–4.1 kC),
  - dielectric constant (at 25 °C, 60 cycles): 2.5–4.0 kC,
  - dielectric strength (10-mm film): 1500 V mm^{-1},
  - power factor (at 25 °C): 0.002–0.02 kC,
  - power factor (at 25 °C, 60 cycles): 0.005–0.02 kC,
  - volume resistivity: 10^{12}–10^{14} Ω cm^{-1};
- ash content: maximum 0.4%.

6.3.2
Dissolution Behaviour

6.3.2.1 Solubility in Water

The solubility of the ethyl ethers of cellulose is dependent upon the amount of etherification (the DS). There following relationships exist:

- Ethylcelluloses with a low degree of substitution – DS 0.5–0.7 – have solubility in 4% sodium hydroxide solution;
- ethylcelluloses having an average DS from 0.8 to 1.3 are characterized by solubility in cold water;

- water insolubility occurs when ethylcelluloses have a very high content of added ethoxyl groups – expressed as degree of substitution – in the range 2.2–2.8.

The solubility of ethylcellulose in water is restricted to a very narrow – and overall to a low- to medium-type (DS ≤ 1.3) – substitution range. Currently, there are no commercially products available that are water soluble. Highly etherified ethylcellulose (DS ≥ 2.2) is completely insoluble in water at any temperature. Some mixtures of EC with organic solvents tolerate up to 10% water in the system. The cellulose derivatives cannot be dissolved in glycerine and propylene glycol.

The highly-substituted ethyl ether of cellulose may be dissolved in:

- organic solvents (including mixtures of polar–nonpolar solvents) such as aliphatic and cyclic alcohols (methanol, ethanol, iso- and *n*-propanol), aliphatic esters (methyl and ethyl acetate), ketones (acetone, methyl ethyl ketone), glycol ethers and acetates, aromatic and chlorinated hydrocarbons (methylene chloride);
- fats and oils and medium-chain triglycerides; EC is dispersed in oil at room temperature and becomes soluble then heated to at least 80–160 °C (176–320 °F) with stirring – the solubility increases with increasing temperature;
- the addition of minerals, salts, sugars, and other hydrocolloids does not improve the solubility of ethylcellulose in water or organic solvents [1,3].

6.3.2.2 **Suitable Solvents**

Effect of Ethoxyl Content (DS)

The solubility of ethylcellulose is dependent on the degree of substitution with respect to the content of added ethoxyl groups. The solubility of EC in water is discussed above. The water-insoluble ethylcellulose grades (DS ≥ 2.2) are – depending on the level of substitution (with respect to ethoxyl content) – more or less readily soluble in organic solvents. The effect of substitution on solubility is illustrated by the data given in Table 6.1. This data shows that there are relatively few single solvents for the lower substitution types (EC products with DS 2.2–2.4). Solubility in all classes of organic solvents improves with increased ethoxyl content until an optimum level is reached within the range of substitution of the B-, C-, and D-types. Above this level of substitution, solubility again becomes more restricted, being limited for the highest levels of possible substitution to highly nonpolar compounds such as the aromatic hydrocarbons [3].

Binary Solvent Systems – Effects of Alcohols [3]

Although there are special applications that require single solvents, for general use they are inadvisable. Almost all ketone, ester, and hydrocarbon solvents for ethylcellulose can be significantly improved by mixing them with relatively small quantities of the lower aliphatic alcohols such as ethanol or butanol. Such binary

Table 6.1 Solubility of ethylcellulose – effect of ethoxyl substitution [3].[a)]

Solvent	Relative solubility[b)]			
	Type A	Type B	Type C	Type D
	DS 2.22–2.41	DS 2.46–2.58	DS 2.58–2.73	DS 2.65–2.81
Toluene (methylbenzene)	G S	**C S**	**C S**	**C S**
Xylene	G I	**C S**	**C S**	**C S**
80 : 20 Toluene – ethanol	**C S**	**C S**	**C S**	**C S**
Ethyl acetate	V H S	V H S	V H S	V H S
Butyl acetate	G S	V H S	V H S	V H S
Acetone	I	V H S	V H S	V H S
Methanol	V H S	H S	I	I
Ethanol	H S	H S	H S	H S
Butanol	I	H S	H S	H S
Isopropanol (91 vol.%)	H S	H S	H S	H S
Ethylene glycol monobutyl ether	G S	—	I	I
Ethylene dichloride	H S	**C S**	**C S**	**C S**
Carbon tetrachloride	G	**C S**	**C S**	**C S**
Methylene chloride	**C S**	**C S**	**C S**	**C S**

Data taken with permission from Reference [3].

a) Testing method: 15 wt% concentration of ethyl cellulose in the solvent.

b) C = clear, G = gelled mass, H = hazy, S = solution, I = insoluble (although some particles might be gelled), and V = very.

mixtures are much better solvents for ethylcellulose than is each component alone, and they have several advantages.

These binary solvents will completely dissolve a much wider range of substitution types of ethylcellulose than will single solvents. For example, toluene alone is a good solvent only for the range of substitution covered by the C-type EC. However, B-type ethylcellulose, although soluble, does not always form a good solution in toluene alone, as evidenced by the fact that it is difficult to duplicate the viscosity in successive solutions of the same concentration, indicating incipient gel formation.

EC with low degree of substitution is poorly soluble in toluene alone. However, a solvent combination of 80% toluene and 20% ethanol will completely dissolve all substituents from 43.5% to 50%. In general, the quantity of alcohol required with esters, ketones, and hydrocarbons to make good binary solvents increases as the degree of substitution becomes lower. For example:

- 90 : 10 toluene–ethanol is a good solvent for EC – type 'B' (48.0–49.5% ethoxyl);
- but to obtain comparable results with the EC – type 'A' and its ethoxyl content of 45.0–47.2%, one needs a ratio of 70 : 30 toluene–ethanol.

The alcohols are especially valuable in improving the solvent power of so-called 'toluene substitutes' – proprietary products derived in many cases by catalytic dehydrogenation of naphthenes and cyclized hydrocarbons. They are characterized by having kauri–butanol values between 40 and 70, and alone show considerably poorer solvency for ethylcellulose than do aromatic hydrocarbons such as toluene and xylene. However, the toluene substitutes become excellent solvents for all substitution types of EC when mixed in a ratio of '70 : 30' of 'toluene-substitute : ethanol'. Some are better solvents than others. In general, the higher the kauri–butanol value is the less ethanol is required to obtain good solvency. Butanol is somewhat less effective than ethanol in these binary mixtures, which is reflected by the fact that viscosities at a given concentration are higher.

A marked reduction in solution viscosity is another advantage gained by using a binary solvent system in which one of the lower aliphatic alcohols is a constituent of the solvent mixture. An evident advantage to be derived from use of a mixed solvent is the possibility of obtaining more concentrated solutions without encountering objectionable gelation or granulation. Control of viscosity in such binary solvent systems is much easier because the solutions are less sensitive to small changes in concentration or temperature. Other advantages exhibited by binary solvents containing an alcohol are improved clarity of solution, freedom from the granulation or gelation encountered when poor or lean solvents are used, and improved tolerance for petroleum thinners.

Solubility in Organic Solvents

The variety of solvents for B-type ethylcellulose is unusually wide. Table 6.2 summarizes the solubility of EC (type B) with DS from 2.46 to 2.58 in organic solvents. Clearly, the aliphatic or straight-chain hydrocarbon class is the only one having essentially no solvent power to dissolve B-type ethylcelluloses. The wide range of solvents greatly simplifies the task of selecting solvents for specific applications.

Particularly in industrial applications, the mixed solvent systems shown in Table 6.2 are very useful because they allow EC solutions with low viscosity and good transparency, and they are also very economical [3].

Dilution with Petroleum Thinners – for Technical Applications

The tolerance of ethylcellulose solutions for dilution with non-solvent petroleum thinners is unusually good – it becomes progressively better with increased ethoxyl content. Such dilution, however, is affected not only by the potency of the active solvent being diluted but also by the degree of substitution of the EC. A poor or lean solvent will obviously tolerate less dilution than a rich or active solvent [3].

6.3.2.3 Viscosity–Concentration Relationships

Like all other cellulose ethers, the thickening effect of ethylcellulose depends on the amount of dissolved substance in solution. In contrast to the water-soluble cellulose derivatives, commercially available EC products are completely

Table 6.2 Solubility of EC (ethylcellulose) in representative organic solvents [3].[a)]

Solvent	Solubility[b)]
Alcohols:	
Methanol	S
Ethanol	S
Propanol	S
Isopropanol	SW
Butanol	S
Amyl alcohol	S
Cyclohexanol	PS
Benzyl alcohol	S
Diacetone alcohol	SW
Chlorinated hydrocarbons:	
Methylene chloride	S
Chloroform	S
Carbon tetrachloride	S
Ethylene dichloride	S
Trichloroethylene	PS
Pentachloroethane	S
Esters:	
Methyl acetate	S
Ethyl acetate	S
Propyl acetate	S
Isopropyl acetate	S
Butyl acetate	S
Amyl acetate	S
Ethylene glycol monomethyl ether acetate	S
Ethylene glycol monoethyl ether acetate solvent	S
Glycol diacetate	PS
Benzyl acetate	S
Cyclohexyl acetate	PS
Methyl formate	S
Ethyl formate	S
Ethyl lactate	S
Butyl lactate	S
Miscellaneous solvents:	
Dichloroethyl ether	PS
Ethylene dioxide (dioxane)	S
Pine oil	S
Hydrocarbons:	
Toluene	S
Xylene	S
Cyclohexane	PS
Tetrahydronaphthalene	S
Turpentine	SW
Aromatic petroleum distillates	PS
Petroleum ether (light petroleum)	I
Hexane	I

Petroleum thinner	I
Ketones:	
Acetone	S
Methyl ethyl ketone	S
Dipropyl ketone	PS
Hexanone	S
Cyclohexanone	S
Methyl cyclohexanone	S
Mixed solvents:	
80 : 20 Toluene–ethanol	S
80 : 20 Xylene–butanol	S
2 : 1 Ether–ethanol	S
15 : 1 Ethanol–camphor	S
70 : 30 Turpentine–butanol	S
Ethers:	
Ethyl ether	S
Ethylene glycol monomethyl ether	PS
Ethylene glycol monoethyl ether solvent	S
Ethylene glycol monobutyl ether solvent	S
Diethylene glycol monoethyl ether	SW
Nitroparaffin:	
1-Nitropropane	S
2-Nitropropane	S
Nitroethane	PS
Nitromethane	I

Data taken with permission from Reference [3].

a) Testing procedure: 15 g ethylcellulose in 85 ml solvent.

b) I = insoluble; PS = partly soluble, but contains gel and granulation; S = soluble (films, obtained from these solutions are clear, although some might appear slightly hazy to others); SW = swollen, but granules not merged into a cohesive gel.

insoluble in water. Therefore, ethylcellulose is dissolved in a mixed solvent system of 80% toluene and 20% ethanol. The viscosity is measured at 5% concentration at a temperature of 25 °C (77 °F) [3].

Figure 6.2 shows typical viscosity–concentration curves for ethylcellulose – type 'B'. There is no proportional slope of the curve. Doubling the amount of EC will increase the viscosity by a multiplying factor – becoming higher, the higher the concentration. Two examples illustrate this. An EC with a viscosity level of 7 (at 5% = 5.6–8 mPa s) has a 10% viscosity of 30–40 mPa s and a 20% viscosity of 500 mPa s. When an ethylcellulose grade of 100 mPa s (at 5% = 80–105 mPa s) is used, the 10% concentration is about 1000 mPa s and the 20% concentration about 30 000 mPa s. The polymeric effect leads to a strong and exponential increase in viscosity at high amounts dissolved.

6.3.2.4 Blending

When the stock of a certain viscosity type of ethylcellulose is not on hand, or a use may be developed for an EC of a viscosity differing from any standard grade,

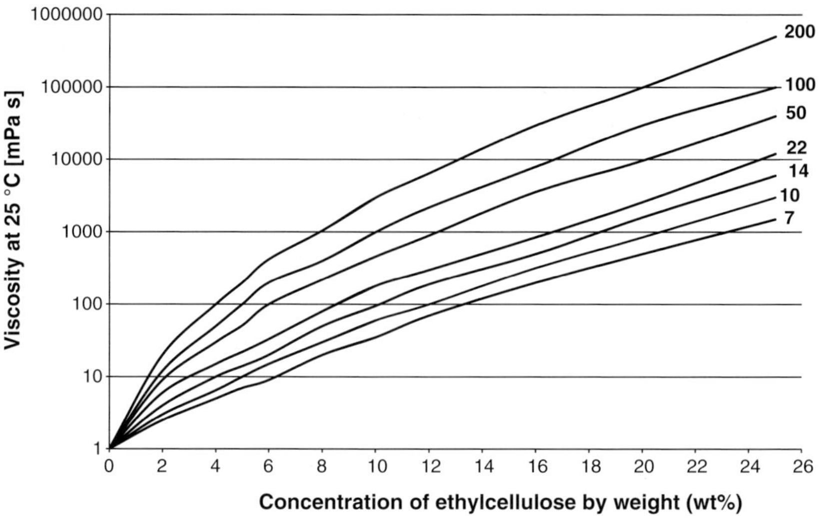

Figure 6.2 Viscosity–concentration curves of EC – type 'B' [3]. Reprinted with kind permission of Ashland Inc.

it is useful to know that different viscosity types can be blended to produce a product of the desired viscosity. The principle of blending is based on the Arrhenius equation, which relates viscosity and concentration. It can be conveniently expressed as follows [3]:

$$\text{Log}\ V_S = \frac{n \bullet \log V_1 + (100 - n) \bullet \log\ V_2}{100}$$

V_S = viscosity sought
n = weight-% of the first component (1) of the mixture
V_1 = viscosity of first component (1) in the mixture
V_2 = viscosity of second component (2) in the mixture

All viscosities must be expressed for a constant polymer concentration and in the same units. Blends can be calculated directly from this equation. However, it is more convenient to use a so-called 'blending chart' (Figure 6.3). With this chart, it is possible to determine, without calculation, the percentage of any two ethylcelluloses of different viscosities that must be blended to secure a desired intermediate viscosity. Likewise, it is possible to determine the viscosity that will result from making any blend.

This chart is set up on semi-logarithmic or logarithmic plotting paper marked off so that the divisions on the horizontal axis represent percentages from '0' to '100', while the divisions on the vertical axis represent viscosity in mPa s or cP for standard 5% solutions of ethylcellulose up to a viscosity level of 256. The given chart extracts the relevant viscosity range (from 32 to 256 mPa s).

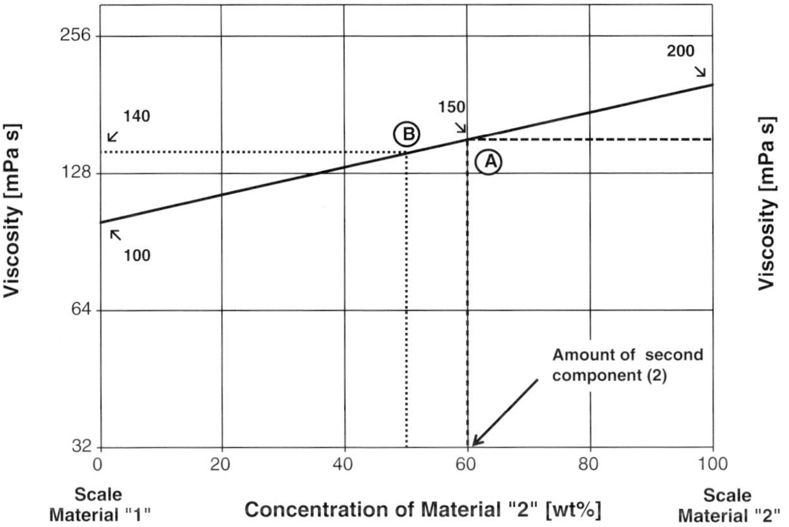

Figure 6.3 Blending chart for ethylcellulose [3]. Reprinted with kind permission of Ashland Inc.

The blending chart is used in the following way: The exact procedure can be described most easily by the use of specific cases. For example, if an ethylcellulose with 150 mPa s viscosity as 5% solution is required the available materials consist of:

- type 100:
  - with a viscosity range of 80–105,
  - an exact viscosity of 97 mPa s in 5% solution);
- type 200:
  - with a viscosity range of 150–250,
  - an exact viscosity of 200 mPa s in 5% solution.

The diagonal line (corresponding to a straight edge) connects type 100 at the 97-mPa s point, on the left-hand side of the chart (V_1), with type 200, on the right-hand side (V_2). The diagonal intersects the horizontal line corresponding to 150 mPa s at the vertical line marked 60% (Point A). This means that 60% of type 200 and 40% of type 100 should be mixed to obtain a 150-mPas ethylcellulose.

Similarly, other blends can be obtained. It is not necessary to draw lines. Merely use a straight edge and read directly. It is important that the percentage read from the diagonal refers to the EC type indicated on the right-hand hand axis (V_2) [3].

The chart can also be used to calculate the viscosity that will result by blending predetermined weights of two different viscosity types. For example, equal weights of the same type 100 and the same type 200, used to illustrate the first example, can be mixed. The vertical 50% line intersects the diagonal at 'Point B' on the horizontal viscosity line, designating 140 mPa s [3].

In general, blending has limitations and some rules have to be followed to guarantee good mixing results:

1) The two viscosity types selected for the blend should be as close together as possible. If widely separated viscosity grades are blended, there is the danger that the resulting mix may yield a lumpy or granular solution. Moreover, the functionality of EC in a certain application depends on the polymer length and, thereby, the molecular weight of the ethylcelluloses used. Choosing close viscosities of the mixing partners ensures a reliable manufacture and a high and homogeneous quality in the final product.
2) Different substitution types should not be blended, because lumpy, granular, or even incompatible solutions can result. The dissolution behaviour of ethylcellulose is a function of its ethoxyl content and degree of substitution.

6.3.3 Gelation and Behaviour with Other Ingredients

6.3.3.1 Overview

Depending on the degree of polymerization, ethylcellulose leads to an increase of the viscosity of oils, medium-chain triglycerides (MCT), and organic solvents.

Gelation is usually not observed. However, there is the possibility of controlled gelation of EC by contacting the cellulose ether with certain vegetable oils, such as linseed oil, soybean oil, corn oil, or cottonseed oil, melted first in this hot and then cooled to room temperature. In the first step, about 3–5% of ethylcellulose is dissolved in a suitable oil at 180–200 °C (356–392 °F); after cooling, a relatively inelastic, rigid gel is formed.

As a hydrophobic substance ethylcellulose shows a good film-forming ability. The films are very tough, resistant, flexible and transparent, and odourless and tasteless.

EC is compatible with many plasticizers. Plasticizers are usually employed with a concentration of 5–30% of the solid content. Most often, esters (such as glyceryl triacetate, triethyl citrate, dibutyl sebacate) are used. Fatty acids (stearic and oleic acid), fatty alcohols (palmityl and stearyl alcohol), and edible oils are also possibilities.

Mixtures of solvents, aliphatic and aromatic substances, may be applied. Mixed solvent systems produce lower solution viscosities and improved film strength than pure solvents. More details can be found in Section 6.3.2.2 ('Suitable solvents').

Because some systems of EC with organic solvents tolerate up to 10% of water, water-soluble polymers may be incorporated into the formulation. In practice, there are heterogeneous coating systems with the non-ionic hydrocolloid HPMC (see Section 6.6 on 'Synergies with other hydrocolloids'). Incompatibilities between ethylcellulose and other food-grade substances such as sugars, salts, minerals, vitamins, flavours, and other food ingredients are not known.

EC is highly compatible with an unusually large number of resins, and plasticizers, oils, and waxes. The high compatibility allows relatively easy development of many economic formulations. Even the addition of small amounts of ethylcellulose to waxes and resins (such as shellac) harden their textures and increase their melting points.

In protective coatings, lacquers, and varnishes, EC reduces the stickiness and accelerates the drying time. Printing inks and plastics with rigid or soft properties can also be formulated easily with the support of ethylcellulose [3].

6.3.3.2 Compatibility with Resins

Resins are used as modifying agents for ethylcellulose formulations in many applications. In coatings and glazes they impart hardness, gloss, adhesion, and an improved resistance to moisture. In plastics, they bring strength, water resistance, and in some products even better dimensional stability. Conversely, ethylcellulose diminishes the stickiness of resins and gives them toughness, thereby increasing their resilience and durability. The positive effects can be achieved, however, only if the resin applied is compatible with the ethylcellulose type used.

The compatibility with resins is also controlled by the ethoxyl content and thus by the degree of substitution (DS) of the ethylcellulose. An EC – type B with 48.0–49.5% ethoxyl (DS 2.46–2.58) – shows the highest compatibility with resins. The tolerance decreases with decreasing ($DS \leq 2.4$) and with increasing degree of substitution ($DS \geq 2.6$). Table 6.3 describes the results of a study of the compatibility of EC with resins as a function of the DS. The research was performed with a mixture ratio 'Resin to EC' of '1 : 1'. Each listed resin represents a distinct class of materials [3].

Table 6.3 Compatibility of ethylcellulose with resins as a function of the DS (degree of substitution) [3].

Solvent	**Compatibility**[a)]			
	Type A	**Type B**	**Type C**	**Type D**
	DS 2.22–2.41	**DS 2.46–2.58**	**DS 2.58–2.73**	**DS 2.65–2.81**
Ester gum 80 (glycerol ester resin)	I	C	C	C
Wood rosin	C	C	C	C
Dewaxed dammar	C	C	C	C
Non-oxidizing alkyd resin	C	C	I	I
Sebacic, oil-modified alkyd resin	C	C	I	I

Data taken with permission from Reference [3].

a) C = compatible and I = incompatible.

Compatibility with B-Type Ethylcellulose [3]

Since ethylcellulose type 'B' with a medium DS has the highest compatibility with resins and is the preferred product group for direct and indirect food applications, this class will be considered in more detail (Table 6.4).

Table 6.4 Compatibility of ethylcellulose (type 'B') with resins [3].

Group	Compatibility resin – EC	Resin	
A	Good – up to 1 : 1	**Natural resins:** batu, run; Congo, run; dammar (dewaxed Batavia); elemi; kauri; manila gum, all grades; mastic gum; pontianak, run, mixed bold; rosin, wood; sandarac; shellac (dewaxed); dark hard pine resin	**Rosin- and terpene-based resins:** methyl ester of rosin; hydroabietyl alcohol; rosin-derived alkyd; ester-gum (glycerol ester of rosin); hydrogenated methyl esters of rosin; pentaerythritol ester of rosin; maleic-modified ester gum resin; maleic-modified glycerol ester of rosin; ester of polymerized rosin; polymerized rosin; hydrogenated rosin; esters of hydrogenated rosin; dark hard-pine resin; zinc resinate
		Alkyd resins: long oil phthalic alkyd; sebacic, oil-modified alkyd resins **Phenolic resins:** oil-soluble pure phenolic resin	**Miscellaneous:** acrylic synthetic fibre; styrene resin
B	Limited – compatible up to 0.3–0.5 : 1	**Natural resins:** white shellac	**Rosin- and terpene-based resins:** hard maleate resin
		Coumarone–indene resins: plastic-grade coumarone-indene resins	**Miscellaneous:** chlorinated paraffin
C	Incompatible – even at 1 : 10	**Natural resins:** accroides (shellac type) Congo ester	**Urea- and melamine-formaldehyde:** urea–formaldehyde resin;

East India Macassar, pale East India Singapore, pale Orange shellac	melamine resin; melamine-formaldehyde resin
Alkyd resins: sebacic, unmodified non-oxidizing alky resin; alkyd resin	**Miscellaneous:** vinyl resin; fatty oil-vinyl toluene copolymer; polymer of bisphenol and epichlorohydrin; mineral-thinner-soluble acrylic ester resin; silicone dimethyl fluid
Rosin-based: distilled methyl ester of all oil; modified pentaerythritol ester of rosin	

Data taken with permission from Reference [3].

The resins are listed in three groups – depending on the tolerance of EC:

[A] resins with good compatibility in all proportions up to '1 : 1' (resin : EC);
[B] resins with limited compatibility to a resin-to-EC ratio of '0.3–0.5 : 1';
[C] resins, incompatible even at ratios as low as '1 : 10' (resin : EC).

In summary, this ethylcellulose (type B, DS $\approx$ 2.5) is generally compatible with:

- resin acids;
- the glycerol, glycol, and diethylene glycol esters of resin acids;
- varnish-type pure phenolics of both the oil-soluble and oil-reactive classes;
- rosin-modified phenolics;
- many natural resins – several of which were compatible only after heat treatment or other special treatment such as dewaxing;
- toluene–sulfonamide–formaldehyde resins for technical applications.

This B-type ethylcellulose shows restricted compatibility with:

- maleic-modified rosin esters,
- urea–formaldehyde resins,
- coumarone–indene resins,
- alkyds modified with phenolics, rosin, or oils;
- cycloparaffins,
- zinc resinates.

This limited compatibility is usually controlled by the degree of resin modification. For instance, compatibility becomes more restricted as the percentage of maleate is increased, as the melting point of the coumarone–indene resins increases, and with increased zinc content in the zinc resinate. Compatibility with modified alkyds becomes less restricted with increased oil or fatty acid modification, or with increased phenolic or rosin modification.

Ethylcellulose (type B, DS ≈ 2.5) is incompatible in all proportions with:

- all pure alkyds based on glycerol phthalate,
- short oil modifications of such alkyds,
- acrylates and methacrylates,
- polystyrene,
- polyisobutylenes,
- vinyl acetate resins.

Table 6.4 represents an overview of the compatibility of medium-substituted ethylcellulose (type B, DS ≈ 2.5) with various resins. Most of these resins cannot be directly used in foods or on the surface of food products. The table covers all possible types of EC application – that is other regulated products, such as cosmetics and pharmaceuticals, as well as packaging materials for foods and technical applications.

Resins of group 'A' give clear films with ethylcellulose in any ratio up to an addition of 50% (ratio = 1 : 1). Resins of group 'B' have only limited compatibility with EC (up to a content of 25–33% of resin use) without sacrificing quality. In group 'C', all resins listed are not compatible with ethylcellulose, even when their concentration in the formulation is below 10%. With these mixtures, no transparent films can be obtained. With some specific resins of borderline compatibility (listed in group 'B'), there may be inconsistencies in compatibility data, depending on minor variations in the resin from lot to lot or other factors not strictly controlled. Consequently, it is possible that other investigators may find discrepancies between their results and those reported here. However, the table gives an accurate idea of the kinds of resins that are compatible (group 'A') or incompatible (group 'C') with EC.

Resins, which are normally incompatible with ethylcellulose alone in a binary system, can be made compatible by the addition of a third component. This third substance acts as a so-called solubilizer – comparable to the function of an emulsifier in emulsions where it helps to combine two different-soluble phases, which cannot be mixed and stabilized under normal circumstances, and would separate very quickly. The resins can be usefully incorporated into compositions containing EC by employing a mutually compatible third ingredient to prevent separation, with its resultant dullness or haze. Solvent plasticizers often serve in this capacity. Cellulose nitrate is also effective.

Compatibility with A-Type Ethylcellulose [3]

Low-substituted ethylcellulose (type A) is used in plastics and protective coatings because of such special properties as its higher melting point, special moulding characteristics, and restricted solubility and compatibility. It is interesting, therefore, to consider the resins that can be used to modify this type of EC.

Resins that are compatible in all proportions, up to '1 : 1' resin : ethylcellulose, are:

- boea, run,
- sebacic, oil-modified alkyd resin,
- wood rosin,
- Zitro resin,
- modified ester of rosin (completely compatible in all proportions, up to '0.3–0.5 : 1' of 'resin : EC').

Many urea–formaldehyde and melamine resins show restricted compatibility with ethylcellulose – type 'A', but are of interest within their limited sphere of compatibility because they are colourless resins and because they possess thermosetting characteristics.

6.3.3.3 Compatibility with Plasticizers

Pure ethylcellulose yields very tough films of excellent tensile strength, flexibility, and elongation characteristics. However, these films lack suppleness. In addition, EC alone softens and flows at too high a temperature ($T \geq 150\,°C$ (302 °F)) to be practical in moulding operations or in other applications requiring good thermoplasticity. Therefore, plasticizing or softening agents are added to ethylcellulose to obtain the appropriate degree of suppleness, to lower the softening point, and to improve thermoplasticity.

The variety of possible plasticizers for EC is unusually large. It includes, with two exceptions, all classes of compounds that have been offered as softening or flexibilizing agents for coatings and plastics. The two exceptions are heat-bodied vegetable oils and highly unsaturated drying oils such as tung, perilla, and oiticica oils.

Notably, fatty acids derived from incompatible oils are compatible with ethylcellulose. In general, it has been observed that 'air-bodying' an oil improves its compatibility over that exhibited by the raw oil. This special processing is carried out after refining the oil and includes treatment with hot air at 100–150 °C (212–302 °F) for several hours.

Table 6.5 lists a large number of plasticizing agents compatible with ethylcellulose. Practically all of the components listed can be used with all substitution types of EC (types A to D). Both vegetable and mineral oils, however, show a more marked tendency to exude from compositions based on the lower-substituted A-type cellulose ether. In addition, solid plasticizers, such as triphenyl phosphate, tend to crystallize from ethylcellulose plastics if used in an excessive proportion – they show a greater tendency to crystallize from A-type formulations than from the other higher-substitution ethylcellulose grades.

Tests for compatibility were made by preparing solutions containing both EC and plasticizer, making flow-outs of the solutions on glass plates, drying them,

Table 6.5 Compatibility of ethylcellulose with plasticizers [3].

Group	Resin	
Compatible plasticizer	**Fatty acids:** linseed oleic ricinoleic stearic and many others	**Miscellaneous types:** acetyl tributyl and acetyl triethyl citrate; amyl, ethyl and butyl oleate; methoxyethyl oleate; butyl and butoxyethyl stearate; butyl and glycol esters of fatty acids; butyl diglycol carbonate; dibutyl tartrate; diisobutyl adipate; tributyrin; camphor; refined mineral oils; lubricating oils – all types
	Fatty alcohols: cetyl myristyl stearyl palmityl and many other	
	Vegetable oils – as raw oil and air-blown: castor oil, corn oil, cotton seed oil, linseed oil, soybean oil	**Phthalate esters – various:** the use of these substances in regulated applications is strongly discouraged because of their health hazards.
	Phosphate esters: tricresyl phosphate, triethyl phosphate, triphenyl phosphate	**Water-sensitive substances:** triethyl citrate; glyceryl di- and -triacetate; esters of glycol, diethylene glycol and glycerol with higher fatty acids (e.g.: glycerol mono- and di-ricinoleate, -laureate, -oleate, -stearate; ethylene glycol mono-ricinoleate,-laureate,-oleate, -stearate)

Data taken with permission from Reference [3].

and examining the dry films. Strangely, despite its wide compatibility, the cellulose ether is actually insoluble at room temperature in almost all plasticizers. For example, if powdered ethylcellulose is added to a liquid plasticizer such as soya oil, linseed oil, or suitable fatty acids, the particles do not even become swollen at room temperature. At about 100 °C (212 °F), a solution is formed and the EC remains in solution when the system is cooled. In commercial practice, ethylcellulose compositions are always compounded by a solvent process or by hot-milling or hot-melt procedures. This unusual property of wide compatibility, along with very limited room-temperature solubility, may explain why EC lacquers and plastics are much more resistant to oils and greases, to plasticizer migration, and to contaminating influences generally than one would expect.

For technical applications it is important to know that ethylcellulose is somewhat softer than nitrocellulose (NC) and cellulose acetate (CA), and therefore requires proportionately less plasticizer than either of these to attain the desired degree of softening for any particular purpose. Both cellulose acetate and nitrocellulose plastics normally require from 30–35% plasticizer by weight for most plastics applications. Plastics based on EC, on the other hand, normally require only 15–20% plasticizer to obtain comparable hardness.

Nitrocellulose lacquers frequently contain as much as one part plasticizer to two parts NC. Lacquer formulations with ethylcellulose seldom contain more than one part plasticizer to four parts EC at comparable hardness.

For the selection of plasticizers (from Table 6.5) for use in foods it is essential to ensure that the substances have approval for direct consumption [3].

Water-Sensitive Plasticizers [3]

Ethylcellulose has been tested and found compatible with several materials with an excessive water sensitivity that detracts from their general usefulness. These water-sensitive softeners are listed in Table 6.5.

Oils as Plasticizers [3]

The oils, both mineral and vegetable, are relatively inexpensive plasticizing agents, and can play a unique and important part in the plasticization of ethylcellulose. Unlike the high-boiling solvent esters, they possess very little active solvent action for ethylcellulose unless heated to temperatures approaching the softening point of EC. The oils possess much less softening effect on the cellulose derivative than do the more active high-boiling esters. Therefore, the oils can be used in the development of compositions requiring a degree of toughness not possible with the high-boiling esters because of their excessive softening action.

Castor oil is unique amongst the vegetable oils compatible with ethylcellulose in that it does not exude from compositions even when it is present in proportions as high as equal parts of oil and EC. All of the other vegetable oils exude when present in such high proportions (1 : 1 ratio). This tendency to exude is lessened somewhat if the oil is air-blown. Oil blends containing castor oil, hydrogenated methyl esters of rosin, or one of the chemical esters as a component of the blend also have a reduced tendency to exude. However, vegetable oils, such as corn, soybean, cottonseed, and linseed, can be incorporated into ethylcellulose compositions up to 30–40% by weight before objectionable exudation is encountered. Linseed and soybean oils exude less readily than do corn and cottonseed oils.

With the exception of castor oil, all of the compatible vegetable oils form rigid non-elastic gels with relatively small amounts of EC (3–5 wt%). Such gels are prepared as hot-melt mixtures by dissolving the EC in hot oil at about 180–200 °C (356–392 °F), and then permitting the mixture to cool at room temperature.

Refined mineral oil can be cited as an example that is processed as plasticizer in technical applications. It is used extensively either as the sole plasticizer or as an important constituent of plasticizer blends in the preparation of extrusion and injection plastics where low water resistance, good dimensional stability, especially under adverse temperature and humidity conditions, as well as high toughness and flexibility even at low temperatures are important requisites. Mineral oils of refined and lubricating grades can be used in amounts up to 20–23% by weight in an EC formulation without excessive exudation being encountered at ordinary temperatures.

It is appropriate at this point to call attention to the toughening type of plasticizers. In addition to vegetable oils, the sebacic acid type of alkyd resins and very long oil-extended alkyds impart suppleness to ethylcellulose plastics with a minimum effect on softening point and toughness. Various mixtures of the vegetable oils with these alkyd resins can also serve as valuable toughening plasticizers that can be used in proportions as high as 50% by weight of the plastic without exudation and stickiness. Such compositions form the basis of cable coatings and calendared coatings.

Plasticizers with Resistance against Oil and Gasoline [3]

There are special uses for ethylcellulose compositions where the wide solubility in solvents and great tolerance for diluents – plasticizers and extenders – are not advantageous properties. One such use is in coatings for automotive and aviation ignition cables, where gasoline resistance is an important requisite. Aviation engines, in particular, are frequently washed with high-test gasoline containing an appreciable amount of aromatic hydrocarbons.

Table 6.6 shows how different EC films – plasticized and unplasticized – respond to exposure to gasoline. The viscosity level is the same for all EC products (100 mPa s at 5% in '80 : 20' toluene–ethanol).

Table 6.6 Gasoline resistance of ethylcellulose films [3].

Types of ethylcellulose	Plasticizer	Condition of film[a)] after 72 h immersion in test solvent[b)]
EC – type A	None	Slight curl
EC – type B	None	Softened
EC – type C	None	Gelled and partly dissolved
EC – type A	Sebacic, oil-modified alkyd resin	Badly curled
EC – type A	Processed castor oil	Soft, curl
EC – type A	Dimethoxyethyl phthalate	Soft, curl, exudation

Data taken with permission from Reference [3].

a) Film: 60 parts of ethylcellulose to 40 parts plasticizer.

b) Tests were made in solvent composed of 60 parts 100-octane gasoline, 5 parts benzene, 15 parts xylene, and 20 parts toluene.

Reference to Table 6.6 shows that the required resistant compositions can best be prepared by resorting to the use of lower-substitution A-type ethylcellulose (DS $\leq$ 2.4) and compounding it with gasoline-resistant plasticizers. The EC (type A) has the lowest solubility in aromatic hydrocarbons and tolerates the least dilution with petroleum thinners.

The following plasticizers are insoluble in petroleum hydrocarbons and oils:

- acetyl triethyl citrate,
- castor oil,
- phthalate (dimethyl-, diphenyl-, dimethoxyethyl-),
- triacetin (glycerol triacetate),
- triethyl citrate,
- sebacic alkyd resin (oil-modified),
- benzyl ether of glycerol.

6.3.4 Physical Properties

As with other water-soluble cellulose derivatives, certain properties of ethylcellulose depend somewhat on the degree of substitution (ethoxyl content). The effect of substitution on softening temperature and melting point, on hardness, on water absorption, on solubility in ethanol, and on solubility in a toluene–ethanol mixture (80 : 20) is represented in Figure 6.4. The influence of substitution on solubility in different organic solvents is discussed in Section 6.3.2.2 ('Suitable solvents'). In this test series, ethylcelluloses with a low viscosity of 50 mPa s (measured as 5% concentration in 80 : 20 toluene–ethanol) are used.

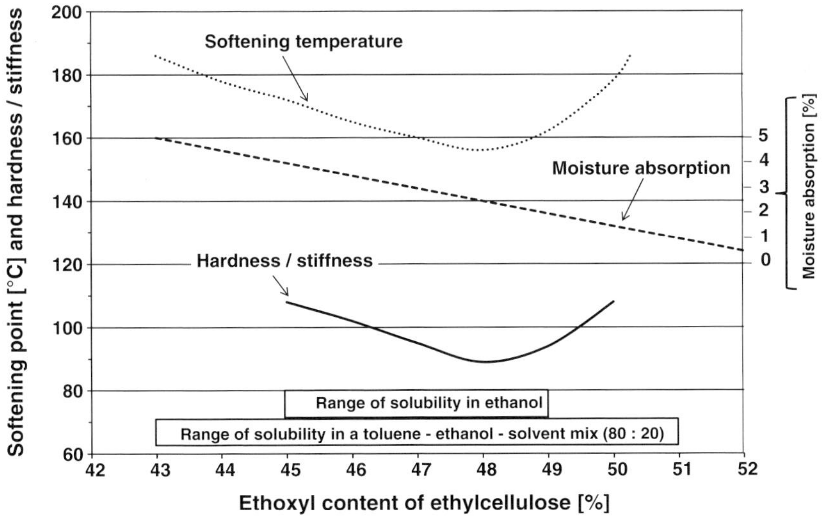

Figure 6.4 Effect of substitution on the physical properties of EC [3]. Reprinted with kind permission of Ashland Inc.

The absorption of moisture by ethylcellulose (at 19 °C (66 °F) and 70% relative humidity) decreases with increasing ethoxyl group content over its range from 43% to 51%. Within this range, the softening temperature and respective melting point and degree of hardness or stiffness show a minimum at an ethoxyl amount of 48%. Ethanol dissolves ethylcelluloses with 45–49% ethoxyl groups, while a mixture of toluene and ethanol in the ratio '80 : 20' expands the solubility to 43–51% ethoxyl.

Conversely, certain other mechanical properties, such as tensile strength, elongation, and flexibility, are not greatly affected by the degree of substitution, but depend largely on the degree of polymerization (DP) and the respective molecular weight (MW), which can be determined by viscosity measurement.

Table 6.7 presents data for the tensile strength, percentage of elongation up to break, and the flexibility of different viscosity grades of ethylcellulose (type B). Figure 6.5 illustrates these data in graphical form. To measure the values shown in Table 6.7 and Figure 6.5, the following method was used:

EC-type B is dissolved in a toluene–ethanol mixture (80 : 20) to prepare a casting solution. By casting, films 0.076-mm (76 μm) thick are made. Drying is carried out for 16 h at 70 °C (158 °F) followed by conditioning for 48 h at 20 °C (68 °F) and 65% relative humidity. Finally, the testing is done.

As visualized in Table 6.7 and Figure 6.5, the values for tensile strength and elongation (at rupture) reach a stable level between viscosity grades of 50 and 100 mPa s (at 5% concentration) and show only a slight increase at higher degrees of polymerization. This means that the length of the cellulose chain is only important for the low-viscosity range (below 55 mPa s at 5%), and only for short-chain EC grades does a chain extension have a positive effect on tensile strength and elongation of the ethylcellulose. Above a critical length – which apparently exists at a viscosity level of 100 mPa s (measured as a 5% solution in a

Table 6.7 Effects of viscosity on the physical properties of EC (ethylcellulose) [3].

Viscosity[a)]		Tensile strength (lb inch^{-2})	Elongation at rupture (%)	Flexibility (double folds)
5% (mPa s)	Intrinsic viscosity (dl g^{-1})[b)]			
8	0.73	6800	10	160
9	0.79	7000	14	200
13	1.04	7300	19	330
24	1.40	8100	28	595
43	1.70	9500	32	1000
94	2.16	10 400	35	1480
184	2.53	10 500	35	2020

Data taken with permission from Reference [3].

a) Viscosity measurement performed in a solvent mixture of toluene–ethanol (80 : 20).

b) Intrinsic viscosity determined by measuring the relative viscosity at different concentrations. Calculation of the reduced viscosity follows through extrapolation of the resulting values to 'zero'. The formula is explained in detail in Section 3.4.1.4.

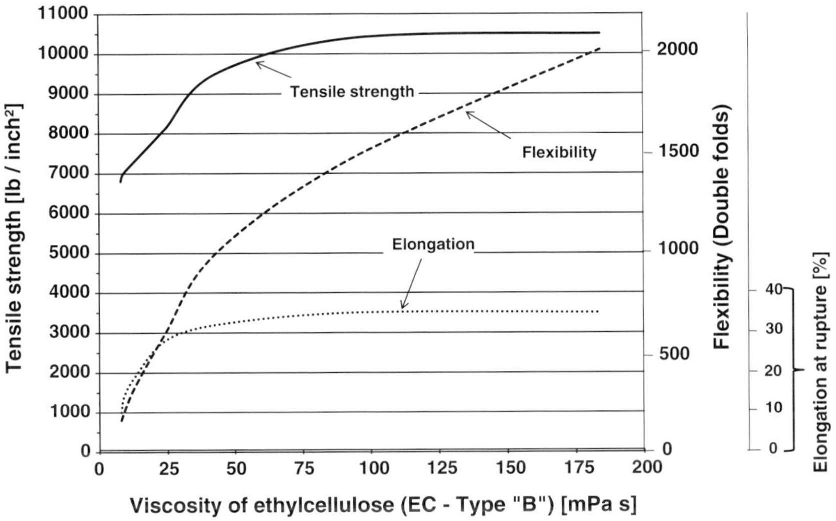

Figure 6.5 Influence of viscosity on the physical properties of EC [3]. Reprinted with kind permission of Ashland Inc.

'80 : 20' toluene–ethanol solvent system) – no further significant improvement of this parameter is observed. The values then reach a plateau. Consequently, only very little additional tensile strength or elongation can be gained by using viscosities higher than about 100 mPa s.

In contrast, the curve of flexibility that is obtained by double-folding of the films exhibits a constant increase. There is no break in the flexibility curve within the presented range shown to indicate what the ultimate folding endurance of EC might be. The missing optimum implies that longer polymer chains are beneficial to the elasticity.

Softening temperature and melting point are not significantly affected by the degree of polymerization.

Within the substitution range offered, ethylcellulose-type 'B' with ethoxyl group contents from 48.0% to 49.5% and a DS from 2.46 to 2.58 is most commonly used because it has the highest compatibility, meets most requirements, and is usually specified due to exhibiting the most generally useful combinations of properties.

The other types of substitution are selected when special requirements must be met.

For instance, low-substituted ethylcellulose (type A) is usually selected for special uses requiring harder, higher melting compositions having better resistance to certain solvents and oils. Moreover, it is the product of choice for injection moulding because it can be injected without surface-lamination or similar mould defects, and yields plastics of better resistance to distortion under heat. A-type EC is used where oil- and gasoline-resistance is important.

Ethylcelluloses of type 'C' and 'D' are useful where extreme dilution with straight-chain hydrocarbon solvents is desirable, and where maximum water resistance is important [3].

6.3.5 Temperature and Viscosity

The viscosity of an ethylcellulose solution is significantly reduced by increasing the solution temperature. For this reason, more concentrated solutions can be prepared at a given viscosity when heat is applied. For example, in the production of films, this 'heat-thinning effect' offers the advantages that less solvent can be used and that drying times can be greatly reduced because less liquid is to be evaporated. In Figure 6.6, the temperature-dependent viscosity profile of an ethylcellulose is illustrated by an example. The product has an average viscosity of about 5.6–8 mPa s (measured as a 5% solution in 80 : 20 toluene–ethanol) and a DS of 2.5. The solvent in this study is a blend of 60% toluene and 40% ethanol. The viscosity is measured with a Stormer viscometer and expressed in the unit of poise; 1 poise corresponds to 0.1 Pas or 100 mPa s [3].

6.3.6 Films

6.3.6.1 Selection of Solvent

The physical properties of ethylcellulose films depend somewhat on the composition of the solvent from which they are deposited. Care must be exercised,

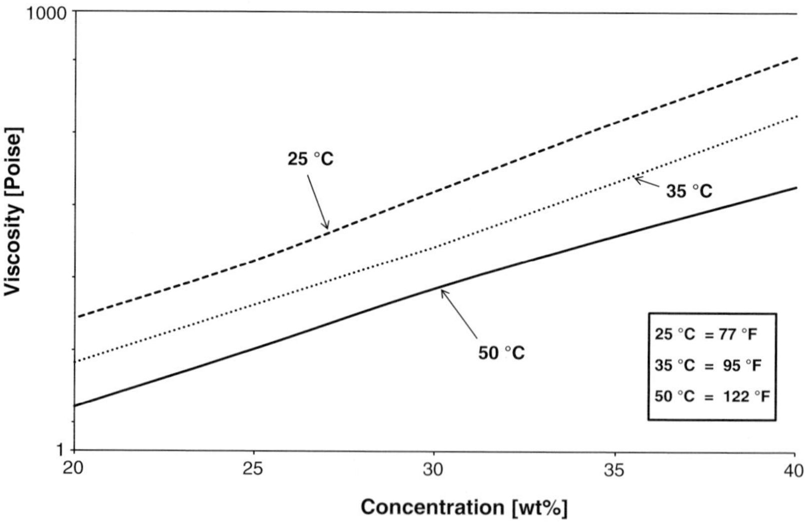

Figure 6.6 Relationship between temperature and viscosity in EC solutions [3]. Reprinted with kind permission of Ashland Inc.

therefore, in proper selection of solvents to ensure the preparation of strong, flexible films. In general, flexible films of maximum tensile strength are obtained when nonpolar solvents such as toluene or xylene (which have no affinity for water) or butyl acetate or methyl isobutyl ketone (which have little affinity for water) constitute a major portion of the solvent at the moment when the film sets to a gel. Brittle films with relatively poorer tensile strength usually result when polar solvents such as acetone or ethyl alcohol, which are miscible with water in all proportions, are present in major amounts at the moment of gelation.

Intensive investigations have shown that a suitable solvent mixture should contain not more than about 30–40% of a low-boiling, water-miscible solvent such as ethanol or acetone. The remainder should consist of solvents such as xylene, toluene, toluene substitute, butyl acetate, or butanol. This means the best films are obtained when the larger proportion (60–70%) of the solvent is nonpolar. The concentration of nonpolar liquid in the film solution is crucial at the moment when the film settles into a gel.

Water in relatively small amounts in a mixed solvent system has a pronounced effect on the physical properties of films. This water may either be present in the solvent initially or may be introduced into the film solution by condensation brought about by the rapid rate of cooling during the drying period.

The foregoing analysis is based on the supposition that most applications are carried out under ordinary atmospheric conditions where no control of relative humidity is exercised. It should be pointed out that strong, flexible films can be obtained from any good solvent mixture, even though they are composed wholly of water-miscible solvents, provided care is taken to use anhydrous solvents and to control the moisture content of the air and the rate of evaporation. This requires special drying rooms or cabinets equipped with air-conditioning apparatus [3].

6.3.6.2 Selection of Plasticizer

Table 6.8 gives evaluation data for several plasticizer–ethylcellulose compositions. The formulations consisted of one part of plasticizer to four parts of ethylcellulose, for a medium-substituted grade of EC (type 'B' with 80–105 mPa s as 5% solution in 80% toluene and 20% ethanol), being tested. Tensile strength, elongation, and flex life were measured for 5-mm thick films of each composition.

6.4 Stability

General Statements

Ethylcellulose is an extremely stable polymer. Since the substance is non-ionic, it has good pH stability (pH 2–13). Ethylcellulose shows high stability when heated

Table 6.8 Evaluation of plasticizers in ethylcellulose films (approx. 5 mm thick – ratio '1 : 4' for plasticizer : EC) [3].

Plasticizer	Double folds	Tensile strength (lb inch^{-2})	Elongation (%)
Dibutyl phthalate	2000	5500	45
Polyethylene glycol di-2-ethylhexoate	1400	5000	50
Raw castor oil	1200	6700	42
Tricresyl phosphate	980	6400	38
Non-oxidizing alkyd resin	910	5850	66
Diamyl phthalate	880	6400	42
Sebacic, oil-modified alkyd resin	600	7150	37
Processed castor oil	600	6700	35
Diisobutyl carbonyl phthalate	560	6400	37
Hydrogenated terphenyl	520	7200	32
β-(*p*-*tert*-Butylphenoxy)ethanol	430	6950	38
Petroleum-derived plasticizer	400	6800	26
Diphenyl phthalate	400	6200	39
para-*tert*-Amylphenol	360	7850	27
Butyl diglycol carbonate	145	2650	13
Styrene resin	110	8200	19
Butoxyethyl stearate	105	2700	19

Data taken with permission from Reference [3].

and very good thermoplastic properties. The glass transition temperature is in the range 129–133 °C (264–271 °F). The softening temperature is about 133–138 °C (271–280 °F) and the melting point has been measured as between 152 and 162 °C (306–324 °F).

Dissolved ethylcellulose is also very resistant to high temperatures. After plasticization (from 135 °C (275 °F) on), EC can be compounded and co-processed with plasticizers, waxes, and other polymers. Compounding improves the flexibility at low temperatures.

Ethylcellulose is inert to treatment with alkaline solutions, even if they are highly concentrated and hot. Dissolved ethyl ether of cellulose withstands concentrated salt solutions and oxidative stress. The shear stability is assessed as high. This and the following statements are extracted from Reference [3].

Colour Stability

Ethylcellulose is virtually colourless and retains this character trait, even under a wide range of conditions and applications. Neither sunlight nor ultraviolet light affect the 'colour'. Therefore, EC can be readily used in combination with dyes and pigments in various products, such as, for example, in protective coatings and plastics, where very high colour stability and close colour control are required.

Compatibility

Ethylcellulose is compatible with an unusually wide range of resins and plasticizers, including oils and waxes. The stability of EC is not affected by these substances and can be considered as high. This wide range of compatibility makes it easy to develop economical formulations in the great variety of uses to which it is adaptable. Small additions of ethylcellulose to waxes toughen their texture and raise their melting point. In varnishes, its use reduces tackiness and increases drying rate. It is quite useful in printing inks. Plastics with rigid or soft properties can be readily formulated. For more information, see Section 6.3.3.

Density

The low density (or specific gravity) of ethylcellulose makes it possible to obtain greater coverage and greater volume per unit weight than with the other cellulose derivatives.

Two examples from the technical area illustrate this effect in coatings and adhesives of similar weight composition:

- Ethylcellulose allows a 45% higher coverage than nitrocellulose.
- EC use results in a 20% greater loading than cellulose acetate.

In plastics, films, and foils, the low density of ethylcellulose gives the same corresponding advantages of greater volume for a given weight.

Highly stable coatings (e.g. protective lacquers) are generally formulated using low-viscosity ethylcellulose grades with 4, 7, or 10 mPa s to achieve the lowest possible thickening effect at the maximum possible solids content. EC types having higher viscosities are always applied when a high toughness and strength of the coating is required.

Electrical Properties

The excellent electrical properties of ethylcellulose (see also Section 6.3.1 on 'Physicochemical parameters'), combined with its good thermal stability and outstanding flexibility and toughness, led to its early and continued use in cable lacquers, where conditions requiring these properties are encountered. It has also been used in plastics for electrical insulation for many purposes.

Flexibility

The great flexibility of ethylcellulose over a wide range of temperatures is one of its most distinctive characteristics. Its retention of flexibility at very low temperatures is especially notable, with many compositions remaining flexible even at −70 °C (94 °F). Ethylcellulose also retains a large measure of its initial flexibility, even after longer exposure to temperatures approaching its softening point (156 °C (313 °F)).

This flexibility is advantageous in many applications, such as protective coatings, adhesives, films, foils, and plastics. It is a property that is desirable when the final products are exposed to very cold and/or very hot temperatures, and when there is the possibility of going through large temperature ranges. The following are areas of application:

- plastics requiring a high resistance at low temperatures;
- furniture lacquers to prevent cold cracking;
- fabric and textile coatings, which must remain flexible in the cold and be resistant to cracks and breaks;
- hot-melt compositions and plastics for sustained toughening action after compounding.

Flammability

The flammability of ethylcellulose is as low as that of any other water-soluble cellulose derivatives. If EC is held in an open flame, it will catch fire and burn. However, formulations with ethylcellulose can be made fire resistant by proper choice of plasticizer and other flame-proofing agents.

Softening Point

The softening or melting point of ethylcellulose is, at 156 °C (313 °F), relatively low. In addition, it is possible, if desired, to make it even lower by proper adjustment of suitable plasticisers. These factors permit ease of working and application, such as in hot-melt application of adhesives, calendar application on cloth, and in injection and extrusion plastics.

Stability to Chemicals and Temperature

Of all cellulose derivatives, none is more stable to chemicals than ethylcellulose.

It is resistant to alkaline solutions, both diluted and concentrated. EC also survives treatment with strong hot alkalis without molecular degradation. Highly concentrated salt solutions are ineffective. Ethylcellulose is a more sensitive to acidic materials than cellulose esters (such as nitrocellulose), which are often used as an alternative.

However, lacquer coatings based on EC satisfactorily withstand the action of dilute acids for limited exposure periods. Dissolved ethylcellulose is resistant to the destructive influence of oxygen under high pressures at room temperature for extended periods of time. In an ozone-containing atmosphere, EC can be kept for one or two weeks with no detectable trace of degradation. Oxygen at temperatures above the softening point of ethylcellulose has a marked degrading effect unless the compositions are stabilized with antioxidants.

The application of heat to ethylcellulose has only slight effects until its melting point is reached. At elevated temperatures ($T \geq 156\,°C$ (313 °F)), discolouration and embrittlement of the product may occur when the ether is not protected by means of antioxidants.

Acidic substances such as high-acid-number resins have a degrading effect on EC, particularly at higher temperatures – and therefore are not recommended to be used in plastics, adhesives, or other hot-melt applications of ethylcellulose.

By use of suitable process temperatures (below the softening point, $T < 150\,^{\circ}C$ ($302\,^{\circ}F$)) in combination with added stabilizers, heat-resistant EC formulations can be created, without destroying the molecule. Such properly formulated and stabilized ethylcellulose compositions have remarkably good resistance to high temperatures, so that plastic compounding, moulding operations, and hot-melt compounding and application can be carried out without degrading the EC. Then, high-temperature-stable coatings for foods and for products such as cables and plastics are possible.

Stability to Light

Light, visible or ultraviolet, has no discolouring effect on ethylcellulose. Clear films and foils transmit virtually the whole spectrum of visible light and varying amounts of ultraviolet light, depending on the stabilizers and plasticizers used in formulation. The characteristics of UV light absorption are significantly affected by the thickness of the film and by the concentration and chemical nature of the stabilizers used. Certain stabilizers reduce the embrittling effect of ultraviolet light on ethylcellulose.

Ethylcellulose, because of its light transmission, together with its flexibility, great toughness, and chemical resistance, is an unusual material when properly used in protective films. Such films can be applied from solution, from hot-melt, by calendering, or by lamination.

Stability upon Action of Light, Oxygen, and High Temperatures

Formulations with ethylcellulose that are not stabilized are subject to oxidative degradation in the presence of sunlight or UV radiation and at elevated temperatures, particularly above the softening point of EC ($T = 156\,^{\circ}C$ ($313\,^{\circ}F$)). Ethylcellulose thus oxidized becomes brittle, and the surface can become badly crazed. Discolouration generally occurs during prolonged exposure at excessive temperatures. Investigations have shown that the harmful effects of oxidation can be inhibited for extended periods of time by the use of antioxidants and by the proper selection of resins, plasticizers, and other modifiers.

The use of antioxidants is the most effective and generally the easiest method of inhibiting oxidative degradation. Ethylcellulose compositions that are subjected to high temperatures during compounding or subsequent use and that may be exposed to ultraviolet light or sunlight should always contain an antioxidant. Acid acceptors should be added to ethylcellulose formulations containing components such as chlorinated resins and plasticizers that form acidic degradation products. Epoxy types are especially useful because of their excellent capacity to absorb free acids. Stabilizers of these types should always be incorporated when ethylcellulose is used in hot-melts, in plastics, and in clear-film applications.

Additional studies demonstrate that substances and compounds that absorb light waves in the range 230–340 nm are effective stabilizers against light-catalysed oxidation.

Usually, the addition of one part of antioxidant or a blend with equal parts of a light absorber or epoxy stabilizer for each 100 parts of ethylcellulose is sufficient to obtain long-term stability of the cellulose ether. Larger quantities of added stabilizers result in very small increased benefits, and as little as 0.25 parts will give marked improvement over unstabilized ethylcellulose compositions.

Outdoor exposures of clear EC films show that a combination of antioxidant, light absorber, and acid acceptor is more effective than a single stabilizer.

Furthermore, most pigments provide protection against light and oxygen-induced degradation during outdoor exposure. In pigmented system where added stability is desired, antioxidants are more effective than ultraviolet light absorbers, because light is excluded by the pigment.

Stability to Water

The types of ethylcellulose discussed in this chapter are all characterized by a high degree of substitution ($DS \geq 2.2$) and are therefore insoluble in water. Owing to the lack of solubility in water, these ethylcelluloses are not affected in their efficiency by the presence of water. EC takes up very little water from moist air or during immersion, and the small amount that is taken up evaporates readily, leaving the ethylcellulose unaltered. Dimensional stability, which is related to water absorption, is unusually good. This property can be especially valuable in many applications.

Thermoplasticity

Ethylcellulose possesses excellent plastic flow characteristics – even at high processing temperatures. Products (foods or plastics) can be processed in heated mixers, on heated two-roll mills, or in extruders without the addition of volatile solvents, thus making manufacturing of certain food-grade compounds possible and plastic compounding more economical. This material characteristic is essential for injection, extrusion, lamination, and calendering operations, as well as for hot-melt compounding and application of adhesives, paper coatings, and casting plastics.

Mechanical Stability (Toughness)

The high tensile strength of ethylcellulose is remarkable. Combined with its excellent flexibility over a wide temperature range, this results in an unusually high toughness and mechanical stability. Toughness, at high and low temperatures, is one of the most useful qualities of ethylcellulose. In this respect, it has marked significant advantages over synthetic resins. In addition, it is superior in this property to other cellulose derivatives. The toughening action that ethylcellulose imparts can be of fundamental importance for protective coatings, adhesives, and plastics.

Taste and Smell – Sensory Stability

Since ethylcellulose is free of taste and smell it is used in coatings for food and in packaging materials such as paper, cardboard, films, and foil that come in contact with food. Plastic materials for packaging foods are another possible application of EC.

6.5 Rheometry

To identify ethylcellulose and distinguish it from other additives, the following methods can be applied [2].

1) **Solubility test:** Ethylcellulose (with DS > 2.3) is virtually completely insoluble in water, glycerol, and propylene glycol. Ethylcellulose with less than 46% of ethoxyl groups is readily soluble in tetrahydrofuran, methyl acetate, chloroform, and in mixtures of aromatic hydrocarbons and ethanol. Ethylcellulose with more than 46% of ethoxyl groups is readily soluble in ethanol, methanol, toluene, chloroform, and ethyl acetate.
2) **Viscosity measurement:** With ethylcellulose, the thickening effect cannot be measured in water in the way that is for all other cellulose derivatives or hydrocolloids. The viscosity value is determined with a 5% concentration of EC in a mixture of toluene–ethanol (80 : 20) at 25 °C (77 °F) using a Ubbelohde viscometer.
3) **Film formation:** 5 g of the sample is dissolved in 95% of a '80 : 20' mixture of toluene–ethanol (w/w). A clear, stable, slightly yellow solution is formed. A small amount of the solution is poured onto a glass plate. The solvent is evaporated under a laboratory fume hood. It leaves a thick, tough, continuous, clear film, which is combustible.
4) **Litmus test:** The pH of a 1% colloidal solution is in the neutral range.

6.6 Synergies with Other Hydrocolloids

Synergistic effects between ethylcellulose and real hydrocolloids do not exist in the same solvent system. The reason for this is the water insolubility of commercially available ethylcellulose grades and the insolubility of hydrocolloids in oil or organic solvents (such as aliphatic alcohols or esters and ketones).

Since some mixtures of EC with organic solvents tolerate up to 10% water in the system, it is possible to incorporate a real hydrocolloid in this organic ethylcellulose solution.

With the aim of controlled release of the active ingredients, a small amount of hydroxypropylmethylcellulose (HPMC) is incorporated into the matrix of an ethylcellulose coating – for example, in practice during manufacture of solid food supplements in the form of tablets. In this respect, it is possible to form water-soluble pores of HPMC in a coating of water-insoluble ethylcellulose. Thus-coated tablets will dissolve very slowly in the human stomach, initially by the HPMC pores becoming hydrated and finally opened. The active substance inside the tablet is only gradually dissolved and thus released slowly.

Procedure for the Preparation of a Tablet Coating Made of Ethylcellulose with Integrated HPMC-Windows (for example)
Ethylcellulose is dissolved in an ethanol–water mixture (90 : 10). The concentration of dry matter (EC and HPMC) should be 10–15% by weight. The amount of ethylcellulose dissolved depends on the molecular weight or degree of polymerization. The shorter the cellulose chain, the higher is the dosage required:

- 10–15% ethylcellulose of viscosity level 7–10 mPa s (measured at 5% concentration in toluene–ethanol (Section 6.5, 'Rheometry' – Point 2))
- 10 wt% ethylcellulose with 20 mPa s;
- optional (only conditionally recommended): 5 wt% ethylcellulose with 45 mPa s.

To reduce the dissolution rate ('very slow release of active substances'), it is advised to increase the dose of a low-viscosity ethylcellulose rather than use the same concentration of a higher-viscosity EC. A high density of shorter polymers on the surface enhances the protective effect of the active ingredients in the core.

It is possible to incorporate a plasticizer in the coating solution (5–30 vol.% of total content of dry matter). As pore former, a hydrophilic polymer with a low molecular weight – such as a low-viscosity HPMC (6–15 mPa s at 2% in water) – is used.

The concentration of the pore former is dependent on the solubility of the active ingredient in the tablet core and on the desired release rate (about 3–10% HPMC based on the amount of ethylcellulose). Coatings on the surfaces of tablets or dragées are usually applied by means of spray-drying processes in a fluidized bed.

6.7 Food Applications with Typical Formulations

6.7.1 Overview of Functions

Ethylcellulose is used in food, nutraceuticals, and dietary supplements for the following functions:

- as a thickening agent for hydrophobic substances (vegetable oils, flavourings, alcohols);

- for the formation of flexible films having high heat resistance and excellent freeze–thaw stability;
- for producing water barrier layers with the target of protecting water-sensitive and water-reactive substances;
- for controlled release of active ingredients (vitamins, minerals, etc.);
- as a binder in granulation processes;
- as a film former and coating agent for tablets and capsules;
- for masking sensory critical active substances with a very bitter, sour, or metallic taste profile.

Higher viscosity ethylcellulose products form thicker films that are stronger and more resistant than the films of low viscosity grades.

Granulation

Granules, associated with ethylcellulose, are suitable as a base material for tablets since they produce very stable products with good solution properties. The granulation of water-sensitive substances with solvents is feasible as well as granulation by extrusion, as ethylcellulose shows thermoplastic properties when heated.

Coatings

For coating products, ethylcellulose is usually dissolved in combination with hydrophilic substances (e.g. HPMC) in a polar–nonpolar solvent mixture. The coating solution is then applied on the surface of the product to be protected by spraying (Section 6.6).

Pressing Tablets

There are two technologies for producing tablets – dry granulation and direct compression. Pressing untreated substances has an economic advantage, so that this method is preferred. Ethylcellulose allows the direct compression of various filling materials and active substances under high pressures with a good dissolution profile in the final product.

The release rate is significantly influenced by the:

- concentration of ethylcellulose,
- total amount of polymer in the system,
- particle size of the EC (granules or fine powder).

The higher the dosage of the ethylcellulose is, and the smaller its particle size, the more resistant the system will be to external stress factors such as solvents.

The release of the active ingredients is slower at a high density of fine particles. Table 6.9 shows the most important applications and suitable ethylcellulose grades.

Table 6.9 Applications for ethylcellulose.

Application	Suitable EC polymers (viscosity)[a]
Coatings for controlled release	• Pure EC with a low or medium viscosity (7–20 mPa s); • ethylcellulose, blended with low-viscosity HPMC
Microencapsulation	Higher viscosities (45–100 mPa s)
Coating of tablets	Low to medium viscosities (7–20 mPa s)
Granulation	Medium viscosities (10–45 mPa s)
Binder/direct compression	Low or high viscosities (7–10 or 100 mPa s)

a) Viscosity is measured as 5% solution in toluene–ethanol (80 : 20) at 25 °C (77 °F) with an Ubbelohde viscosimeter.

6.7.2
Coating and Encapsulation with Ethylcellulose

An example of the use of ethylcellulose is the coating of systems for encapsulation and for masking of substances with undesirable organoleptic properties. Table 6.10 presents a typical formulation.

6.8
Non-food Applications

For ethylcellulose, there are various possible applications in technical goods and other regulated products [3].

Table 6.10 Formulation for masking a bitter-tasting substance [3].

Ingredient	Amount (%)
Bitter-tasting substance	1
Ethylcellulose (EC 45)	10.5
Emulsifier Tween 80	0.5
Solvent methanol	88

Data taken with permission from Reference [3].

Processing guideline:

- ethylcellulose, the bitter-tasting substance and emulsifier are dissolved in the solvent;
- the homogeneous solution is spray-dried at about 90 °C (194 °F);
- the resulting powder does no longer has a bitter taste.

1) **Pharmaceutical applications:**
   Ethylcellulose is one of the few water-insoluble polymeric excipients. The substance is used for and in:
   - coating of tablets;
   - development of systems with controlled release of active ingredients;
   - microencapsulation;
   - masking of ingredients with undesirable flavours;
   - granulation processes.

   From a technological perspective, the transition from medical drugs to solid food supplements is fluid – whereas the choice of suitable solvents and carrier systems has nearly no limits in the manufacture of pharmaceutical products it is very restricted in food supplements.
   - **Coating for tablets:** Ethylcellulose with an approval for direct consumption produces durable coatings for tablets that offer good adhesion and function to control the rate at which the ingested medicine is released. Typically, the tablet is spray-coated from a solvent system. Pharmaceutical manufacturers also treat products with an EC-based aqueous polymer dispersion that is spray-coated onto the tablets, thus avoiding solvent systems that are sometimes undesirable. In either case, the coating masks bitter tastes and improves stability. Patents for further instructions are available.
   - **Binder for tablets:** The use of ethylcellulose as a binder and filler in tablets of vitamin, mineral, and prescription drugs for both humans and animals has a long tradition. The oil-soluble cellulose ether helps to produce a stable, attractive tablet without significantly affecting the dissolution and release rate.
   - **Encapsulation:** Microencapsulation of active ingredients with ethylcellulose offers pharmaceutical manufacturers another means to control the solubility and release rate. The coacervation process with a solvent is a possibility, which has been filed in several patents.
2) **Personal care – cosmetics:**
   Ethylcellulose is globally approved for cosmetic applications. An important feature is the masking that gives rise to controlled release of active ingredients. Applications are lipstick and nail polish, where EC can create a lasting brilliance. The stabilization of perfumes and flavour substances is another option. Further typical applications are the thickening of body creams and lotions. Recent developments are towards water-resistant liquids such as sunscreen products.
3) **Emulsions:**
   EC-containing emulsions can be formulated for environmental concerns and low costs. Ethylcellulose emulsions have been developed for use as a permanent sizing agent for textiles to replace starches and various other water-soluble gums.

4) **Inks:**
   a) **Screen-process inks:**
   The use of ethylcellulose in screen-process inks has grown substantially. Ethylcellulose is soluble in organic solvents that are not injurious to lacquer stencils, and it has wide compatibility with many resins and plasticizers. Compositions are usually based on one part by weight of ethylcellulose and a few parts of plasticizer and resin. A heavy loading of pigment and inert ingredients are possible.
   b) **Magnetic inks:**
   Because of the high dielectric constant of ethylcellulose it can also be used for formulating magnetic inks. These have unique adhesion holdout properties.
   c) **Gravure and flexographic inks:**
   Ethylcellulose is being used in both gravure and flexographic inks. It contributes good scuff resistance, pigment dispersion, and holdout features, and controls viscosity. Compatibility with ink resinates is good. Pigments are easily incorporated by adding ethylcellulose colour chips or flushed colours, or by any of the usual methods of grinding. Typically, one part of the pigments is mixed with three parts by weight of ethylcellulose. EC contributes significantly to the formation of strong friction-resistant films. Ink films have a reduced stickiness with improved adhesion on difficult surfaces through ethylcellulose. Additionally, inks with EC also dry faster. Flexographic inks are commonly formulated with shellac, or combinations of shellac and hard resins, in alcohol solutions. Ethylcellulose is added to promote adhesion and toughness and provide other features, the same as with rotogravure inks.
5) **Ceramics:**
   In ceramic products, ethylcellulose finds utility for adjusting the rheological properties, and it acts as a binder. It generates the necessary green strength in raw (unfired) ceramics during thermal curing and burns without leaving any residue.
6) **Conductive paste (electronics):**
   Ethylcellulose is a component of electronic conductive pastes and inks. It permanently controls the rheology (viscosity) and acts as a binder or stabilizer for metallic pigments. Combustion takes place without leaving any residue.
7) **Casting plastics:**
   Hot-melt casting compositions consist of resins, plasticizers, waxes, and mineral fillers toughened with ethylcellulose. This type of plastic can be cast in bronze, brass, iron, aluminium, wood, or plaster moulds. Some mould materials, such as wood and plaster, require a sealer coat. Remarkably accurate surface reproduction is possible with these hot-melt plastic castings. They are used for the preparation of dies, jigs, and tools in the aircraft industry, and hold promise of wide use in other industrial applications.

8) **Grinding of pigments:**
Ethylcellulose has been found to be an excellent base for pigment grinding. Because of its low melting point, EC, plasticized or unplasticized, can be made into a colloid readily on either the two-roll or the Banbury mill. Plastic thus produced is very tough and sticky while hot, and is well adapted to pulling apart pigment agglomerates. The high temperature at which ethylcellulose can be processed without decomposition ensures uniform grinding and dispersion of pigment particles.

9) **Films and foils:**
In film form, ethylcellulose is well suited in terms of properties and appearance as a wrapping material and as electrical insulation. Unplasticized film can be produced with a high degree of toughness and flexibility. A plasticizer-free film of 0.03 mm (0.001 inch), for example, can be flexed over 2000 times on a double fold tester before failure occurs. Even at −70 °C (94 °F) the film has a high degree of flexibility. Tensile strength is on the order of 100–160 g cm^{-2} (7000–11 000 lbs $inch^{-2}$), while elongation is at 10–35%. The water absorption capacity is low, even in very humid atmospheres, where moisture absorption is about 2%. This limited water uptake has no appreciable effect on the dimensional stability of the film. In fact, ethylcellulose films are not noticeably affected even after long immersion in water, when a drying step follows.

10) **Adhesives:**
In adhesives, ethylcellulose contributes to:
- flexibility at low temperatures;
- strength and firmness;
- broadening of the critical melting temperature range with a resultant decrease in plastic flow;
- an increase in melting point of mixtures;
- a decrease in sweating of plasticizers;
- better control of tackiness in adhesive films.

Hot-melt adhesives containing ethylcellulose have a fast initial tack while still in the liquid or molten state – but then harden quickly upon cooling. Furthermore, there is no residual solvent odour or taste. Solvent-based ethylcellulose adhesives are useful in places where ethylcellulose films or plastics must adhere to other surfaces. EC also provides easy weldability of packaging materials and high stability of sealing seams directly after their production.

11) **Hot-melt applications:**
Hot-melt applications of ethylcellulose have received considerable attention. This is because EC is a product with an unusual combination of properties, which makes it easily adaptable to this mode of application. It can be made stable to heat. It has excellent thermoplasticity. It dissolves readily in many hot resins, plasticizers, oils, and wax mixtures. Of particular interest is the fact that it imparts to such mixtures a remarkable toughness. Furthermore, all hot-melt compositions

are economical to apply. No volatile solvents are involved. Application is usually a simple, one-operation procedure.

Various compositions are now in use. Some applications are relatively well known and have been in use for years. Others are new, have interesting possibilities, and are expanding rapidly. Examples of well-known uses are shellac substitutes, potting compounds used in electrical insulation work, and decalcomania transfer compositions toughened with ethylcellulose. Numerous possibilities exist for the hot-melt coating of paper with resin and plasticizer mixes toughened with the cellulose derivative.

Compositions based on EC can be tailored to specific requirements such as non-blocking, flexibility over a wide temperature range, high gloss, or pale colour. Ethylcellulose hot-melt strip coatings provide protection during both shipping and storage.

12) **Plastics:**
Ethylcellulose-based plastic belongs to the group known as thermoplastic materials – that is, those that can be shaped by pressure while hot and that, when cool, retain the impressed form. A plastic made from EC has the usual operating advantages of other thermoplastic products – for example, re-use of scrap; adaptability to fast, low-cost injection and extrusion processes; ready workability with hand or machine tools; and easy finishing.

An outstanding characteristic of ethylcellulose plastic is its high impact resistance at low temperatures. It has the lowest specific gravity of any of the cellulosic thermoplastics now available, and, with the exception of nitrocellulose, it shows the lowest water pick-up of these thermoplastics. It has the highest alkali resistance and also has fair resistance to acid.

EC plastic easily meets all ordinary requirements of practical strength properties at temperatures between −57 and +77 °C (−70 and 170 °F). The use of ethylcellulose plastic is indicated where hardness, combined with toughness, close dimensional tolerance, ability to hold dimensions during immersion in water, and high impact strength at low temperatures are desired.

Military rocket tapes and moulded parts can be made from a premium grade of A-type EC. This is a result of the low ash content, which is critical to the shelf life of the product.

13) **Special lighting/fluorescent lamps:**
In the coating of fluorescent lamps, ethylcellulose is used to adjust the rheological properties, and as a binder. The EC will be removed in the final heating process, where it burns without residue and the triphosphor layer adheres to the glass tubes.

14) **Speciality coatings, paints, and lacquers:**
Satisfactory formulation of ethylcellulose lacquers and their good performance depend largely on the selection of the appropriate kind and proportion of modifying agents to meet the requirements of each use.

Ethylcellulose has been found to yield lacquers of excellent toughness and flexibility. These properties are retained over a wide temperature range, extending from low to high. Ethylcellulose lacquers, for example, can be formulated to be unusually resistant to cold-checking. Furthermore, they can be created to retain their colour extremely well on exposure to sunlight.

Tests and experience indicate that properly formulated EC lacquers are as durable as conventional nitrocellulose coatings. The specific properties of ethylcellulose products make them a desirable choice for use over metal, wood, leather, and rubber.

In general, the solvent system for EC lacquers consists of 80% aromatic hydrocarbons and 20% alcohols. The hydrocarbon portion is composed of toluene and xylene or mixtures of these two substances. The remaining alcohol fraction is composed of ethanol, isopropanol, and butanol or mixtures of these alcohols.

For some coating applications, such as those on asphalt tile and polystyrene plastics, it is necessary to use special solvent combinations instead of the cited mixtures. These special solvent compositions may give higher viscosities and somewhat hazy solutions, but the lacquers dry to clear films and can be used without difficulty if the lower solids necessary with these solvents can be tolerated.

In conclusion, ethylcellulose provides water-resistant and cold- and heat-resistant coatings that require only low maintenance; such coatings are also suitable for marine applications. The possibilities for application of ethylcellulose in paints, coatings and lacquers are very varied and are described in more detail in the following examples:

a) **Hart lacquers for rigid surfaces:**
Such lacquers will give satisfactory service where a tough, crack-resistant coating is desired on a rigid surface, even when it is exposed to the extremes of outdoor temperature. For such applications, ethylcellulose is modified with hard resins and plasticizers. The resistance to cracks and fractures remains even under extreme temperatures.

b) **Tough lacquers:**
Many lacquer applications on non-rigid surfaces require, as basic desired properties, toughness, wear-resistance, good adhesion, resistance to discolouration and alkali, and flexibility. Such properties are required in lacquers for paper, linoleum, cellophane, surgical tape, wet and dry sandpaper, and lacquers for the decoration of textiles. Ethylcellulose in combination with suitable plasticizers and stabilizers can fulfil these functions in lacquers.

c) **Bronzing lacquers:**
Ethylcellulose is also used in copper and aluminium bronze lacquers. These EC compositions are substantially free from discolouration and gelation. Correct selection of the resin is an important aspect in the preparation of a coating formulation. In an experimental series, dewaxed dammar (wood resin) developed less colour in the lacquer than did other resins tested. Citric acid (1% based on EC) is an effective stabilizer against the development of any green colour.

d) **Lacquer for polystyrene plastic:**
The solvent combination is an important factor in any successful coating for polystyrene plastic. In an ethylcellulose lacquer for polystyrene plastic, alcohols must constitute the major portion of the solvent, although a small amount of an active solvent for the plastic is desirable to get some 'bite' into the moulded part. A suitable active substance would be an ester, ketone, or aromatic hydrocarbon.

e) **Lacquer for rubber:**
Whether the emphasis is on gloss, adhesion, or flexibility, ethylcellulose coatings can be formulated to give the desired properties. EC lacquers have met these requirements when applied to either cured or uncured rubber, and yield a finish with outstanding adhesion and high flexibility.

f) **Speciality wood finishes:**
Ethylcellulose lacquers containing high proportions of low-cost resins and a small proportion of high-viscosity EC are economical, air-drying coatings. Although the films may be somewhat soft, their flexibility and toughness are sufficient to make them interesting for applications such as drawer coatings or backboards for chests.

g) **Water-white wood finish:**
When ethylcellulose films are exposed to ultraviolet light, they tend to bleach rather than discolour. This tendency indicates that they have a place in water-white wood finishes. Low-substituted ethylcellulose (type A) without a plasticizer produces a water-white composition hard enough to stand some rubbing and polishing. A mixture of 80% nitrocellulose and 20% EC gives a lacquer that also shows reasonably good colour retention.

h) **Alkali-resistant lacquer:**
Because there is increasing interest in maintenance paints that have good alkali resistance, work has been carried out on the use of stabilized ethylcellulose compositions in this field. It was found that the use of epoxide-type stabilizers, plus antioxidants, produces EC coatings of very high alkali resistance.

Ethylcellulose lacquers applied to steel panels were found to be unaffected after 100 days at room temperature in either a 5% caustic solution or a 1% ammonia atmosphere. A more spectacular test showed these coatings to be unaffected even though they remained immersed in a 70% caustic solution at 76 °C (170 °F) for 19 days.

i) **Paper lacquers and coatings:**
Because of the varied special properties that ethylcellulose makes possible in paper lacquers, it is used widely in this field. The following are achievable with this material:

- light initial colour,
- good colour retention under sunlight and ageing exposures,
- heat-sealing,
- good gloss,
- resistance to blocking,

- resistance to heat discolouration,
- flexibility over a wide range of temperatures,
- alkali resistance.

Reconstituted cigar wrappers are flexible, yet tough, when they are formulated with ethylcellulose.

j) **Solvent-based strip coatings:**
In recent years, there has been much interest in temporary protective coatings that can later be stripped from the surface they protect. The major volume of these coverings has been deposited from hot-melts, but a growing market is developing for solvent-based products.

k) **Flow-back high gloss lacquer:**
Other interesting ethylcellulose lacquers are flow-back paper coatings. These formulations are designed to give especially high gloss when the coating is reflowed by application of heat at about 177 °C (350 °F) for a few seconds after the lacquer is applied and dried in a normal manner.

l) **Lacquers, topcoats, varnishes – the final layer:**
The use of ethylcellulose in certain types of varnishes gives them the inherent toughness and quick-drying properties associated with the cellulose derivatives. Addition of EC to some varnishes shortens the drying time appreciably, increases toughness, reduces the amount of metallic drier necessary, decreases surface tack, and improves resistance to rapid temperature changes.

There are two different procedures for adding ethylcellulose to varnishes. The simplest and most foolproof is the 'coldcut' method. EC is first dissolved in its usual solvent combination, then added to a varnish that has been thinned with a solvent rich enough to hold the ethylcellulose in solution. If the varnish contains the usual solvent of mineral spirits or other aliphatic hydrocarbon, the EC will precipitate on being added to the varnish.

The 'hot-melt'-procedure involves the addition of ethylcellulose flakes to the hot varnish. Careful temperature control is necessary to avoid degradation of the cellulose ether.

The quantity of EC added will control the properties of the coating, but 10%, based on resin and oil, is adequate for most purposes. Medium-substituted ethylcellulose types of low to medium viscosity (types B and C with 5% from 10 to 200 mPa s) are used, with the C-types having a slight advantage in this application. The viscosity level of EC controls the amount of extra bodying desired in the varnish.

6.9 Nutritional Properties

In 2004, the Scientific Commission (SCF) for review of the European Food Safety Authority (EFSA) came to the following conclusions:

- Ethylcellulose was examined on the basis of existing safety data for the closely related cellulose derivatives. Like other approved cellulose-based substances, the ethyl ether of cellulose is considered as safe.
- The SCF committee confirmed that during the manufacturing process, the steam treatment and the subsequent drying step, all volatile residues – including the unused substitution educt ethyl chloride – are completely removed.
- Owing to the highly hydrophobic nature of ethylcellulose and its high molecular weight (>500 kDa), the ethyl ether of cellulose passes the human gastrointestinal tract without significant change after oral ingestion. The energy content of EC is 0 kcal g^{-1}.
- Therefore, the scientific panel decided to categorize ethylcellulose in the group of modified cellulose products for which no maximum daily quantity is defined ('ADI not specified') [4].

6.10
Legislation

Legal situation in Europe (EU)

Ethylcellulose has been evaluated with positive results by the European Food Safety Authority (EFSA) (see 'Official Statement' [4). Since 2006, ethylcellulose (E 462) has been officially listed as a food additive in Annex I of the EC Directive 95/2. Thus, EC is generally approved in food products as a technological additive with a dose of 'quantum satis'. The new Directive 2006/52/EC (amendment of Directive 95/2/EC) has been valid since 26 July 2006, and had to be implemented into national legislation of the member states by 15 February 2008. Ethylcellulose may be used for the encapsulation of flavours and for the coating of vitamins. It finds application in ink for printing on fruits and baked goods. In addition, ethylcellulose is incorporated as a hydrophobic component in packaging paper that comes into contact with water- and fat-containing food.

In the USA, ethylcellulose has been approved by the FDA for many decades as a direct and indirect food additive in protective coatings of food. An example of such an application is the fixation of flavours.

After the general approval of ethylcellulose as food additive in Europe, European food law has a real advantage over US law. This fact has to be taken into account in the development of food products with EC being used globally.

While in North America the declaration of ethylcellulose may be made with 'Modified cellulose', in the European Union the chemical name or the E number must be mentioned on the label. An alternative naming is currently being discussed and checked by the relevant EU committees.

FDA Status of Ethylcellulose (USA)

In the USA, the use of food grade ethylcellulose is regulated in the United States Food Chemicals Codex (FCC). These EC types also meet the monograph requirements in the National Formulary, current edition. Furthermore, ethylcellulose is included in the CTFA (Cosmetic, Toiletry and Fragrance Association) ingredients dictionary.

The low- and medium-viscosity EC-types with low and medium degree of substitution (types A, B, and C; see Section 6.3.1) are in accordance with the requirements of the US Food Administration (US FDA, FDA) for use in food applications and products, which may come into contact with food. The requirements for ethylcellulose are determined in FDA regulation No. 21 and can be found in different sections, depending on the specific application.

Table 6.11 gives an overview of FDA regulations for ethylcellulose.

Legal Situation in Other Countries

Cellulose derivatives are widely accepted. In countries and areas outside of the USA and the European Union, local food legislation needs to be checked.

Table 6.11 FDA status of ethylcellulose for regulated applications [3].

Regulation – No. 21 CFR . . .	Application
175.105	Adhesives
175.300	Resinous and polymeric coatings
175.390	Zinc–silicon dioxide matrix coatings
176.170	Components of paper and paperboard in contact with aqueous and fatty foods
176.180	Components of paper and paperboard in contact with dry foods
177.1210	Closures with sealing gaskets for food containers
182.90	Substances migrating to food from paper and paperboard products – ethylcellulose
573.420	Binder or filler in dry vitamin preparations to be incorporated into animal feed
73.1 (b)[a)]	• Diluents in colour additive mixtures for marking food: - inks for marking food supplements in tablet form, gum, and confectionery; - inks for marking fruits and vegetables; • diluents in colour additive mixtures for colouring shell eggs.
73.1001[a)]	Diluents in colour additive mixtures for drug use exempt from certification: • ingested drugs; • inks for branding pharmaceutical forms; • externally applied drugs
172.868[a)]	Ethylcellulose – for use as a binder and filler in dry vitamin preparations, as a component of protective coatings for vitamin and mineral tablets, and as a fixative in flavouring components

Data taken with permission from Reference [3].

a) These regulations are only fulfilled by special EC grades especially developed for pharmaceutical purposes and applications that involve contact with food products.

Specifications of these EC types: (i) chemical composition: 48.0–49.5% ethoxyl; (ii) degree of substitution (DS) = 2.46–2.58; (iii) viscosity range = 7–100 mPa s as 5% solution in 80 : 20 toluene–ethanol.

Products having wide regulatory approval (FDA, USP, EU) such as MC and CMC (carboxymethylcellulose) are more frequently used. Ethylcellulose has the newest approval of all cellulose-based stabilizers, being now generally allowed according to European Food law. This will help to implement EC in the food laws of other countries.

References

1 Balser, K. and Iseringhausen, M. (1975) Cellulose ether, *Ullmann's Encyclopedia of Industrial Chemistry* (eds E. Bartholomé *et al.*), vol. **9**, 4th edn, Verlag Chemie, Weinheim, pp. 192–212.

2 European Union (2008) Commission directive 2008/84/EC of 27 August 2008, laying down specific purity criteria on food additives other than colours and sweeteners. *Official Journal of the European Union*, **51**, L253, 94–102. Available at http://eur-lex.europa.eu/LexUriServ/LexUriServ.do?uri=OJ:L:2008:253:0001:0175:EN:PDF.

3 Ashland Inc. (2012) Ethylcellulose: Physical and Chemical Properties (PRO_250-42A_3-02) – Technical Information, recipes and recommendations. http://www.ashland.com/products/aqualon-ethylcellulose.

4 EFSA (26 April 2004) Opinion of the scientific panel on food additives, flavourings, processing aids and materials in contact with food (AFC) on a request from the Commission related to ethyl cellulose as a food additive. *EFSA Journal.* doi:10.2903/j.efsa.2004.35. http://www.efsa.europa.eu/en/efsajournal/pub/35.htm.

7
Hydroxypropylcellulose

Synonyms: hydroxypropyl ether of cellulose, cellulose-hydroxypropyl-ether, cellulose 2-hydroxypropyl ether, HPC

E number:	E 463
CAS number:	9004-64-2
FDA code:	21 CFR 172.870

Related hydrocolloids: HPMC

7.1
Manufacturing

Hydroxypropylcellulose (HPC) is defined as a cellulose product obtained directly from natural occurring strains of fibrous cellulose, which are partially etherified with hydroxypropyl groups [1].

HPC with a molar degree of substitution of 4 was brought to the market in the United States of America at the end of the 1960s.

Hydroxypropylcellulose is produced by the reaction of alkali cellulose with the reagent 1,2-propylene oxide at high temperatures and pressures. The reaction can be carried out in an organic solvent or in an excess of the reagent. Since HPC has a coagulation and flocculation point in water (>40–45 °C (104–113 °F)), the substance can be cleaned with hot water. In the cleaning step, undesired by-products such as unwanted salts and ethers are removed. The purification is necessary to achieve the status of food, which requires 99.5% pure HPC.

Figure 7.1 shows the production process for hydroxypropylcellulose.

7.2
Chemistry

The HPC molecule is a polymer with a base of β-D-glucose units, where the free hydroxyl groups are partially or completely substituted with hydroxypropyl

Cellulose and Cellulose Derivatives in the Food Industry: Fundamentals and Applications, First Edition.
Tanja Wüstenberg.

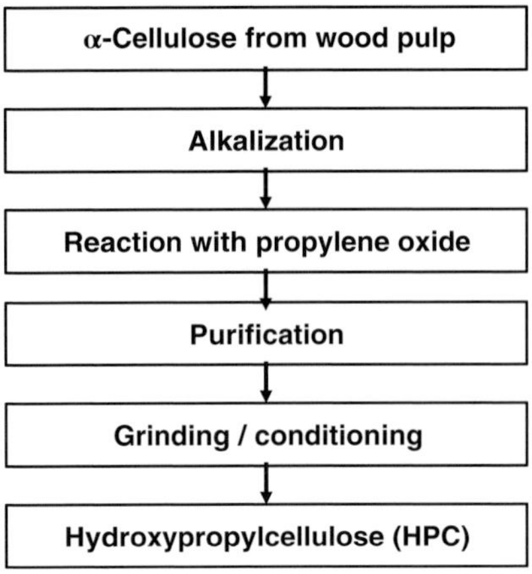

Figure 7.1 Flow diagram of the production of HPC.

groups. In the reaction with propylene oxide, secondary OH groups are formed. These secondary groups can react further. The probability of the etherification of already substituted HP groups is high. The molar substitution maximum is 4.6.

The hydroxypropyl ether of cellulose has the general chemical formula:

$$\left[C_6H_7O_2(OH)_x(OCH_2CHOHCH_3)_y(OCH_2CH[R_w]CH_3)_z\right]_n$$

where:

the degree of substitution (DS): $x + y + z = 3$;
the molar substitution (MS): $y + z(1 + w) \leq 4.6$;
R is a substituent with the number of w hydroxypropoxyl groups.

An alternative notation is [1]:

$$\left[C_6H_7O_2(OR^1)(OR^2)(OR^3)\right]_n$$

where R^1, R^2, and R^3 can be:

H
$CH_2CHOHCH_3$
$CH_2CHO(CH_2CHOHCH_3)CH_3$
$CH_2CHO[CH_2CHO(CH_2CHOHCH_3)CH_3]CH_3$.

One of many possible structural formulas of the hydroxypropylcellulose is shown in Figure 7.2. In this case, the molar substitution is 3.0. The degree of polymerization is 'n'. At the C2-atom, there is a monomeric hydroxypropyl substitution. At the C6, the HP-substitution is dimeric.

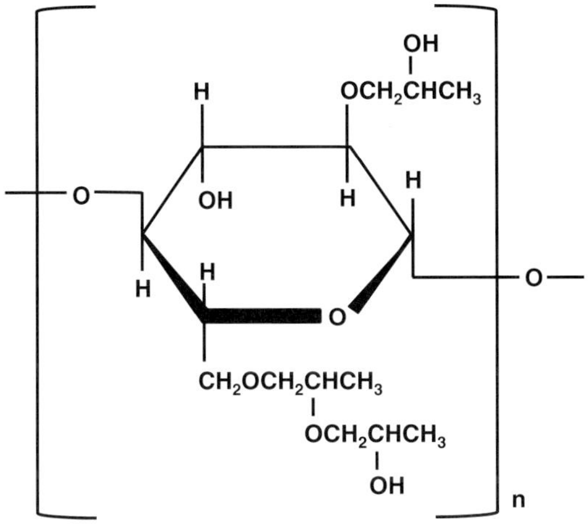

Figure 7.2 Structural formula of hydroxypropylcellulose (HPC).

The following molecular weights are calculated:

- $162.14\,g\,mol^{-1}$ for a non-substituted anhydroglucose unit;
- $336.37\,g\,mol^{-1}$ for a tri-substituted β-D-glucose module (DS = 3.0).

Macromolecules with approximately 100–3000 anhydroglucose monomers weigh about $30\,000–1\,000\,000\,g\,mol^{-1}$.

The content of hydroxypropoxyl groups ($–OCH_2CHOHCH_3$) is at least 80.5%. This corresponds to a maximum of 4.6 hydroxypropyl groups per glucose unit in relation to the dry substance [1].

7.3 Rheology

7.3.1 Physicochemical Parameters and Properties

Hydroxypropylcellulose has the following product parameters and physical and chemical properties [2]:

- appearance: white to grey-white or cream-coloured to pale yellowish coarse granular or fibrous powder;
- molecular weight for regulated products (25 °C (77 °F), Brookfield LVT):
  - low viscosity:
    - type A: 200–600 mPa s (at 10% conc.) – MW $80\,000\,g\,mol^{-1}$,
    - type B: 75–150 mPa s (at 5% conc.) – MW $95\,000\,g\,mol^{-1}$;

- medium viscosity:
  - type C: 150–400 mPa s (at 5% conc.) – MW 140 000 g mol^{-1},
  - type D: 150–400 mPa s (at 2% conc.) – MW 370 000 g mol^{-1};
- high viscosity:
  - type E: 4000–6500 mPa s (at 2% conc.) – MW 850 000 g mol^{-1},
  - type F: 1500–3000 mPa s (at 1% conc.) – MW 1 150 000 g mol^{-1};

- ash content (as Na_2SO_4): maximum of 0.2% (free of silica);
- moisture content (as packed): maximum 5%;
- pH value (1% solution at 20 °C): 5.0–8.0;
- degree of substitution (as molar substitution, measured in mol): 3.4–4.4;
- solubility in water: clear and smooth solutions at temperatures below 38 °C (100 °F);
- solubility in organic solvents: dissolves easily in many polar solvents to give clear and smooth solutions;
- biological oxygen demand (BOD_5): 14 000 ppm;
- bulk density: 500 g l^{-1} (varies with type);
- softening temperature: 100–150 °C (212–302 °F);
- burnout temperature: burns out completely at 450–500 °C (842–932 °F) in N_2 or O_2;
- density, with respect to specific gravity (2% solution at 30 °C (86 °F)): 1.010 g cm^{-3};
- refractive index (2% solution): 1.337;
- pH stability: 2–11;
- freezing point: 0 °C (32 °F);
- surface tension (at 20 °C (68 °F)) of a 0.1% solution: 44 mN m^{-1} (dyne cm^{-1});
- interfacial tension (at 25 °C (77 °F) of a 0.1% solution in water versus refined mineral oil: 12.5 mN m^{-1} (dyne cm^{-1}) (in comparison, the value for water is 31.6 mN m^{-1});
- bulking value in solution: 0.334 l kg^{-1} (0.04 gal lb^{-1}).

7.3.2 Dissolution Behaviour

7.3.2.1 Suitable Solvents

Pure hydroxypropylcellulose is soluble in water between 0 and 38 °C (32 and 100 °F). Aqueous HPC solutions are exceptionally clear and smooth flowing. They show pseudoplastic behaviour with decreased viscosity under shear, exhibiting almost no thixotropy, and return to the original viscosity once the shear force is removed [3]. At temperatures above 40–45 °C (104–113 °F), HPC is insoluble in water and already dissolved HPC precipitates from aqueous solutions. This precipitation is also called 'flocculation' and is thermoreversible. The precipitation of HPC is visible through a clouding of the solution, a significant viscosity reduction, and the occurrence of highly swollen fibres or clumps. When cooled below 40 °C (104 °F), the precipitated cellulose ether goes back into solution. This relationship is shown in figure 7.3.

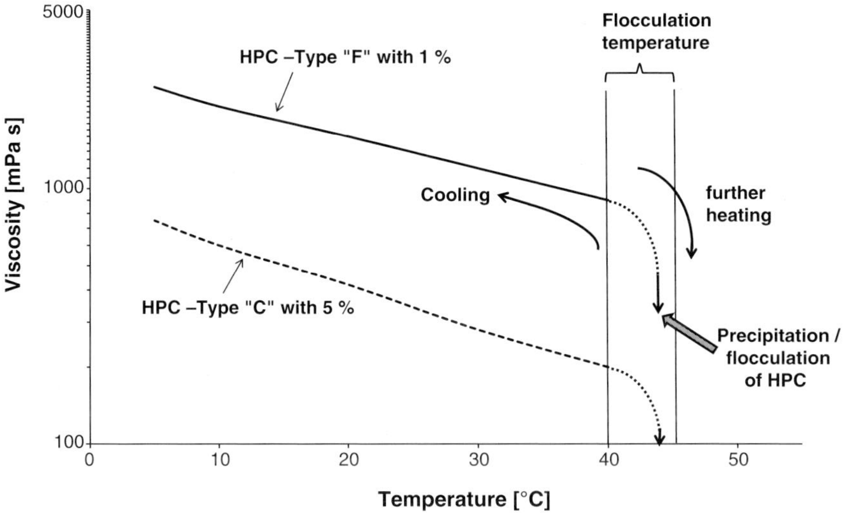

Figure 7.3 Effect of temperature on the viscosity of an aqueous HPC solution [2]. Reprinted with kind permission of Ashland Inc.

In contrast to methylated cellulose ethers, hydroxypropylcellulose never forms a gel. In addition, the addition of ingredients – competing with HPC for the available water – such as sugar and polyelectrolytes cannot induce gelation. These substances, however, have an effect on the flocculation temperature, moving it to lower temperatures (see Sections 7.3.3 and 7.4.1). The hydration speed of HPC is slow. Like all cold-water-soluble hydrocolloids, powders of hydroxypropylcellulose also tend to form lumps when they are stirred too quickly into cold water ($T = 0–35\,°C$ (32–95 °F)).

Methods used to avoid unwanted aggregates are:

- pre-wetting with hot water at 50–60 °C (122–140 °F),
- dispersing in vegetable oil,
- pre-mixing with non-polymeric materials such as sugars and salts or inert ingredients such as cooking starches, flours, pigments.

The pre-wetted or 'diluted' HPC can then be easily transferred into cold water and homogeneously incorporated into the recipe. Alternatively, the HPC must be introduced slowly in water below 35 °C (95 °F) with constant stirring. This solution procedure can create a foaming effect due to the surface activity of HPC.

All HPC types dissolve very well in many polar organic solvents by forming clear, smoothly flowing solutions – both at room temperature and at high temperatures. Unlike the thermal reaction in water, HPC shows no tendency to precipitate or flocculation when heated in organic liquids. As a general rule, the more polar the solvent is, the better the solubility of HPC will be. Methanol, ethanol, propylene glycol, and dioxane (ethylene dioxide) are the best solvents for HPC.

Another possibility is the addition of HPC to hot waxes or molten fats. Acetylated monoglycerides, glycerides, poly(ethylene glycol), and propylene glycol in the molten state are suitable for solubilizing powdered HPC. Hydroxypropylcellulose is very compatible with many waxes and oils, having a high molecular weight and melting at high temperatures, and can be used to improve the properties of such materials. The addition of HPC to these systems increases the viscosity, elasticity, and hardness and thus the resistance of surface coatings to cracking and breaking [2].

The solubility of hydroxypropyl cellulose in alcohol–water mixtures depends on the composition of the mixture with respect to its alcohol content. The addition of alcohol to an aqueous HPC solution generally leads to an increase of the viscosity and to an increase of the flocculation point (the temperature at which the HPC precipitates). The reason for this is the improved solubility of HPC in organic solvents. At high alcohol concentrations in solution (at least 45 vol.% ethanol, methanol, or propylene glycol), the HPC solution can be heated to the boiling point. In these systems, the HPC does not flocculate but passes through a phase of maximum viscosity at approximately 50–70 °C (122–158 °F). The best solubility and thus the greatest thickening effect is achieved with the EtOH–water mixture [2].

The solution behaviour of HPC in alcohol–water mixtures is described in detail in the next section (see Figures 7.4 and 7.5 below). Table 7.1 gives an overview of the general solubility.

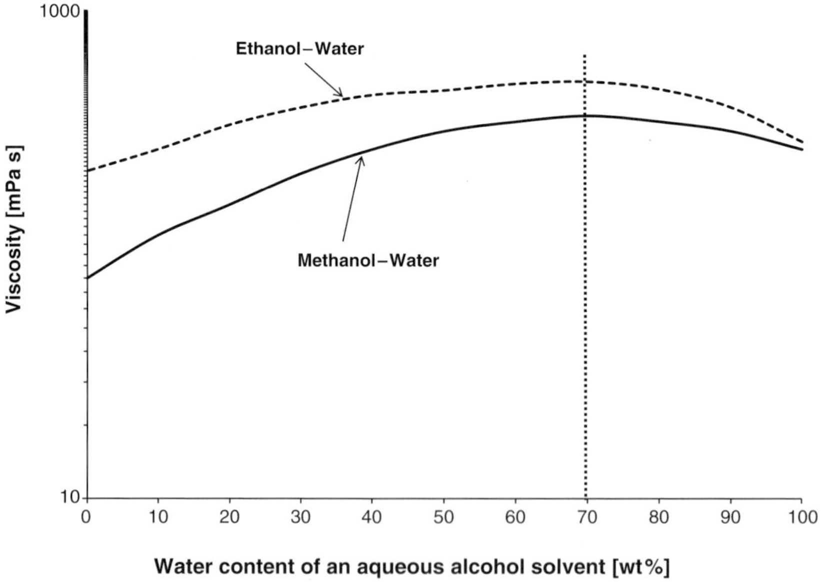

Figure 7.4 Viscosity of HPC in aqueous alcohol solutions [2]. Reprinted with kind permission of Ashland Inc.

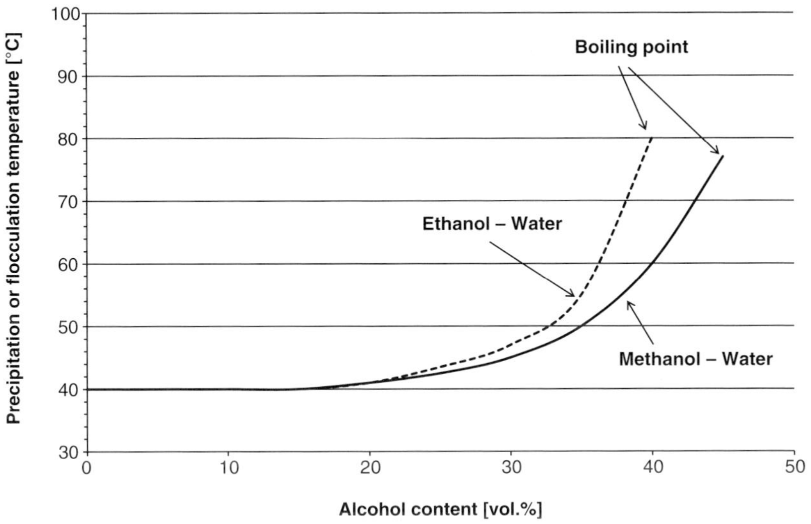

Figure 7.5 Precipitation (or flocculation) temperature of HPC in aqueous alcohol solutions [2]. Reprinted with kind permission of Ashland Inc.

7.3.2.2 Dissolution Curves in Water and Aqueous Alcohol Mixtures

Dissolution of HPC in Pure Water [2]

The viscosity of an aqueous HPC solution decreases as the temperature of the solvent is increased. This effect is known and is considered normal for polymers in solution. As shown in Figure 7.3, for two different types of HPC, the viscosity is reduced by about half when the temperature is raised by

Table 7.1 Solvents for hydroxypropylcellulose[a] [2].

Clear and smooth solutions in:	Cloudy rough solutions in:	Insoluble in:
Water	Lactic acid	Linseed oil
Concentrated acetic acid (glacial), E 260	Acetone	Soybean oil
Methanol	Butyl acetate	Sunflower oil
Ethanol	Methyl acetate	Glycerol
Propylene glycol	Tertiary butanol	Mineral oils
Isopropanol (95%)	Isopropanol (99%)	Toluene
Glycerol (propanetriol) : water (3 : 7)		Xylene
Dioxane (ethylene dioxide)		
Acetone : water (9 : 1)		
Ethylene dioxide (dioxane)		
Methylene chloride : methanol (9 : 1)		
Toluol : ethanol (3 : 2)		

Data taken with permission from Reference [2].

a) Note: not all mentioned solvents are food-approved. Please check before application!

about 15–20 °C (27–36 °F). This effect occurs consistently up to the temperature of flocculation/precipitation at 40–45 °C (104–113 °F).

For pure hydroxypropylcellulose, the solubility in water is limited to the cold-temperate range 0–38 °C (32–100 °F). When an aqueous HPC solution is further heated, there is a precipitation or flocculation. The transparent and smoothly dissolved HPC then precipitates as swollen filaments or lumps, and a strong decrease in viscosity is observed. The process is completely reversible – when cooled below 40 °C (104 °F), precipitated HPC dissolves again.

Dissolution of HPC in Aqueous Alcohols [2]

The viscosity of hydroxypropylcellulose in aqueous alcoholic solutions varies with the solvent composition. The viscosity goes through a maximum value at a mixing ratio of seven parts of water to three parts of alcohol (by weight). Figure 7.4 illustrates the course of solubility. The HPC solution can be prepared by two different methods. The first possibility is to incorporate dry HPC powder directly into the alcohol–water mixture. Alternatively, the HPC can be first dissolved in water or in the alcohol, and then the second solvent component is added.

Owing to the presence of alcohol in the system, the solubility of hydroxypropylcellulose is improved compared to that in pure water. This results in an increase in flocculation temperature and in the thickening power. In some cases, HPC can be heated to boiling without precipitating. With high alcohol concentrations (e.g. at least 45% by volume of ethanol, methanol, and propylene glycol) dissolved HPC can be heated to the boiling point without flocculation. Between 50 and 70 °C (122 and 158 °F) it passes through a phase of maximum viscosity. The ethanol–water mixture offers the greatest solubility and viscosity.

The factor by which the flocculation is increased depends on the nature of the alcohol and its concentration in the system.

Figure 7.5 shows the relationship between alcohol content and the precipitation temperature for aqueous solutions with ethanol and methanol. In systems with more than 45% ethanol or methanol, the HPC solution can be boiled without precipitation occurring. Propylene glycol behaves like methanol and shows the same trend for the concentration-dependent curve of the flocculation temperature. Other water-miscible organic liquids, which are capable of dissolving HPC, also increase the temperature of precipitation.

7.3.2.3 Surface and Interfacial Tension

Hydroxypropylcellulose is a surface-active polymer. Water solutions display greatly reduced surface and interfacial tension. Because of this, HPC functions as an aid in both emulsifying and whipping. These properties, coupled with protective colloid action, enable it to perform a dual role in the following systems:

- oil-in-water emulsions – stabilizer and emulsification aid;
- foamed systems – stabilizer and whipping aid.

Table 7.2 Surface and interfacial tensions of HPC solutions [2].

HPC (wt%)	Surface tension (dyne cm^{-1} (mN m^{-1}))	Interfacial tension versus refined mineral oil, (dyne cm^{-1} (mN m^{-1}))
0 (= water)	74.1	31.6
0.01	45.0	—
0.1	43.6	12.5
0.2	43.0	—

Data taken with permission from Reference [2].

Table 7.2 illustrates the reduction of surface and interfacial tension of water solutions containing HPC. All viscosity types have essentially the same effect and a concentration of HPC as low as 0.01% produces close to the maximum reduction in surface tension [2].

The low surface tension of water-based solutions containing HPC tends to promote foaming and air entrainment. If this presents problems, a water-dispersible antifoaming agent can be used and should be added to the water prior to solution preparation. Recommended defoamers are silica-type products, lauryl and octyl alcohols, and blends of oils, fatty acids, waxes, and non-ionic surfactants. Concentrations are generally in the range 25–200 ppm. However, it is advisable to consult the manufacturers of the antifoaming agents for their recommendations for a particular system [2].

7.3.3 Gelation and Behaviour with Other Ingredients

Despite the similarity and affinity with methylcellulose and HPMC (hydroxypropylmethylcellulose), hydroxypropylcellulose does not form gels upon heating. Temperatures above 40–45 °C (105–113 °C) lead to precipitation of HPC from aqueous solutions with a phase separation and the formation of highly swollen fibres or clump-like aggregates. The precipitation or flocculation is completely reversible (reversed by cooling). The presence of other substances in the HPC solution may strongly influence the flocculation temperature. However, gelation is not possible.

When HPC is dissolved in an alcohol–water mixture that contains at least 45% methanol or ethanol, the solution can be heated to boiling, without the HPC being precipitated.

Hydroxypropylcellulose has a high compatibility with other substances. Because of its dual solubility, HPC can be processed with water-soluble as well as with solvent-soluble polymers. If other substances are used in high concentrations, this can lead to precipitation or salting out of the HPC, because there is not enough solvent available to keep the cellulose ether in the hydrated state. The behaviour of hydroxypropylcellulose with dissolved inorganic salts in water varies, depending on the type of electrolyte and its concentration. At higher

Table 7.3 Compatibility of hydroxypropylcellulose with inorganic salts[a),b)] [2].

Salt	Salt concentration (wt%)			
	2	5	10	50
Ammonium nitrate	+	+	+	—
Calcium chloride	+	+	+	—
Disodium phosphate	—	—	—	—
Ferric chloride	+	+	+	—
Sodium acetate	+	+	—	—
Sodium carbonate	+	—	—	—
Sodium chloride	+	+	—	—
Sodium nitrate	+	+	+	—
Sodium sulfate	+	—	—	—
Sucrose	+	+	+	—

Data taken with permission from Reference [2].
a) Testing method: Medium viscosity HPC (2% = 125–450 mPa s) is added as 2% solution to the specific salt solution. The compatibility is measured visually. The final HPC concentration was approx. 0.1 wt% in all cases.
b) + = compatible; − = not compatible.

amounts of salt, there is a tendency for HPC to salt out. Then the molecule precipitates in the solution as an extremely swollen product. The phenomenon of salting out is associated with a decrease in viscosity. In addition, clouding of the solution is observed. In borderline cases, the salting-out effect is not immediately visible, but occurs only after a certain holding time (by up to 48 h) [2].

Table 7.3 shows an overview of the compatibility of dissolved HPC with different polyelectrolytes in different concentrations (the testing method is given in the table).

HPC is compatible with most natural hydrocolloids and synthetic water-soluble polymers. A good tolerance of hydroxypropylcellulose was found with methylcellulose (MC), sodium-carboxymethylcellulose (Na-CMC), hydroxyethylcellulose (HEC), guar gum, gelatine, sodium alginate, locust bean gum, and sodium caseinate.

Furthermore, HPC can be used in formulations that are preserved with the following food approved preservatives: sodium benzoate, sodium propionate, methyl and propyl hydroxybenzoate, and sorbic acid and its potassium, sodium, and calcium salts.

The compatibility with plasticizers such as propylene glycol and glycerol, together with the good thermoplastic properties of HPC, is advantageous especially in hot extrusion and moulding processes. By the addition of lubricants in the formulation, the adherence with respect to sticking to the surfaces of the machines is minimized. Stearates and lecithins are most suitable for food productions.

Interactions can occur between surfactants and HPC. There is no general rule or mechanism for it. By the substitution with hydroxypropyl groups, the HPC molecule is more lipophilic than other water-soluble cellulose ethers. This makes it compatible with a large number of anionic, cationic, non-ionic, and amphoteric surfactants. With some ionic surfactants (e.g. sodium and ammonium lauryl sulfate), the flocculation point of HPC can be increased. For non-ionic surfactants, this effect is not observed.

When using known solvents, HPC can be processed with water-insoluble polymers such as zein, shellac, or ethylcellulose. Films and coatings made of this heterogeneous mixture are characterized by a homogeneous appearance and very good quality [2].

7.4 Stability

7.4.1 Process Stability

The stability of hydroxypropylcellulose against process-related changes of temperature, pH value, and shear force is high. The HPC molecule reacts to these parameters by a temporary and reversible change in its physical state or consistency. On removal of the stress factor, the starting value and original texture are regained. Under normal conditions in food, there is no degradation of the viscosity or functionality.

As with most polymers, with increasing temperature the viscosity of the solution decreases. HPC shows a stronger response to an increase in temperature than, for example, methylcellulose, HPMC, or CMC. If the temperature rises by about 15 °C (27 °F) – for example from 20 to 35 °C (from 68 to 95 °F), the viscosity of the HPC is halved, regardless of the polymer concentration and its molecular weight.

The drop in viscosity is logarithmically proportional to the temperature increase up to about 40–45 °C (104–113 °F), when the precipitation or flocculation point is reached. Precipitation from the aqueous solution is completely reversible and, by cooling below 40 °C (104 °F), the homogeneous solution with the specific initial viscosity is reconstituted. The phenomenon of precipitation is visible by the appearance of turbidity in the solution. In addition, a significant decrease in viscosity is measured. The solution separates – the HPC emerges as an extremely swollen polymer. The transition from dissolved state to flocculated state takes place for hydroxypropylcellulose without the formation of a gel.

The way in which HPC precipitates from aqueous solutions after a temperature increase depends on:

- molecular weight of the polymer;
- presence of sugars, polyelectrolytes, and other water-soluble substances;
- additional application of shear forces.

Table 7.4 Flocculation temperature of HPC solutions with different additives [2].

Composition of the solution	Temperature of precipitation (flocculation point)	
	(°C)	(°F)
1% Hydroxypropylcellulose (HPC) – reference	41	106
1% HPC + 1% NaCl	38	100
1% HPC + 5% NaCl	30	86
0.5% HPC + 10% sucrose	41	106
0.5% HPC + 20% sucrose	36	97
0.5% HPC + 30% sucrose	32	90
0.5% HPC + 40% sucrose	20	68
0.5% HPC + 50% sucrose	7	45

Data taken with permission from Reference [2].

Low viscosity HPC grades precipitate as highly swollen single fibres. High viscosity products begin as larger lumps. The addition of CMC or specific surface-active substances reduces the tendency to agglomerate with increasing temperature, so that even higher molecular weight HPC precipitates in finer forms. The flocculation point is shifted to higher temperatures through the addition of organic solvents because they improve the solubility of HPC. The temperature of precipitation is reduced in the presence of relatively high concentrations of other dissolved substances, since they compete with HPC for the available water. The size of the effect is controlled by the type and amount of dissolved substances [2].

Table 7.4 shows the effect of salt and sucrose on the flocculation point of HPC (1% = 1500–3000 mPa s).

The table shows that the flocculation temperature decreases continuously with increasing concentration of dissolved salt or sugar. At high doses of these highly hydrophilic substances, it can already take place at room temperature. Rheologically, HPC solutions flow smoothly and in non-Newtonian fashion. They show no or only a small structure or thixotropy. At high shear rates, the solutions behave pseudoplastically and react with a temporary viscosity loss during the shear treatment. Upon removal of shear forces, the viscosity returns to its initial value, the polymer is not destroyed. The shorter the cellulose chain is (low degree of polymerization with respect to low molecular weight), the lower the reversible shear thinning will be. Low shear rates reduce the temporary reduction in viscosity, too. The highest stability against shear force is found for low molecular weight HPC types at low shear stress.

Hydroxypropyl cellulose is a surface-active polymer. In aqueous solutions, a strong reduction of the interfacial tension is measured. The effect is stronger than that of methylcellulose and at least comparable to that of HPMC.

Therefore, HPC acts as an emulsifier and foaming agent in multiphase mixtures such as oil-in-water emulsions and foams. The required dosage is low (0.01–0.5% HPC). Through the addition of lauryl and octyl alcohols to the solvent water, an undesired foam formation during shear treatments (stirring, homogenizing, pumping) can be prevented.

As a neutral molecule, hydroxypropylcellulose is very pH stable. Dissolved HPC does not change its viscosity in the range pH 2–11. For long storage periods in very acidic or alkaline media, the polymer can be decomposed in the presence of oxidizing agents (peroxide, caustic soda). The proof of degradation of the cellulose chain is a significant loss of viscosity. In comparison with other hydrocolloids and cellulose derivatives, hydroxypropylcellulose has a higher stability against chemically and biologically induced hydrolysis. When light, oxygen, and enzymes can be excluded, ready-made HPC solutions of pH 6–8 are very long lasting at low temperatures [2].

7.4.2 Hygroscopicity

Like most water-soluble substances, hydroxypropylcellulose absorbs moisture from the atmosphere or environment. The amount of absorbed water depends on the relative humidity and the temperature of the environment. The maximum water content of all types of HPC is 5%; it is usually between 2% and 3%. It is recommended to store the material in tightly closed containers and in a dry environment to avoid an increase of the moisture content. Basically, HPC has a low affinity for water. At any given relative humidity (RH), it has lower equilibrium humidity than most other water-soluble polymers. The equilibrium moisture content of HPC at 50% RH and 23 °C is 4%. At the same temperature, it increases to 12%, if the ambient air has a RH of 84%.

7.4.3 Polymer Degradation

Long-term studies have indicated that the molecular stability of HPC is controlled significantly by the initial molecular weight and the storage conditions. Low- and medium-viscosity types retain 97% of their viscosity even after three years of storage at room temperature with regular contact with ambient air. High-viscosity HPC products are susceptible to viscosity losses with storage time. During the first year, the viscosity decreases by about 10%, and after two years 20% thickening loss can be expected. The stability can be increased by storing the HPC at room temperature, but only rarely exposing it to the atmosphere. It is recommended to 're-test' (measurement of the viscosity) after 12 months when using long-chain hydroxypropylcelluloses and then to repeat this test at a three-month interval, to prevent quality changes in the end application due to possible viscosity degradation.

7.4.4 Combustion

Hydroxypropylcellulose has excellent binding properties. It is – particularly in technical applications – often used as a temporary binder. Examples include ceramics, coating or glazes, refractory materials, and powder-coated metal products. The HPC polymer is evaporated or burnt without residue at high temperatures of at least 250–500 °C (480–930 °F) in an oxidizing, reducing, or inert atmosphere. The extremely low ash content of the HPC and the complete absence of organic residues after combustion ensure that a virtually contamination-free product results after heat treatment.

7.4.5 Thermoplasticity and Film Formation

In General

Hydroxypropylcellulose is a thermoplastic polymer that is suitable for all extrusion and hot forming processes. The thermoplasticity is independent of the molecular weight with respect to the cellulose chain length. HPC forms very flexible, stable and durable films when drying out of an aqueous solution. The films are not sticky, even at high humidity. They form a permanent barrier to oils and fats. In addition, the films can be sealed or welded hot due to the thermoplastic properties of HPC. Film formation is explained in Section 7.7.2 ('Films and coatings').

Thermoplasticity [2]

Hydroxypropylcellulose shows excellent thermoplastic properties and can be processed by virtually all current fabrication methods used for plastics production. Injection- and compression-moulding, blow moulding, injection foam moulding, vacuum-forming, and extrusion of films, sheets, foam profiles, and filaments can be carried out on conventional equipment when using pelletized moulding powder.

All HPC types are – regardless of their molecular weight – thermoplastic and can be moulded and extruded easily. In general, the low-viscosity products are preferred for injection- and blow-moulding where strength, stiffness, hardness, dimensional stability, and shape are very important. The medium- and high-molecular HPC grades are recommended for the extrusion of systems where greater flexibility and tensile strength or elongation properties are desirable. When fillers or extenders are used, long-chain HPC types are selected because of their superior binding properties and their ability to provide good flexibility and strength in respect to toughness – even if up to 90% of filling material is incorporated.

Although it is not necessary from a functional point of view, small amounts of plasticizers are used to ensure a smooth uniform flow of the melt and a homogeneous end product. The addition of plasticizers gives a softer and more flexible character to the finished products. Generally, propylene glycol, glycol, poly(ethylene glycol), and trimethylolpropane are particularly suitable as plasticizer.

Lubricants are added to the formulations of mould castings with HPC to achieve easy removal from the mould and to prevent the HPC laminating itself. Lamination leaves a more or less thin film remaining on the mould surface. The undesired result is that the moulded product 'peels' when it is detached from the mould after the curing time. Glycerol monostearates, lecithin, and various other stearates are effective lubricants. Silicones are used in technical applications. Antioxidants may be added to the low-viscosity HPC types to prevent colour changes or losses and thermal degradation. Ascorbic acid is an effective antioxidant in food products.

To achieve the required functional properties, to change the physical characteristics, or to reduce costs, fillers and extenders can be incorporated into the HPC recipes. In general, any filler material may be used that survives the melting temperatures of HPC. As filler in food products, starches, pigments, flavours, and various foods are possibilities that have been successfully evaluated in casting and extrusion systems. Depending on the particle size and bulk density of the filler, loadings of 40–95% by weight can be realized with HPC.

Twin-screw extruders are recommended to prepare the pelletized forming powder.

A one-screw compounder can be used, if it is equipped with a nylon screw or with a metering screw, which has a long deep-positioned feed zone.

7.5 Rheometry

The composition of HPC can be detected by chemical cleavage of the ethers, followed by gas chromatography.

Hydroxypropylcellulose can be differentiated from other cellulose ethers by the dissolution properties. Hydroxypropylcellulose is soluble in cold water up to approximately 38 °C (100 °F). In addition, HPC is transparently soluble in cold and hot polar organic solvents – in this behaviour it differs from methylcellulose, HPMC, and CMC. In the temperature range 40–45 °C (104–113 °F) HPC is no longer soluble in water. Already dissolved HPC precipitates, with the transparent solution becoming cloudy, and there is a sharp drop in viscosity with a visible phase separation. Hydroxypropylcellulose does not form gels upon heating, in contrast to the methylated cellulose ethers. Its precipitation is reversible and can be reversed when cooled below 40 °C (104 °F) [2].

7.6 Synergies with Other Hydrocolloids

Mixtures of the non-ionic cellulose ether HPC with the anionic sodium-carboxymethylcellulose (Na-CMC) show a synergistic effect concerning the viscosity. The resulting viscosity of a solution containing 0.5% HPC and 0.5% medium-viscosity CMC is almost twice as high as mathematically expected. The synergistic viscosity win of HPC–CMC–mixtures is no longer observed at pH values below 3 and above 10. The presence of even small amounts of salt (e.g. 0.5% NaCl) can compensate this effect, because CMC as anionic substance is relatively sensitive to salts and, in a salty medium, it cannot develop its full viscosity.

With hydroxyethylcellulose (HEC), a neutral molecule, this synergistic effect on viscosity is not observed – the thickening corresponds to the calculated values [2].

Compared to all other hydrocolloids, HPC was found to be the least effective in reducing flavour intensity [4].

7.7 Food Applications with Typical Formulations

Hydroxypropylcellulose (HPC) is a non-ionic cellulose ether with an unusual combination of properties such as [5]:

- solubility in organic solvents at any temperature;
- hydration in cold water;
- insolubility in hot water without gelation;
- thermoplasticity;
- interfacial activity with simultaneous thickening and stabilization of multiphase mixtures;
- maintenance of flavour intensity.

7.7.1 Foam Products

The focus of applications is dairy-free whipping products (NDC, 'non-dairy whipping creams'). In whipped food products, HPC fulfils the following functions:

- good stability of foams during storage at room temperature;
- short, lightweight foam structure (fluffy);
- improved texture and mouthfeel;
- reduction/suppressing of synaeresis, no shrinking after whipping;
- resistance against overwhipping and phase reversal (emulsion breakage);
- production of real veg(etari)an foams without the use of milk proteins.

Tables 7.5 and 7.6 illustrate the mode of operation of hydroxypropylcellulose in foams.

Table 7.5 Formulation for a non-dairy whipping product without extra protein supplement [5].

Phase	Ingredient	Amount (%)
A	Hydroxypropylcellulose (150–400 mPa s, 2%)	0.4
	Water (50–65 °C) (122–149 °F)	10.0
	Polysorbate 60	0.3
B	Water (cold)	48.7
C	Sugar	15.0
	Salt	0.1
	Colour	Optional
	Cream flavour	Optional
D	Partially hydrogenated palm kernel oil with lecithin	25.0
	Mono-/diglycerides	0.1
	Polyglycerol esters of fatty acids	0.4
Sum		100%

Data taken with permission from Reference [5].

Processing guideline:

1. HPC is dispersed in hot water of 50–65 °C (122–149 °F). After 5 min, the polysorbate is stirred into the dispersion. Stirring is continued until a homogeneous dispersion is achieved.
2. Phase (A) is blended with cold water (Phase B) at room temperature. For complete dissolution of the HPMC, a stirring time of 2–3 h is required.
3. The dry ingredients (Phase C) are mixed and added to the water phase (AB) under stirring (about 5 min.). The mixture is heated to about 60 °C (140 °F).
4. The fat and the emulsifiers (Phase D) are melted slowly. The fat phase is heated to 60 °C (140 °F).
5. The fat-emulsifier-phase (D) is added slowly to the water-phase (ABC) under stirring.
6. Sterilization at 140 °C (284 °F) for 4 s or HTST-processing
7. Homogenization then takes place in two stages at about 125 and 25 bar (1,812 and 363 psi).
8. Finally, cooling to 5 °C (41 °F) and filling.

7.7.2
Films and Coatings

Water-Soluble Films

Another application field of hydroxypropylcellulose is edible coatings for food products.

Flexible packaging films can be produced with conventional extrusion techniques. Coatings can be applied on products either by extrusion or from aqueous or organic solvent systems.

Table 7.6 Recipe for whipping products with soy proteins [5].

Ingredient	Amount (%)	Amount (%)
Hydroxypropylcellulose (150–400 mPa s, 2%)	0.4	—
Hydroxypropylcellulose (1500–6500 mPa s, 2%)	—	0.28
Water	58.5	58.62
Isolated soy protein concentrate (ISP)	3.0	3.00
Polysorbate 60	0.3	0.30
Partially hydrogenated palm kernel oil with lecithin	22.0	22.00
Mono-/diglycerides	0.1	0.10
Polyglycerol esters of fatty acids	0.4	0.40
Sugar	15.0	15.00
Salt	0.1	0.10
Flavour	0.2	0.20
Sum	100%	100%

Data taken with permission from Reference [5].

Processing guideline:

1. Addition of HPC to hot water (approx. 60 °C (140 °F)) and vigorous agitation for 30 min. The cellulose ether dissolves, when the water temperature goes below 38 °C (100 °F). Alternatively, ⅓ of the water can be heated to 60 °C (140 °F). Then the HPC is dispersed in the hot water. Subsequently, the remaining water is added cold under stirring. Agitation is continued for at least 30 min.
2. The soy protein is added to the HPC solution and stirred for about 30 min. until the protein is dissolved.
3. All fats and emulsifiers are melted together at low heat.
4. The fat-emulsifier-mix is added slowly to the HPC-soy-water-phase under stirring.
5. Premixing of sugar, salt and flavour. The dry mix is incorporated into the main phase (from step 1–4) under constant agitation.
6. Homogenization.
7. Cooling to 5 °C (41 °F) for several hours, to achieve the fat crystallization
8. Whipping with maximum speed in a high-shear-mixer for approx. 2 min.
9. Cooling of the foam.

The HPC films (Table 7.7) are characterized by a high gloss and good flexibility. The hydrocolloid improves the toughness, hardness, and resilience of films and coatings. This is evident by a reduced cracking and less breakage. Hydroxypropylcellulose also acts as a binder and gives the films both heat sealability and thermal plasticity in extrusion processes. The films act as a barrier against oil and oxygen. Even with high humidity or moisture, they do not tend to be sticky, but stay dry [2,5].

Table 7.7 Formulation for water-based film coatings without plasticizer [5].

Ingredient	Amount (%)
Hydroxypropylcellulose (300–600 mPa s, 10%)	10
Water	90
Sum	100

Data taken with permission from Reference [5].

Processing guideline:

1. 1/5 of the water amount is heated to 90 °C (194 °C);
2. HPC is dispersed in the hot water;
3. 3/5 of the remaining water amount is added cold to the HPC dispersion; under agitation, the HPC is dissolved slowly during the cooling phase;
4. the remaining water can be used to adjust the viscosity to a specific value (approx. 150 mPa s);
5. the HPC solution is used to coat food products.

Suitable Plasticizers for Pure HPC Films [2]

Since HPC films are inherently very flexible, it is not necessary to add a so-called plasticizer or softener to cast films. In films produced by extrusion, plasticizers perform important tasks such as:

- ensuring lubrication of the nozzle;
- reduction of the melt viscosity;
- improving the uniformity of the melt;
- reduction of the heat seal temperatures;
- reduction of film hardness;
- increasing the flexibility;
- improving the extensibility.

The concentration of the plasticizer (as pure substance or mixture) should not exceed 5%, but usually only 2% or less is used. Higher dosages of plasticizer impair the properties of HPC films. The tensile strength, elasticity, and ductility are reduced by overdosing. The films show increased stickiness at high humidity. The transparency decreases significantly so that the films will be cloudy and dull. The shape or dimensional stability deteriorates and the solution properties are negatively affected.

The plasticizers listed below are suitable for HPC films having a concentration of up to 5%:

- propylene glycol (1,2-propanediol, E 1520),
- glycerol (1,2,3-propanetriol, E 422),
- poly(ethylene glycol) (PEG),
- trimethylolpropane (TMP, 2-ethyl-2-(hydroxymethyl)-1,3-propanediol).

To increase the film resistance at high humidity, for example to reduce the tackiness in a humid atmosphere, lubricants can be used, such as:

- poly(ethylene glycol) (having a molecular weight of 1000–20 000 g mol^{-1}),
- propylene glycol monostearate,
- glycerol monostearate,
- silicones (in technical applications).

HPC as Additive for Other Films and Coatings [2]

Owing to its good film-forming properties and its wide range of solubility, hydroxypropylcellulose is a useful additive and modifier for other films and coatings. The solubility of HPC in polar organic solvents allows its addition to many organic-solvent-soluble resins, while its broad compatibility ensures homogeneous films of improved quality. The addition of HPC to films and coatings not only increases their flexibility but also improves their resilience with respect to toughness and heat seal properties. The water resistance is reduced as well as the tendency to crack and fracture. Owing to the solubility of HPC in cold water (T=0–38 °C (32–100 °F)) and in polar organic solvents, there are various possibilities for the production of solutions that can be used for the casting of films and for coating operations.

As HPC is insoluble in hot water (T> 40 °C (104 °F)), it is essential to pay attention to temperature control during the drying of aqueous solutions to prevent so-called 'blushing', which is the formation of turbidity. This can occur at higher temperatures and can have a negative affect on more than just the visual properties of the film.

Water-Insoluble Films and Coatings [2]

Films and coatings of HPC can be made water insoluble by the addition of resins. This works when the resin reacts with the freely available hydroxyl groups of HPC with crosslinking. Several resins are able to respond to this crosslinking reaction with OH groups. The degree of cure and the degree of solubility of the final coating system generally depend on the following factors:

- reactivity of the insoluble resin;
- temperature during the curing process;
- pH value of the system;
- amount of crosslinking resin used.

It is this context, high temperatures, low pH values, and higher concentrations of the resins improve curing and increase the water resistance and the rigidity with respect to dimensional stability.

These crosslinked compositions usually have a greater durability in water, but they are not completely insoluble in water or resistant to water. They swell in the presence of water. Owing to the partial absorption of water, the films and coatings are less resilient (especially less extensible). They thus have a reduced wet strength.

7.8 Non-food Applications

The applications of hydroxypropylcellulose in other regulated products and technical applications are manifold. Table 7.8 gives an overview of the key applications and properties.

7.9 Nutritional Properties

Hydroxypropylcellulose is physiologically inert. Like all native celluloses, it will neither be absorbed nor metabolized by the human body. Since the substance is considered to be non-toxic, no ADI value has been set. HPC acts as a soluble dietary fibre and thus has a positive effect in the digestive system. At very high levels of intake, a laxative effect is observed. However, in typical food applications this concentration is not achieved. Table 7.9 lists nutritional information of HPC.

7.10 Legislation

European Union (EU)

In the European Union, hydroxypropylcellulose is generally approved as a technological additive for food under the E number 463. There is no maximum limit – the dosage is 'quantum satis'. The function of the HPC can be described with the term 'thickening agent' or 'stabilizer'. In Europe, the labelling term 'modified cellulose', which is allowed in the USA, is not sufficient. The declaration of HPC is made with the chemical name or the E number.

United States of America (USA)

Food-grade HPC complies with the requirements of the US FDA for direct addition to food for human consumption, as specified in the Code of Federal Regulations, Title 21, Section 172.870. This highly purified HPC also conforms to the specifications for hydroxypropylcellulose set forth in the Food Chemical Codex (FCC) [2].

In the USA, HPC can be found on food labels and product specifications as 'modified cellulose'.

Table 7.8 Applications of hydroxypropylcellulose [2].

Product group	Specific applications	Properties of HPC
Adhesives	Solvent-based	Thickening
	Hot-melt	Thermoplasticity
Aerosols/foams	Food – whipped toppings	Stabilizer, foaming aid
	Cosmetics - emulsions	
	Technical – solvent-based	Film-former, binder
Binder	Burnout types	Ready burnout, low residue, solvent-soluble
	Electrical insulators	
	Ceramic glazes	
	Alcohol core-wash-compounds	Thickener, binder, suspending agent, solvent-soluble
	Matrix board manufacture	
Coatings	Edible food coatings	Glaze, oil- and oxygen-barrier
	Textile and paper coatings	Solvent-soluble film-former, barrier against oil and fat, heat-sealable
	Film coatings	
Cosmetics	Hair styling products	Film former, alcohol-soluble thickener
	Alcohol-based preparations	
	Perfumes and eau de colognes	
	Emulsion creams, lotions, shampoos	Emulsion stabilizer, thickener, surface activity/foam formation
Encapsulation	Micro- and macro-encapsulation	Solubility, edible, formation of a flexible film, fast release
Extrusion	Films and sheets	Binder, thermoplastic, water- and solvent-soluble
	Profiles and filaments	
Foods	Whipped toppings	Stabilizer, whipping aid
	Edible coatings for nuts and candies	Protective coating and oil barrier
	Glazes for confections	High gloss and colour coatings
	Reformed food products	Binder for moulding and extrusion
Moulding	Injection-, compression-, and blow-moulding	Binder, thermoplastic properties, water- and solvent-soluble
Paint removers	Acid-based scrape-off and flush-off	Thickener, acid resistant
Paper	Coatings	Solvent-soluble, flexible, thermoplastic film barrier
Pharmaceuticals	Tablet binder Tablet coating Modified release Liquids and semi-solids	Aqueous and solvent solubility, thermoplastic binder; non-ionic, pH stable, thickener, suspending agent, diffusion barrier, flexible films, surface active

Plastic foams	Foamed sheets, tubes, rods	Thermoplasticity
PVC processing	Suspension polymerization	Surface-active protective colloid
Printing	Inks, based on water, alcohol and glycol	Thickener, binder, suspending aid
Miscellaneous	Cleaners (acid-based)	Thickener, acid-resistant
	Polishes, based on water and solvents	Thickener, stabilizer, suspending aid

Data taken with permission from Reference [2].

Table 7.9 Nutritional values of hydroxypropyl cellulose [2].

Ingredient	Amount (in 100 g)
Energy content	0 kcal
Fat	0 g
Fibre (total)	97 g
Soluble dietary fibres	97 g
Carbohydrates/sugar	0 g
Protein	0 g
Vitamins (sum)	0 g
Sodium	11 mg
Potassium	1 mg
Calcium	6 mg
Iron	1 mg
Magnesium	1 mg
Water	3 g
Ash	>0.25 g

Data taken with permission from Reference [2].

Canada

In Canada, hydroxypropylcellulose is listed in the Food and Drugs Act and Regulations as a food additive that may be used as emulsifying, gelling, stabilizing, and thickening agent [4].

Other Countries

Please check the local legislation before first use.

References

1 European Union (2008) Commission directive 2008/84/EC of 27 August 2008, laying down specific purity criteria on food additives other than colours and sweeteners. *Official Journal of the European Union*, **51**, L253, 94–102. Available at http://eur-lex.europa.eu/LexUriServ/LexUriServ.do?uri=OJ:L:2008:253:0001:0175:EN:PDF.

2 Ashland Inc. (2012) Klucel™ hydroxypropylcellulose: physical and chemical properties (PC-11229). Information from the document library at http://www.ashland.com/products/klucel-hydroxypropylcellulose.
3 Imeson, A. (2010) *Food Stabilizers, Thickeners and Gelling Agents*, Wiley-Blackwell, Oxford.
4 Stephen, A.M., Philips, G.O., and Williams, P.A. (eds) (2006) *Polysaccharides and Their Applications*, 2nd edn, Taylor & Francis Group, Boca Raton, FL, pp. 147–179.
5 Ashland Inc. (2012) Hydroxypropylcellulose (Klucel, Aerowhip) – Information and formulations for film coatings and non-dairy toppings (65.011-E1, VC-630A). Formulations from the document library at http://www.ashland.com/products/aerowhip-hydroxypropylcellulose.

8
Hydroxypropylmethylcellulose

Synonyms: HPMC, methylhydroxyprop(ox)yl cellulose, MHPC, 2-hydroxy-propyl ether of methylcellulose, hypromellose

E number:	E 464
EINECS:	Not listed
CAS number:	9004-65-3
FDA code:	21 CFR 172.874

Related hydrocolloids: methylcellulose (MC), methylethylcellulose (MEC), hydroxypropylcellulose (HPC)

8.1
Manufacturing

Hydroxypropylmethylcellulose (HPMC) is a mixed ether. It is defined as a cellulose product derived directly from naturally occurring vegetable fibres, which is partially etherified with methyl groups, with a small amount of substituted hydroxypropyl groups [1].

During manufacturing, simultaneously to methylene chloride, small amounts of propylene oxide are also added. The flow chart in Figure 8.1 shows the production process.

The hydroxypropylmethylcelluloses basically have the same character as the pure methylcelluloses. The introduction of the hydroxypropyl group into the molecule of MC improves the clear solubility and increases the temperature of the gel point, which is also described in the literature as 'flocculation point' in this context. In addition, the thermoplastic properties and solubility in organic solvents can be enhanced by the hydroxypropyl groups.

From a technical point of view, the reaction of cellulose with propylene oxide can be controlled very well despite the high reaction temperatures. During this reaction, salts and other substances are formed as undesirable by-products. These need to be removed to obtain food-grade quality with a maximum 0.5% of residual salt content. The purification is carried out by washing the HPMC

Cellulose and Cellulose Derivatives in the Food Industry: Fundamentals and Applications, First Edition.
Tanja Wüstenberg.

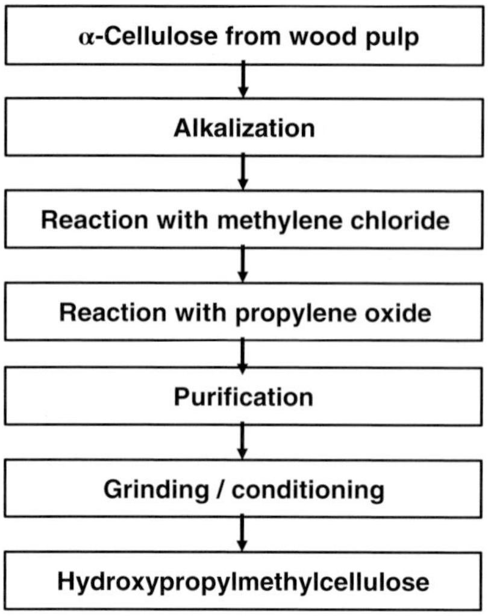

Figure 8.1 Production of hydroxypropylmethylcellulose (HPMC).

several times with hot water. More information on the manufacturing process is given in Chapter 5 (especially Sections 5.1.1 and 5.1.2, where the production of cellulose ethers is explained in detail).

8.2 Chemistry

The molecule of hydroxypropylmethylcellulose has a cellulose backbone made of β-D-glucose units with a $(1 \rightarrow 4)$-linkage. The three free hydroxyl groups are partially etherified with methyl groups. The hydroxypropyl and methyl groups can be attached to both the naturally occurring hydroxyl groups of the cellulose and to newly formed, meaning already substituted, hydroxypropyl groups through an ether bridge. The molar substitution (MS) may therefore differ from the degree of substitution (DS). The HPMC molecule is shown in Figure 8.2. The methyl-DS is 1.5 in this case. The hydroxypropyl-MS is 0.75. The total DS is 1.75.

The general chemical formula for HPMC is very complex because multiple substitutions are possible as introduced functional groups replace the former OH groups of the glucose monomers. It is generally [1]:

$$[C_6H_7O_2(OR^1)(OR^2)(OR^3)]_n$$

where R^1, R^2, and R^3 can be:

H
CH_3
$CH_2CHOHCH_3$

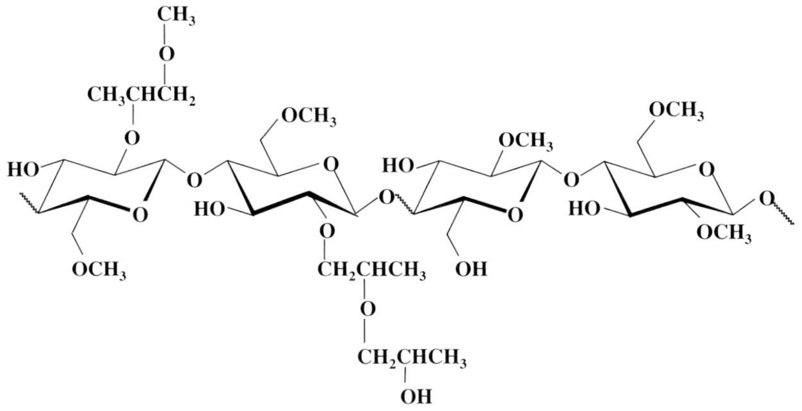

Figure 8.2 Molecule of hydroxypropylmethylcellulose (HPMC).

$CH_2CHO(CH_2CHOHCH_3)CH_3$
$CH_2CHO[CH_2CHO(CH_2CHOHCH_3)CH_3]CH_3$.

The degree of substitution (DS) of water-soluble HPMC grades is between 1.5 and 2.0. The gravimetric content of methyl groups (abbreviated 'Me' → $-OCH_3$) is 19–30%. The content of hydroxypropyl groups (abbreviated as "HP' → $-OCH_2CHOHCH_3$) is 3–12% (in each case based on the dry matter). The molecular weight is between 13 000 to 200 000 g mol^{-1}. This correlates with a degree of polymerization of 70–1100 glucose monomers [1–3].

8.3 Rheology

Chapter 5 gives further information on methylated cellulose derivatives. See in particular Section 5.3 (Rheology), which describes the rheological principles for MC and HPMC; Section 5.3.1 covers the general swelling and dissolution behaviour, and Section 5.3.3 deals with gelation.

8.3.1 Dissolution Behaviour

Hydroxypropylmethylcellulose having a medium to high DS (1.5–2.0) is very soluble in cold water (0–30 °C (32–86 °F)). The solubility of HPMCs is improved compared with that of pure methylcelluloses. Their hydration takes place at higher temperatures (see Table 5.7 in Section 5.3.4).

Transparent colloidal solutions of low to high viscosity are formed, depending on the degree of polymerization and the substance concentration.

Like powdered methylcellulose, HPMC also tends to form lumps when transferred to cold water because of its high speed of hydration. The methods used for aggregate-free incorporation, described in Section 5.3.4, can also be applied to HPMC products. Clear solubility and transparency can be increased by short-term cooling to 0–5 °C (32–41 °F).

Hydroxypropylmethylcellulose (DS 1.5–2.0) is insoluble in:

- hot water at temperatures above about 55 °C (131 °F);
- ethanol (100%) – but soluble in mixtures of EtOH with water (up to 90 : 10);
- fats, oils, MCT.

Good solubility of HPMC is achieved in polar organic solvents. In addition, hydroxypropylated methylcelluloses are soluble in esters, fatty acids, and propylene glycol. Dissolution is also possible completely anhydrously, when heating takes place.

In the presence of sugars, the viscosity increases disproportionately. At sugar concentrations of 45% and higher, there is no longer complete hydration of the HPMC. Salt (NaCl) and polyelectrolytes are tolerated in large amounts. Section 5.3.2 deals with the solution behaviour of the closely related methylcellulose.

While the type and level of substitution of a hydroxypropylmethylcellulose is the critical factor in heat-induced gelation, the amount of substituted groups and the ratio of 'methyl : hydroxypropyl' (Me : HP) does not matter at temperatures that are colder than the respective gelation temperature. As shown in Figure 8.3, the viscosity of a HPMC in pure water at room temperature is only a function of

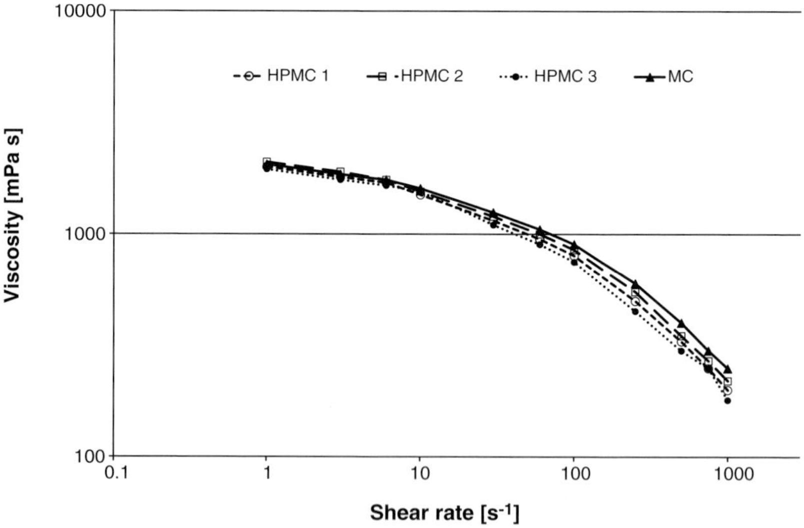

Figure 8.3 Viscosity of isoviscous HPMC grades with different substitution types (2% HPMC with a viscosity of 4000 mPa s, measured at 20 °C (68 °F); Brookfield LVT).

shear. Iso-viscous HPMC types with the same cellulose chain length (in this measurement, 2% = 4000 mPa s) demonstrate comparable viscosity data after shear treatment. The substitution level (measured in % by weight of substituted Me- or HP-groups) has no effect at cold temperatures [4].

8.3.2 Gelation of Hydroxypropylmethylcellulose

Based on Section 5.3.3 ('Gelation of methylcellulose'), we discuss here only the peculiarities that arise on the introduction of hydroxypropyl groups into the methylcellulose molecule. HPMCs also show the reversible gelation that is typical of methylated cellulose ethers upon heating of their solutions. Differences exist in the temperature of gel formation and the strength of the formed gels.

The amount of substituted hydroxypropyl and methyl groups determines the:

- start temperature of the gelation,
- strength of the resulting gel,
- stability of the gel with respect to its re-melting temperature.

The methyl group is the gelling component in a HPMC molecule. Hydroxypropyl substituents increase the hydrophilicity of the molecule and stabilize the hydration shell around the macromolecule. Consequently, they increase the gelling temperature. By changing the ratio of methyl groups to hydroxypropyl groups, the gelation temperature of HPMC types can be varied within wide limits [5].

Upon cooling, a phase transition into the viscous state occurs. The value of the solution viscosity is the same value as before heating. A hysteresis has to be taken into account, since gel formation starts at a higher temperature than the re-melting of the gel. The course of gelation is shown in Figure 5.19.

European Union (EU) food law allows HPMC types with:

- 19–30% methyl (Me),
- 3–12% hydroxyprop(ox)yl (HP).

In comparison, the tolerance range of pure methylcellulose is set at 25–33% methyl ($-CH_3$) and not more than 5% hydroxyethoxyl ($-OCH_2CH_2OH$). This means that low- as well highly methylated HPMC products are possible. Highly methylated HPMC (29% Me) with a high HP-content (10%) gels at about 58–64 °C (136–147 °F) into a semi-solid gel. With a low HP-content (6%) and the same Me-content, the gelation to a semi-solid gel takes place at 62–68 °C (143–154 °F) instead. By reducing the methyl-content to 22% (with a medium HP-content of 8%), the gelation temperature can be adjusted to 70–90 °C (158–194 °F). The gel strength is low; it results in a soft gel. An overview of the MC and HPMC types currently commercially available is presented in Table 5.4. Other factors influencing the gelation of methylated cellulose ethers are elucidated in Section 5.3.3 ('Gelation of methylcellulose'). As with the pure methylcelluloses,

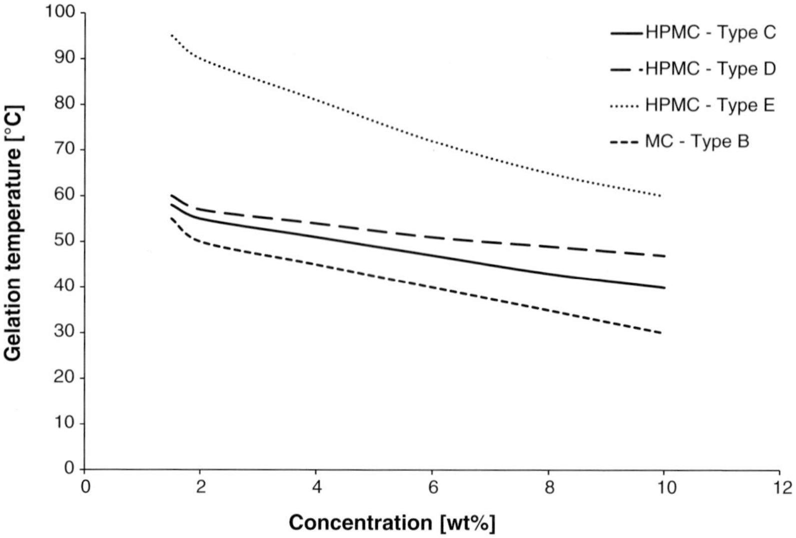

Figure 8.4 Concentration-dependent gelation temperature for different HPMC types.

there is a proportional relationship between the concentration of HPMC, its viscosity level, and its gelling behaviour. The following applies: the higher the concentration and the viscosity of the polymer, the lower is the gelation temperature.

Figure 8.4 shows schematically the relationship between the concentration of a methylated cellulose ether and its gelation temperature The actual shape of the curve (inclination angle) is determined by the viscosity level and the content of methyl- and hydroxypropyl groups.

In each application case, it must be considered that the total solids content and residual water content of the recipe decisively determine the level of the gelling temperature and the gel strength.

8.3.3
Behaviour of HPMC with Other Ingredients

Hydroxypropylmethylcellulose is a non-ionic polymer having a high compatibility with other ingredients. Because of the close relationship with pure methylcelluloses, the reader is referred to Section 5.3.7 for further reading. The HPMC molecule has a very heterogeneous structure. There are hydrophilic anhydroglucose monomers and substituted hydrophilic hydroxypropyl groups as well as substituted hydrophobic methyl groups. Thus, the HPMC prefers to locate at the interfaces of multiphase systems. The molecule has emulsifying properties and can support traditional emulsifiers. Unwanted interactions between HPMC and other ingredients of the formulation are rare. In most cases, either the high water-binding capacity or the enormous speed of absorption is the cause of negative interactions.

HPMC may be used in sugar-containing formulations of up to 45% sugar. Molecules with low levels of methyl groups tolerate higher sugar concentrations than higher substituted products. The tolerance to sodium chloride, phosphates, mineral salts, and other polyelectrolytes is high. A recipe may – depending on the substitution type of the HPMC used – contain a maximum of 10–20% salt (NaCl). Only beyond this salt content is the effectiveness of the HPMC limited and the viscosity development reduced. The presence of di- and trivalent ions may reduce the temperature of gelation to room temperature. Since no charged groups are present, there it is no complex formation, and no precipitation of the cellulose derivative. An example illustrates the effect of divalent ions. An HPMC gels in pure water at 58 °C (136 °F). An addition of 2% sodium carbonate lowers the gelling temperature to 35 °C (95 °F). With 4% Na_2CO_3 in the solution, gel formation starts already at 20 °C (68 °F). An addition of 10% calcium sulfate ($CaSO_4$) decreases the gelation temperature of this HPMC by only 4 °C (about 7 °F).

Acids and alkalis have only a minimal effect on the neutral molecule of the HPMC, in which there are no charged groups. The stability in the pH range 2–12 is very good.

The compatibility of HPMC with animal and vegetable proteins is high. It is best to ensure by a suitable process technology that the proteins are present at their optimum functionality.

In neutral milk beverages (pH ~7.0), phase separation was observed occasionally after stabilization with MC and HPMC. The casein–(HP)MC balance is very sensitive and reacts to small concentration fluctuations; consequently, the use of HPMC or MC is not advised in this application. HPMC may be incorporated in formulations having low water and high fat contents. Short-chain molecules with a low thickening effect and a low water requirement are particularly suitable for these conditions.

Like methylcellulose, HPMC can be used to reduce the oil absorption of food products during frying. The film-forming and gelling properties of HPMC allow – with a specific surface treatment of the product – a barrier to be built. The barrier layer prevents the water vapour pressure leaving from inside and reduces the penetration of the oil from the outside. This increases the yield and the product is nutritionally more ('less fat') valuable, providing the same taste as the traditional method of preparation. Owing to the faster hydration, the use of HPMC is easier because the cooling step can be omitted. However, the effectiveness of a pure methylcellulose is better owing to the lower gelation temperature and higher gel strength. If sufficient cooling of the methylcellulose spray- or dipping-solution or of the liquid batter cannot be guaranteed, a high methylated HPMC should be used as an alternative to MC.

HPMC may be used in alcoholic products for thickening and stabilization. Water-clear solubility is obtained in systems with up to 80% alcohol content.

Previous research projects and application tests have not shown any negative interactions of HPMC with other hydrocolloids.

8.4 Stability

The storage stability of powdery cellulose ethers and their microbiological stability are explained more precisely in Sections 5.4.1 and 5.4.2, respectively.

Hydrated hydroxypropylmethylcellulose is highly stable.

Most manufacturing processes from raw materials to food products are associated with changes in temperature and shear force impact. This leads to reversible changes in the hydrated HPMC molecule. The presence of acids and alkaline substances has no significant effect in the pH range 2–12. The neutral molecule is still stable and shows an increase in viscosity corresponding to the chain length. Rheologically, an HPMC solution is characterized by a non-Newtonian flow behaviour below the gelation point. The solution reacts pseudoplastically or thixotropically – in other words, by shear forces, such as stirring, kneading, homogenizing, and pumping the viscosity is reduced. The stronger and longer the shear stress is, the lower is the resultant viscosity ('shear thinning'). This effect is reversible. After a rest period, the initial viscosity is regained.

The substitution type has no impact at temperatures up to 35 °C (95 °F) – all water-soluble cellulose derivatives (MC, CMC, HPMC) show the same behaviour when applying shear forces. An essential attribute of the hydroxypropylmethylcelluloses is their surface activity. Even pure methylcellulose has this property, but to a lesser extent. The measurement of solutions of various hydrocolloids shows that highly substituted HPMC grades with a high HP-content have the largest surface-active effect. The values are listed in Table 8.1. Water is the 'zero'-reference without any foaming power. The lower the dynamic surface tension is, the higher is the foaming capability.

The measured surface tension correlates with the type and amount of substituted groups. Carboxymethyl groups have no effect as the example of

Table 8.1 Surface tension of different cellulose ether solutions.

Hydrocolloid	Carboxymethyl content (%)	Methyl content (% Me)	Hydroxypropyl content (% HP)	Surface tension (γs) (dyne cm^{-1})
Water	—	—	—	72
Cellulose gum (CMC)	High	—	—	68
Methylcellulose (MC)	—	High	—	53–59
HPMC – type C	—	High	High	45–48
HPMC – type D	—	High	Low	51–55
HPMC – type E	—	Medium	Medium	47–55
HPMC – in general	—	Medium to high	Low to high	45–55

cellulose gum (CMC) shows. The introduced methyl groups render the MC and HPMC molecule surface-active. The biggest effect is delivered by a low- to medium-viscosity HPMC with a high proportion of HP-groups and a high methyl group content. The influence of hydroxypropyl groups on this product characteristic is massive.

The difference between various highly substituted HPMC molecules (type C, D, or E) is usually only measurable in pure solutions. In food formulations, the effects are often no longer significant.

This interfacial effect has to be taken into account in the preparation of solutions. The reduction of surface tension by HPMC is an advantage whenever foams and other multiphase mixtures should be stabilized permanently. HPMC acts as a foam generator. Traditional foaming agents such as caseinates can be omitted, so that vegetarian and vegan products can be created. If egg is not working as a foam generator, because the recipe contains fatty substances, HPMC can take over this function. Application examples are meringue products with cocoa powder, nut oils, or coconut milk.

The temperature stability of dissolved HPMC is given from − 50 to +135 °C (−58 to +275 °F). In acidic products, its viscosity is stable at pasteurization temperatures ($T < 100$ °C (212 °F)). When very acidic food products with pH ≤2.5 are sterilized (at $T > 121$ °C (250 °F) and 2 bar (29 psi)), a slight decrease in HPMC viscosity is detected. This irreversible loss can be compensated by a higher dosage of around 10–20%.

Dissolved HPMC does not show phase separation at freezing temperatures, unlike the solutions of most other hydrocolloids. The solubility of HPMC increases with decreasing temperature, which is reflected in an enhanced transparency of its solutions. HPMC molecules retain, even at very cold temperatures, their affinity for water and are always surrounded by a hydration shell. The water is frozen out slowly, so that very small ice crystals are formed. Hydroxypropylmethylcellulose acts as crystallization control with a permanent stabilization of food products.

The gels formed during heating are also characterized by a good stability. The gels are thermoplastic and therefore also suitable for extrusion processes at high temperatures and pressures.

When HPMC solutions are dried under cold conditions (below their specific gelation temperature), transparent elastic films are produced that are resistant at room temperature. These films form a permanent barrier to fats and oils and dissolve slowly when exposed to cold water. Upon heating, the films gel, and become permeable to water vapour.

8.5 Rheometry

The composition of HPMC can be determined by chemical cleavage of the ethers followed by gas chromatography (GC). The methylated cellulose ethers

(MC, HPMC) can be differentiated from other hydrocolloids through the reversible gelation of their solutions at warm temperatures. To determine the type of substitution, the temperature of gelation and the gel strength are considered. For both methods, a 2% solution is prepared and heated indirectly in a test tube. When the solution changes from transparent to opaque-white, gelation has begun. The solution should be turned completely into a gel in order to be sure of its consistency. Pure methylcellulose gels below 50 °C (122 °F) and forms sliceable gels. HPMC needs higher temperatures (above 58 °C (136 °F)) to form a gel. The gels are of a soft to semi-firm consistency.

8.6
Synergies with Other Hydrocolloids

So far, neither synergies nor incompatibilities have been found between HPMC and other hydrocolloids.

8.7
Food Applications with Typical Formulations

The relationship between HPMC and methylcellulose, with the resulting similar properties, leads to the fact that the products are exchangeable with each other. For food applications, formulations of 'methylcellulose' can also be used with HPMC, additionally (see Chapter 5, Section 5.7).

8.7.1
Foams

Tables 8.2 and 8.3 gave some example formulations.

Table 8.2 Formulation for egg-free dried foams (type 'baiser'/meringues).

Ingredient	Amount (g)
HPMC – type C with 6–50 mPa s	5–10
Powdered sugar	250
Water	200
Additional taste provider (nut oils, coconut flakes, cocoa, etc.)	Quantum satis

Processing guideline:

1. Preparation of 200 ml of a 2.5% HPMC solution by applying the 'hot-cold technology';
2. slow addition and incorporation of 250 g of powdered sugar;
3. addition of flavouring substances and homogeneous distribution in the mass;

4. whipping at high speed until it forms a thick foam;
5. skim small amounts of foam and place on a greased baking tray;
6. bake at 115 °C (240 °F) until the 'meringues' are dry (about 1 h).

Alternatively, the dry mix method can be applied. A homogeneous premix of 5 g HPMC and 50 g of sugar is stirred into cold water (for about 2 min). Thereafter, alternately pitch briefly and add a further quantity of sugar until a firm foam results. Then the flavour carrier is incorporated. The remaining procedure follows steps 4–6.

Table 8.3 Formulations for dairy-free whipping products (vegetarian cream).

Ingredient	Amount (%)	
	30% Fat[a)]	10% Fat
Water	46.75	73.78
Hydrogenated vegetable oil	32.00	9.96
Glucose-fructose-syrup + glucose syrup	16.00 + 4.00	—
Sugar	—	14.96
HPMC – type D with 50 mPa s	0.60	0.75
Polysorbate 60 (Tween 60)	0.30	0.30
Polysorbate 80 (Tween 80)	—	0.02
Polyglycerol esters of fatty acids (PGE)	0.20	—
Sorbitan monostearate	—	0.13
Alginate (low viscosity, with high gel strength)	0.15	0.10
Sum	100	100

a) The formulation must be diluted with 50% cold water (= 2 parts of cream : 1 part of water).

Processing guideline:

A. Preparation of the water phase:
1. dispersing the HPMC in hot water ($T >$ 95 ° C (203 °F)) with shear force;
2. addition of the alginate with further stirring;
3. cooling to 30 °C (86 °F);
4. addition of the sugar syrup;
5. heating to 40 °C (104 °C) and addition of the water-soluble emulsifiers (Polysorbate, PGE).

B. Preparation of the fatty phase by melting the fat at 65 ° C (150 °F).

C. Mixing the water phase and the fat phase:
1. heat the water phase to 75 ° C (167 °F);
2. slow addition of the fat phase to the water phase;
3. homogenization ('down stream') of the hot mixture at 200 °C (392 °F) with 20 bar;
4. cooling to 8 °C (46 °F), ripening of the fat for 72 h.

Table 8.4 Applications of HPMC in coatings, films, and confectionery.

Application	Product	Viscosity	Gel strength	Concentration (%)
Bakery glaze	HPMC	Low	Soft–semi-solid	0.1–0.2
Buttercream frosting	HPMC	High	Soft	0.6–0.8
Buttercream icing	HPMC	Low	Soft	0.1–0.2
Candy coatings	HPMC	Low	Semi-solid	0.5–6.0
Caramel	HPMC	Low	Soft	0.3–0.5
Chocolate nougat	HPMC	Low	Semi-solid	0.3–0.5
Icing – very thin/flat	HPMC	Low–medium	Soft–semi-solid	0.1–0.2
Icing – low density	HPMC	Low	Soft–semi-solid	0.2–0.75
Piping gel (not sliceable)	HPMC	High	Soft	0.25–0.5

8.7.2 Films and Coatings with HPMC

8.7.2.1 Overview

HPMC shows various functions and benefits in films and coatings:

- suppression of sugar crystallization, reduced brittleness;
- increased elasticity and improved film forming;
- good and clean flavour profile, enhanced release of flavours;
- thermal process stability;
- creamy texture, improved spreadability, consistency control;
- better freeze–thaw stability;
- compatibility with low water activity and high fat contents;
- lubricant – simplified production and lower abrasion losses;
- influence over the solubility of coatings (controlled release).

Table 8.4 lists the use of HPMC in confectionery, films, and coatings.

8.7.2.2 Typical Formulations

Tables 8.5–8.8 give some typical formulations for coatings, glazes and films.

Table 8.5 Formulations for films (adhesive solutions and dry products).

Ingredient	Amount (%)	
	Spray solution as an adhesive or glaze	Films (dry)
Water	98	90–95
HPMC – type C with 15 mPa s, alternative: methylcellulose (MC)[a], type B with 15 mPa s	2	5–10
Colours + flavours	Optional	Optional
Spices, herbs, seeds	—	Optional

a) Note: MC requires a longer hydration time and a lower hydration temperature than HPMC.

Processing guideline:

1. Weighing of HPMC (or MC) – the amount depends on the desired viscosity and film thickness;
2. divide the water amount into two equal parts – one set is cooked, the other amount is cooled;
3. disperse the HPMC or MC powder in the boiling water ($T \geq 90\,°C$ (194 °F)) with moderate stirring;
4. addition of cold water ($T = 10\,°C$ (50 °F));
5. slowly stir until the solution is clear/transparent and viscous;
6. addition of colours, flavourings, spices, and so on to the cellulose ether solution and then stirring until a homogeneous phase is formed;
7. pour the solution on glass plates and then dry the films at cold temperatures below the gelation temperature (HPMC max. 50 °C (122 °F), methylcellulose up to max. 40 °C (104 °F)); the drying takes several days.

Table 8.6 Formulation for 'warm' glazes for baked goods.

Ingredient	Amount (%)
Powdered sugar	77.0
Water (cold)	22.7
NaCl	0.2
HPMC – type E with 100 mPa s	0.1
Sum	100.0

Processing guideline:

1. Premix the HPMC with approx. 20 g of sugar;
2. add the dry mixture to the water and then boil the sugar syrup;
3. mix the rest of the sugar quantity and the salt; slow addition to the hot syrup; moderate stirring until a homogeneous solution is obtained;
4. keep the glaze warm to about 50 °C (122 °F) and use on warm pastries (e.g. dough nuts).

Table 8.7 Formulations for 'cold' glazes for baked goods.

Ingredient	Amount (%)			
	Sugar glaze/ frosting	Vanilla-buttercream-icing	Chocolate-buttercream-icing	Chocolate glaze, fat-free
Powdered sugar	81.9	74.0	64.7	41.0
Granulated sugar	—	—	—	28.9
Water	10.0	—	—	19.5

(continued)

Table 8.7 (Continued)

Ingredient	Amount (%)			
	Sugar glaze/ frosting	Vanilla-buttercream-icing	Chocolate-buttercream-icing	Chocolate glaze, fat-free
Margarine/shortening	—	15.0	12.4	3.4
Condensed milk	—	9.7	9.7	—
Glucose syrup (77% dry matter) with high fructose content (55%)	8.0	—	—	—
Chocolate, liquid, unsweetened	—	—	11.9	—
Cocoa, de-oiled	—	—	—	5.8
Vanilla	—	1.1	1.1	—
NaCl	—	0.1	0.1	—
MCC + guar	—	—	—	1.0
Agar			—	0.2
HPMC – type C with 15 mPa s	0.1	—	—	0.2
HPMC – type E with 100 mPa s	—	0.1	0.1	—
Sum	100.0	100.0	100.0	100.0

Table 8.8 Formulation for frosting.

Ingredient	Amount (%)
Powdered sugar	81.8
Water	10.0
Fructose syrup	8.0
HPMC – type C with 15 mPa s	0.2
Sum	100.0

Processing guideline:

1. pre-dissolve the HPMC in hot water; 3.6 g HPMC is dispersed in 180 g hot water ($T > 85\,°C$ (185 °F)) and brought into solution by cooling to 25 °C (77 °F);
2. 270 g of sugar and sugar syrup are weighed in a mixing bowl;
3. 30.6 g of HPMC solution are incorporated in the sugar mixture under slow stirring until a homogeneous frosting develops;
4. the finished product can be frozen until ready for use;
5. prior to application, the frosting must be completely thawed.

8.7.3 Fillings

The formulation of various fillings is shown in Tables 8.9–8.12.

Table 8.9 Formulation for a fat-free chocolate cream filling for cookies.

Ingredient	Amount (%)
Powdered sugar	50.6
Glycerol	10.6
Maltodextrin	10.6
Cocoa, de-oiled	9.4
Water	8.7
Fructose syrup	6.9
Instant starch (E 1412)	3.0
HPMC – type E with 100 mPa s	0.2
Sum	100.0

Table 8.10 Formulation for chocolate mousse and lemon mousse fillings.

Ingredient	Amount (%)		
	Chocolate mousse	Chocolate mousse, gelatine-free	Lemon mousse
Vegetarian creamer (liquid)	37.8	37.8	37.7
Water	19.4	20.4	19.2
Sugar	19.3	19.3	19.2
Whipping base (56% palm kernel oil, 20% sugar, 20% sugar syrup, 4% sodium caseinate)	12.1	12.1	12.0
Cocoa powder, de-oiled	9.7	9.7	—
Polydextrose	—	—	7.8
Lemon flavour, natural	—	—	2.0
Gelatine	1.5	—	1.5
Citric acid, crystalline	—	—	0.4
HPMC – type D with 50 mPa s	0.2	0.7	0.2
Sum	100.0	100.0	100.0

Table 8.11 Formulations for fruit fillings for cakes and dough pockets.

Ingredient	Amount (%)			
	Apple	Apple, low calorie	Cherry	Blueberry
Fruit, fresh or frozen	59.06	68.9	38.9	56.4
Fruit juice	—	—	33.0	11.5
Water	16.0	16.8	—	11.5
Sugar	15.5	—	15.0	10.6

(continued)

Table 8.11 (*Continued*)

Ingredient	Amount (%)			
	Apple	Apple, low calorie	Cherry	Blueberry
Polydextrose	—	11.5	—	—
Fructose syrup	7.0	—	10.0	8.5
Modified waxy maize starch (E1422)	2.0	2.0	2.0	1.0
Colour, natural	Optional	—	Red: 0.03	Optional
Sodium alginate	0.1	0.16	—	—
Citric acid, crystalline	Maleic acid: each 0.04		0.5	0.2
Aspartame	—	0.1	—	—
HPMC – type E with 4000 mPa s	—	—	0.6	0.3
HPMC – type E with 100 000 mPa s	0.3	0.5	—	—
Sum	100.0	100.0	100.0	100,0

Table 8.12 Formulation for a lemon cream filling for cakes.

Ingredient	Amount (%)
Water	53.0
Sugar	21.7
Vegetable fat	13.7
Buttermilk powder	6.0
Egg yolk	1.8
Starch (E 1422)	1.5
Lemon powder	1.5
Citric acid, crystalline	0.4
Salt	0.2
HPMC – type E with 100 000 mPa s	0.2
Sum	100.0

8.7.4 Bakery Products with HPMC

8.7.4.1 Traditional Baked Goods

Tables 8.13–8.17 give the formulation for several traditional bakery products.

Table 8.13 Formulation for wheat bread with reduced carbohydrate content.

Ingredient	Amount (%)
Water	35.2
Soy flour	19.2
Wheat gluten	15.0
Wheat flour	10.0

Vegetable fat	5.0
Yeasts	5.0
Soy protein isolate	4.0
Linseed	2.0
Dextrose	1.5
Baking soda	1.0
Salt	1.0
HPMC –type D with 50 mPa s	1.0
Lecithin	0.1
Sum	100.0

Table 8.14 Formulation for a cake with fruit pieces.

Ingredient	Amount (%)	Amount (g)
Sugar	25.0	150
Butter	25.0	150
Whole egg	25.0	150
Wheat flour	15.2	91.2
Wheat starch	6.1	26.6
Emulsifier mix of glycerol monostearates (GMS), polyglycerol esters of fatty acids (PGE) and acetic acid ester	2.4	14.4
Baking powder	0.6	3.6
HPMC – type D with 450 mPa s	0.3	1.8
Xanthan[a)]	0.2	1.2
Salt	0.2	1.2
Sum	100	600
+ 20% Raisins or other fruit pieces, based on the dough weight		

a) Mixture of HPMC and xanthan to achieve the optimal result in texture, volume, and the distribution of the fruit pieces.

Processing guideline:

1. Mixing of the dry ingredients;
2. mixing of butter and eggs in a separate bowl;
3. incorporation of the dry mixture into the butter–egg mix; stir for about 5 min at medium speed;
4. addition of the raisins to the dough; stir for 1 min at low speed;
5. filling of 450 g of dough into a loaf shape;
6. baking for 60 min at 160 °C (320 °F);
7. cooling the baked cake on a metal grid/sieve.

Table 8.15 Formulations for tortillas.

Ingredient	Amount (%)
Corn flour	56.425
Water	43.500
HPMC – type E with 4000, 15 000 or 100 000 mPa s	0.075
Sum	100.000

Table 8.16 Formulations for reduced-fat pancakes, waffles, and pastry dough.

Ingredient	Amount (%)		
	Pancakes, fat-reduced	Waffles, fat-reduced, frozen	Pastry dough, for microwave
Water	—	—	25.0
Milk (1.5% fat)	70.20	56.4	—
Wheat flour	16.85	27.0	49.8
Vegetable oil + butter (salted)	—	—	15.0 + 5.0
Whole egg (powder)	5.62	2.0	—
Maltodextrin	4.22	4.6	1.0
Sugar	—	4.0	—
Whey powder, sweet	1.40	1.5	2.0
Salt	0.56	0.4	1.2
Sodium carbonate (soda ash)	0.34	0.3	—
Baking powder (salt-free)	0.28	3.0	—
Pectin LM	0.28	—	—
HPMC – type D with 50 mPa s	0.25	0.3	—
HPMC – type E with 4000 mPa s	—	—	1.0
Monoglycerides, water-dispersible	—	0.5	—
Sum	100	100	100

Table 8.17 Formulations for muffins with reduced fat content.

Ingredient	Amount (%)		
	With blueberries	Muffin – pure	With cinnamon
Water	25.5	31.6	8.5
Wheat flour	27.4	30.7	28.4
Sugar	16.5	18.5	29.4
Blueberries, drained	13.4	—	—
Maltodextrin + polydextrose	—	—	4.2 + 4.2
Whole egg, fresh	8.2	9.2	13.2
Vegetable fat	3.4	3.8	2.8
Glycerol	—	—	3.1
Skimmed milk powder	2.0	2.3	—
Baking powder	1.3	1.5	1.1
Vanilla	0.8	0.9	0.2
Monoglycerides	0.5	0.6	0.6
Salt	0.3	0.3	—
Lecithin	0.3	0.3	0.4
Cinnamon	—	—	0.5
MCC – powdered cellulose	—	—	3.0
HPMC – type D with 50 mPa s	—	0.3	—
HPMC – type E with 100 000 mPa s	0.4	—	—
Methylcellulose – type B with 15 mPa s	—	—	0.4
Sum	100	100	100

8.7.4.2 Gluten-Free Products

For gluten-free products, formulations are given in Tables 8.18–8.21.

Table 8.18 Formulations for allergen-free breads.

Ingredient	Amount (%)	
	Variant 1	Variant 2
Water (max. 30 °C (86 °F))	42.35	44.30
Rice flour ('long grain')	26.00	4.90
Rice starch	20.00	—
Corn flour	—	19.40
Potato starch	—	19.40
Sugar	5.00	4.00
Milk powder	—	4.00
Sun flower oil	2.50	—
Active dry yeasts	1.50	2.00
HPMC – type E with 4000 mPa s	1.40	1.00
Salt	1.25	1.00
Sum	100	100

Processing guideline:

Variant 1:

1. Blend all dry ingredients – except for the yeasts – together in a mixer with a paddle attachment;
2. dissolve yeasts in warm water;
3. add the yeast-slurry and the oil to the dry ingredients in the mixer;
4. mix all ingredients together for about 10 min; the dough will be somewhat sticky and will not form a cohesive ball of dough like traditional wheat dough does;
5. cover your hands with a little soybean oil and remove dough from mixer;
6. using rice flour as a dusting agent, form the dough into desired shapes/loaves;
7. cover the loaves with a damp cloth; proof for 60 min at 35 °C (95 °F);
8. preheat the steam oven to 200–220 °C (392–428 °F); slash the top of the bread loaves with a sharp knife;
9. bake for approx. 15 min or until the bread is brown and sounds hollow when tapped;
10. cool completely before packaging; bread can be kept overnight without becoming stale.

Variant 2:

- See 'variant 1';
- When using fresh yeast, the amount of water is to be reduced by 1.5%. The swelling time of active dry yeasts is approx. 5 min. This is to avoid the yeasts

coming into direct contact with the salt – when transferring the starter culture to the dry mixture;
- filling weight for one bread is approx. 460 g;
- proofing for 30–40 min at 30 °C (86 °F);
- baking at 200–230 °C (392–446 °F) for 25–30 min;

Table 8.19 Formulations for gluten-free buns (heated in baking tins).

Ingredient	Amount (%)	Amount (g)
Water (max. 30 °C (86 °F))	44.5	300
Rice flour	22.3	150
Buckwheat flour	13.4	90
Corn flour	4.5	30
Corn starch	4.5	30
Lemon juice	2.7	18
Shortening (margarine)	1.8	12
Yeasts	1.4	9
Baking powder	1.4	9
Sugar beet syrup	1.8	10.6
HPMC – type E with 4000 mPa s	0.8	5.4
Salt	0.9	6
Sum	100	670 g

Processing guideline:

The procedure is similar to that for the production of gluten-free bread. The rolls can be filled and baked in muffin pans.

Table 8.20 Formulations for different gluten-free cake doughs.

Ingredient	Amount (%)		
	Apple cake (without egg)	Biscuit cake	Choux pastry
Water	11.56	4.7	32.8
Corn starch	18.75	—	6.5
Potato starch	18.75	—	P-Flour: 6.6
Corn flour	8.9	13.0	6.5
Rice flour	—	13.0	—
Egg	—	41.8	37.2
Shortening (margarine)	22.25	—	6.5
Sugar	18.75	26.0	2.7
Vanilla sugar	—	1.4	Flavour: 0.13
HPMC (type E with 4000 mPa s)	1.0	0.1	1.0
Salt	0.04	—	0.07
Sum	100	100	100

Table 8.21 Formulation for extruded rice noodles – gluten-free pasta.

Ingredient	Amount (%)
Rice flour	Quantum satis
Water	25.0
Mono- and diglycerides of edible fatty acids	1.5
HPMC – type E with 4000 mPa s	0.6

8.7.5 Chilled and Frozen Dairy Products and Desserts

8.7.5.1 Functions and Overview

HPMC is used in dessert products for various reasons:

- good water absorption and retention, no synaeresis or dehydration, prolonged standing performance of toppings;
- emulsifying properties, stabilization of multiphase systems;
- excellent foaming and stabilization of the air bubbles, increased 'overrun' (short-chain MC/HPMC molecules are absorbed in the interfaces between water and air, and then gel there because of their high concentration);
- high stability even during freeze–thaw cycles, the control of crystallization leads to the formation of small ice crystals, creamy melt-away when products are consumed frozen;
- control of viscosity, texture/consistency, and mouthfeel;
- compatible with other hydrocolloids;
- also effective in applications with high fat content.

Table 8.22 gives an overview of the application of HPMC and methylcellulose in dessert products.

Table 8.22 Use of hydroxypropylmethylcellulose and MC in desserts.

Application	Product	Viscosity	Gel strength	Conc. (%)
Pudding, cook-up type	HPMC	Medium	Soft	0.2–0.4
Vanilla cream (type 'custard')	HPMC	High	Soft	0.3–0.5
Ice cream, sorbets, sherbets	MC or	Medium	Firm	0.1–0.5
	HPMC or	Low	Semi-firm	
	HPMC	Medium–high	Soft	
Cheese cake (cold preparation)	MC or	Medium	Firm	0.3–0.4
	HPMC	Low–medium	Soft	

(*continued*)

Table 8.22 (*Continued*)

Application	Product	Viscosity	Gel strength	Conc. (%)
Instant puddings	MC or HPMC	Medium Medium–high	Firm Soft	0.3–0.6
Mousses	HPMC	Low	Semi-firm	0.2–0.7
Whipped toppings (RTU)	HPMC	Low + medium	Semi-firm + soft	0.2–0.5
Whipping cream concentrates	HPMC	Low	Semi-firm	0.3–0.4

8.7.5.2 Typical Formulations

Tables 8.23–8.27 give typical formulations for desserts, chilled and frozen dairy products.

Table 8.23 Formulation for ice cream.

Ingredient	Amount (%)
Milk (3.5% fat content)	63.48
Sugar	23.84
Skimmed milk powder (NFDM)	7.03
Sugar syrup with high fructose content	4.80
Vanilla extract	0.54
HPMC – type D with 50 mPa s	0.31
Sum	100.0

Table 8.24 Formulations for vanilla puddings, egg cream ('custard'), and chocolate desserts.

Ingredient	Amount (%)				
	Vanilla pudding	Vanilla pudding, reduced fat	Custard cream, 30% reduced egg	Chocolate pudding reduced-fat, type 1	Chocolate pudding reduced-fat, type 2
Milk, skimmed	56.0	53.1	72.0	54.0	51.1
Water	20.9	25.3	—	20.0	24.8
Sugar syrup (high fructose content)	19.6	18.0	—	19.4	19.0
Eggs, whipped	—	—	15.7	—	—

Sugar	—	—	8.9	—	—
Cocoa, de-oiled	—	—	—	3.0	1.4
Flavour 'chocolate'	—	—	—	—	0.4
Modified waxy maize starch (E1442)	2.3	2.3	2.8	2.3	2.0
Salt	0.5	0.5	—	0,4	0,4
Sodium stearoyl lactylate (SSL), E 481	0.1	0.1	—	0.1	0.1
Vanilla flavour or extract	0.2	0.2	0.2	0.2	0.2
HPMC – type E with 4000 mPa s	0.4	0.4	—	0.5	0.4
HPMC – type E with 100 000 mPa s	—	0.1	0.4	0.1	0.2
Sum	100.0	100.0	100.0	100.0	100.0

Table 8.25 Formulations for lemon sherbets.

Ingredient	Amount (%)	
	Standard	Fat-free variant
Water	57.25	60.82
Powdered sugar	20.00	20.00
Lemon juice concentrate	8.00	8.00
Maltodextrin	5.00	5.00
Glucose	5.00	5.00
Vegetable oil	3.50	—
Emulsifier – mono-/diglycerides of edible fatty acids	0.30	—
Citric acid	0.30	0.30
Pectin	0.17	0,20
Locust bean gum	0.17	0.20
Lemon flavour	0.05	0.08
HPMC – type E with 100 000 mPa s	0.26	0.40
Sum	100.0	100.0

Table 8.26 Formulations for yogurt, fermented.

Ingredient	Amount (%)	
	Standard	Yogurt preparation, sugar-free + reduced fat
Milk (3.5% fat content)	91.2	—
Water	—	74.1
Maltodextrin	—	10.2
Sugar	5.8	—
Sweetener (aspartame)	—	0.1
Starch (modified waxy maize starch)	1.5	3.0
Skimmed milk powder	1.0	12.0
Cultures for stirred yogurt	0.2	0.1
HPMC – type E with 100 000 mPa s	0.3	0.5
Sum	100.0	100.0

Table 8.27 Formulation for fat-reduced sour cream.

Ingredient	Amount (%)
Fat substitute – modified corn starch	30.9
Sour cream (20% fat)	27.0
Water (18 °C (64 °F))	17.8
Polydextrose	10.8
Skimmed milk powder	10.8
Maltodextrin	1.0
Lactic acid (85%)	0.4
Monoglycerides	0.3
Sodium stearoyl lactylate (SSL)	0.2
Sour cream flavour	0.2
Salt	0.2
Propylene glycol alginate	0.1
Xanthan gum	0.1
HPMC – type E with 100 000 mPa s	0.2
Sum	100.0

8.7.6
Confectionery

The formulation for some confectionary products is shown in Tables 8.28–8.30.

Table 8.28 Formulation for nougat.

Part	Ingredient	Amount (%)	Processing guideline
A	Powdered sugar	9.3	1. Dry blend HPMC with some powdered sugar and disperse into water;
	Water (7 °C (44 °F))	5.7	2. add egg white solids and mix well;
	Egg white, dried	0.5	3. add remaining powdered sugar and whip for 10 min;
	HPMC – type D with 50 mPa s	0.4	
B	Glucose syrup	37.3	4. dissolve sugar in water:
	Granulated sugar	24.0	add glucose syrup,
	Water	7.5	boil to 130 °C (260 °F);
			5. add syrup in a thin stream to the whip using a low speed and a flat beater;
C	Cocoa powder	3.6	6. mix the powders together;
	Malt powder	3.6	7. incorporate the dry blend to the mix from part '1+2' under moderate stirring.
	Skimmed milk powder	3.7	
	Powdered sugar	2.7	
D	Fat	1.7	8. melt the fat at low temperature and add to the mixture with slow mixing and minimum time for dispersion;
			9. spread onto cooling table.
			10. cutting, after sufficient cooling time.
	Sum	100.0	

Table 8.29 Formulation for soft caramel.

Ingredient	Amount (%)
Sweetened condensed milk	25.4
Glucose syrup with 42 DE	23.4
Granulated sugar	14.0
Brown sugar	14.0
Unsalted butter	11.1
Water (20 °C (68 °F))	9.3
Sorbitan monostearate	1.5
Salt	0.9
HPMC – type E with 100 mPa s	0.4
Sum	100.0

Table 8.30 Formulations for fruit bites with cream cheese.

Ingredient	Amount (%)
(A) Cream cheese layer:	
Cream cheese (Philadelphia type, full fat)	80.7
Yogurt, natural	12.0
Instant starch (E 1422)	3.0
Maltodextrin	3.0
Sugar	1.0
HPMC – type D with 450 mPa s	0.3
(B) Fruit layer:	
Pineapple pieces (canned), chopped	70.0
Apricots puree (canned, drained)	18.0
Sugar	9.0
Pre-gelatinized rice starch	2.0
HPMC – type D with 220 000 mPa s	1.0
Sum (A or B) – each	100.0

Processing guideline:

A. Preparation of the cream cheese layer:

1. HPMC is pre-mixed with sugar and stirred into the yogurt. Allow to stand for 2 min and mix again for 30 s.
2. The remaining dry ingredients are mixed together.
3. Add the cream cheese and the HPMC-yogurt-slurry together and mix well until homogeneously blended.
4. Add the dry mix (step 2) and stir gently.

B. Preparation of the fruit layer:

5. Mix the pineapple pieces and mashed apricots together.
6. Blend the starch, sugar, and HPMC together and mix into the fruit (hydration time 5 min). Mix gently for 30 s.

C. Production of the 'bites':

7. Spread the cream cheese evenly over a greased or non-stick base to give a layer (approximately 1–2 mm thick).
8. Spread the fruit mix evenly over the cream cheese in a layer (approximately 1–2 mm thick).
9. Cut the layers into 3–5 mm squares and deep freeze.
10. Once frozen, batter, crumb, and flash fry.
11. Re-freeze.

8.7.7 Reformed Products with HPMC

Formulations for croquettes and for filled dough pockets are shown in Tables 8.31 and 8.32, respectively.

Table 8.31 Formulations for croquettes [6,7].

Ingredient	Amount (%)	
	Variant A	Variant B
Puree from fresh potatoes	87.4	79.0
Potato flakes (dried)	6.0	11.0
Milk powder	0.8	—
Salt	0.8	1.0
HPMC – type D with 450 mPa s	—	0.5–1.0
5% Solution of HPMC – type E with 100 mPa s	5.0	—
Breadcrumbs	—	5.0
Water	—	3.0–3.5
Sum	100.0	100.0

Data taken with permission from References [6,7].

Processing guideline:

Variant A:

1. Preparation of a 5% HPMC solution with the 'hot-cold method';
2. fresh potatoes are peeled, cooked, and mashed;
3. the dry components are weighed and mixed;
4. formation of the croquettes;
5. frying for 3 min at 180 °C (356 °F).

Table 8.32 Formulations for filled dough pockets ('steamed buns') [8].

Ingredient	Wt%, based on flour amount
Wheat flour	100.0
Sugar	10.0
Modified starch	5.0
Shortening	4.0
Salt	1.0
Baking powder	1.0
Dry yeasts	1.0
Water	~67.0
CMC – 30 000 or 40 000	0.2
HPMC – type D with 50 mPa s	0.2

Data taken with permission from Reference [8].

Processing guideline:

1. Mix the dry ingredients until a homogeneous mixture is obtained (e.g. 1–2 min with a Hobart mixer at low speed.);
2. addition of the fat component; incorporation into the dry mix for several minutes at low speed;

3. addition of the water quantity; mixing for several minutes at medium speed;
4. replacement of the mixing attachment with the kneading attachment;
5. kneading for 10 min at low speed;
6. divide the dough into small portions ('pockets', rolls);
7. rest period (for yeast fermentation) for 1 h at 38–40 °C (100–104 °F) and high humidity;
8. steaming (10–12 min).

8.7.8
Soups, Sauces, Salad Dressings, and Marinades

Table 8.33 shows the formulation for dry marinades. By adding high-viscosity HPMC to a traditional mixture, an improved dry marinade can be made. Pieces of meat get a good gloss and the water loss during frying is reduced. Compared to wet marinades, there is the advantage of a simple change of taste style by rapid dry cleaning of the production equipment and environment.

Table 8.33 Formulations for dry marinades.

Ingredient	Amount (%)		
	Coating with herbs	Glaze with lemon-butter	Coating for seafood
Maltodextrin	58.2	55.9	44.0
Dextrose	16.3	6.8	20.9
Chicken broth	11.3	—	—
Wheat flour	—	5.9	—
Vegetable oil	—	5.8	Olive oil: 8.0
Flavour – type ...	—	Butter: 5.8	Sea food: 7.7
Modified starch (E 1422)	6.5	2.9	4.6
Paprika	1.6	—	1.5
Garlic powder	1.6	3.4	3.9
Lemon powder	—	3.0	—
Herbs	Parsley: 0.8	Dill: 0.5	Parsley: 0.4
Salt	0.8	6.8	3.9
Black pepper	0.75	0.6	Lemon pepper: 1.2
Cayenne pepper	0.24	—	—
Green onions, dried	0.5	—	—
Onion powder (fine)	—	0.9	1.5
Hydrolysed yeasts	—	0.8	—
Orange powder	—	—	0.9
HPMC – type E with 100 000 mPa s	1.50	0.9	1.5
Sum	100.0	100.0	100.0

The basic recipe for a dry marinade is as follows:

5% HPMC (type E with 100 000 mPa s),
5% starch,
5% spices,
20% dextrose,
65% maltodextrin.

The formulations for creamy sauces and soups (Table 8.34) and mayonnaise substitutes (Table 8.35) are shown here.

Table 8.34 Formulations for creamy sauces and soups – viscosity build up during cooling.

Ingredient	Amount (%)
Water (hot, 80–95 °C (176–203 °F))	87.8
Oil	6.0
Skimmed milk powder	2.0
Modified waxy maize starch (E 1422)	2.0
Salt	1.3
Glutamate (MSG) – optional	0.4
HPMC – type D with 220 000 mPa s	0.5
Sum	100.0

Table 8.35 Formulations for mayonnaise-substitutes – fat-reduced instant products (without homogenization).

Ingredient	Amount (%)	
	18% Fat	3% Fat
(A) Dry blend:		
Sugar	8.6	8.6
Instant starch, drum-dried (E 1422)	5.0	5.0
Salt	1.5	1.5
Maltodextrin	1.0	1.0
Full egg, dried	1.0	1.0
MCC – cellulose gel with CMC	0.8	—
Mayonnaise flavour	0.6	0.6
HPMC – type D with 4000 mPa s	0.3	0.3
HPMC – type E with 100 000 mPa s	0.1	0.3
Xanthan	0.3	0.3
(B) Liquid phase:		
Water	55.5	73.1
Vegetable oil	18.0	3.0
Lemon juice, pasteurized	3.3	3.3
Vinegar (5%)	3.0	1.5
Mustard	1.0	0.5
Sum	100.0	100.0

8.7.9 Beverages

8.7.9.1 Functions and Overview

HPMC fulfils in drinks and beverages the following tasks:

- suspension of particles over a wide temperature range (from frozen or cold to hot) without sticky mouthfeel;
- crystallization control in frozen slushies/smoothies and formation of small crystals with 'smooth' melting behaviour;
- rapid construction of foams and their stabilization in instant cappuccinos;
- solubility in alcohol–water mixtures, thickening of transparent alcoholic drinks and creamy liqueurs;
- increase in viscosity of milk shakes, improved mouth feel and creaminess without loss of flavour;
- high compatibility with vitamins and minerals in highly enriched beverages (nutraceuticals) because of the neutral molecule character;
- thermally stable at pasteurization and sterilization (UHT) temperatures;
- very good acid stability (down to pH 1).

Table 8.36 summarizes how HPMC is used in beverages.

8.7.9.2 Typical Formulations

Tables 8.37–8.40 give some typical formulations.

8.7.10 Flavour Concentrates

8.7.10.1 Functions and Overview

HPMC and methylcellulose show the following features and benefits in flavour emulsions:

- stabilization of the emulsion;
- interfacial activity;

Table 8.36 Application of HPMC in beverages.

Application	Product	Viscosity	Gel strength	Conc. (%)
Cappuccino – RTD/instant	HPMC	Low	Soft	0.2–0.3
Chocolate milk/cocoa	HPMC	High	Soft	0.1–0.2
Frozen slushies	HPMC	Low–medium	Soft	0.1–0.3
Suspension of particles	HPMC	Low	Soft	0.7–1.0
Alcoholic beverages	HPMC	Low–medium	Soft–semi-firm	0.1–0.5

Table 8.37 Formulations for hot prepared drinks from powder mixtures.

Ingredient	Amount (%)	
	Hot soup	Hot chocolate
Water (60–80 °C) (140–176 °F)	90.0	90.0
Dry mixture	9.9	9.9
HPMC – type E with 100 mPa s	—	0.1
HPMC – type E with 100 000 mPa s	0.1	—
Sum	100.0	100.0

Table 8.38 Formulations for frozen beverages ('slushies') – 'piña colada' type, alcohol-free.

Ingredient	Amount (%)
Water	60.1–60.35
Pineapple juice	15.9
Coconut cream	7.9
Sugar syrup with 42 DE	7.9
Fructose syrup	7.9
HPMC – type E with 100 or 15 000 mPa s	0.05–0.3
Sum	100.0

Table 8.39 Formulations for alcoholic beverages.

Ingredient	Amount (%)
Water	To 100%
Alcohol	Max. 40%
Other liquid ingredients such as cream and so on	Up to 30%
HPMC – type C or D with 50 mPa s	0.50–2.00
HPMC – type C or E with 4000 mPa s	0.05–0.20
HPMC – type E with 100 mPa s	0.50–2.00
Sum	100.0

- high efficiency – low amounts comparable to alternative emulsifying hydrocolloids;
- improved dispersion of the individual components;
- high loading of the resulting powder with oil and its safe enclosure;
- encapsulation of flavours;
- inclusion of – from the sensory perspective – negative acting substances (fish oils, bitter-tasting nutraceuticals etc.);
- *disadvantage*: the comparatively low dry matter of the MC/HPMC emulsions mostly does not allow economical results using spray-drying processes.

Table 8.40 Formulations for foamy instant beverages with coffee and tea (cold).

Ingredient	Amount (g)			
	Coffee	Coffee	Coffee	Tea
Water, iced	147 ml	147 ml	150 ml	150 ml
Sugar	—	—	6.0	6.0
Skimmed milk powder	1.5	1.5	—	—
Vegetable creamer	—	—	8.0	8.0
Instant coffee powder	1.5	1.5	1.0	—
Instant black tea powder	—	—	—	1.0
HPMC – type E with 100 mPa s	—	0.2	0.2	0.2
HPMC – type E with 4000 mPa s	0.1	—	—	—
Sum	100.0	100.0	100.0	100.0

Table 8.41 Use of methylcellulose and HPMC in applications with flavour oil.

Application	Product	Viscosity	Gel strength	Concentration (%)
Flavour oil emulsions	MC or	Low	Firm	0.5–3.0
	HPMC	Low	Semi-firm	
Encapsulation of flavour oils	MC or	Low	Firm	3.0–8.0
	HPMC	Low	Semi-firm	

Table 8.41 shows the use of HPMC and methylcellulose for the stabilization of flavour oils.

8.7.10.2 Typical Formulations

Typical formulations for flavour emulsions are given in Tables 8.42 and 8.43.

Table 8.42 Formulations for emulsions for spray drying.

Ingredient	Amount (%)
Water	55–90
Substance (oil) to be encapsulated	1–5
Maltodextrin	5–25
Emulsifier	1–3
HPMC – type C with 5–15 mPa s	3–12
Sum	100

Table 8.43 Formulations for liquid emulsions to protect flavour oils.

Ingredient	Amount (%)
Water	80.5–88.7
Substance (oil) to be encapsulated	0.3–1.5
Maltodextrin	5
Emulsifier	1–3
MC – type B with 15 mPa s	3–5
HPMC – type C with 5–50 mPa s	2–5
Sum	100.0

8.8 Non-food Applications

For technical applications, cosmetics, and pharmaceutical products, there is also the possibility to apply HPMC and pure methylcellulose reciprocally. The options are described in Chapter 5 on methylcellulose (see Section 5.8).

8.9 Nutritional Properties

The nutritional values of HPMC are the same as for methylcellulose (see Table 5.49 in Section 5.9.1). Hydroxypropylmethylcellulose is considered as safe. No ADI has been defined. As soluble fibre, HPMC is not digested or resorbed by the human organism.

In recent years, health benefits were found in the consumption of the long-chain (high-viscosity) HPMC types. Nutritionally unwanted or harmful substances in food – such as unhealthy fats – can be absorbed into the polymer and funnelled out of the body through this HPMC carrier. Furthermore, delayed sugar resorption is observed, thus enabling the glycemic index (GI) to be controlled by the use of hydroxypropylmethylcellulose. In large quantities, HPMC has a laxative effect.

8.10 Legislation

European Union (EU)

Hydroxypropylmethylcellulose is used as a technological additive to achieve a particular effect in food. The description of HPMC on the label can be made as 'stabilizer' or 'thickening agent'. The E number is E 464. In the European Union, HPMC is generally approved for food without a maximum limit ('quantum satis'). Exceptions are untreated products and those food products that may not be modified by additives.

United States of America

Food additives are regulated in the Unites States by the Food and Drug Administration (FDA) and the U.S. department of Agriculture (USDA). Of the chemical additives, only MC and sodium CMC are generally recognized as safe (GRAS). HPMC manufacturers have requested the extension of GRAS status to HPMC. The other cellulose derivatives of importance – HPC, HPMC, MEC, and EC – are approved under part 172 Title 21 of the CFR. The compounds and their specific approvals for HPMC can be found in section 172.874. There are other specific approvals for cellulose derivatives such as those that define their use in 'Adhesives and Coatings for Food Use' (21 CFR 175.300) or specific food products such as 'Artificially Sweetened Fruit Jellies' (21 CFR 150.141). The USDA has regulatory authority over meat products in the United States. The approvals for cellulose ethers in meat products are different to those for other foods. Only sodium CMC and MC are approved as extenders and stabilizer in meats. Applications for admission of HPMC are made by the cellulosics manufacturers. Finally, in addition to these approvals for CMC and methylcellulose, HPMC is listed in the 'USDA Standards and Labelling policy book' as an appropriate ingredient when used in a manner consistent with the policy book's regulations [4].

In the USA, HPMC is also authorized to be labelled as 'modified cellulose' or 'carbohydrate gum'.

Canada and Australia

In Canada, hydroxypropylmethylcellulose is listed in the 'Food and Drugs Act and Regulations' as food additive that may be used as emulsifying, gelling, stabilizing, and thickening agent.

In Australia, HPMC has been approved for inclusion in the 'Food Standards Regulations' under Section A 10, group 1, and referencing modifying agents [4].

Other Countries and Areas

In other countries and areas, HPMC needs to undergo local approval before being applied. As the product is intensely examined in Europe and the USA, successful EU and US-FDA approvals will be helpful for implementation in other markets.

References

1 European Union (2008) Commission directive 2008/84/EC of 27 August 2008, laying down specific purity criteria on food additives other than colours and sweeteners. *Official Journal of the European Union*, **51**, L253, 94–102. Available at http://eur-lex.europa.eu/LexUriServ/LexUriServ.do?uri=OJ:L:2008:253:0001:0175:EN:PDF.

2 Scherz, H. (1996) *Hydrocolloids: Stabilizers, Thickening and Gelling Agents in Food Products*, Food Chemistry/Food Quality, vol. 2, Food Chemical Society GDCh, Behr's Verlag GmbH, Hamburg.

3 Balser, K. and Iseringhausen, M. (1975) Cellulose ether, *Ullmanńs Encyclopedia of Industrial Chemistry* (eds E. Bartholomé *et al.*), vol. 9, 4th edn, Verlag Chemie, Weinheim, pp. 192–212.
4 Stephen, A.M., Philips, G.O., and Williams, P.A. (eds) (2006) *Polysaccharides and Their Applications*, 2nd edn, Taylor & Francis Group, Boca Raton, FL, pp. 147–179.
5 Belitz, H.-D., Grosch, W., and Schieberle, P. (2009) *Food Chemistry*, 4th edn, Springer-Verlag, pp. 294–297.
6 Ashland, Inc. (2012) Benecel high-purity methylcellulose and hydroxypropyl methylcelllulose, physical and chemical properties (56001-E4 REV). From the document library at http://www.ashland.com/products/benecel-methylcellulose-mc-and-hydroxypropylmethylcellulose-HPMC.
7 Murray, J.C.F. (2000) Cellulosics, in *Handbook of Hydrocolloids* (eds G.O. Phillips and P.A. Williams), Woodhead Publishing Limited, Cambridge, pp. 219–230.
8 Ashland Inc. (2012) Steamed Buns with Benecel^{TM} HPMC and Aquasob^{TM} Cellulose Gum (CMC) (Formula 01-1051). Formulation from the document library at http://www.ashland.com/products/benecel-methylcellulose-mc-and-hydroxypropylmethylcellulose-HPMC.

9
Methylethylcellulose

Synonyms: MEC, EMC, ethylmethyl ether of cellulose, ethylmethylcellulose, methylethyl ether of cellulose, cellulose ethyl methylether

E number: E 465
CAS number: 9004-59-5 and 9004-69-7
FDA code: 21 CFR 172.872

Related hydrocolloids: methylcellulose, HPMC

9.1
Manufacturing

Methylethylcellulose (MEC) is a mixed ether of cellulose that is made of natural plant cellulose obtained by alkalization and etherification with methyl chloride and ethyl chloride. Another possible reaction with ethylene oxide is very difficult to control. Methylethylcellulose gels or flocculates at higher temperatures than pure methylcellulose does. This property is used in the purification of the substance from undesired by-products of the etherification. MEC is washed with warm water above the gelation or flocculation temperature. Figure 9.1 shows, schematically, the manufacturing process for the production of methylethylcellulose.

Because of its close relationship with methylcellulose (MC), the product is referred to in practice as methylethylcellulose, MEC. In some older literature, the name ethylmethylcellulose, EMC, can be still found. In the 90 years of the last century, the production volumes of methylethylcellulose amounted to about 60 tons per year worldwide. There was only one manufacturer.

The MEC product, which is no longer commercially available, had a degree of substitution (DS) 'methyl' of 0.7 and a DS 'ethyl' of 0.3. Rheology, salt tolerance, and pH stability are comparable to the properties of methylcellulose [1]. The general difference to MC is the solution behaviour and the gelling properties, which are markedly differently for methylethylcellulose, depending on the strength of the etherification. Details can be found in Section 9.3 ('Rheology').

Cellulose and Cellulose Derivatives in the Food Industry: Fundamentals and Applications, First Edition.
Tanja Wüstenberg.

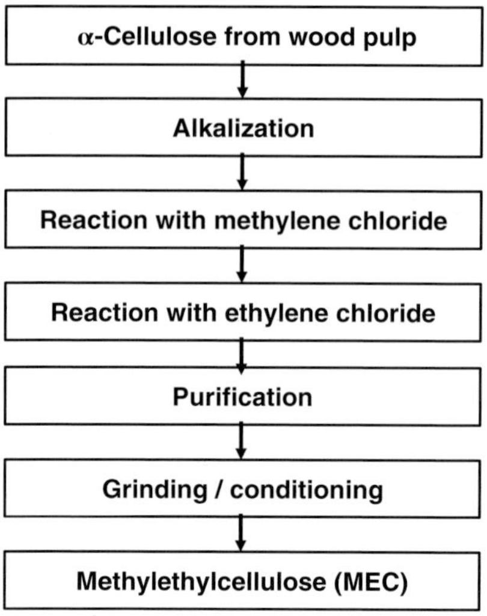

Figure 9.1 Flow diagram for the production of methylethylcellulose.

The low production volumes are based on the fact that the product is not used in technical applications. In addition, in the food industry, the number of applications is limited. The properties of the MEC can be provided by other cellulose derivatives, which are easier to use with the same functionality.

Since 1998, MEC has been out of production due to the low use levels, difficulties in the application, and lack of expansion opportunities in the food sector.

9.2 Chemistry

A long chain of ß-D-glucose monomers forms the basic structure of methylethylcellulose. The three free hydroxyl groups of the monosaccharide molecule glucose are substituted with ethyl or methyl groups. The chemical formula of the MEC molecule is:

$$[C_6H_7O_2(OH)_x(OCH_3)_y(OC_2H_5)_z]_n$$

The following applies:

- number of ethoxyl groups: $z = 0.57–0.8$;
- number of methoxyl groups: $y = 0.2–0.4$;
- number of free OH groups: $x = 3 - (y + z)$;
- degree of substitution: $DS = y + z$;
- degree of polymerization: n = number of glucose monomers.

The structural formula is shown in Figure 9.2.

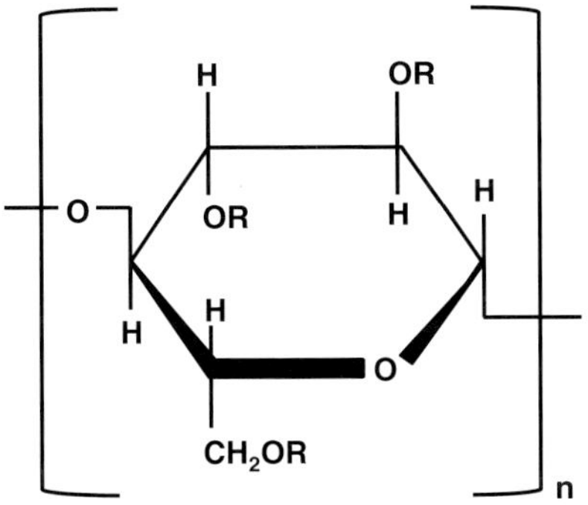

Figure 9.2 Structural formula of methylethylcellulose (R = H, CH_3, or C_2H_5).

The molecular weight of a non-substituted unit is 162.12 g mol^{-1}. A degree of substitution of 0.77 (DS_{Methyl} = 0.19 and DS_{Ethyl} = 0.58) increases the molecular weight to 181 g mol^{-1}. With a DS of 1.2 (DS_{Methyl} = 0.3 and DS_{Ethyl} = 0.9), the monomer weighs 191 g mol^{-1}. The polymers have a molecular weight of 30 000–40 000 g mol^{-1}, which corresponds to a chain length of about 150–200 glucose units.

Methylethylcellulose contains (based on the dry substance):

- 3.5–6.5% methoxyl groups (-CH_3),
- 14.5–19.0% ethoxyl groups (-C_2H_5)
- 13.2–19.6% alkoxyl groups in total (calculated as methoxyl) [2].

9.3 Rheology

9.3.1 Dissolution Behaviour

Methylethylcellulose first came onto the market as a cotton-like product. The cotton-like MEC had to be soaked overnight in cold water to be hydrated. Only after this swelling step could it be incorporated into food formulations. To simplify the application, a powdered MEC was developed. The viscosity of commercial products was in the range 10–60 mPa s (at a concentration of 2.5%).

In cold water, methylethylcellulose forms colloidal solutions. As with all cellulose ethers, the water solubility is dependent on the degree of substitution. Low-substituted MEC forms turbid solutions with some 'muddy' consistency in cold water and weak gels when heated. Higher-substituted methylethylcellulose (with more than 20% substituted methyl and ethyl groups in total) shows a better solubility at lower water temperatures.

The resulting solutions are clear. The highly etherified MEC types produce solid gels with increasing temperature.

9.3.2
Gelation and Behaviour with Other Ingredients

Like all methylated cellulose ethers (MC, HPMC (hydroxypropylcellulose)), methylethylcellulose forms thermally reversible gels on heating its aqueous solutions, too. On cooling, the more or less solid gels melt back to viscous solutions and return to their original viscosity. The gelling temperature and the gel strength are dependent on the degree of substitution and on the concentration of methylated cellulose ether.

Highly substituted MEC (about 20% in total alkoxyl) forms solid gels at temperatures above 50 °C (20 °F). Methylethylcellulose with low DS shows weak gelation, giving soft gels at 65–70 °C (150–160 °F). The temperature of gel formation and the properties of the resulting gels are related to the dry matter concentration. Generally, the presence of other ingredients reduces the gelling temperature and increases gel strength.

Compatibility with other components of the food formulation is high. The water quantity required for hydration of the MEC is low due to the short polymer chains. One effect of this is that high amounts of sugars and salts can be tolerated by the MEC system. Foam generation and stabilization is the main characteristic feature of the MEC, a feature that is still obtained in the presence of fats and oils, proteins (e.g. egg white or soy protein), and other hydrocolloids. Foams, which return to the liquid state during longer standing times, can be whipped again to be fixed. The whipping capability, the so-called 'overun', of methylethylcellulose is very high. A low-viscosity MEC (10–60 mPa s at 2.5%) can generate a volume of about 400% foam as a 2.5% solution. The qualities of the foams from MEC are comparable to those from whipped egg white – the pore structure is very fine and the whipping volume is equal.

9.4
Stability

Owing to its neutral molecular nature, dissolved methylethylcellulose is very stable to acid and salt additions. The viscosity of the solution is not significantly affected.

Moreover, temperature changes have no irreversible consequences. The freeze–thaw stability is very good, as with all other cellulose ethers. MEC can stabilize frozen food products and influence crystal formation positively, so that a creamy texture with fine ice crystals results. Heating leads to the formation of reversible gels of different strength, which melt back again to the viscous state on cooling. Low-viscosity MEC can be used in UHT processes, since it forms suitable soft gels. The disadvantage of MEC with low DS is the limited solubility. MEC with a high degree of substitution (20% imported alkoxyl groups) is highly and sufficiently soluble. However, it cannot be heated in UHT processes, since its gels are too firm and would block the plates or tubes of heat exchangers by clot formation. The negative interaction of solubility and gelation leads to a strong limitation of possible applications.

The use of shear forces leads to the formation of foams from aqueous or alcoholic solutions because of the surface activity of methylethylcellulose. The whipping capacity is very high and is inversely proportional to the chain length of the MEC. The lower the viscosity of the MEC (which means, the shorter the cellulose chain) the higher the so-called 'overrun' will be. Its function as a foam generator is retained in systems with high fat contents. Proteins do not interfere or reduce the whipping behaviour of the MEC. In terms of volume and pore structure, the foaming power of methylethylcellulose is comparable to egg white. The advantage of MEC is in the plant-based raw material and the ability to function smoothly in the presence of substances that interfere with egg white. MEC foams are very fat-stable. Foam products with MEC can also be whipped again by renewed shear treatment after storage-induced liquefaction.

9.5 Rheometry

Methylethylcellulose can be distinguished from other hydrocolloids and some cellulose derivatives (MCC (microcrystalline cellulose), CMC (carboxymethylcellulose), HPC (hydroxypropylcellulose)) by its solubility in cold water, its foaming power, and its heat of gelation. MEC is more surface active than methylcellulose and forms weaker gels upon heating (approximately 65–70 °C (150–160 °F)). The solubility of methylethylcellulose in cold water is limited compared to that of pure HPMC, MC, or HPC. The resulting solutions are cloudy and sometimes have a 'muddy' consistency. Another distinguishing feature of MEC against a methylcellulose or HPMC is its solubility in ethanol. Foams are very stable in high-fatty systems through the use of methylethylcellulose and also remain stable in the presence of proteins.

9.6 Synergies with Other Hydrocolloids

Synergistic effects of hydrocolloid mixtures including MEC are not known. Currently, this cellulose ether is not produced on an industrial scale. A resumption of commercial production is doubtful due to the aforementioned reasons. Methylethylcellulose, therefore, is not included in research projects on positive

Table 9.1 Formulation for a whipping cream ('whipped topping').

Ingredient	Amount (%)
Vegetable oil or fat	20–35
Milk or soy protein	1–4
Methylethylcellulose (10–40 mPa s, 2.5%)	0.3–0.5
Methylethylcellulose (40–60 mPa s, 2.5%)	0.2–0.4
Emulsifiers	0.4–1
Sugar or glucose syrup	6–20
Salt, flavour, colour	To achieve the desired taste
Water	To 100% (approx. 40–70)
Sum	100%

Processing guideline:

The procedure is identical to the production of whipping products with HPC (see Section 7.7.1 on 'Foam products').

interactions of gelling agents and stabilizers. MEC has been replaced mainly by low-viscosity HPMC with medium to high methyl content.

9.7 Food Applications with Typical Formulations

The excellent foam stabilization by methylethylcellulose – which is also exhibited in high fat and high protein containing systems – is responsible for its potential use in:

- dairy-free whipping products (a typical formulation is given in Table 9.1),
- cakes,
- meringues without egg,
- dietary and egg-free food products.

Since MEC is no longer produced for commercial purposes, only one recipe is presented in this chapter. In principle, the formulations for foams in Chapter 7 on hydroxypropylcellulose can be transferred to MEC (see Section 7.7.1 on 'Foam products').

9.8 Non-food Applications

Methylethylcellulose is not used outside the food sector. In technical products and regulated non-food applications, the properties of the MEC are covered by other cellulose derivatives (like MC, HPMC, or HPC).

9.9
Nutritional Properties

Like all other cellulose ethers, methylethylcellulose is also physiologically inert and is not metabolized by the human body because of its derivation from cellulose. MEC is a soluble fibre, which is considered as toxicologically safe. An ADI value has not been specified.

9.10
Legislation

European Union (EU)

In the EU, methylethylcellulose grades have a general authorization as additives with technological function. The products have the E number E 465 and may be used in all food products in unlimited dosage ('quantum satis'). Excepted from this application are untreated and those food products that may not be modified by additives. MEC must be declared on the food label. The description can be made as with 'filler', 'carrier', 'stabilizer', or 'thickening agent'.

United States of America (USA)

In the USA, methylethylcellulose is registered in the 'US Code of Federal Regulations (CFR)' Title 21 under Part 172 as number 872 (= 21 CFR 172.872). Additionally, MEC is registered in the Food Chemical Codex, a compendium of internationally recognized standards for determining the purity and quality of food ingredients, including processing aids, preservatives, flavourings, colorants, stabilizers, thickeners, and nutrients.

Canada

In Canada, methylethylcellulose is listed in the 'Food and Drugs Act and Regulations' as a food additive that may be used as emulsifying, gelling, stabilizing, and thickening agent [1].

References

1 Stephen, A.M., Philips, G.O., and Williams, P.A. (eds) (2006) *Polysaccharides and Their Applications*, 2nd edn, Taylor & Francis Group, Boca Raton, FL, pp. 147–179.

2 European Union (2008) Commission directive 2008/84/EC of 27 August 2008, laying down specific purity criteria on food additives other than colours and sweeteners. *Official Journal of the European Union*, **51**, L253, 94–102. Available at http://eur-lex.europa.eu/LexUriServ/LexUriServ.do?uri=OJ:L:2008:253:0001:0175:EN:PDF.

10
Sodium Carboxymethylcellulose

Synonyms: cellulose gum, Na-CMC, CMC, sodium salt of the carboxymethyl ether of cellulose

E number: E 466
CAS number: 9004-32-4
FDA code: 21 CFR 182.1745

Related hydrocolloids from the group of cellulose derivatives:

enzymatically hydrolysed carboxymethylcellulose, crosslinked sodium carboxymethylcellulose, croscarmellose.

Hydrocolloids of other substance classes with similar functionality:

highly esterified pectin (HM pectin).

10.1
Manufacturing

Sodium carboxymethylcellulose (CMC) with food approval is defined as the sodium salt of a carboxymethyl ether of cellulose obtained directly from natural strains of fibrous plant material [1].

CMC was discovered shortly after World War I and has been produced on a large scale since 1935.

The cellulose ether CMC is produced by treatment of cellulose with aqueous sodium hydroxide solution and subsequent reaction with monochloroacetic acid or its sodium salt. In a parallel reaction, the by-products sodium chloride (table salt) and sodium glycolate are formed out of the etherification chemicals with free sodium hydroxide.

The purification and the removal of these salts for the production of high-purity sodium carboxymethylcellulose is carried out with solvents, preferably water. At the end of the carboxymethylation process, the CMC must be

Cellulose and Cellulose Derivatives in the Food Industry: Fundamentals and Applications, First Edition.
Tanja Wüstenberg.

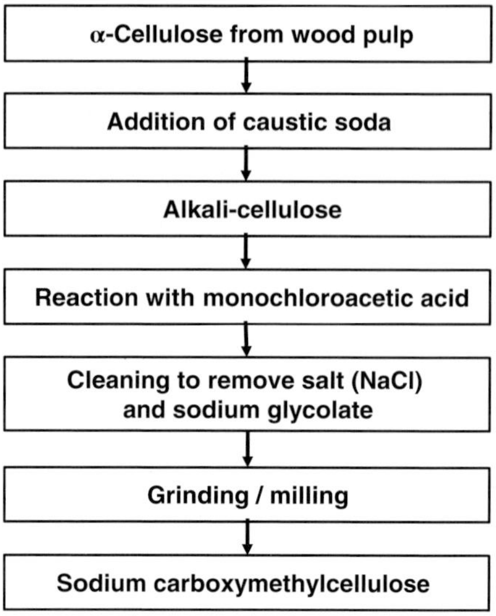

Figure 10.1 Flow diagram of the CMC production.

neutralized as a rule, since it usually includes a slight excess of sodium hydroxide. Although the neutral point of Na-CMC is in the weakly alkaline range at a pH of 8.25, the pH is generally adjusted to 7–7.5. If the pH of still moist carboxymethylcellulose is adjusted or neutralized to 6 or lower then the dried CMC has limited solubility. The solutions are turbid, show streaks, and contain insoluble gel particles. When the pH value is adjusted to 4 or below before drying, the resulting CMC is insoluble in water. In the final process step, the CMC is dried and milled to the desired particle size.

Food quality material contains at least 99.5% pure CMC and a maximum 0.5% of residual salts – as the sum of sodium chloride and sodium glycolate. The latter is limited to a maximum of 0.4% in the dry matter. The maximum total sodium content is 12.4% based on the dry matter. According to European food law, the degree of substitution is the range of 0.2–1.5 carboxymethyl groups per anhydroglucose unit. The lower limit for the molecular weight is 17 000 g mol^{-1}, which correlates with a degree of polymerization (DP) of approximately 100 [1].

The US Food Chemical Codex requires a minimum viscosity of 25 mPa s and a degree of substitution (DS) of 0.2–1.5 for food-grade CMC products.

Figure 10.1 shows the method for production of carboxymethyl cellulose.

10.2 Chemistry

The CMC molecule is a polymer with a base of β-D-glucose units, whose free hydroxyl groups are partially substituted by carboxymethyl groups. During the reaction with monochloroacetic acid, no secondary OH groups are formed. The DS of commercial CMC grades for food products is between 0.6 and 1.25 (average DS = 1.0).

Carboxymethyl cellulose and its sodium salt have the general chemical formula:

$$\left[C_6H_7O_2(OH)_x(OCH_2COONa)_y\right]_n$$

where:

- n is the degree of polymerization;
- x is the number of remaining free hydroxyl groups ($x = 1.50–2.80$);
- y is the degree of substitution, that is, the amount of substituted OH-groups ($y = 0.2–1.50$);
- $x + y = 3.0$, since in each glucose monomer three free OH groups are available.

The reaction of cellulose with monochloroacetic acid and the structural formula of CMC are presented in Figure 10.2. In this case, the average substitution is 1 (DS = 1.0).

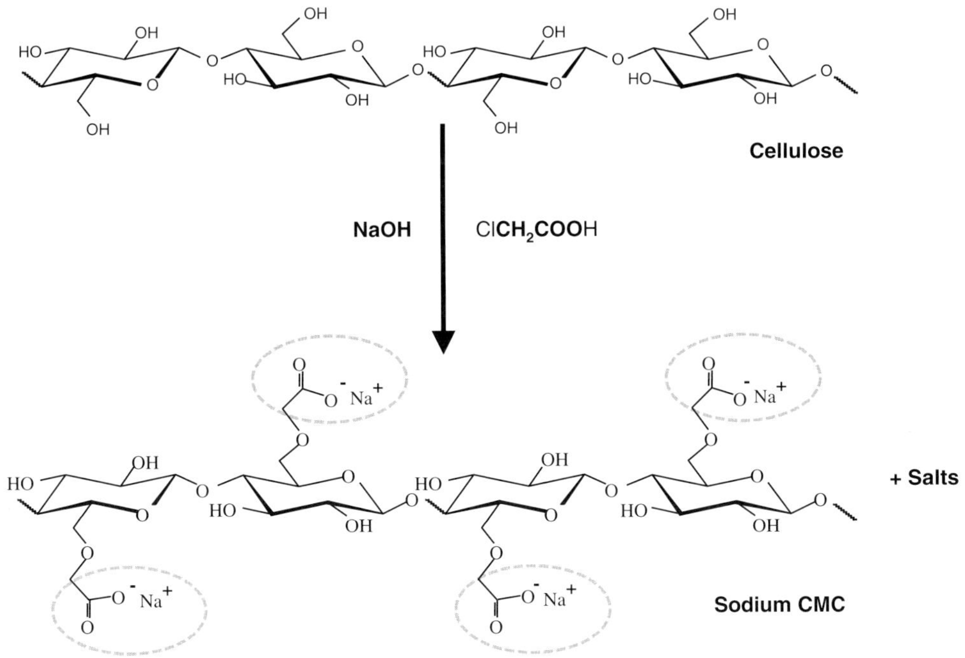

Figure 10.2 Reaction to form a molecule of Na-CMC from cellulose.

Table 10.1 Relationship between viscosity, molecular weight, and DP (degree of polymerization) [2].

CMC – viscosity (mPa s), (2%, Brookfield)	Molecular weight ($g\,mol^{-1}$)	Polymerization degree (DS = 0.7)
Low (<100)	90 000	400
Medium (<5000)	250 000	1100
High (>20 000)	700 000	3200

Data taken with permission from Reference [2].

Table 10.2 Relationship between degree of substitution and sodium content of a CMC.

Substitution degree (DS)	0.1	0.2	0.3	0.4	0.5	0.6	0.7	0.8
Sodium content (wt%)	1.35	2.58	3.69	4.72	5.66	6.53	7.33	8.08
Substitution degree (DS)	0.9	1.0	1.1	1.2	1.3	1.4	1.5	
Sodium content (wt%)	8.77	9.42	10.02	10.59	11.12	11.62	12.09	

The molecular weight of one carboxymethylated anhydroglucose unit with a substitution of 0.2 is 178.14 $g\,mol^{-1}$. At a DS of 1.5, a substituted glucose building block weighs 282.18 $g\,mol^{-1}$. Food-grade macromolecules have a molecular weight of at least 17 000 $g\,mol^{-1}$. This corresponds to a degree of polymerization (DP) of about 100 [1].

Table 10.1 illustrates the relationship between viscosity, molecular weight, and DP.

Table 10.2 shows the relationship between the degree of substitution of a CMC and its sodium content (Na). The approximate factor is 'DS : Na = 1 : 10'.

10.3
Rheology

10.3.1
Dissolution Behaviour

10.3.1.1 General Correlations

Polysaccharides with Carboxyl Groups

Polysaccharides with carboxyl groups as alkali salts are readily soluble in neutral and alkaline aqueous systems. Sodium-CMC belongs to this group as well as alginates and pectins. The molecules are relatively stretched by repulsion of the carboxylate ions and show for the same reason no intermolecular interactions. The viscosity of these solutions is high, but depends on pH value. At $pH \leq 3$

gelation or precipitation occurs by elimination of electrostatic repulsion and dimerization of undissociated carboxyl groups via hydrogen bonds [3].

Specific Facts on Na-CMC

Sodium carboxymethylcellulose (with a DS of 0.7–1.5) dissolves very well in cold and hot water. Through the excellent solubility of medium to highly substituted CMC types over the wide temperature range 0–100 °C (32–212 °F), the cellulose ether CMC can be differentiated from other cellulose derivatives, such as the methylated cellulose ethers (MC, HPMC (hydroxypropylmethylcellulose), MEC (methylethylcellulose)) and HPC (hydroxypropylcellulose), which form gels or flocculate at warmer temperatures, and accordingly are no longer soluble at elevated temperatures.

The optimum solubility of cellulose gum is between 25 and 40 °C (77 and 104 °F). Higher temperatures cause a reversible decrease in viscosity.

Like all cellulose ethers, carboxymethylcellulose (with a DS $\geq$ 0.6) forms colloidal solutions in water that are clear and transparent.

Commercially available CMC types have a degree of substitution between 0.4 and 1.5.

Since the 'merger' of adjacent hydroxyl groups of different cellulose chains significantly determines the solubility of the molecule, the DS of the carboxymethylcellulose has a strong influence on the physical properties of the solutions that can be observed for all cellulose ethers.

The degree of substitution thus determines the behaviour of the CMC in water. The higher the content of carboxymethyl groups introduced, the higher the solubility is and therefore the smoother the solutions are. In this context the uniformity of the substituents is to be mentioned as a second factor of influence. Ideal hydration is achieved by using a cellulose gum with high DS and high uniformity of the introduced groups (see upper molecule in Figure 5.6). The solubility decreases with more block-wise arrangement of the carboxymethyl groups. For example, it is possible to synthesize a CMC with a relatively high DS of 0.8 and a strict block-wise distribution of substituents that as a consequence is not completely soluble in water. At low degrees of substitution, the largest share of etherification occurs in the amorphous and the crystalline surface regions of the cellulose chain. This creates a CMC molecule with a large original or untreated, that is, native chain area, which shows much association with adjacent cellulose chains, resulting in low water solubility. Carboxymethylcellulose with degrees of substitution below 0.3 is not soluble in water – only alkaline media come into use as solvents. With increasing DS, an increasingly number of carboxymethyl groups is incorporated into the cellulose chain, whereby the chain–chain association is interrupted and a partially soluble polymer is formed.

CMC having a DS from 0.3 to 0.6 is only partially soluble. It swells in water and shows a rough texture with visible fibres. For a DS of 0.7, the crystalline zones of the chain are sufficiently often interrupted by 'foreign groups' to become a highly water-soluble material.

Degrees of substitution from 0.6 to 1.0 provide complete dissolution of the CMC in water. The resulting solutions are fibre-free, they flow homogeneously and smoothly.

With substitution degrees above 1.0, the relative concentration of the native hydroxyl groups is so small that no or only a very little association of neighbouring chains takes place [4].

By increasing the DS to 1.0–1.4, the stability and solubility of the cellulose ether arises even under extreme conditions. This phenomenon is used in systems with low water contents, low pH values, and high electrolyte contents.

The 'staying power' and stability of CMC solutions is inversely proportional to the solubility. The solutions of highly substituted CMC grades (with DS of 1.2 or higher) have a high gloss, very smooth fibre-free flow with minimal thixotropy, a high salt tolerance, good stability in low-water products, and low staying power. They tend to lose shape and flow away while low-substituted CMC types provide fibrous salt-sensitive solutions having a yield point.

Cellulose gum has a very high rate of hydration in water. In addition, the water absorption capacity of CMC is excellent. High-viscosity types can absorb up to '150 g of water per gram of CMC'. Salts reduce the absorption capacity. The water retention over the storage time is good – the highest performance can be found in medium-viscosity CMC. It can physically 'bind' water the longest and keeps it better fixed in competitive situations (e.g. moist filling–dry dough) than high-viscosity CMC types or other hydrocolloids such as guar, xanthan, and instant starches.

The presence of salts and polyelectrolytes can affect the solubility of cellulose gum in water. The effect of the ions is dependent on the type and quantity as well as the degree of substitution and molecular weight of the CMC. In addition, the preparation method can have a significant impact.

The cellulose ether CMC is not soluble in:

- pure ethanol,
- pure iso-propanol (IPA, isopropyl alcohol),
- vegetable oils,
- water-miscible non-solvents such as glycerol and propylene glycol.

Dispersions of CMC in these liquid media can be used to stir the hydrocolloid aggregate-free in cold water. Wetting of the CMC particles with a vegetable oil or alcohol prevents too rapid hydration. In addition, the speed of viscosity development is reduced by slowing the rate of dissolution, which especially gives advantages for cold-water soluble, high-viscosity hydrocolloids in terms of better solution quality. The mixing ratio is one part of CMC powder to three to ten parts of liquid. The solid dispersion, also called a 'slurry', is then fed to the water phase while stirring.

For very high-viscosity CMC types, this slurry-method is the only way to produce a smooth, homogeneous, and 100% functional solution. During an alternative slow stirring into water, the viscosity would build up so quickly that the agitator would no longer produce a turbulent flow but would only rotate in the centre at its maximum radius, while in the outer zones of the vessel a solid mass 'is standing'.

Although carboxymethylcellulose is insoluble in organic solvents, it can be dissolved in solvent mixtures of water and water-miscible solvents such as ethanol and acetone. For example, Na-CMC can be completely solubilized in an ethanol–water mixture, when the maximum alcohol content is 40 vol.%. Then, the CMC forms a transparent solution with full viscosity. Above 40 vol.% ethanol, turbidity occurs and the thickening effect is limited (visible decrease in viscosity). For acetone–water mixtures the limiting values are slightly lower than for ethanol (35–40 vol.%).

CMC, being pre-dissolved in pure water, tolerates a higher amount of ethanol or acetone than a dry CMC powder, which is directly added into the solvent mixture. Short-chain carboxymethylcellulose also tolerates higher concentrations of organic solvents in mixed systems with water than high-viscosity CMC products. The behaviour of highly substituted carboxymethylcellulose in solvent mixtures is discussed in detail in Section 10.3.1.7 ('Influence of solvent mixtures').

The production of a dry mixture of CMC with other dry formulation components, such as sugars, dextrins, flours, cooking starches, pigments, and so on, is an alternative way to reduce the hydration rate of the cellulose ether and to suppress the formation of lumps in water.

The viscosity of the solution is controlled not only by the concentration of CMC but also depends on the degree of polymerization or molecular weight of the cellulose gum used.

With increasing dosage, and length of the polymer chain, the thickening effect is also increased. As with most hydrocolloids this leads to the development of so-called 'superstructures'. There is no proportional relationship between input quantity and final viscosity. A rule of thumb says that doubling the concentration (*C*) of CMC increases the viscosity (*V*) in water by a factor of 5–10:

$$\text{Rule of thumb}: \quad 2C_{\text{CMC}} \rightarrow (5-10)V_{\text{System}}$$

The dissolution time is determined by the particle size or granulometry of the cellulose ether. Granular CMC products with relatively large particles need a longer hydration time for complete dissolution. Granules have the advantage that they do not dust and that they can easily – and without other excipients – be dissolved in water without formation of any disturbing aggregates. In the manufacture of dry mixes, coarse CMC tends to separate. For functional compounds, it is therefore recommended to use CMC in powder form.

Calcium Salts of CMC

In contrast to the fully water-soluble sodium salt of carboxymethyl cellulose, there is also a calcium-based salt available that is insoluble in water. The negatively charged carboxymethyl groups ($-COO^-$) of two different molecules or of

The difference between a pseudoplastic and a thixotropic flow is that the latter is a time-dependant shear-thinning.

10.3.1.5 Relationship between DS, DP, and Flow Characteristics of CMC

For carboxymethylcelluloses, pseudoplasticity generally decreases with increasing degree of substitution (DS). At DS $\geq$ 1.0, CMC forms solutions with very little interchain association between adjacent cellulose chains ('chain–chain interaction'), so that these solutions have no or only minor hysteresis loops at increasing and decreasing shear rates.

The thixotropic flow properties of CMC are the result of interchain hydrogen bonding between adjacent areas of relatively unsubstituted anhydroglucose units. In CMC, which has comparatively relatively large regions of unsubstituted hydroxyl groups on the sugar rings, there is a strong tendency for the single chains to develop interchain hydrogen bonds, resulting in a gel-like network. The solutions of these CMC-types show a rough, gel-like flow.

Conditions for the production of CMC grades with high thixotropy are:

- very uneven distribution of substituted carboxymethyl groups (CM) (blockwise arrangement with CM-rich and CM-free zones);
- low degree of substitution (up to DS $\leq$ 0.7).

With increasing degree of substitution, the distribution of substituents is statistically more homogeneous. The tendency for interchain association and to form intermolecular hydrogen bonds decreases anti-proportionally with increasing DS. With an average carboxymethyl group contents from 0.7 onwards, commercial CMC products are created with pseudoplastic non-thixotropic rheology. As consequence of a high degree of uniformity of substitution, their solutions flow smoothly and are free of fibres [4].

10.3.1.6 Effect of Shear on CMC Solutions

CMC is generally used to thicken, suspend, stabilize, gel, or otherwise modify the flow characteristics of aqueous solutions or suspensions. Preparation and use of its solutions involve a wide range of shearing conditions. It is important that the user of CMC understand how the rheology is affected by the degree of substitution (DS) and degree of polymerization (DP) [5].

Pseudoplasticity [5]

Small amounts of CMC dissolved in water greatly modify solution properties. The most obvious immediate change is an increase in viscosity. A particular CMC solution will vary in viscosity as different physical forces are applied to the solution.

These physical forces may be high, intermediate, or low shear stresses. For example, rolling, spreading, or vigorous mixing provides high shear stress. After a liquid film has been formed, gravity and surface tension control flow. The forces of gravity and surface tension are conditions of low shear stress. Pouring a liquid from a bottle is an example of an intermediate shear stress. If a solution of high-viscosity cellulose gum appears to be viscous when poured from a bottle, it will behave as a thin liquid when applied as a lotion. When, however, high shear stress is removed, it will instantly revert to its original high viscosity. Such flow behaviour is referred to as pseudoplastic flow, or time-independent shear-thinning, a form of non-Newtonian flow. It differs from time-dependant shear-thinning, called thixotropic flow.

Solutions of CMC exhibit pseudoplasticity because the long-chain molecules tend to orientate in the direction of flow as the applied force (shear stress) is increased. The resistance to flow (viscosity) is thereby decreased. When a lesser force is applied, the apparent viscosity is higher because random orientation and entanglement of non-aligned molecules causes increased resistance to flow. When shear rate is plotted against shear stress (Figure 10.4), a Newtonian fluid gives a straight line passing through the origin (zero point). A pseudoplastic liquid, such as a CMC solution, gives a curved line.

A plot of apparent viscosity against shear rate as in Figure 10.5 produces a horizontal straight line for a Newtonian fluid and a curved line for a pseudoplastic one.

Solutions of low-molecular-weight CMC grades have a low viscosity and generally are less pseudoplastic than solutions of higher molecular weight products. At very low shear rates, all sodium carboxymethylcellulose gum solutions approach Newtonian flow behaviour (Figure 10.6).

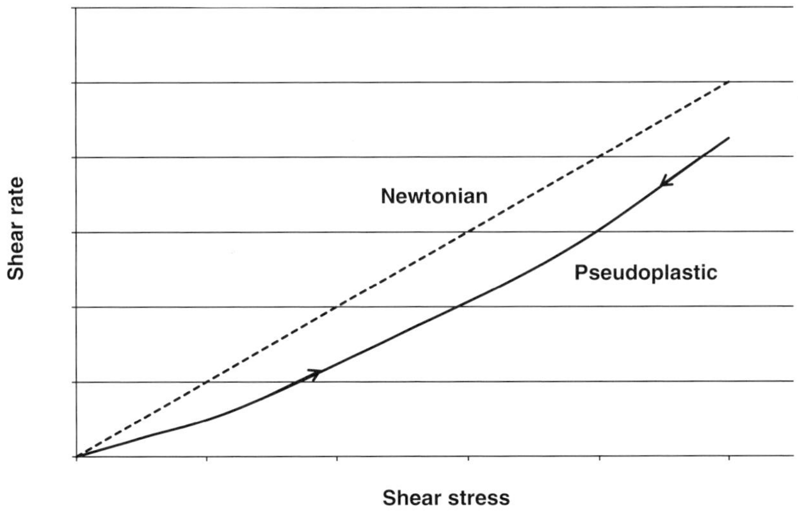

Figure 10.4 Shear rate versus shear stress for Newtonian and pseudoplastic liquids [2]. Reprinted with kind permission of Ashland Inc.

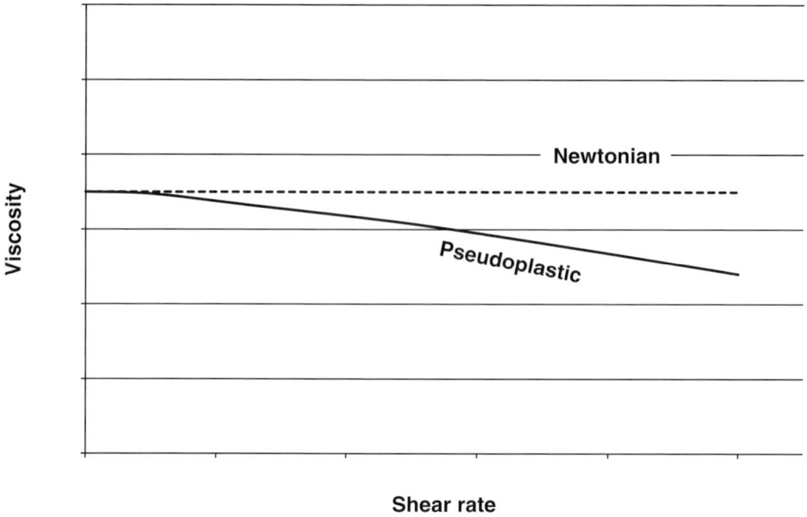

Figure 10.5 Viscosity versus shear rate for Newtonian and pseudoplastic liquids [2]. Reprinted with kind permission of Ashland Inc.

Thixotropy [5]

When long-chain polymers associate intermolecularly, they tend to develop a three-dimensional (3D) structure and exhibit thixotropy. Generally, solutions of medium- and high-viscosity CMC of DS between 0.9 and 1.2 are pseudoplastic rather than thixotropic. In contrast, medium- and high-viscosity cellulose gums

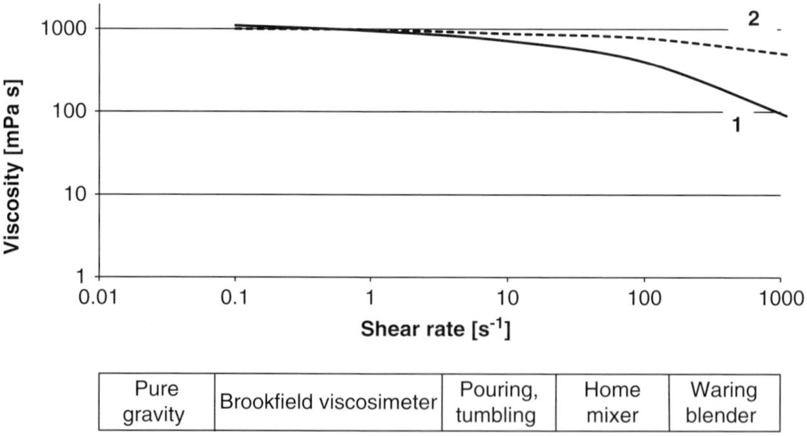

Figure 10.6 Effect of shear rate on apparent viscosity of Na-CMC-solutions with DS 0.7 (1 = 1% high-viscosity CMC; 2 = 10% low-viscosity CMC) [2]. Reprinted with kind permission of Ashland Inc.

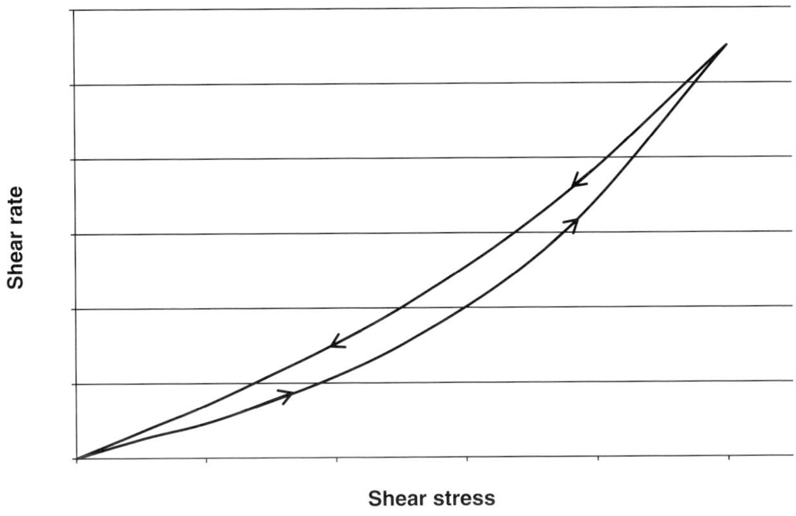

Figure 10.7 Thixotropic flow [2]. Reprinted with kind permission of Ashland Inc.

of DS 0.4–0.7, especially those that are slightly less uniformly substituted, show thixotropic behaviour in solution.

Thixotropy is a time-dependant viscosity change. It is characterized by an increase in viscosity when a solution remains at rest for a period of time. In certain cases, the solution may develop gel strength or even set to an almost solid gel. If sufficient shear stress is exerted on a thixotropic solution, the structure is broken and viscosity is reduced. Thixotropic solutions show hysteresis loops when shear rate is plotted against shear stress (Figure 10.7). The increased shear stress required to break the thixotropic structure reduces the resistance to flow.

If a solution has gel strength, a spur is seen in the hysteresis loop (Figure 10.8). This indicates the stress necessary to break the gel structure and cause the solution to flow with its normal apparent viscosity. Whereas a pseudoplastic solution instantly reverts to its at-rest viscosity after shear removal, a thixotropic CMC solution requires time for return to its at-rest state.

Thixotropic solutions are often desirable or essential for certain uses of cellulose gum, such as for suspension of solids. High- and medium-viscosity types of CMC (with DS 0.4–0.7) generally exhibit thixotropic behaviour.

The factor responsible for the thixotropic nature of most cellulose gum solutions is the non-uniformity of substitution. These naked regions or 'gel centres' tend to associate, forming a 3D network or structure that produces an increase in viscosity. Shear forces can break the structure. Some researchers conclude that thixotropy in CMC solutions arises because of the presence of small quantities of unsubstituted crystalline residues in the CMC. These regions would be present as fringed micelles that could form crosslinking centres, entrapping various amounts of molecularly dispersed CMC and effecting formation of a 3D structure.

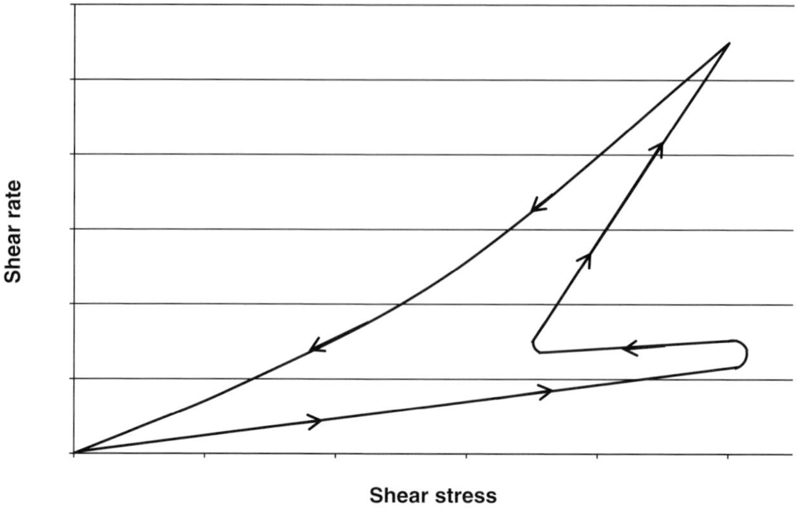

Figure 10.8 Extremely thixotropic flow with gel strength [2]. Reprinted with kind permission of Ashland Inc.

Thixotropy is concentration dependant. The more hydrocolloid is in solution, the more crowding occurs, which enhances the magnitude of the thixotropic increase. The reason why high-viscosity low-DS type cellulose gums generally display thixotropy is because these species have the greatest amount of unsubstituted or insoluble regions in the cellulose backbone.

Some CMC applications require a polymer solution having few or no thixotropic properties. Evenly substituted, high-DS polymers of medium and high viscosity are satisfactory here.

Thixotropic CMC grades have a rough irregular flow, non-thixotropic CMC solutions flow very uniformly, fibre-free, and smoothly.

10.3.1.7 Influence of Solvent Mixtures

The behaviour of highly substituted CMC in heterogeneous or mixed solvent systems such as a glycerine–water mixture is similar to that of CMC in pure water. The course of the viscosity curve in the solvent mixture is identical to the viscosity development in water. However, the viscosity of the individual fractions in the solvent mixture controls the resulting viscosity and thickening performance of the CMC.

Example

If a solvent system of glycerol–water in the ratio 60 : 40 is used, which has a mixing viscosity of 10 mPa s, then the viscosity thereof is about tenfold higher than in pure water. When CMC is homogeneously dispersed or dissolved in the glycerol–water mixture, the final viscosity is also a factor of ten higher than in the analogous dissolution in pure water.

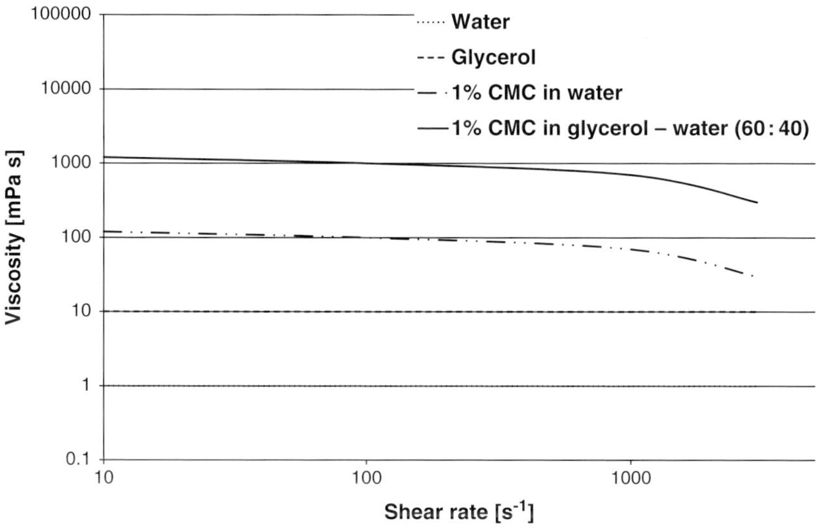

Figure 10.9 Effect of solvent mixtures on the dynamic viscosity of CMC [2]. Reprinted with kind permission of Ashland Inc.

Figure 10.9 shows the effect of solvent mixtures on the thickening effect of CMC as a function of the shear rate.

Table 10.3 provides an overview of the tolerance of carboxymethylcellulose in ethanol–water mixtures. In this series of tests, 95% ethyl alcohol is added slowly with vigorous stirring to a 1% CMC solution. The CMC has a degree of substitution of 0.7.

In these experiments, the tolerance of the CMC to ethanol could be improved by the subsequent addition of the organic solvent to the already completely dissolved CMC.

Short CMC molecules tolerate higher amounts of ethanol in solvent mixtures than long-chain CMC types.

Table 10.3 Compatibility of CMC with ethanol in solvent mixtures [5].

Viscosity level 'CMC'	Volume ratio of ethanol to CMC solution (1%)	
	First visible haze	First significant precipitation
Low	2.4 : 1	3.6 : 1
Medium	2.1 : 1	2.7 : 1
High	1.6 : 1	1.6 : 1

Data taken with permission from Reference [5].

10.3.1.8 Effect of Dissolved Substances on the Viscosity

The effect of dissolved substances such as salts or polar non-solvents on the viscosity of CMC solutions is not only controlled by their concentration but also significantly determined by the order of addition of hydrocolloid (CMC) and dissolved substance.

As illustrated in Figure 10.10, the viscosity development of a CMC is only slightly affected when the hydrocolloid is completely dissolved in pure water first before other substances are added in liquid or solid form to the CMC solution. The viscosity drop is minimized compared to the situation in which the CMC is added to the solute in solution. Conversely, when dry CMC is incorporated into heterogeneous solutions of salts, acids, and/or alkalis, significantly lower final viscosities have to be expected.

For brine this is probably due to an immediate ionic interaction between the salt and the polymer carboxylate group, effectively shielding the anionic charge and substantially inhibiting hydrogen bonding between the anionic site and water molecules [4].

Furthermore, the extra (and already present) solutes prevent the crystalline regions of the CMC molecules of the cellulose backbone being broken up, which would make them available for accretion of water molecules.

The effect of dissolved substances occurs on smaller dimensions, when the substituents (carboxymethyl groups) are evenly distributed in the cellulose chain, because then there are fewer crystalline zones present [2].

Specifically, this is found in higher substituted CMC grades. Thus, a CMC with DS 0.9 is much more stable to salts and acids than a CMC having an average DS of 0.7.

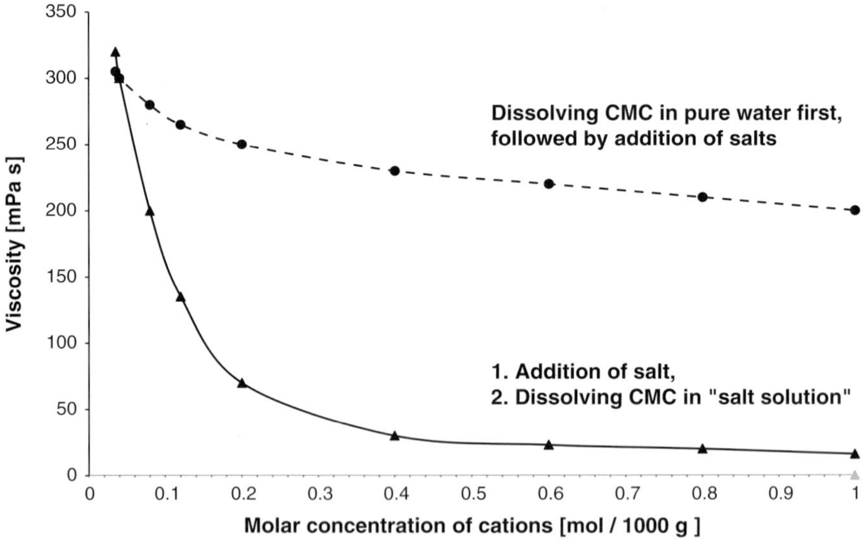

Figure 10.10 Effect of salts and their order of addition on the viscosity of a 1% solution of sodium CMC [2]. Reprinted with kind permission of Ashland Inc.

The course of the solid curve in Figure 10.10, when releasing CMC in a heterogeneous system, shows similar curves in the presence of:

- sodium chloride (NaCl);
- other monovalent chloride salts (KCl or LiCl);
- sodium chloride with caustic soda (NaOH) at pH 10;
- sodium sulfate (Na_2SO_4);
- sodium diphosphate ($Na_4P_2O_7 \cdot 10\,H_2O$) at pH 9.5–9.8 [2].

CMC shows similar behaviour in the presence of acids (H_3O^+ ions). Again, the time point of the addition of acid – besides its concentration – is a critical factor for the development of viscosity. This relationship is considered in detail in Section 10.4.3.

To avoid the problem of reduced viscosity in these situations, it is recommended that the CMC be dispersed in water first and allowed to hydrate. Following polymer hydration, the salt is added to the desired concentration. The probable reason why this is effective is because as CMC hydrates it collects water molecules around itself. These dipoles orientate with respect to the anionic carboxylate groups, partially satisfying and partially shielding them from the cations that are added later [4].

10.3.1.9 Blends of CMC Types

In principle, two grades of CMC with different viscosities may be mixed to obtain a viscosity that is within the range of the two grades. Since the viscosity is an exponential function of concentration, the viscosity of a mixture cannot simply be assumed to be the arithmetic mean.

Example

To obtain a target viscosity of 900 mPa s for 2% CMC solution, 28% of material A (of 2% with 1800 mPa s) have to be mixed with 72% of material B (of 2% with 700 mPa s).

The mixing ratios are often determined graphically and visually by the means of so-called 'blending charts'. An example is in presented in Section 5.3.2.6 (Figure 5.18).

If no blending chart sheet is currently available, the following formula can be used to determine the mixing ratio approximately. From the Arrhenius equation the following relationship between viscosity and polymer concentration can be derived [2]:

$$\text{Log}\, V_S = \frac{n \log V_1 + (100 - n) \log V_2}{100}$$

V_S	=	viscosity searched
n	=	wt% of the first component (1) of the mixture
V_1		viscosity of the first component (1) in the mixture
V_2	=	viscosity of the second component (2) in the mixture

Comments:

- All viscosities must be expressed for a constant polymer concentration and in the same unit of measurement.
- The chemical composition of CMC types used should be similar (comparable DS).
- The viscosities of selected CMC grades should be as close to each other as possible.
- In the application, the viscosity often plays a secondary role alongside other functional properties. The stability is – in many cases (such as beverages, sauces, emulsions/foams, etc.) – not only a function of the viscosity but also a function of the number of CMC molecules present and of their chain length.
- In practice, it is inadvisable to mix different types of CMC to generate isoviscous products because the production safety thereby decreases and basically an unnecessary risk will be taken.

10.3.2
Gelation of Carboxymethylcellulose

Possibilities and Principles for Gelation

As a pure solution in hot or cold water, cellulose gum does not show classical gelation like other cellulose derivatives (MC, HPMC) or gel-forming hydrocolloids (such as starch, gelatine, alginates, carrageenans, pectins) do.

In contrast to methylcellulose or hydroxypropylmethylcellulose, when using CMC there is no increase in viscosity with respect to gelling when heating their aqueous solution. The opposite is the case – high-viscosity CMC solutions exhibit a (albeit thermoreversible) decrease in viscosity when exposed to heat.

In principle, it is not possible to generate sliceable gels in pure water with an average substituted carboxymethylcellulose (DS of 1.0) alone.

Theoretically, there are four options by which to induce a gel formation with sodium CMC:

1) **Using of a 'jelly' CMC:**
   With high concentrations (from 2% on) of low-substituted CMC (with DS ≤ 0.6, optimum DS = 0.3–0.4), very viscous gel-like solutions can be produced using high shear forces, but these gels still have a viscous character and retain their spreadability.

2) **Ionic interaction:**
Ionic polymers can be crosslinked by di- or trivalent counterions. The addition of di- or multivalent cations or salts may cause gelation of the cellulose ether at room temperature. Trivalent cations such as Al^{3+} (as acetate, citrate, or sulfate), Fe^{3+}, and Cr^{3+} are particularly suitable to induce ionic interaction. The resulting gels are not stitch resistant, but are spreadable paste-like, transparent to opaque, and stable at freezing temperatures, and when heated. In the food sector, this interaction is not or hardly used due to the nature of the chemical elements and the high quantities required.
3) **Intermolecular H-bonding:**
Hydrogen-bonded gels can be obtained by lowering the pH of an aqueous solution of cellulose gum. Dispersing CMC in dilute hydrochloric acid (0.1 M HCl) causes formation of a network by replacement of the sodium in CMC with hydrogen. The anionic carboxyl groups ($-COO^-$) are protonated due to the excess of oxonium ions (H_3O^+) and neutral carboxyl groups (–COOH) result. Between the –COOH groups of different CMC molecules, hydrogen bonds are formed and a 3D network structure can develop. Moreover, the neutralized carboxyl groups are less soluble than the negatively charged ions. The hydrogen bonds additionally reduce the solubility of CMC in water. An elastic hydrogel occurs.
4) **Complex coacervation:**
Complex coacervate gels can be formed by electrostatic interaction between a polyanion (like CMC) and a polycation. The basic principle of this gelation type is that polymers with opposite charges stick together, forming soluble and insoluble complexes, depending on pH value of the solution and concentration of solubilized hydrocolloid. For example, anionic CMC can be complexed with cationic chitosan to create such a gel.

Gels from Low-Substituted CMC Grades

The only option by which to produce gels with CMC even without addition of salt is the use of a carboxymethylcellulose with a very low degree of substitution (DS) and additional application of very strong shear forces. The concentration of CMC must be high (at least 2%). The DS must be below 0.6. The gelation is probably the result of disaggregation of the fringed micelles in the crystalline regions of the cellulose chain, which provides more potential crosslinking points. As consequence, 3D network formation is possible. More crystalline regions are present in the low-substituted CMC grades than in the high-substituted types. A US patent (from Nijhoff 1968) uses a CMC with DS 0.35–0.4 for the preparation of gels by high shear forces during homogenization processes. Examples are low-calorie spreads, dressings, and desserts [5].

The gels also differ from those of classical gelling agents such as gelatine and κ-carrageenan. They are of a more viscous nature and spreadable rather than cut resistant. The principle is reminiscent of the use of microcrystalline cellulose, which must also be activated by strong shear forces. In addition, the availability of such carboxymethylcelluloses is low to non-existent, because manufacturers have specialized in the production of fully water-soluble products (from DS 0.6).

Gelation of CMC with Trivalent Cations

For special, and usually non-food, applications the formation of CMC gels with trivalent metal ions is described in more detail here. These gels can have various textures – the whole range from soft, flowable and pourable, to very firm. The latter, though, also remain pasty and spreadable or malleable and never reach gelatine-like sliceable consistency. In the following, the use of aluminium ions in the gelation of CMC is described in detail. Alternatively, trivalent iron (Fe^{3+}) and chromium (Cr^{3+}) are suitable.

The role of chromium(III) in the human body is discussed controversially. There is evidence that Cr^{3+} could have a role in carbohydrate and fat metabolism of mammals. This evidence is currently being studied. Earlier indications that the popular dietary supplement chromium(III) picolinate has a beneficial effect on body composition could not be confirmed. In recent studies with animal cells a mutagenic and carcinogenic effect of this compound has been found. The data currently available suggest that chromium deficiency is extremely unlikely to occur. In addition, a toxic effect from higher doses of Cr(III) is improbable because of the extremely low solubility of Cr(III) hydroxide (6.7×10^{-31}). It is therefore only slightly absorbed in the human gut. In the USA, the recommended intake of Cr^{3+} has been reduced from 50–200 μg per day to 35 μg per day for adult men and 25 μg per day for adult women. Chromium(VI) compounds are extremely toxic. They are mutagenic and damage human DNA. Chromium(VI) enters the body through the respiratory track and damages the lung tissue. The toxic effect increases with the insolubility of the salt. In Europe, its use is legally strongly restricted. Owing to these health risks in combination with missing advantages, the application of chromium for CMC gelation in regulated and technical applications is not recommended.

The salt-induced gelling of cellulose gum is demonstrated by using aluminium ions.

The gradual addition of aluminium ions to a carboxymethylcellulose leads to uniform crosslinking and thus to gelling of CMC.

The network formation is controlled by several factors:

- concentration of polymer (CMC) in the solution;
- ratio of amount of cations to the amount of anionic carboxymethylate groups;
- pH value;
- chain length of the polymer (MW, DP).

Such gels can be stored for more than four years without a change in texture, synaeresis, or a colour change or discolouration.

The rate of gelation and the character of the resultant gel can be controlled by temporary chelation of the trivalent metal ions with anions suitable for this purpose, such as, for example, citrates and maleates. In addition, temperature control and adjustment of pH of the solution influence the gelation. Aluminium-CMC gels are resistant to non-chelating acids. In alkaline media, however, they dissolve slowly.

Cellulose gum gels can be produced that are not changed by prolonged heating and withstand temperatures up to 90 °C (194 °F) for 1-h duration without damage.

If a very long storage time is desired, an additional preservative should be used.

By using slowly soluble monobasic aluminium acetate ($(HO)_2Al(C_2H_3O_2)$) or dissolved aluminium sulfate ($(Al_2(SO_4)_3)$, various gels can be created.

Tables 10.4 and 10.5 present the relationships between concentration of CMC, rate of gelation, and texture of the resulting gel. A CMC with DS 0.7 was used.

Aluminium acetate has the molecular formula is $C_6H_9AlO_6$. There are three salts: neutral aluminium triacetate, basic aluminium diacetate, and basic aluminium monoacetate. The latter is used here.

In summary a high absolute concentration of CMC (≥1%) at a relatively high relative concentration of aluminium ions (CMC : Al^{3+} = 5 : 1) leads to formation of a solid, stiff gel in 1–2 h. By increasing the mixing ratio in favour of CMC to over 10 : 1, the time of gel formation is extended and the gels become softer. In addition, a decreasing polymer concentration slows down the speed of gelation.

In general, the use of a high concentration of a short-chain, low-viscosity CMC, and therefore the selection of less thixotropic grades, leads to more elastic gels.

High amounts of aluminium salts give gels that are brittle and fragile. The characteristic gel properties such as elasticity and brittleness can be modified by the addition of non-ionic hydrocolloids, proteins, salts, and sugars [5].

Table 10.4 Gelation of CMC with aluminium ions (from basic aluminium acetate[a)]) [5].

CMC – DS 0.7		Aluminium salt – concentration			
Viscosity level	**Concentration in solution (% solids)**	**On CMC (%)**	**On total weight (%)**	**Gel texture**	**Time to gel (h)**
High	1.0	20	0.2	Rigid	1–2
High	0.5	20	0.1	Rigid	2–4
High	0.5	7	0.035	Soft	12–24
High	0.5	7	0.035	Soft	1
Medium	2.0	15	0.3	Rigid	0.5–1
Medium	1.5	7	0.011	Firm	2–4
Low	4.0	20	0.8	Rigid	2

a) Aluminum acetate is produced by the reaction of aluminium hydroxide with acetic acid.

Table 10.5 Gelation of CMC with aluminium sulfate solutions ($Al_2(SO_4)_3$) [5].

CMC – DS 0.7			Solution from aluminium sulfate and citric acid[a)]				
Viscosity level	Concentration solution (%)	pH[b)]	Citrate : Al^{3+} (molar ratio)	pH[c)]	$Al_2(SO_4)_3$ addition (%)[d)]	Time to gel formation	Gel strength
High	1.0	7.0	0.33	2.0	16	Immediate	Firm
High	1.0	7.0	0.5	2.0	16	Immediate	Soft
High	1.0	5.5	0.5	6.5	16	1 h	Semi-firm
High	1.0	5.5	0.5	7.0	16	2.5 h	Soft
High	0.5	5.5	0.5	6.5	32	10–15 min.	Soft
High	0.25	6.0	0.5	6.5	64	3–4 d	Very soft
DS 1.2 (medium)	2.0	5.5	0.3	7.0	8	3 h	Semi-firm

a) The aluminium sulfate and citric acid must be combined in solution before addition to the gum solution. Solution concentration is 3.2% aluminium sulfate (as $Al_2(SO_4)_3{\cdot}18\,H_2O$).
b) Solutions are generally neutral. The lower pH values were obtained by addition of hydrochloric acid (HCl).
c) Citric acid–aluminium sulfate solutions are normally at pH 2.0. The neutral pH values are obtained with caustic soda (NaOH). The NaOH is added to the solution just prior to its use.
d) Weight of $Al_2(SO_4)_3$ added as a percentage of the dry gum weight.

10.3.3
Behaviour of CMC with Other Ingredients

Generally, carboxymethylcellulose is highly compatible with other components of the formulation. Dissolved CMC is compatible with most water-soluble non-ionic, anionic and cationic polymers and hydrocolloids. Positive or negative interactions are always to be expected when CMC competes with other substances for the free water available in the system. In principle, Na-CMC is more tolerant to strongly hydrophilic substances such as sugars than other cellulose ethers are. Low-viscosity cellulose gums with a high degree of substitution have the highest tolerance in aqueous media, because their water demand is low. The longer the chain of cellulose with low DS is, the more strongly the solubility of CMC is influenced by other substances.

The following sections explain in detail the effects of several ingredients and additives that are part of food formulations. The interactions of salts of polyelectrolytes with CMC are special and unique compared to their impacts on other hydrocolloids, and depend on various factors which are discussed below.

Sugar

In principle, carboxymethylcellulose shows a higher tolerance to high sugar concentrations than other cellulose ethers (MC, HPMC).

CMC produces in sugar solutions a significantly higher viscosity than would be expected from the viscosities of the individual components. The thickening effect of a CMC–sugar mixture in water is higher than the increase in viscosity through single components (synergism). This synergistic effect on viscosity is probably caused by the formation of hydrogen bonds between the sugar molecules and the free hydroxyl groups of the carboxymethylcellulose chain. The kind of impact CMC has on sugar solutions is determined by the order of addition and the method of preparing the solution.

The greatest thickening result is obtained when the sugar is added to an already completely dissolved CMC.

When dry carboxymethylcellulose powder is dissolved in a sugar solution, the viscosity development may not only be delayed but also reduced, because there is not sufficient free water available for complete hydration of CMC. The CMC will then possibly not be fully dissolved and partially still exist in the system in form of dry particles. The solution appears opaque and cloudy.

Special CMC products with a very high degree of substitution ($DS \geq 1.2$) are more tolerant of 'water shortage' and may even be dissolved satisfactorily in solutions with a low content of free water. These carboxymethylcellulose grades offer a similar viscosity profit in mixed solvent systems of water and sugar to that detected in glycerol–water mixtures (see Section 10.3.1.7 on the 'Influence of solvent mixtures' [5].

The increased tolerance of these highly substituted CMC products occurs not only in sugar solutions but also in the presence of salts or polyelectrolytes.

Sugar can be used for premixing powdered CMC to avoid undesired aggregate or lump formation during transfer of the CMC into water.

CMC can be used to thicken sugar syrups with more than 50% sugar (dry matter) – without precipitation of sugar crystals. In sugar syrup with 60% sugar, the limit of complete solubility is reached when 1% CMC of a viscosity level of 10 000 mPa s (measured at 2% CMC concentration at 25 °C (77 °F) with a Brookfield spindle) is added. CMC grades with higher thickening effect must, accordingly, be dosed lower. CMC products having a shorter cellulose chain can be used in higher concentrations.

Figure 10.11 illustrates the effect of sucrose on the viscosity of an aqueous carboxymethylcellulose solution. In this experiment, the CMC is premixed with sugar and then dissolved in water at neutral or acidic pH value. The thickening effect is not proportional. The higher the concentration of CMC, the higher is the factor with which the viscosity of CMC in pure water has to be multiplied to obtain the specific viscosity of CMC in the sugar solution. The reason for this is – as with CMC alone, the so-called 'polymeric effect'.

CMC operates in sugar-containing systems as a crystallization inhibitor. In supersaturated sugar solutions, carboxymethylcellulose delays the crystallization of sugar. In addition, the growth of sugar crystals (rate of crystallization) is slowed down, so that there are fewer large crystals. Low-viscosity low-substituted CMC types are more effective as crystallization inhibitors than the long-chain grades. High-viscosity cellulose gums are preferred if water loss by vaporization or evaporation is to be counteracted. In addition, dried sugar films (e.g. as coatings of bakery and confectionery products) are more stable with the addition

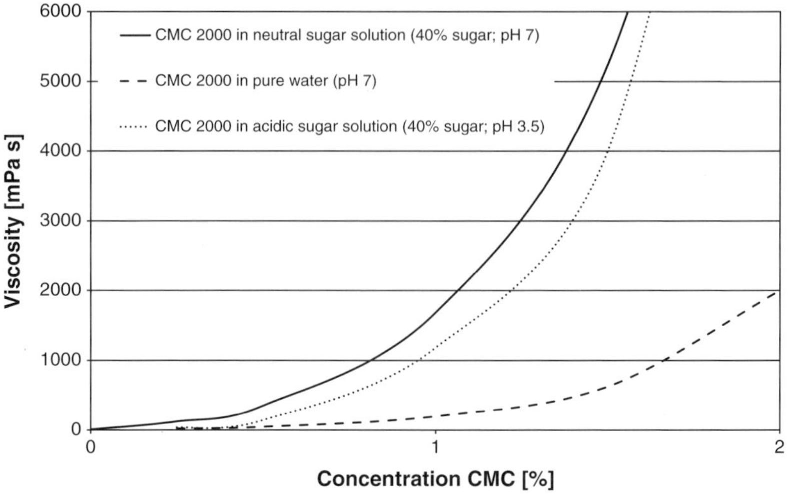

Figure 10.11 Course of CMC viscosity with sugar and acid addition (reduction of pH).

of small amounts of short-chain cellulose gums. They break less quickly and have finer crystal sizes giving a smooth glaze.

Salts and Polyelectrolytes

The effect of salts on CMC depends on the:

- degree of substitution and degree of polymerization of CMC;
- nature with respect to valence of the added ions and their concentration;
- production process and the kind of contact of the CMC with the ions;
- type and duration of applied shear force;
- pH of the system.

As general rules [5]:

- Monovalent cations usually form soluble salts with carboxymethylcellulose. They have very little effect on the viscosity of the solution, on clarity, and other properties at low to moderate salt levels.
- Divalent cations show different reactions and often lead to borderline situations – flocculation or turbidity is possible due to complexation. The viscosity of these solutions is generally reduced, since part of the CMC is 'inactive', being complexed as a less soluble salt with the cations.
- Trivalent cations often react with CMC to form insoluble salts.
- Dissolved carboxymethyl cellulose is less sensitive to salt addition than dry CMC powder, which is incorporated in saline solutions (Figure 10.10).
- The higher the degree of substitution of the cellulose gum, the lower is the effect of salts (see Figures 10.12–10.15 below).

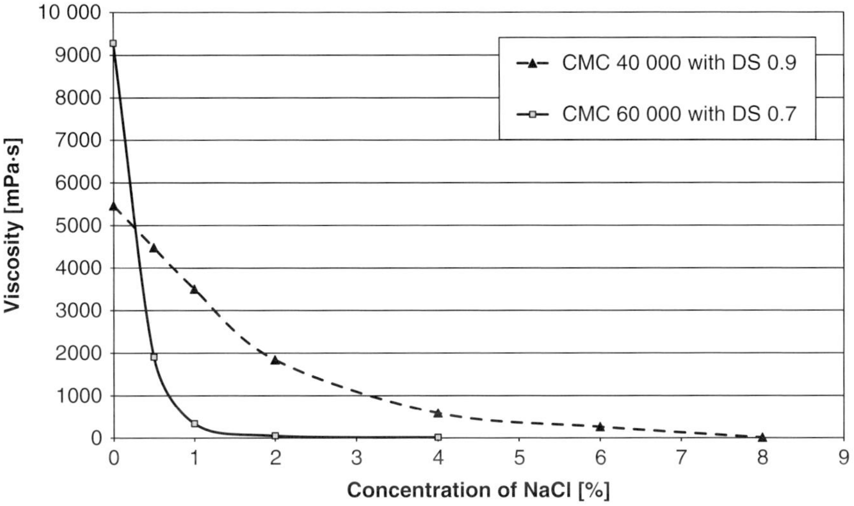

Figure 10.12 Effect of salt (NaCl) on viscosity of cellulose gum.

- Low-viscosity CMC grades show higher compatibility over a wider concentration range than medium- and high-viscosity carboxymethylcellulose.
- The additional application of shear forces improves the viscosity development (shear treatment ↑ → viscosity ↑).

Table 10.6 establishes a relationship between polymer composition, salt concentration, and mechanical shear forces in the presence of monovalent ions (NaCl).

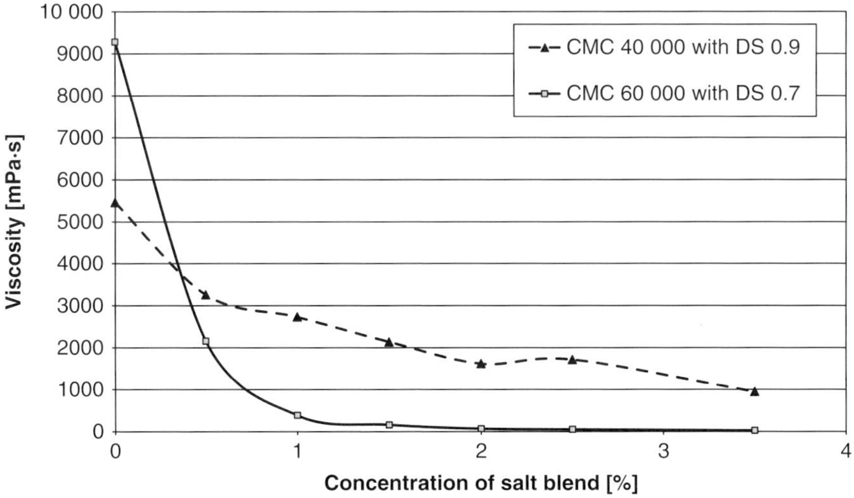

Figure 10.13 Influence of a salt mixture on the viscosity of a CMC solution (1%). Salt mixture in wt%: 7% $NaNO_2$ + 71% NaCl + 22% Na_2HPO_4.

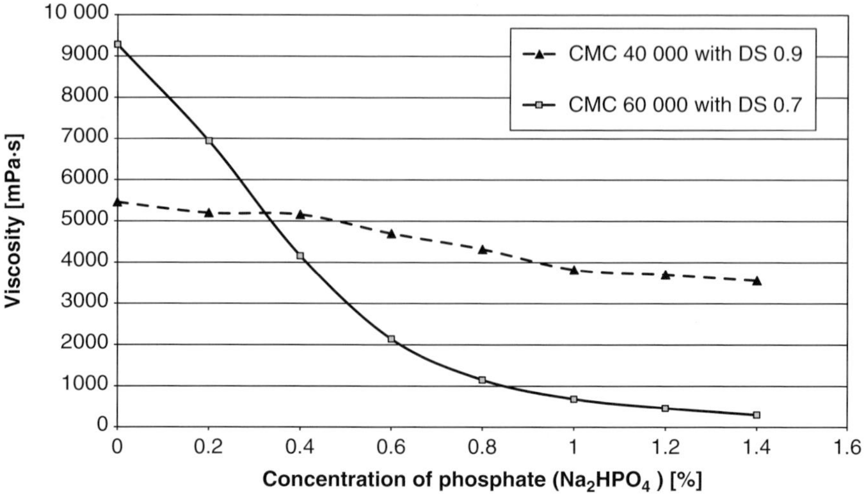

Figure 10.14 Influence of phosphate (Na_2HPO_4) concentration on viscosity of a CMC solution (1%).

The CMC grades used differ with respect to degree of substitution (DS) and degree of polymerization (DP) or molecular weight. The resulting viscosities are determined with a Brookfield spindle viscometer at 25 °C (77 °F). The carboxymethylcellulose is incorporated in dry form into the salt solutions.

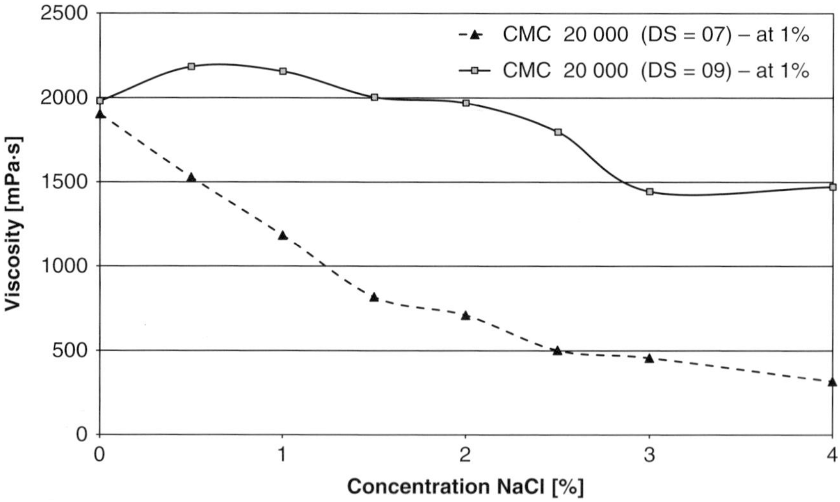

Figure 10.15 Impact of the degree of substitution on the salt stability of isoviscosity CMC grades.

Table 10.6 Factors influencing the disaggregation of CMC [5].

CMC type	Viscosity (mPa s)					
	Anchor stirrer			Waring blender		
	Distilled water	4% NaCl	Saturated NaCl	Distilled water	4% NaCl	Saturated NaCl
DS 0.4 (high viscosity)	900	11	6	4000	65	16
DS 0.7 (high viscosity)	1680	140	45	760	1040	2440
DS 0.9 (high viscosity)	1680	570	165	760	750	1720
DS 0.9 (medium viscosity)	215	160	225	125	95	235
DS 1.2 (medium viscosity)	175	80	180	100	55	140

Table 10.7 shows the compatibility of CMC with dissolved inorganic salts.

A 1% solution of a medium- to high-viscosity CMC (type 2000 mPa s at 1%) is prepared; 1 g of this 1% solution is added to 15 g of a salt solution containing 10% or 50% (saturated) concentration of salt. Thus, 1 g of CMC is faced with 150 or 750 g of salt.

Singly charged cations (Na^+, K^+) usually react with cellulose gum to form water-soluble salts.

The viscosity of CMC is largely controlled by the order of addition of CMC and salt. If the cellulose ether is sufficiently dispersed in water and completely hydrated first, then a subsequent dose of monovalent ions has only a minimal impact on viscosity. CMC dissolved and fully hydrated is, therefore, relatively salt stable.

When CMC is added dry to a salt solution, the cations present suppress the development of viscosity. The reason for this is that the ions inhibit the loosening of the polymers, the so-called disaggregation, and thus reduce the rate of dissolution. More information can be found in Section 10.3.1.8 ('Effect of dissolved substances on the viscosity').

Highly substituted CMC (DS of 1–1.25) is more 'resistant' to the influence of salts than CMC grades with a low degree of substitution. At constant DS, short-chain carboxymethylcelluloses tolerate higher amounts of added ions or salts than medium- and long-chain CMC products. The dilution effect is 100% reversible. After removal of the ions, the viscosity develops accordingly.

This salt sensitivity is used in the manufacture of injection brines. In the highly concentrated solution for injection, the CMC does not thicken the aqueous phase when salt concentrations (as the sum of sodium chloride, phosphates, and nitrates) of 5–15% are present. The solution has a low viscosity and can be

Table 10.7 Compatibility[a)] of CMC with inorganic salt solutions [2,5].

Salt	10% Solution	50% or Saturated solution
Aluminium nitrate	*P*	*P*
Aluminium sulfate	*P*	*P*
Ammonium chloride	**C**	**C**
Ammonium nitrate	**C**	**C**
Ammonium sulfate	**C**	*P*
Calcium chloride	**C**	*P*
Calcium nitrate	**C**	*P*
Chromic nitrate	*P*	*P*
Disodium phosphate	**C**	**C**
Ferric chloride	*P*	*P*
Ferrous chloride	*P*	*P*
Ferric sulfate	*P*	*P*
Potassium ferricyanide	**C**	**C**
Potassium ferrocyanide	**C**	**C**
Magnesium chloride	**C**	**C**
Magnesium nitrate	**C**	**C**
Magnesium sulfate	**C**	**C**
Sodium carbonate	**C**	**C**
Sodium chloride	**C**	**C**
Sodium dichromate	**C**	**C**
Sodium metaborate	**C**	**C**
Sodium perborate	**C**	**C**
Sodium sulfate	**C**	*P*
Sodium sulfite	**C**	**C**
Sodium thiosulfate	**C**	**C**
Silver nitrate	*P*	*P*
Stannic chloride	*P*	*P*
Zinc chloride	*P*	*P*
Zinc nitrate	*P*	*P*
Zinc sulfate	*P*	*P*

Data taken with permission from References [2,5].
a) **C** = compatible, *P* = precipitation (not compatible).

mixed with other hydrocolloids. After injection, the CMC builds up viscosity slowly because there is a much lower salt concentration present in the application system (e.g. in the meat piece). The formulation for brines is presented in Section 10.9.6.2 ('Injection brines').

Figure 10.12 shows the effect of the reversible salt dilution of CMC in saline solutions of different concentrations. In addition, it is shown how the degree of substitution influences the salt stability. The data are based on the following experimental setup:

CMC granules (concentration of 1%) are mixed slowly into a solution of NaCl. After sufficient hydration and dissolution time, the viscosity is measured with a Brookfield spindle.

As illustrated in Figure 10.12, an extreme effect is observed using low-substituted CMC (DS $\leq$ 0.7). The degree of polymerization (molecular weight) is no longer relevant from a salt content of 0.5% sodium chloride and higher.

Figures 10.13 and 10.14 show how a 1% CMC solution reacts in contact with salts other than sodium chloride. The experimental setup is the same as for Figure 10.12.

Figure 10.15 gives an idea of the effect of the substitution degree when using isoviscosity CMC grades. The experimental setup corresponds to that in Figures 10.12–10.14. This study confirms that a higher degree of substitution results in a much improved salt stability.

The salt sensitivity of cellulose gums provides the opportunity for process control.

Over the storage period, the viscosity of CMC builds up in salt solutions, too. The cellulose ether is not destroyed by transfer into the brine, but the NaCl solution will slow down its hydration, because the amount of free water is low and the osmotic pressure is high. This effect can be initiated by sodium chloride, phosphates like Na_2HPO_4, and nitrates (e.g. $NaNO_2$) or also by mixtures of these salts.

Divalent cations do not form crosslinked gels with the anionic CMC, but the cellulose derivative can be complexed. Such complex formation leads to a reduction in viscosity and can also be seen by the presence of a haze. Cations of calcium, magnesium, barium, cobalt, magnesium, ferrous(II), and manganese(II) induce these reactions with CMC.

Precipitation is observed only with heavy metals.

CMC products with a very high degree of substitution (DS $\geq$ 1.2) are only slightly affected by moderate concentrations of divalent cations, when these ions are added only after the dissolution of CMC in water [5].

Trivalent cations lead to the precipitation of CMC as insoluble salt or to its gelation (see also Section 10.3.2.3). The formation of aggregates is caused by crosslinking, ionic bonding, or complex formation.

This precipitation is triggered by the trivalent ions Al^{3+}, Cr^{3+}, and Fe^{3+} as well as by salts from silver, copper, ferric, ferrous, zinc, lead, and zirconium.

The effect of trivalent cations on dissolved CMC can be used for the preparation of gels. The slow addition of trivalent ions – such as aluminium ions (Al^{3+}) – gives a uniform crosslinking of individual CMC molecules with each other through their carboxymethyl groups.

The strength of the gels increases with:

- higher concentration of CMC;
- higher molecular weight and chain length of the CMC;
- increasing concentration of trivalent metal ions;
- decreasing pH.

Since the crosslinking salts do not occur naturally in foods at all, or are not present in sufficient quantities, and are not allowed to be applied for human consumption, the resulting gels have only indirect technical uses. Applications are gel pads for cooling or keeping warm.

Fats/Oils

The compatibility of cellulose gum with fats and oils is good. Short-chain and medium-viscosity CMC can also be incorporated into products with high fat and low water contents.

Owing to its film-forming properties, dissolved CMC leaves an oily impression on the tongue. In combination with the high water binding capacity and good water retention, this sensory behaviour makes CMC a suitable replacement for fats and oils in many recipes.

CMC is not soluble in oil. Therefore, CMC–oil dispersions can be used as low-viscosity but highly loaded precursors in food manufacturing (e.g. emulsion preparation).

Films of dried CMC solutions are used as a barrier against fat in some applications (see Section 10.4.4 on 'Films of CMC').

Milk and Soy Proteins

CMC shows unique behaviour in the presence of proteins. Proteins are polymers of different amino acids, which occur in many different forms. From their numerous properties, the sensitivity of proteins to pH value changes and heat effects are the most important characteristics, because they severely limit the capabilities and application possibilities of the polypeptides. Through its anionic molecular character, CMC can react with several proteins to form soluble and stable complexes when the protein is in its isoelectric pH range. Although the effect on the system is controlled primarily by the pH, the interactions also depend on the type of protein and its concentration, the processing temperatures, and the type and concentration of CMC. At low temperatures and at pH values below 3 and above 6, conventional CMC grades react with milk proteins (caseins) to afford insoluble complexes that may be removed from the system as precipitated agglomerates. In the pH range 3.0–5.5, stable soluble complexes are formed.

The maximum stability is characterized by an abnormally high viscosity compared to the viscosities of the individual components.

As long as the pH of the system is greater than the isoelectric point of the protein, there is a synergistic activity driven by ionic interactions between the anionic CMC chain and the cationic protein chains, generating a higher viscosity than otherwise expected [4].

CMC–casein complexes are relatively sensitive to shear forces. During stirring, a reversible decrease in viscosity is observed. The complexes are stable to heat, but a slightly reduced viscosity is observed while heating up. The caseins denature to a much lesser extent than in systems without CMC [6].

To extend the functionality of carboxymethylcellulose, special acid-stable variants of CMC have been developed in recent years, which are highly effective even at pH 3.0 and lower.

The origin of the protein does not matter – vegetable raw materials such as wheat and soy protein are also stabilized like animal proteins from milk (sodium and calcium caseinate, whey protein), egg white, or gelatine.

The reaction between water-soluble proteins and sodium carboxymethylcellulose is primarily anionic in nature, although other stabilizing forces are conceivable.

The absence of interactions between proteins and non-ionic hydrocolloids or uncharged cellulose ethers (such as methylcellulose, HPMC, or HPC) is sound evidence for ionic, namely, anionic–cationic, interactions between Na-CMC and acidic proteins.

The mechanism of protein stabilization by CMC therefore presupposes the existence of cationic groups in the proteins, as is the case, for example, with the acidification of milk to pH values of 4.3 and less. At pH values below the isoelectric point (IEP) of the respective milk proteins (pH $[IEP_{Caseins}] = 4.9–5.4$; pH $[IEP_{Whey\ proteins}] = 4.2–5.1$), their amino groups ($–NH_2$) are protonated through excess oxonium ions (H_3O^+) and thus have a positive charge ($–NH_3^+$). Between these proteinogenic cationic amino groups and the anionic carboxymethyl groups ($–COO^-$) of CMC, electrostatic forces of attraction are generated. At pH 4, about 25% of the carboxymethyl groups of CMC with DS 0.8 are still ionized so that there are enough negatively charged groups left to interact with the cationic proteins. Increasing the pH to 5 will increase the number of ionized carboxymethyl groups in this specific CMC to about 50% and thus enhance the stabilization effect for proteins, which in turn also have fewer positive charges at a higher pH value.

It is supposed that the relatively small CMC molecules form a protective layer on the surface and around the proteins and thus prevent repulsion of the positively charged proteins, which would cause their precipitation as larger and visible aggregates.

Figure 10.16 illustrates the operation of stabilizing proteins with carboxymethylcellulose.

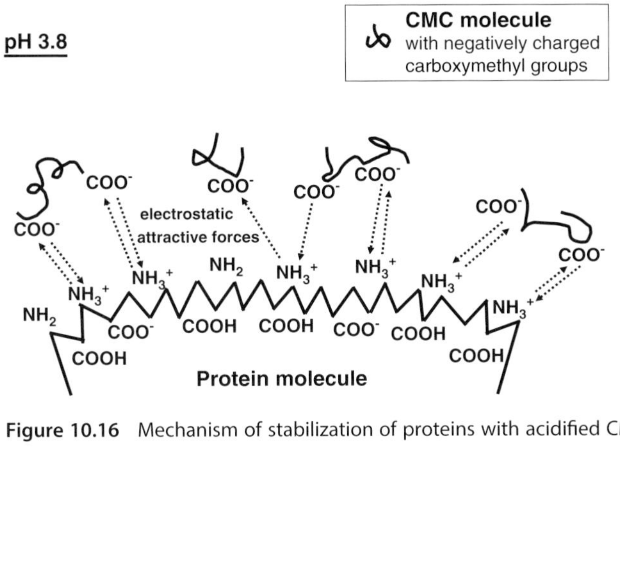

Figure 10.16 Mechanism of stabilization of proteins with acidified CMC.

The shorter the cellulose chain of CMC, the better is the protection effect. However, the effect of CMC as a protective colloid for proteins is completely independent of the lengths or shape of the polymers and the resulting viscosity. Optimal stabilization of whey drinks is achieved at higher concentrations (1.5–3%) of low-viscosity CMC grades (2% = 30 mPa s). Medium-viscosity cellulose gum (2% = 100–2000 mPa s) with a DS of 0.7–1.0 stabilizes cationic milk and soy proteins in acid dairy and non-dairy drinks.

The following examples explain the effect of CMC on the solubility of soy proteins: A 0.5% solution of a high-viscosity CMC with DS 0.7 provides solubilization of an equal amount of soy protein in its isoelectric pH range, as evidenced by a large increase in apparent viscosity (Table 10.8). This interaction between cellulose ether and soy protein is dependent on both the degree of substitution and the degree of polymerization of Na-CMC. To stabilize a system with 0.5% soy protein, or rather maintain the protein in solution, a 0.5% solution of a long-chain CMC with DS 0.4 is required. Using a medium-viscosity CMC with DS 1.2, the concentration must be increased to 1.5% to achieve a comparable solubilizing effect. A low-viscosity CMC with DS 0.7 does not solubilize the soy protein; it would aggregate and precipitate. This indicates that the chain length of the cellulose backbone has an important role in the physicochemical protein stabilization with CMC. If the acidic protein system is sufficiently stabilized at room temperature by CMC, and stable CMC–protein complexes are formed, subsequent heating neither insolubilizes the protein nor irreversibly dissociates the complex [5].

Table 10.8 elucidates the relationship between pH, protein solubility, and CMC addition. The measurements are based on using a high viscosity, non-thixotropic carboxymethylcellulose with DS 0.7.

The reduction of CMC viscosity with decreasing pH in pure water (Table 10.8) is typical and caused by the irreversible neutralization or protonation of the anionic carboxymethyl groups. The viscosity decrease turns into a strong increase in thickening when the pH is reduced further to 3.0 or below (see Section 10.4.3.4).

Table 10.8 Interaction between CMC and soy proteins [5].

pH	Viscosity of an aqueous solution (mPa s)		
	0.5% Soy protein	**0.5% CMC**	**0.5% Soy protein + 0.5% CMC**
6.9	3	180	160
6.5	3	160	160
5.9	3	160	170
5.6	Precipitation	125	330
4.7	Precipitation	96	990
4.2	Precipitation	82	920
3.5	Precipitation	62	600

The protective effect of CMC at acidic pH below 5.6 can be seen in the high viscosity values which verify the solubilization of soy protein by complexation. At higher concentrations of soy protein, a suspending action is evident for the gum and/or CMC–protein complex [5].

The reaction of caseins with CMC is similar to the interactions between soy protein and CMC. Carboxymethylcellulose inhibits precipitation of casein in its isoelectric pH region and produced high viscosities by complexing. Casein is apparently more readily solubilized by CMC than is soy protein. The increase in viscosity is more pronounced with sodium caseinate than with calcium caseinate as the pH value decreases [5].

In summary, carboxymethylcellulose has the ability to solubilize proteins in their isoelectric pH ranges. For example, acidified milk proteins and soy proteins are protected against the effects of heat. Flocculation of the caseins, whey, and soy proteins is prevented by CMC due to steric repulsion forces.

Other hydrocolloids do not show these positive interactions with proteins or are restricted. Guar cannot keep proteins at the IEP in solution, and heat protection is not given, too. Sole stabilization by guar is insufficient. It is based on reduction of the reaction rate by increasing the viscosity of the aqueous phase. The same goes for xanthan, starches, and locust bean gum. It is known that κ-carrageenan is able to stabilize milk proteins excellently (especially α-S_2- and κ-casein) at very low concentrations when the liquid has a neutral pH range of 6.5–7.5. With acidified proteins, the hydrocolloid exhibits, indeed, a solubilizing effect in the IEP area, but no heat protection.

Most hydrocolloids do not offer a permanent and viscosity-independent stabilizing effect through electrostatic interactions. If the system is controlled only on the basis of a viscosity increase of the aqueous phase, then mechanical separation, flocculation, and phase separations often occur when diluting concentrates or during the thermal treatment of foods.

Pectin (highly esterified grades) is traditionally used as a stabilizer in acidified milk products. The stabilizing mechanism of HM pectin is similar to that of CMC – via electrostatic attraction forces. The acidic groups in pectin are stronger ($pK_a = 2.9$–3.3) than in CMC ($pK_a \approx 4.5$). These values do not exactly reflect the whole polymer composition because the total amount of acidic groups available is important. The following relationships have to be considered. HM pectin has less free carboxy groups than LM pectin, but the latter gels in the presence of calcium ions from milk and therefore is not suitable for liquid dairy products. CMC products with lower DS have fewer free carboxymethyl groups than high-DS grades. At pH 4, more than 90% of the free carboxyl groups in HM pectin are ionized, while at pH 5 all free carboxyl groups (>99%) are ionized. From the ionization status, a higher pH would be better for the pectin–protein interaction, but at $pH \geq 5$ pectins are no longer stable and start to degrade. Pectin chain degradation occurs due to β-elimination, whereby the glycosidic bonds of the pectin backbone are cleaved, consequently reducing the molecular weight. Therefore, the stabilization of acidified milk proteins with HM pectin is limited to a narrow pH interval. The ionization values for CMC are 25% of free carboxymethyl groups at pH 4 and 50% at pH 5. At pH 4 and above, both pectin and CMC have enough ionized groups to interact with cationic proteins. At pH <4, it is important to use a CMC grade with a high degree of substitution (DS ≥ 0.85). CMC has greater ability to stabilize proteins at pH >5 than pectin.

The above considerations show that CMC is a valuable alternative to HM pectin in acidified dairy products of liquid consistency. After introduction of the alternative name 'cellulose gum' for CMC, the cellulose ether is increasingly used in food products, particularly in dairy applications. The advantage of CMC is found in the more pH-independent mode of action. This makes a subsequent correction of acidity for sensory reasons possible. In addition, CMC can be used in homogenization processes with high pressures of 250 bar (3625 psi) without problems such as molecular degradation.

Egg White and Gelatine [5]

Carboxymethylcellulose is used commercially to stabilize egg white for drying or freezing processes. The functionality of the egg protein is enhanced by CMC.

Gelatine can be modified in its physicochemical properties by addition of CMC. Gelatine – being in principle only hot-water soluble as a native extract product – even becomes soluble at lower temperatures by 'mixing in' of the cold-water-soluble cellulose derivative.

Water-Soluble Polymers – Other Hydrocolloids (Except Pectins)

CMC reacts compatibly with most water-soluble polymers (hydrocolloids) over a wide concentration range.

Generally, low-viscosity carboxymethylcellulose tolerates higher concentrations of other polymers than does long-chain CMC types, because its water demand is low(er).

The quantity of available water is usually the limiting factor for the solubility and the resulting functionality of CMC as well as the effectiveness of the other hydrocolloids employed.

The compatibility of cellulose gum with other non-gelling thickeners, such as xanthan, guar, and LBG is good. With guar, CMC can cause a viscosity profit (synergistic effect).

In starch systems, CMC is used to increase viscosity economically, as well as control synaeresis at all temperature levels (from frozen to boiling). Cellulose gum provides viscosity to starch–water slurries prior to gelatinization of the starch [5].

With gel-forming hydrocolloids, such as starches, alginates, agar, and κ-carrageenan, a positive effect on the stability of the resulting gel by CMC is observed. For more in-depth information see Section 10.6 'Synergies of cellulose gum with other hydrocolloids'. Negative interactions between CMC and HM pectin were found, which are described in the subsequent subsection. Further incompatibilities of CMC with other hydrocolloids are not known.

HM Pectin

Negative interactions can occur between CMC and highly esterified pectins. The anionic CMC molecules react with the negatively charged molecules of HM pectin to form insoluble complexes, which then precipitate. As mentioned in Chapter 1, pectin has – like CMC – dissociated carboxyl groups ($-COO^-$), which are partially esterified with methanol. Because of the similar structure of the side groups of these hydrocolloids, electrostatic repulsion forces arise that result in the precipitation of the substances. Since the function of both hydrocolloids and their effect as a protective colloid for cationic proteins is very similar, it is recommended to only use them separately and not in combination.

Other Cellulose Derivates Such as Non-ionic Cellulose Ethers and MCC

CMC can be combined with non-ionic cellulose derivatives (such as methylcellulose, HPMC, MEC, HPC, HEC) without limitation.

Negative interactions and undesired 'reactions' are not known and are not expected due to the inactivity or inertia of neutral cellulose.

When isoviscosity products (e.g. carboxymethylcellulose and hydroxypropylcellulose) are mixed together, a synergistic effect in viscosity can be observed in the mixture. It does not matter whether the two cellulose derivatives are first dry blended together as powders and then dissolved as powder mix or if already hydrated cellulose ether solutions are brought together.

The presence of salts and electrolytes in the system reduces the synergistic effect.

Table 10.9 shows the resulting viscosities when mixing anionic CMC with non-ionic cellulose derivatives (HEC, HPC). In both examples, a significant gain in viscosity is observed.

Colloidal MCC is produced by the co-processing of microcrystalline cellulose with carboxymethylcellulose with the aim of increasing the swellability or water absorption capacity of MCC significantly. CMC thus improves the functionality of the microcrystalline cellulose in aqueous systems, while the largely native cellulose structure of MCC ensures the 'stability' of the CMC to heat. Through this positive effect, diverse application options for MCC–CMC mixtures are possible.

Table 10.9 Synergistic effects on the viscosity of blends from CMC with non-ionic cellulose ethers [5].

Polymer	Viscosity of a 1% solution at 25 °C (77 °F) (mPa s)	Viscosity of a '1 : 1' blend (1%), solution at 25 °C (77 °F) (mPa s)	
		Calculated	Measured
High-viscosity CMC (DS 0.9)	1500	1650	3200
High-viscosity HEC	1800		
High-viscosity CMC (DS 0.9)	1500	1570	3280
High-viscosity HPC	1640		

10.4 Stability

10.4.1 Microbiological Attacks

CMC has high biological and physicochemical stability.

It is more resistant to microorganisms than many other water-soluble hydrocolloids. Its solutions, though, are not immune to microbial attack.

By moderate heat treatment (pasteurization) many microorganisms can be destroyed without affecting the properties and functions of the CMC significantly.

Heat treatment at 80 °C (176 °F) for 30 min, or alternatively 1-min at 100 °C (212 °F), is generally sufficient to achieve adequate storage stability.

If pure CMC solutions are to be stored, the use of a preservative is recommended to prevent degradation of viscosity. The preserving substances should be added directly after preparation of the solution to stabilize it against possible subsequent entry of cellulase-containing microorganisms.

Cellulases are hydrolytic enzymes that cleave the cellulose chain and thus reduce the viscosity dramatically. Even the presence of minimal microbial traces, which contain highly active cellulases, causes in non-conserved systems – even after subsequent addition of a preservative – rapid degradation of the cellulose chain and an irreversible loss of the thickening power (see also Section 5.4.2 on the microbiological stability of methylcellulose). Cellulase enzymes in water solutions can usually be destroyed by heating for 15 min at 50 °C (122 °F) and pH 9–10.

In the past, problems caused by cellulases were rarely detected.

Potential contaminant sources with some cellulase activity are:

- insufficiently purified xanthan,
- non-blanched vegetables.

Suitable preservatives for cosmetics and medicines include:

- sodium benzoate
- sodium propionates
- sorbates (Na and K salts)
- phenol
- methyl and propylparabens
- formaldehyde [2,5].

Only sorbates, benzoates, and propionates are suitable for human consumption.

In the production of pure CMC solutions for re-use, the heating method is generally preferred.

10.4.2
Chemical Degradation

Under certain conditions, aqueous CMC solutions are exposed to chemical degradation. A permanent loss of viscosity occurs when long-chain CMC molecules are 'cut'. The irreversible loss of viscosity is accelerated by increasing the temperature, as with most chemical processes. In addition, a reduction of the pH value has a positive effect on the degradation.

Some manufacturers have developed acid-stable versions of carboxymethylcellulose that show improved properties over conventional CMC products in dissolved form in acidic systems. These CMC grades are characterized by a good clear solubility, high transparency, and constant viscosity over the storage period even in highly acidic applications (with pH < 3.0).

An oxidative degradation of CMC can be observed under alkaline conditions in the presence of oxygen. The extent of viscosity loss is enhanced by additional heating and/or by ultraviolet light.

Preventive actions are in this case the addition of an antioxidant, exclusion of oxygen, and prevention of a highly alkalized system.

CMC solutions have the best stability and a long shelf life under the following conditions:

- protection or absence of microorganisms and active enzymes (cellulases);
- neutral pH (pH 6.5–8.0);
- prevention of elevated temperatures over an extended period of time;
- exclusion of oxygen and sunlight [2].

10.4.3
Functional Stability

Cellulose gum is a very stable and highly functional additive when comparing its dose–response relationship to that of other hydrocolloids.

The solubility and stability of CMC is influenced by the following factors:

- amount of free water;
- presence of other polymers (other gelling and thickening agents, as well as proteins and polysaccharides);
- application of shear forces;
- presence of salts;
- occurrence of temperature changes or increases.

The extent of the effect is dependent on the:

- degree of polymerization (chain length, molecular weight);
- degree of substitution (DS);
- uniformity of substitution;
- time point of the action or the production method.

The effect is reversible in many cases – which allows various possibilities for process control.

Influence of Concentration on the Viscosity of CMC

The viscosity is largely determined by the length of the polymer molecules.

At higher concentrations, viscosity increases as an exponential function of concentration. A rule of thumb is that a doubling of the concentration causes a tenfold increase in viscosity [5].

An increase in the concentration of CMC has a disproportionate effect on the solution viscosity. Doubling the amount used can lead to thickening by a factor of ten in the colloidal solution. The reason for this lies in the polymeric nature of the CMC molecules. The macromolecules interact with each other to form network-like structures. However, even at the highest concentrations, in pure aqueous solutions the gel formation observed with methylated cellulose ethers never occurs.

The increase in reduced viscosity upon further dilution of dilute solutions of CMC is due to increased negative charge of the molecule. In concentrated solution, there is little tendency for the counter ions (sodium) to move out of the sphere of influence of the charges on the polymer molecules. Upon dilution, the cations tend to move into the aqueous interpolymer regions, leaving a net charge on the molecules. As dilution continues, the charge density on the chains increases, and the chains continue to uncoil (Figure 10.17) [5].

Effect of Shear Forces on CMC

Mechanical forces occurring during stirring, kneading, homogenization, and pumping affect the viscosity.

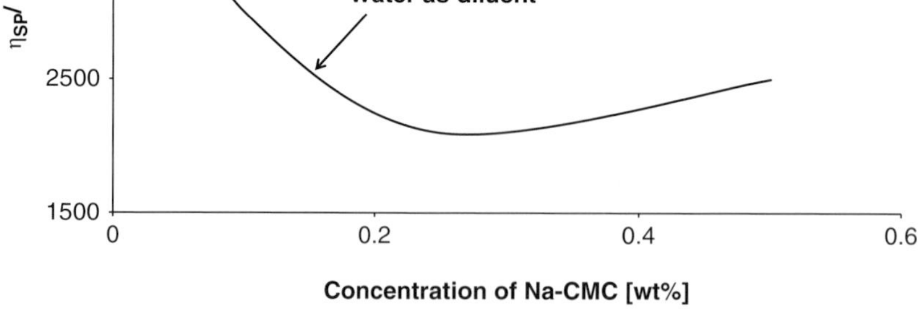

Figure 10.17 Viscosity data for dilute solutions of sodium CMC [2]. Reprinted with kind permission of Ashland Inc.

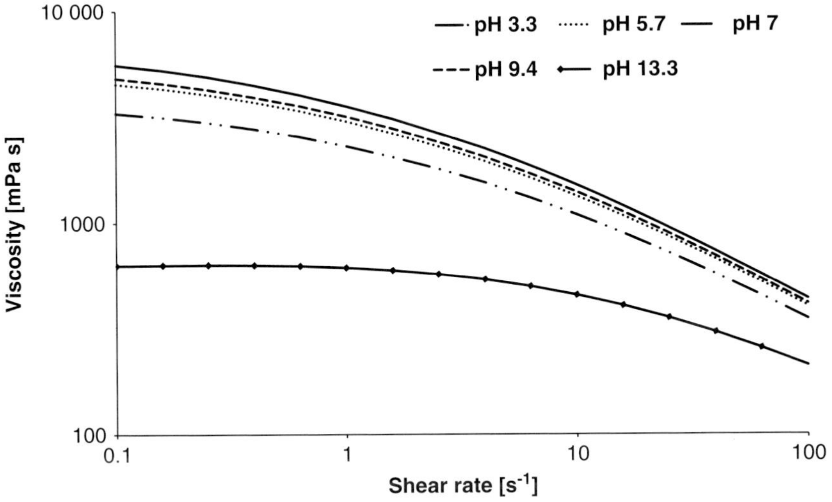

Figure 10.18 Viscosity as a function of shear rate and pH value.

Generally, solutions of medium- and long-chain CMC grades with a DS of 0.9–1.2 indicate pseudoplastic flow behaviour.

When there are lower degrees of substitution (0.4–0.7) for the same polymer length, one observes thixotropy.

By shear stress, the viscosity decreases. This effect is fully reversible in the pH range 3.8–7.0. After a rest phase, the initial viscosity returns, time delayed.

Figure 10.18 illustrates how the viscosity of a high-viscosity CMC (with 20 000 mPa s at 2%) changes under shear force application. The CMC is used in a 1% aqueous solution, which is subsequently brought to the corresponding pH value by addition of sodium hydroxide solution or diluted citric acid. Thereafter, shear treatment is performed.

As expected, the highest viscosity is measured at neutral pH (6.8–7.5).

A moderate alkaline pH of 9.4 affects the thickening power a little less than a weak acid system (at pH 5.7).

The higher the acid concentration, and thereby the lower the pH of the CMC solution, the lower is the viscosity measured.

In a strongly acidic medium (pH < 2.8), the viscosity would rise again. The reason for this acidic-thickening is the formation of the free, but insoluble acid form of carboxymethylcellulose.

The highest viscosity reduction is obtained in a highly alkaline pH range (pH 13–14). Through the transfer of CMC in a strongly alkaline system (pH > 10), the cellulose chain is degraded irreversibly. This is expressed in a massive loss of viscosity.

For further information on rheology, see Section 10.3.1.3 on rheological profiles.

Effect of Changes in Temperature Changes (Cold/Heat) on CMC

Cellulose gum is soluble in water at any temperature between 0 and 100 °C (32 and 212 °F). The higher the temperature is, the better is the solubility, and the lower is the resulting viscosity as a consequence of temperature increase.

As is generally true with aqueous cellulose ether solutions, with the notable exceptions of MC, HPMC, and HPC, the viscosity decreases linearly with increasing temperature on a semilogarithmic plot [4].

The solutions of all CMC types are liquid and thin-flowing at temperatures above 90 °C (194 °F), the chain length of the cellulose ether no longer has an influence.

Figure 10.19 illustrates this property of reversible thermal viscosity reduction.

CMC shows no heat flocculation when boiled in pure water. Aqueous CMC solutions – without any other additives – never form gels. The behaviour of CMC is reversible at pasteurization temperatures ($T \leq 100$ °C (212 °F)). Upon cooling and thus lowering the temperature of the CMC solution, the initial viscosity is restored completely.

By boiling in acidic systems ($T \leq 100$ °C (212 °F) and pH > 3.5), cellulose gum is not destroyed or severely degraded. However, at low pH values (<3.8) a viscosity loss of about 10–20% per heating cycle is encountered. This irreversible reduction can, and at times must, be compensated by a higher starting dose. When sterilizing food ($T > 100$ °C (212 °F)), degradation of the cellulose chain is measured in particular with high-viscosity, long-chain CMC products. This is associated with a more or less strong viscosity loss. Depending on the intensity of the heating procedure, CMC molecules with slight to moderate thickening effect are formed. Low-viscosity CMC is better protected against heat-related degradation.

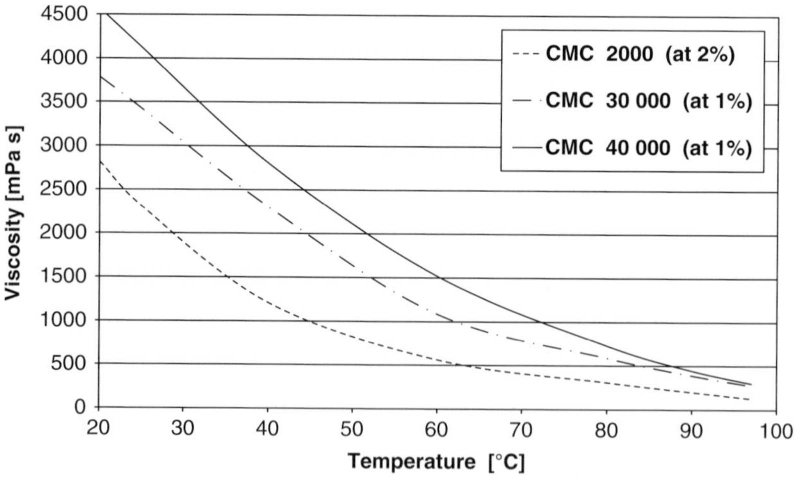

Figure 10.19 Course of viscosity of various CMC types with increasing temperature.

Through the general heat dilution, CMC allows a very good – fast and economic – heat transfer in pasteurization and UHT processes. Compared with other heat-gelling hydrocolloids, the use of CMC herein has significant advantages in heat transfer. Foods, particularly dairy products, receive a gentle treatment.

The thermal properties of carboxymethylcellulose allow its use as a processing aid.

Cooking high-viscosity low-substituted CMC grades (with DS $\leq$ 0.7) in the presence of salts (NaCl, phosphates) is one way to reduce the 'fill viscosity', set for high-efficiency fast-pulsing filling processes, irreversibly upon subsequent thermal preservation and hence obtain a liquid final product. A high viscosity during filling is important in order to dispense cleanly even at very high pulsing.

Cellulose gum is a very stable additive, even at low temperatures. The functionality and water retention are also maintained at temperatures below the freezing point. The freezing point of water is not influenced and is still at 0 °C (32 °F). The cellulose ether does not exert osmotic pressure like sugars and salts, so there is no freezing point depression. The water activity (a_W value) is hardly affected. In some applications a slight positive effect of CMC on the water activity in terms of a 1–5% reduction is observed (e.g. a CMC-free sausage with an a_W value of 0.92 changes its value to 0.89 on addition of 0.3% CMC – a reduction of 3.3%). In most cases, however, hydrocolloid use requires additional water, so that CMC compensates for this by keeping the water activity more or less neutral.

CMC acts as a permanent crystallization control and lowers the speed of crystal formation. The water is frozen out only very slowly. This results in smaller ice crystals. Food products with CMC can go through many freeze–thaw cycles and are characterized by an improved consistency and a smooth mouthfeel. These foods are not as rough and sandy in texture as products that have been stabilized with hydrocolloids that are more frost-susceptible.

Effect of pH Value on CMC

In the 1980s, CMC was described in research papers as being very sensitive to acid and therefore rated as unsuitable for use in acidic systems.

In principle, carboxymethylated cellulose derivatives indicate a higher sensitivity in the presence of acids (H_3O^+ ions) due to their anionic substituents ($-COO^-$) than do other cellulose ethers with neutral substituents such as methylcellulose, HPMC, ethylcellulose, HPC, and MEC. However, in the last 30 years, the manufacture of more acid tolerant CMC grades has been made possible – by changing and improving the pulp quality and by modifying the production processes. Generally, the acid tolerance of cellulose gum has been increased significantly. In addition, the manufacturers of these cellulose derivatives have developed special acid-stable CMC products.

The strength of acids is described by the so-called pK_a value. Stronger acids have a greater tendency to lose their protons and have lower pK_a values than weaker acids. The pK_a of the carboxymethyl group in CMC is 4.5. In comparison, the pK_a of galacturonic acid in pectin is 3.5, which means that the acidic groups in pectin are stronger than those in CMC. At pH 5, about 50% of the carboxymethyl groups of a CMC with a DS of 0.8 are ionized. Addition of acid and thus decreasing the pH to 4 causes partial neutralization of the anionic COO^- group into the neutral COOH group. Only 25% of the carboxymethyl groups of this CMC are still ionized at pH 4. The values for freely available anionic groups depend on degree of substitution. A lower DS means fewer acidic groups for a specific CMC product.

CMC has good stability over a wide pH range (pH 3.5–12). The best effect with carboxymethylcellulose is achieved in systems of pH 3.8–7.0. The maximum viscosity of CMC is measured at pH 6.5–8.5. Above pH 9, there is a slight decrease in viscosity. In the more alkaline region above pH 10, the viscosity decreases sharply and the cellulose chain is cleaved irreversibly. Below pH 6, a slightly stronger decrease in viscosity is observed. At pH values below 3.8, the functionality of CMC is still present but is, however, somewhat reduced. CMC is insoluble in systems with pH 3.0 or lower. Then the less-soluble free-acid carboxymethylcellulose predominates and the viscosity usually increases significantly. Finally, the polymer is precipitated. A medium-viscosity CMC with DS 0.7 shows very little degradation upon standing overnight at pH 2.

In preparing acidic CMC solutions, it is best to dissolve the CMC in neutral water first. If a solution of cellulose gum is prepared prior to the addition of acid, a higher viscosity is obtained than when dry CMC is dissolved in acidic solution. CMC may undergo acid-catalysed hydrolysis in acidic systems, especially at higher temperatures, resulting in permanent loss of viscosity [5].

Through applying an optimized process technology, the acid tolerance of carboxymethylcellulose can be increased, whereby applications with pH values of 2.5–3.0 are possible (e.g. emulsions and concentrates for beverages). Decisive for this is the primary dissolution of CMC with subsequent addition of diluted acids in the second step. Acid-stable CMC grades offer advantages particularly in clear-soluble low-viscosity final products in terms of flexible manufacturing, higher loading, and improved quality.

Figure 10.20 and 10.21 indicate the influence of pH on the viscosity development of CMC grades with different chain lengths and different degrees of substitution.

As the graphs show, the effect of acid or alkali on CMC types with short cellulose chains can be classified as low. The viscosities of the solutions are fairly constant. Even with medium-long average chain lengths, there are only moderate variations in viscosity over the pH range. For high-viscosity CMC molecules, which are long-chain polymers, with a low degree of substitution, the pH value is an absolutely critical parameter that controls the development of viscosity and thus the thickening effect.

If a liquid system is to maintain a constant thickening, with respect to stabilizing over a wide pH range, a higher concentration of a low- to medium-viscosity CMC with DS ≥ 0.8 is required to achieve the best possible product and production safety.

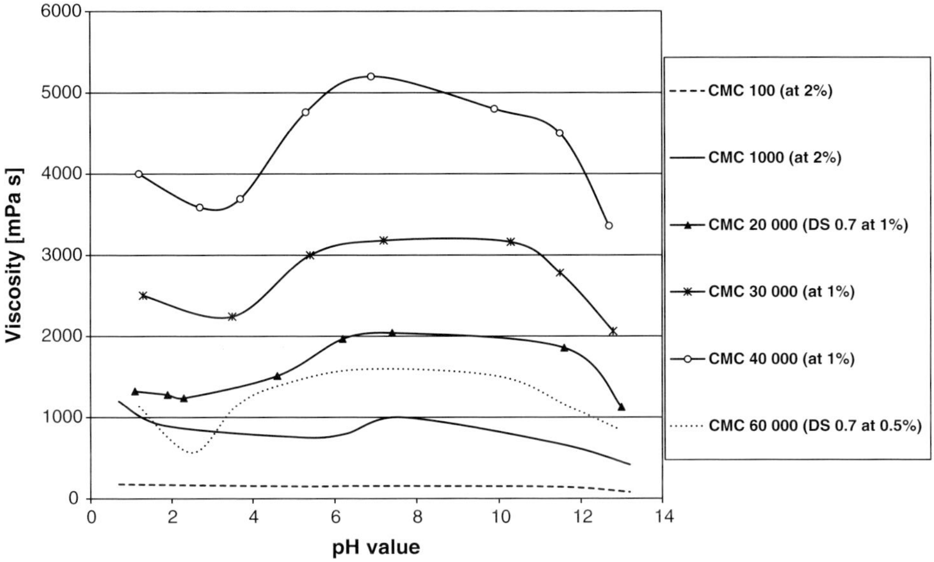

Figure 10.20 Viscosity profile for various types of CMC over the pH range shown.

Figure 10.21 illustrates the combination effect of salt addition (0.5% NaCl) and change in pH. The measurements are based on the following experimental setup:

1) preparation of a dry premix of 1% CMC and 0.5% NaCl;
2) dissolution of the hydrocolloid–salt mixture in tap water at 20 °C (68 °F);

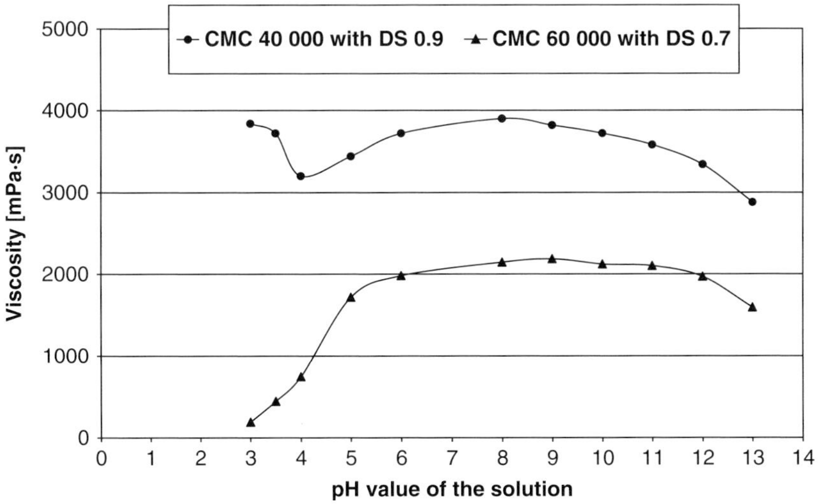

Figure 10.21 Change of viscosity as a function of the pH value with the addition of salt.

3) stirring for 20 h on a rotating laboratory shaker;
4) adjustment of the pH with sodium hydroxide solution (3% resp. 40%) and citric acid (10% resp. 40%);
5) measurement of viscosity after 15 min waiting time.

Again, the degree of substitution has a larger influence on the viscosity development than the degree of polymerization. The addition of salt has an extremely negative effect on the thickening volume of low-substituted long-chain CMC types. More detailed information on the interactions of CMC with salts can be found in Section 10.3.3.2.

Influence of Calcium Ions (Ca^{2+}) on CMC

Di- and trivalent metal ions can support the gelation of CMC. A very high concentration of these metal ions can also lead to flocculation and precipitation of the anionic CMC molecule. The natural concentration of calcium ions in milk ($1.2\,g\,l^{-1}$) is not critical for the function of CMC. To achieve gelling, several times higher Ca^{2+}-ion levels are required. The phase separation, observed after the stabilization of neutral milk with CMC, is probably caused by dehydration of the caseins. During the microbiological acidification of milk with lactic acid bacteria, carboxymethylcellulose disturbs the formation of the milk curd by binding the calcium ions being released from the casein micelles. As consequence, a 3D structure cannot be built up. Therefore, CMC is only suitable for stabilizing or thickening of yogurt when added post-fermentative.

Effect of Proteins on CMC

The anionic nature of CMC leads to interactions with many acidic (cationic) proteins with the formation of soluble and stable complexes. By CMC, a permanent stabilizing effect occurs without any increase in viscosity. The stabilization is not caused by gelation. No internal network structure is formed. CMC solubilizes proteins in their isoelectric pH regions, and protects them against the effects of heat. The consequence is a reduction in the flocculation caused by electrostatic interaction forces. The behaviour of carboxymethylcellulose with proteins is presented in detail in Section 10.3.3.4.

Impact of Salts and Polyelectrolytes

The interactions of CMC with salts are described in detail in Section 10.3.3 on the 'Behaviour of CMC with other ingredients'.

Effect of Moisture on CMC

CMC absorbs moisture from the atmosphere. Figure 10.22 shows the effect of relative humidity on the equilibrium moisture content of three different DS levels [5].

Conclusions from the analytical measurement data are:

- The higher the relative humidity (RH) of the atmosphere is, the more moisture is absorbed by the dry CMC powder.

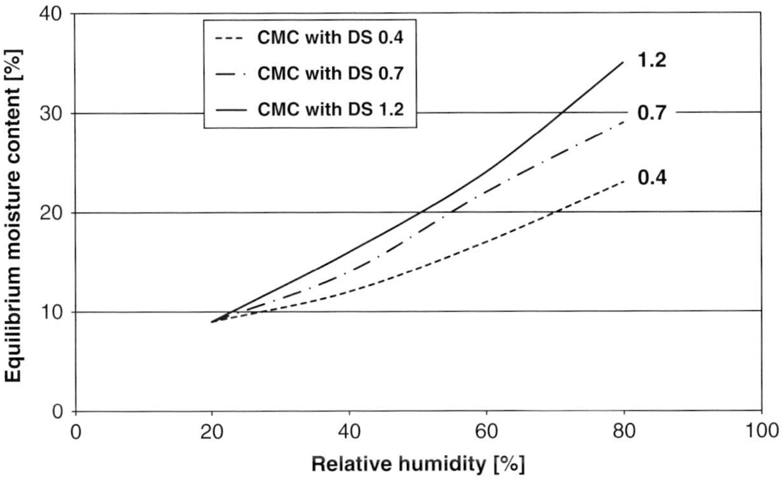

Figure 10.22 Effect of relative humidity on equilibrium moisture content of CMC at 25 °C (77 °F) [2]. Reprinted with kind permission of Ashland Inc.

- With increasing DS and, thus, enhanced general solubility of the CMC product in water the equilibrium moisture content increases when the RH rises.

10.4.4 Films of CMC

Water-Soluble Films

CMC is seldom used to prepare free or unsupported films. However, its ability to form strong, oil-resistant films is of great importance in many applications.

Films can be obtained by evaporating the water from aqueous CMC solutions. These fairly flexible films are unaffected by oils, greases, and organic solvents. Their typical properties are given in Table 10.10. The prepared films were 2-mm thick and contained about 18% moisture [2].

Table 10.10 Typical properties of CMC films [2].

Property	CMC type – viscosity		
	Low (100 mPa s at 2%)	Medium (1000 mPa s at 2%)	High (30 000 mPa s at 2%)
Tensile strength in kg cm−2 (in psi)	563 (8000)	915 (13 000)	1056 (15 000)
Elongation at break (%)	8.3	14.3	14.3
Flexibility (double folds)	93	131	513
Electrostatic charge	Negative	Negative	Negative
Refractive index	1.515	1.515	1.515
Specific gravity	1.59	1.59	1.59

Data taken with permission from Reference [2].

Suitable drying temperatures are between 50 and 90 °C (122 and 194 °F). If the applied heat is too strong, keratinization is observed. The CMC melts to yellow plastic-type flakes. The films of cellulose gum are transparent and relatively flexible. They have a residual water content of 5–20%. Like CMC itself, the films made from CMC dissolve slowly in cold and in hot water. Finally, they are completely solubilized and disappear.

Water-soluble colours can be added to the film solutions before drying. In comparison to films, obtained from MC or HPMC solutions, the CMC films are not hot-water-stable. Owing to the gelation at warm temperatures, methylcellulose and HPMC films are resistant in hot aqueous systems, and thus maintain their barrier functions against greases and oils. When humidity is high, cellulose gum films behave hygroscopically – they take up water vapour to the equilibrium moisture content and thus become sticky.

In the presence of fats, oils, lubricants, and fat-based organic solvents or alcohols (pure ethanol or iso-propanol (IPA), glycerol, propylene glycol), the films are stable and form barriers. Generally, however, CMC films do not provide a barrier to cold or hot water vapour or steam.

To increase the flexibility of the films different plasticizers may be incorporated into the CMC casting solution before the drying process. With 10–30% glycerol or propylene glycol in the formulation, the elongation can be improved by 40–50%. Furthermore, the folding endurance is greatly increased up to 10 000 double folds. Plasticizers (food-grade and without food approval) that have proved effective with CMC are [2,5]:

- diglycerol,
- ethanolamines,
- ethylene glycol
- 2,3-butylene glycol,
- glycerol,
- glycerol α-monomethyl ether,
- glycerol monochlorohydrin,
- 1,2,6-hexanetriol,
- 2-methyl-2-nitro-1,3-propanediol,
- nono-, di-, and triacetin (1,2,3-propanetriol triacetate, E 1518, $C_9H_{14}O_6$, glycerin triacetate),
- 1,3-pentanediol,
- poly(ethylene glycol) (with a molecular weight of 600 or less),
- propylene glycol,
- trimethylolpropane.

Generally, films of methylcellulose and HPMC are less brittle than comparable films from CMC. For these reasons, CMC is rarely used to make plasticizer-free, load-bearing or flameproof films.

Insoluble Films for Non-food Applications

CMC films can be insolublized by crosslinking the hydroxyl or carboxy groups of the gum with a polyfunctional resin, such as polyamide, melamine–formaldehyde, or dimethylol urea.

An aqueous solution of CMC and the resin can be air dried at room temperature, or at elevated temperatures, to obtain insoluble films of good strength and flexibility.

Clear films with varying degrees of water insolubility can be prepared by controlling the amount of added resin and the time and temperature at which the film is cured.

Water-insoluble films can also be produced from a dry film of CMC by treating it with a solution of salt, usually of a trivalent cation, that can replace the sodium ions [5].

10.5 Rheometry

To identify CMC products and to distinguish them from other hydrocolloids and cellulose derivatives, the following methods are applied [1]:

1) **Solubility:**
   - completely soluble in cold water (from DS 0.6 completely soluble, fibre-free, and smooth);
   - insoluble in 100% ethanol;
   - transparent, soluble in ethanol–water mixtures up to 40% alcohol by volume.
2) **Foam test:**
   A 0.1% solution of the sample is shaken vigorously. There is no foam development. There is the possibility of differentiating between methylcellulose, HPMC, HPC, and MEC.
3) **Precipitation:**
   A 5% solution of copper sulfate or aluminium sulfate (5 ml) is added to a 0.5% solution of the sample (5 ml). With CMC grades, a precipitate is formed. All other cellulose ethers, gelatine, tragacanth, and locust bean gum do not show a visible reaction.
4) **Colour reaction:**
   A CMC sample (0.5 g) is added to water (50 ml) and dissolved with stirring, until a clear solution is achieved. In a test tube, 1 ml of the solution is diluted with 1 ml of water and then mixed with five drops of 1-naphthol TS.

   In the inclined tube, sulfuric acid (2 ml) is poured carefully along the side, so that a lower layer is formed. The interface is coloured purple.

10.6
Synergies of Cellulose Gum with Other Hydrocolloids

CMC shows positive interactions with some hydrocolloids.

Since the mechanisms for gelling hydrocolloids differ from those for thickeners, both groups are considered separately.

10.6.1
Behaviour with Other Thickeners

A synergism is found in mixtures of carboxymethylcellulose with guar gum. The resulting viscosity of blends will be '+50% to +100%' higher than expected mathematically. The best effect is achieved when isoviscous CMC and respective guar gum types are used. The recommended mixing ratio is '1 : 1'.

With the combination of CMC with non-ionic cellulose ethers (MC, HPMC, HPC, and HEC), viscosity profits from '+25% to +100%' are measured when each of these cellulose derivatives is applied in the same ratio (1 : 1). An example is given in Section 10.3.3.8 (see Table 10.9).

Mixtures of xanthan and CMC do not have unpredictable interactions. With a high proportion of xanthan gum and a low volume of CMC, the system 'reacts' from a rheological point of view as a pure xanthan addition with a corresponding thickening and a yield point. When the mixing ratio is shifted towards a higher proportion of CMC, then the typical flow behaviour of a pure CMC solution is observed.

While xanthan gum as strong thickener induces gelation in combination with other non-gelling thickening hydrocolloids such as locust bean gum (LBG) and konjac glucomannan, no gelation occurs when co-processing the linear non-branched CMC with xanthan.

A '1 : 1 – mixture' of CMC and LBG provides an improved property profile compared with both hydrocolloids alone. The cold-water solubility and functionality is better than with pure LBG. The heat resistance of CMC is increased by blending. Particularly at low shear rates, the viscosity is more stable. There is, however, no gel formation.

Figure 10.23 shows how a mixture of CMC with guar gum behaves in relation to the measured viscosity. The actual thickening performance is higher than that theoretically calculated. For various CMC or guar types, different curves of blending viscosity are produced. It is worth conducting an individual measurement series in order to find the optimum mixing ratio for the application concerned. Generally, a mixture should contain about 30–70% CMC and 70–30% guar gum to achieve a significant viscosity gain. It is recommended to select isoviscous products as mixing partners.

10.6.2
Interactions with Traditional Gelling Agents

Cellulose gum can be used in combination with cold gelling hydrocolloids, such as κ-carrageenan, alginate, and agar–agar. The synergistic effect on gel quality is

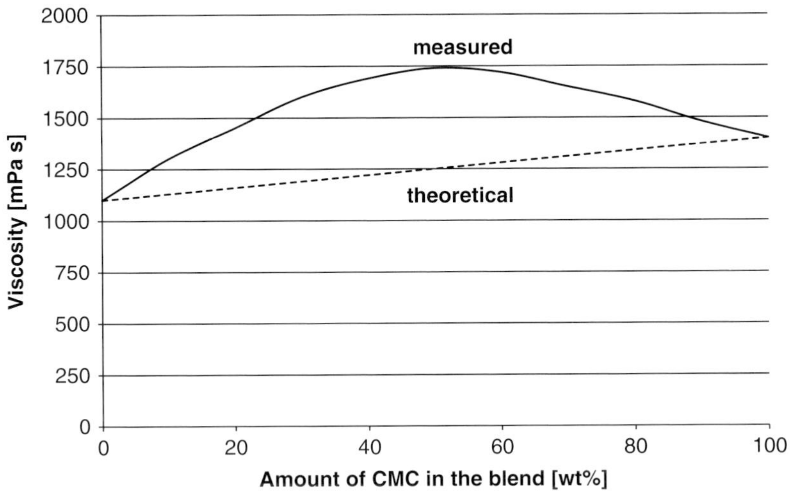

Figure 10.23 Synergism between guar gum and cellulose gum.

also produced by the interaction of cellulose derivative with starches. The kind of starch does not matter. The principle can be applied to all raw materials (corn, potato, tapioca, etc.), regardless of the amylose content and the pretreatment or modification. The positive interactions occur through the CMC molecules being incorporated in the 3D network of the gel. There they can intercept water that is released from the gelling hydrocolloids over the storage period. The cellulose ether acts as a 'sponge' in this heterogeneous system and reduces synaeresis.

By combining a classic gelling agent with Na-CMC, the quality of the gel is improved – it has a higher elasticity and prolonged stability. The gel dries out much more slowly and does not break or only breaks after a significantly longer shelf life. In addition, CMC acts as crystallization control in freezing processes and thereby enhances the performance of the food product under ice-cold conditions and after freeze–thaw cycles.

10.7 Functional Properties of Carboxymethylcellulose

Na-CMC has some very interesting properties and can therefore significantly improve the quality of food. In addition, in technical applications as well as in cosmetic and pharmaceutical products, the following characteristics of CMC have a positive effect:

- control of rheology and viscosity in aqueous systems, adjusting of texture and consistency;
- high water-binding capacity, good water retention (function as a 'sponge'), reduced synaeresis, prolonged freshness and shelf life of food products;

- solubility in cold and hot water from 0 to 100 °C (32 to 212 °F);
- excellent freeze–thaw stability (crystallization control);
- delayed retrogradation of starches;
- possibility to stabilize products with minimal impact on viscosity (stabilization without thickening);
- protective colloid for proteins exposed to acid and heat;
- film-forming properties (barrier function, improved mouthfeel);
- increased plasticity and elasticity of dough;
- improved machinability due to reduced adhesion and thus a simplified cleaning of surfaces;
- replacement of fat with conservation of the 'full fat'-mouthfeel;
- high efficiency at low dosage (0.015–0.5%);
- good compatibility with other hydrocolloids;
- free of odour or taste, no influence on colour;
- solutions are free of fibres and show a smooth flow;
- 100% soluble fibre;
- energy content (measured): 0 kcal g^{-1} (according to the latest EU Regulation: 2 kcal g^{-1}).

10.8 Use of Purified CMC in Regulated Applications

Since its commercial introduction in the USA in 1946, sodium carboxymethylcellulose has found an ever-increasing number of applications.

The many important functions provided by this polymer make it a preferred thickener, suspending aid, stabilizer, binder, and film-former in a wide variety of regulated applications. Table 10.11 presents an overview of possible uses of CMC in cosmetics, food, and pharmaceutical products.

Table 10.11 Options to use CMC in regulated products [2].

Types of uses	Specific application	Properties utilized
Food products	Frozen desserts, soft-serve	• Controls ice crystal growth; • improves mouthfeel; • body and texture
	Pet food	• Water binder, • gravy thickener, • extrusion aid, • binder of fines
	Protein foods	• Retains water, • improves mouthfeel
	Baked goods	• Batter viscosifier; • improves moisture retention; • enhances texture

	Beverages	• Suspending aid; • rapid viscosifier; • improves mouthfeel and body; • protein stabilizer in acidified drinks
	Desserts, icings, toppings	• Odourless and tasteless; • thickens; • controls sugar crystal size; • improves texture; • inhibits synaeresis
	Low-calorie foods	• No caloric value, • thickens, • imparts body and mouthfeel
	Syrups	• Clear/transparent, • thickens • imparts favourable mouthfeel and body
	Dressings, sauces	• Thickener and suspending aid; • imparts body and mouthfeel
	Animal feed, extrusion products	• Lubricant • binder • film-former
Cosmetics	Toothpaste	• Thickener, • flavour stabilizer, • suspending aid, • binder
	Shampoos, foamed products	• Suspending aid, • thickener, • foam stabilizer, • high water-binding capacity
	Creams, lotions	• Emulsion stabilizer, • film-former • thickener
	Gelled products	• Thickener, • gelling agent, • film-former
	Denture adhesives	• Wet tack, • long-lasting adhesion
Pharmaceutical products	Ointments, creams, lotions	• Stabilizer, • thickener, • film-former
	Jellies, salves	• Thickener, • gelling agent, • protective colloid, • film-former
	Tablet binder, granulation aid	• High strength binder
	Bulk laxatives	• Physiologically inert, • high water-binding capacity
	Syrups	• Thickener
	Suspensions	• Thickener • suspending aid

Data taken with permission from Reference [2].

10.9 Food Applications with Typical Formulations

Because of its versatile properties CMC is used in a wide range of products. The viscosity data in the presented formulations refer to 2% CMC solutions, measured in mPa s (cP) with a spindle viscometer (Brookfield) at 25 °C (77 °F).

10.9.1 Bakery and Cereal Products

Tables 10.12–10.19 show formulations for bakery and cereal products.

Improved quality of traditional wheat flat bread with guar:

- Higher yield, longer shelf-life, less breakage in folding/rolling;
- 2.4% guar (on dry basis) can be replaced by:
  - 0.5–1% guar +0.3–0.5% CMC,
  - ratio for 'guar–CMC blends': 4 : 1 to 8 : 5.

Table 10.12 Formulations for traditional wheat bread and rye bread.

Ingredient	Amount (g), based on the total amount of flour	
	Wheat bread	Rye bread (70 : 30)
Wheat flour (type 550)	100	30
Rye flour (type 1150)	—	70
CMC – 1000 or 10 000	0.5–1.5	0.5–1.5
Water	62	80
Yeasts	4	2
Acidulants	—	0.7
Salt	2	2
Fat	1	—

Processing guideline:

- The CMC is homogeneously mixed with the dry ingredients (flour, salt).
- After activating the yeasts, the liquid ingredients of the formulation are added to the dry mix.
- Dough is prepared by kneading.
- After proofing of the dough, the bread can be filled into moulds or baked as freely altered variant.

Table 10.13 Formulations for bread rolls/buns [7].

Ingredient	Amount (%), based on moist dough
Wheat flour (type 550)	57.2
Water	36.6
Yeasts	2.8
Sugar	1.1
Salt	0.9
Ascorbic acid (as a solution containing 1 wt%)	0.3
Emulsifier (mono-/diglycerides)	0.2
CMC – 10 000	0.9
Sum	100.0

Data taken with permission from Reference [7].

Table 10.14 Formulation for tortillas, wraps, and flat breads based on wheat flour.

Ingredient	Amount (g)
Wheat flour	100
Salt	1.5
Baking soda	1.2
Water	80–120
Fat/oil	11–14
CMC 20 000 or 40 000 (amount based on wet dough)	0.1–0.2%
Guar (2–4 times the amount of CMC)	Addition recommended

Table 10.15 Formulation for a dry mixture for tortillas or flat breads based on corn flour [8].

Ingredient	Amount (g), based on flour
Corn flour *(Masa)*	100
CMC 40 000	0.3–0.5
Potassium sorbate	0.36
Fumaric acid	0.30

Data taken with permission from Reference [8].

Table 10.16 Formulation for a cereal power-bar.

Ingredient	Amount (%)
Rice krispies	77.00
CMC 1000 or 2000	0.05
Maltodextrin	9.00
Water	13.95

Data taken with permission from Reference [8].

Table 10.17 Formulations for instant noodles [7].

Ingredient	Amount (%), based on flour amount
Wheat flour	100.0
Water	32.0
Salt	1.5
Polyphosphate	0.1
Potassium and sodium carbonate	Each 0.1
CMC – 1000	0.3

Data taken with permission from Reference [7].

Table 10.18 Formulation for fruitcake [6].

Ingredient	Amount (g)
Wheat flour	28.2
Sugar	21.0
Fat	22.0
Baking powder	0.5
Whole egg	23.0
CMC 40 000	0.3
Water (additional – compared to the standard without CMC)	5.0
Sum	100.0

Processing guideline:

- A homogeneous mixture of the dry ingredients (flour, salt, baking powder, CMC, guar) is very important.
- The liquid ingredients are then incorporated into the dry mix.

Processing guideline:

- CMC powder and maltodextrin are dry mixed (alternatively, a CMC granulate can be used);
- dissolve the dry mix in water until a homogeneous solution is achieved;
- the liquid binder is added to the krispies and homogeneously distributed;
- drying at 40 °C (104 °F).

Table 10.19 More formulations for cakes and cookies [7].

Ingredient	Amount (g)				
	Cookies, reduced-fat (10% fat)	Sponge cake (stirred-type)	Sponge cake (fluffy type)	Sponge cake, reduced sugar	
Wheat flour	52.0	28.9	28.2	28.2	28.2
Whole egg	—	23.2	35.1	35.14	35.24
Margarine	10.0	18.7	—	—	—
Oil	—	4.3	—	—	—
Sucrose	20.0	20.8	31.6	15.9	—
Salt	0.3	—	0.34	—	—
Skimmed milk powder	—	0.9		—	—
Baking soda	2.6	0.8	0.5	0.5	0.5
Water	14.0	2.3	4.2	4.2	4.2
Mono-/diglycerides	0.2	—	—	—	—
CMC – 10 000	0.6	0.1	0.06	0.06	0.06
Vanilla flavour	0.3	—	—	—	—
Isomalt	—	—	—	15.9	—
Acesulfame	—	—	—	0.1	—
Lactitol	—	—	—	—	15.9
Polydextrose	—	—	—	—	15.9
Sum	100.0	100.0	100.0	100.0	100.0

Data taken with permission from Reference [7].

10.9.2
Dairy Products – Ice Cream, Desserts, and Cheese Preparations

The formulation for various dairy products is shown in Tables 10.20–10.26.

Table 10.20 Formulations for milk-based ice cream [6,7].

Ingredient	Amount (%)
Water	61.1–61.26
Butterfat	12.0
Fat-free milk solids (MSNF)	11.5
Sugar	15.0
Stabilizer	Guar: 0.12–0.2 (or κ-carrageen: 0.03)
CMC with 10 000–40 000 mPa s (2%), with substitution degree (DS) 0.7–0.95	0.12–0.2
Sum	100.0

Data taken with permission from Reference [7].

Table 10.21 Formulations for sorbets/sherbets, water ices, and ice lollies [7].

Ingredient	Amount (%)				
	Sherbet type A	Sherbet type B	Water ice type A	Water ice type B	Water ice type C
Milk components (mixture of fat + non-fat)	—	—	5	—	—
Sucrose	25	13	9	15.5	25
Dextrose	5	—	—	5.5	7
Glucose syrup	—	22	22	—	—
CMC (10,000–20 000 mPa s, 2%, acid stable)	0.2	0.25	0.2	0.25	0.25
Fruit puree/fruit juice	20	15	15	—	—
Citric acid	0.4	0.5	0.5	0.4	0.4
Colour/flavour			Optional		
Water	49.4	49.25	48.3	78.35	67.35

Data taken with permission from Reference [7].

Table 10.22 Formulations for stirred yogurt with high viscosity ('white line').

Ingredient	Amount (%)		
	Yogurt, standard	Yogurt, reduced-fat	Yogurt, low-fat
(A) Preparation of yogurt			
Milk with full-fat-content (3.5% fat)	98	—	—
Milk, reduced-fat (1.5% fat)	—	97	—
Milk, low-fat (0.1% fat)	—	—	96
Skimmed milk powder (SMP)	2	3	4
(B) Stabilization of yogurt			
Yogurt, fermented (pH 4.3–4.6)	99.4–99.6		
CMC – granules with 2000 to 20 000 mPa s (2%) and DS 0.7–0.95	0.4–0.6		

Processing guideline:

- The addition of CMC should be after fermentation. Upon addition of the CMC to the milk before the fermentation, the anionic cellulose ether disturbs the formation of the milk curd, by binding the calcium ions.
- CMC granules are stirred slowly into the yogurt.
- In the production of spicy yogurt with herbs or curd/quark, powdered CMC can be premixed with the spices and herbs before being stirred under the white mass.

Table 10.23 Formulations for fruit yogurt, stirred-type ('red line').

Ingredient	Amount (%)
(A) Fruit preparation	
Mashed fruit (fresh or frozen)	50–51
Sugar	35
Water	13
Citric acid (1 : 1) to adjust pH value	QS to pH 3.9 ± 0.2
CMC – powder with 2000 to 20 000 mPa s (2%) and DS 0.7–0.95	1–2
(B) Fruit yogurt	
Yogurt (pH 4.3–4.6)	75–85
Fruit preparation (A)	15–25

Table 10.24 Formulations for chocolate desserts and vanilla pudding [7].

Ingredient	Amount (%)	
	Chocolate dessert	Vanilla pudding
k-Carrageenan	0.1–0.2	0.15
CMC – 2000 to 20 000 mPa s (2%), with substitution degree (DS) 0.7–0.95	0.1–0.2	0.15
Cocoa powder (alkalized)	2.0	—
Skimmed milk powder	2.0	2.0
Modified starch (acetylated distarch adipate)	1.5	2.0
Glucose syrup	3.0	—
Sugar	14.0	8.0
Cream, 30–40% fat	7.0	—
Milk (full fat)	70.2	87.46
ß-Carotene	—	0.06
Vanilla flavour (optional: addition of vanilla bean)	—	0.18
Sum	100	100

Data taken with permission from Reference [7].

Table 10.25 Formulations for fresh cheese preparations (spreads).

Ingredient	Amount (%)	
	Italian fresh cheese preparation, reduced-fat	Fresh cheese preparation, low-fat
Fresh cheese (40% fat in DM)	80.0	—
Fresh cheese (from skimmed milk)	—	78.8
Water	—	20.0
Yogurt (low-fat)	13.0	—
Olive oil[a)]	3.5	—
Coconut oil[a)]	1.5	—
CMC – 30 000 to 40 000 mPa s (2%)	0.5	0.8
Italian herbs and spices	1.0	—
Salt	0.5	0.4
Sum	100.0	100.0

a) High proportion of unsaturated fatty acids, but less cholesterol (about 35% less than in pure cream cheese with the same fat content).

Table 10.26 Formulations for processed cheese preparations and the respective cheese spread.

Ingredient	Amount (%)		
	Processed cheese, standard	Processed cheese with whey protein addition	Processed cheese without protein addition
Cheese (Edam, Gouda, Emmental, Maasdam, Chester)	55.5	50.0	50.0
Butter	9.0	8.0	8.0
Whey protein concentrate (WPC)	3.5	3.5	—
Melting salts such as: citrates E 330–333, lactates E 325–327 phosphates E 339–341 and E 450–452	1.8	1.8	1.8
CMC – 20 000 mPa s (DS 0.7)	—	0.5	—
CMC – 30 000 mPa s (DS 0.82–0.95)	—	—	0.8
Water	27.7	33.7	36.9
Cheese (re-work)	2.5	2.5	2.5
Sum	100.0	100.0	100.0

10.9.3
Syrups and Fillings

For the formulation of syrups and fillings see Tables 10.27 and 10.28, respectively.

Table 10.27 Formulations for syrups for ice cream, pancakes, and pudding desserts [7].

Ingredient	Amount (g)		
	Maple syrup standard	Maple syrup with CMC	Maple syrup sugar-reduced
Water	—	26.31	58.72
Sugar syrup (42 DE)	99.55	73.0	—
Fructose syrup	—	—	40.0
Natural and artificial maple flavour	0.275	0.265	0.27
Colour – caramelized sugar	0.075	0.075	0.075
Artificial sweetener	—	—	0.01
Potassium sorbate (25% solution)	0.1	0.05	0.05
CMC with 1000–2000 mPa s (2%), with DS 0.82–0.95	—	0.3	0.875
Sum	100.0	100.0	100.0

Data taken with permission from Reference [7].

Table 10.28 Formulations for a filling/covering for bakery goods (minimum of 40 °Brix).

Ingredient	Amount (%)
Mashed fruits (fresh or frozen)	20–50
Sugar	35
Water	14–44
Citric acid (1:1) to adjust pH value	QS to pH 4.0 ± 0.2
CMC[a)] – powder with 1000 to 40 000 mPa s (2%) and DS 0.7–0.95	0.2–0.5
Sum	100.0

a) For bake-stable fillings for cakes, the addition of methylcellulose or HPMC is recommended (amount 0.5%, HPMC is used at sugar contents of 35% and more).

10.9.4
Beverages (RTD, Concentrates, Emulsions)

Because of its performance, carboxymethylcellulose is incorporated into many beverages. Advantages are:

- protection protein,
- stabilization without thickening,
- transparency of the solutions,
- viscosity increase also in acidic systems,
- neutral taste,
- slightly creamy mouthfeel.

CMC occurs in ready-to-drink products, and also in instant beverages or powdery products. The cellulose ether may be added to the finished concentrate or to the crude emulsion. Table 10.29 shows an overview of possible applications of cellulose gum in beverages.

Table 10.29 Application of CMC in beverages – overview.

Category	Product	CMC type[a)]	Concentration (in the final product) (%)	Effect
Protein-containing drinks from milk and pure vegetable products (RTD)[b)]	Buttermilk, yogurt, and other acidic beverages based on milk	CMC 2000 or CMC 20 000, each with DS 0.7–1.0	0.2–0.5	Thickening
		CMC 100	Minimum 1%	Protein protection
	Whey drinks (pH 3.8–4.3)	CMC 30 or 100	1.5–2.5	Stabilization, protein protection, mouthfeel
	Beverages with soy protein (pH <5)	CMC 100 + CMC 2000/ 20 000	0.5 + 0.3	
	Milk shakes		0.1–0.3	

(continued)

Table 10.29 (*Continued*)

Category	Product	CMC type[a)]	Concentration (in the final product) (%)	Effect
		CMC 1000 or CMC 2000		Thickening, crystallization control
	Cocoa drinks (1.5% fat)	CMC 100 + κ-carrageen	1.0 + 0.04	Protein protection, stabilization
Emulsions and basic concentrates	Concentrates for fruit-containing beverages with sugar (50 °Brix, pH 3.8, acid content 1–2%)	CMC 30	1.5–2.5	Stabilization, mouthfeel
	Emulsion for sugar-free beverages ('light')	CMC 100	1.5–2.5	
	Sports drinks (isotonic, high content of polyelectrolytes)	CMC 100	1	
	Fruit juice beverages (RTD)[b)]	CMC 30 000 or CMC 40 000	0.05–0.25	Thickening, stabilization
Powder mixtures for instant beverage	Instant beverages, fruit-based (cold or hot): 20% powder, 80% water	CMC 20 000 or CMC 30 000	0.1–0.4 (Powder contains 0.5–1.5% CMC)	Thickening, stabilization, mouthfeel
	Coffee drinks (hot preparation)	CMC 2000 or CMC 10 000	0.25–0.5	
	Cold instant beverages based on coffee or tea	CMC 100	0.5–1	Stabilization, mouthfeel
Alcoholic drinks	Liqueurs without cream, transparent (with max. 40% alcohol)	CMC 100 CMC 1000 CMC 10000	0.5–2 0.1–1 0.1–1	Stabilization, mouthfeel
	Cream liqueurs (3–20% alcohol)	CMC 1000 to CMC 10 000	0.1–1 0.1–1	Viscosity, stabilization, protein protection (optional)
	Cream liqueurs with milk proteins	CMC 1000 + κ-carrageen	0.1 + 0.1	Protein protection, stabilization, thickening
	Eggnog (liqueur)	CMC 1000	0.1–0.5	
	Wine	CMC 30	0.05–0.1 $g\,l^{-1}$	Crystallization inhibitor

a) The number indicates the viscosity of a 2% CMC solution – measured with Brookfield LVT at 25 °C (77 °F).
b) Ready-to-drink.

Tables 10.30–10.34 give the recipe and formulation of various drinks.

Table 10.30 Basic recipe for an acidic milk drink – acidification of milk with fruit juice [6].

Ingredient	Amount (%)
Milk (Full-fat or reduced fat)	53.75–53.85
Orange juice (11 °Brix)	41.3
Sugar	4.7
CMC – 10 000 or 20 000	0.15–0.25
Sum	100.0

Table 10.31 Formulations for acidic milk drinks with fruit components – beverages based on acidified milk in form of whey, buttermilk, or yogurt [6].

Ingredient	Concentration (%)						
	Whey drinks		Yogurt drinks		Buttermilk drinks		
Sweet whey, fermented to pH 4.0 ± 0.2	74	50	—	—	—	—	—
Yogurt	—	—	32–42	30–40	—	—	—
Buttermilk	—	—	—	—	40	40	89.6
Water	—	33	38–48	41.8–51.5	41	41.5–41.8	—
Sugar	—	9–9.5	11	10	10	10	10
Fruit concentrate	—	6	8	8	8	8	8
Fruit juice (with sugar)	24–24.5	—	—	—	—	—	—
CMC 30 or 100	1.5–2	1.5–2	1	—	1	—	—
CMC 2000 or 20 000	—	—	—	0.2–0.5	—	0.2–0.5	0.3–0.4
Sum	100	100	100	100	100	100	100

Processing guideline:

Powdered CMC is premixed with the sugar and then stirred into the milk; dissolution time is about 15 min while stirring.

Processing guideline:

1. Powdered CMC is premixed with the sugar and then transferred into the liquid phase. Alternatively, CMC granules are used, which are added directly to the milk base. Stirring for 15 min;
2. indirect heating to 70 °C (158 °F) in a heat exchanger;
3. homogenization (two-stage) at 150–200 bar (2175–2900 psi);
4. pasteurization at 85 °C (185 °F) for 15 s;
5. cooling to 20 °C (68 °F) and aseptic filling.

Table 10.32 Formulations for fruit preparations (FP) with about 40 °Brix as an additive for milk-based drinks.

Ingredient	Concentration (%)	
	For beverages with low viscosity	For beverages with medium viscosity
Mashed fruits (fresh or frozen)	50	50
Sugar	30	34
Water	10–14	14–15.4
Citric acid (1:1)	To adjust pH to 3.9 ± 0.2	
CMC with 30 or 100 mPa s (2%)	6–10	—
CMC with 2000 mPa s (2%)	—	0.6–2
Dosage FP on milk basis (%)	15–25	15–25
Concentration (%) in beverage	Minimum 1.5	0.15–0.3

Table 10.33 Formulation for chocolate drinks (RTD, instant powder mix) [7].

Ingredient	Concentration (%)		
	Beverage (A)	Beverage (B)	Powder mix (C)
Milk, reduced-fat (1.5% fat)	90.46	—	—
Milk, fat-free (0.1% fat)	—	91.16	—
Skimmed milk powder	—	—	53.5
Sugar	4.5	4.5	36.0
Glucose syrup	2.0	2.0	—
Cocoa powder (de-oiled)	2.0	2.0	9.7
CMC with 100 mPa s (2%)	1.0	—	—
CMC with 2000 mPa s (2%)	—	0.3	0.8
κ-Carrageenan (highly purified)	0.04	0.04	—
Sum	100	100	100

Data taken with permission from Reference [7].

Processing guideline:

Powdered CMC is premixed with sugar and then added to the fruit.

Processing guideline:

For chocolate beverages (RTD) – variants A and B:

- CMC is premixed with the dry components, and then dissolved in milk.

Processing guideline for instant chocolate drinks – variant C:

- 16 g powder mixture is stirred in 84 g water (hot or cold).

Table 10.34 Formulations for fruit juice-containing drinks (soft drinks) [7].

Ingredient	Concentration (%)	
	Beverage with orange juice[a)] (pH 3–3.5)	Guava juice beverage[a)] (pulp stabilization), pH 3.2–3.7
Water	84.3	59.69
Orange juice concentrate (65 °Brix)	5	—
Guava pulp	—	30
Sugar	10.5	10
Citric acid, crystalline	0.15	0.15
Calcium chloride	—	0.01
CMC – 20 000 or 30 000 mPa s (2%)	0.05	0.15
Sum	100	100

Data taken with permission from Reference [7].
a) Note: Consider the local food laws!

Processing guideline:

CMC is premixed with sugar in the first step. Then the dry blend is dispersed into the liquid phase and dissolved under moderate stirring.

Isotonic Electrolyte Drinks

Sports drinks are highly loaded with minerals and vitamins and are characterized by a very low viscosity (water-thin). Stabilization is realized by the addition of 1% of a low-viscosity CMC (2% = 100 mPa s). The best functionality of CMC is achieved by producing a highly concentrated stock solution and subsequent dilution with the electrolyte-containing system.

Instant Beverages

The Formulation for Each of Several Instant Beverages is Given in Tables 10.35 and 10.36

Table 10.35 Formulations for fruit-based and flavoured instant beverages [7].

Ingredient	Concentration (%)		
	Fruit-containing beverage powder	Beverage powder with flavour – type 1	Beverage powder with flavour – type 2
Sugar, finely ground	77–78	88.7–89.1	90.7–91.2
Fruit powder	20	—	—
Citric acid, crystalline	1	6.5	4
Sodium citrate	0.5	2.5	2.2

(continued)

Table 10.35 (*Continued*)

Ingredient	Concentration (%)		
	Fruit-containing beverage powder	**Beverage powder with flavour – type 1**	**Beverage powder with flavour – type 2**
Flavour	Optional	1	1
Colour	Optional	0.1	0.1
Minerals + vitamins	Optional	Optional	Optional
Clouding agent for opacity	Optional	Optional	Optional
CMC – 20 000 or 30 000	0.5–1.5	0.8–1.2	1.5–2
Sum	100	100	100
Preparation of the final beverage:			
Proportion of powder mixture (%)	15–20	10–15	9–12
Proportion of water (hot or cold) (%)	80–85	85–90	88–91

Data taken with permission from Reference [7].

Table 10.36 Formulations for hot chocolate drink, coffee and cappuccino beverages – type 'instant' and RTD [7].

Ingredient	Concentration (%)			
	Variant A[a)] (powder mix)	**Variant B[a)] (powder mix)**	**Variant C (RTD)**	**Variant D (RTD)**
Sugar	50.5–50.75	56	8	5.6
Non-dairy cream[b)]	37	13	2.2	—
Milk powder, fat-free[b)]	—	18	2	8.4
Soluble coffee (instantized)	12	10.7	1.8	—
Cocoa powder (de-oiled)	—	2	0.3	1.6
Flavour + colours	Optional	Optional	Vanilla powder: 0.02	—
CMC – 2000, 10 000 or 20 000	0.25–0.5	0.3	0.05–0.1	0.07–0.1
Water	—[a)]	—[a)]	To 100 (~85.6)	To 100 (~85.3)
Sum	100	100	100	100

Data taken with permission from Reference [7].

a) Preparation: 30 g of powder mix are dispersed in 170 g of hot water at 60–80 °C (140–176 °F).

b) When using milk powder, phosphates must be added additionally to ensure an appropriate protein protection.

Alcoholic Beverages

Tables 10.37 and 10.38 Give the Formulation for Some Alcoholic Beverages

Table 10.37 Formulation for liqueur (low-viscosity, 20% DM) [7].

Phase	Ingredient	Concentration (%)
A	Sucrose solution	28.0
	Alcohol (96 vol.%)	18.0
B	CMC – with 10 000 to 30 000 mPa s (2%)	0.2–0.4
	Water	53.6–53.8
C	Flavour	Optional
	Colours	Optional
	Sum	100

Data taken with permission from Reference [7].

Table 10.38 Formulation for creamy liqueur with cream and milk proteins (50–55% DM).

Ingredient	Concentration (%)
Water (added as a cream with 30% fat content)	33.8
Fat (as cream)	30
Sugar	20
Alcohol	13
Milk proteins	3
Flavour + colours	Optional
κ-Carrageenan	0.1
CMC 1000	0.1
Sum	100

Processing guideline:

1. Mixture of alcohol and sucrose solution (A);
2. dissolve the CMC in water with a fast mixer (B);
3. mixing phase (A) and (B);
4. addition of colours and flavourings (C);
5. filling in bottles;
6. pasteurization.

Processing:

CMC is premixed with sugar and then dissolved in cream.

10.9.5 Soups, Sauces, Salad Dressings, and Marinades

Tables 10.39–10.48 list the formulation for various products under this heading.

Table 10.39 Formulations for creamy soups.

Ingredient	Amount (%)
Water	93.4–93.6
Whey protein	2.5
Corn starch	2.5
NaCl	0.7
Spices	0.3
CMC – 10 000 or 20 000	0.4–0.6
Sum	100.0

Table 10.40 Formulations for instant sauce – type 'Béchamel'.

Ingredient	Amount (%)
(A) Dry mixture:	
Milk powder	55.0
Fat powder	27.5–30.0
Pregelatinized corn starch	6.0
NaCl	6.0
Nutmeg	0.5
CMC – 2000 to 20 000	2.5–5.0
(B) Finished sauce:	
Water	80
Dry blend	20

Table 10.41 Formulations for tomato ketchup [6].

Ingredient	Amount (%)	
	Standard (DM approx. 33%)	Reduced tomato content
Tomato concentrate (30 °Brix)	40.0	25.0
Water	30.0	44.5
Sugar	20.0	20.0
Vinegar (target pH 3.7–3.9)	7.5	7.5
NaCl	1.5	1.5
Spices	0.5	0.5
CMC – 30 000 or 40 000	0.5	1.0
Sum	100.0	100.0

Table 10.42 Formulations for a dry mixture for a pizza topping.

Ingredient	Amount (%)
(A) Dry mixture:	
Tomato powder	55.0–56.0
Sugar	32.0
NaCl	7.5
Citric acid, crystalline	0.7
Onion flavour	0.7
Tomato flavour	0.8
Spice mix[a)]	0.8
CMC – 10 000 to 30 000	1.5–2.5
(B) Ready topping:	
Water	80
Dry mixture	20

a) 45% Pepper, 30% oregano, 15% marjoram, and 10% thyme.

Table 10.43 Formulation for salsa sauce (Mexican-type, hot and sour, DM 10–20%).

Ingredient	Amount (%)
Water	30.8–31.3
Tomato pieces	30.0
Tomato concentrate	10.0
Pepper (red and green)	16.0
Onions (garlic optional)	8.0
Spices	1.0
Vinegar (pH 4.0 ± 0.2)	2.0
NaCl	1.0
CMC – 30 000 or 40 000	0.7–1.2
Sum	100.0

Table 10.44 Emulsions – salad cream, type 'mayonnaise', with 30% fat content without egg.

Ingredient	Amount (g)	
	Standard	Type 'freeze–thaw stable'
Water	38.2	39.9
Sunflower oil	30.0	30.0
Vinegar (5%)	13.0	13.0
Lactic acid (10%)	4.0	4.0
Sugar	4.0	4.0

(continued)

Table 10.44 (*Continued*)

Ingredient	Amount (g)	
	Standard	Type 'freeze–thaw stable'
Maltodextrin (DE 20)	4.0	4.0
Salt	2.0	2.0
Mustard	2.0	2.0
Pea protein	1.5	—
CMC – 30 000	1.3	0.8
Methylcellulose – type A with 50 000 mPa s	—	0.3
Sum	100.0	100.0

Table 10.45 Basic formulation for salad dressings (variations are given in Table 10.46).

Basic formulation	Amount (%)
Oil (sunflower)	40.0
Water	35.0
Sugar	11.0
Vinegar	10.0
NaCl	2.75
Egg yolk	1.0
CMC – 30 000 or 40 000	Minimum 0.25
Sum	100.0

Table 10.46 Variations for 'salad dressing' from the basic recipe (Table 10.45).

Type of dressing	Amount (%)
French dressing	25% Oil; 12% tomato concentrate (30 °Brix); 1% each of paprika powder, mustard powder, and onions
Yogurt dressing	30% Yogurt; 26% water; 15% oil; 4% lemon juice; 3% herbs and spices
Reduced-fat dressings	5% Oil • Replacement of the oil content by water, vinegar, tomato concentrate (30 °Brix) resp. yogurt; • increase in the CMC concentration (0.5%)

Processing guideline:

Traditional process of making mayonnaise; CMC can be premixed with the dry materials, or dispersed in oil.

Table 10.47 Overview for marinades.

Type of marinade	CMC type	Function
Neutral marinades based on oil	Medium- to high-viscosity CMC	Film-formation, water retention
Acidic marinades (emulsions)	Low-viscosity CMC (30–100 mPa s, 2%) with a concentration of 1.0–3.0%	Stabilization without thickening effect (no or only a slight viscosity increase), film-formation
	Medium-viscosity CMC (10 000–40 000 mPa s, 2%) with a concentration of 0.3–0.8%	Thickening/stabilization, film-formation, water retention

Table 10.48 Formulations for dips.

Ingredient	Amount (g)
Fresh cheese (10–20% fat)	40.0–50.0
Skimmed milk	40.0–50.0
Spices + herbs	4.0
Citric acid (1 : 1)	2.0
Onions	2.0
Sugar	0.5
NaCl	0.5
Cheese flavour (your choice)	0.5
CMC – 1000 to 20 000	0.5–1.5
Sum	100.0

10.9.6 Meat Products

10.9.6.1 Scalded/Boiled Emulsified Meat Products – Sausages

With carboxymethylcellulose, the quality of emulsified meat products such as boiled sausages can be enhanced. The sensory properties can be improved and the shelf life extended. The sausages have a good bite ('knack'), increased juiciness, and the tendency to synaeresis and drying out is reduced. Medium-viscosity CMC grades (viscosity at 2% from 30 000–40 000 mPa s) with a higher degree of substitution (DS = 0.8–1.0) is particularly suitable. The quantities used are 0.1–0.4%; this means 1–4 g kg^{-1} CMC on moist minced meat, depending on protein and fat content. Owing to the cellulose gum addition, milk proteins may be waived. Moreover, the amount of salt can be reduced. The flow diagram in Figure 10.24 (upper flow chart) shows how to integrate carboxymethylcellulose into the traditional production process for boiled sausages. Alternatively, a single-stage process ('all-in-one') is possible (see Figure 10.24, lower flow chart). In this rapid method, a dry mixture of salts (NaCl, nitrates, phosphates), the

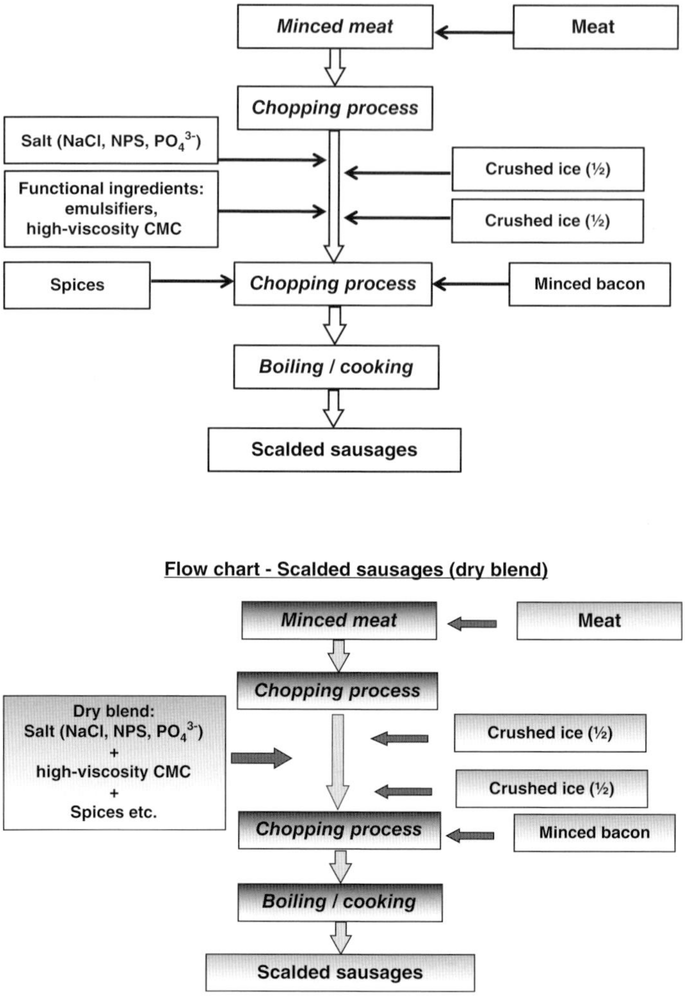

Figure 10.24 Processing of scalded/cooked sausages with CMC.

high-viscosity CMC, and the spices is prepared. This dry blend is then fed to the minced lean meat in the bowl chopper after the first addition of crushed ice (50%). A third possibility is the production of a fat emulsion with CMC. This is made a day before sausage production, cold stored, and then processed like a conventional OW-emulsion the next day.

Table 10.49 shows the formulation for Frankfurter-type sausages.

Table 10.49 Formulation for Frankfurter-type sausages (based on chicken MDM (MDM = mechanically deboned meat)) [7].

Ingredient	Amount (%)
(A) Stabilizer blend:	
Soy protein concentrate	93.4
Mono- and diglycerides	0.6
Carrageenan	3.0
CMC – 40 000[a)]	3.0
Sum	100
(B) Stabilizer–fat emulsion:x	
Stabilizer blend (A)	7.7
Chicken fat (3 mm cubes)	46.0
Water/ice	46.3
Sum	100
(C) Composition of the sausage:	
Chicken MDM	70
Stabilizer–fat emulsion (B)	26
Nitrite salt (0.6%)	1.8
Spices	2.2
Sum	100

Data taken with permission from Reference [7].
a) The alternative is a blend of locust bean gum (1.2%), xanthan gum (0.3%), and guar gum (1.5%).

Processing guideline:

1. Prepare the stabilizer blend by dry mixing of all ingredients listed in (A).
2. The procedure for making the emulsion is as follows:
   a. add water/ice and stabilizer blend into the bowl chopper and run on high speed until homogenous and lump free;
   b. add chicken fat (cut at 3 mm) and chop at high speed until a thorough emulsion is obtained;
   c. age emulsion at 4 °C (39 °F) for 24 h.
3. The procedure for making the Frankfurter sausages is as follows:
   a. place mechanically deboned chicken meat (MDM) in the bowl chopper and run at high speed, adding nitrite salt and spices;
   b. add emulsion to the bowl chopper and chop until thoroughly mixed; maintain the temperature below 6–8 °C (43–46 °F);
   c. fill sausage mixture into 23-mm casings;
   d. cook at 78 °C (172 °F) until a core temperature of 69 °C (156 °F) is reached;
   e. cooling and storage in a cooling cabinet/refrigerator.

10.9.6.2 Injection Brines

The reversible salt sensitivity of long-chain CMC with a low degree of substitution (below 0.7) is used to produce injection brines for meat products (see Section 10.3.3.2). Tables 10.50 and 10.51 give the formulation for two different injection brines.

Processing guideline:

1. Preparation of the brine;
2. the hams are injected to 30% uptake using a multi-needle injector;
3. a vacuum tumbling process follows; total tumble time is 1 h; the tumbling is performed for a 20 min period, followed by 10 min rest;
4. the hams are packed in a cooking bag, cooked at 78 °C (172 °F) until a core temperature of 69 °C (156 °F) is achieved, and then cooled to 5 °C (41 °F).

Table 10.50 Formulation for an injection brine compound with CMC and xanthan.

Ingredient	Concentration (%)
Salts (NaCl, phosphates, carbonates)	86
Xanthan	7
CMC 50 000 with DS = 0.6	7
Sum	100
Amount of dry mixture in the liquid brine	3 (+97% water)
Injection amount, based on meat	30

Table 10.51 Formulation for an injection brine compound with CMC and carrageenan.

Ingredient	Concentration (%)
Water	79.35
Salt, amino acids	5.4
Spray-dried glucose syrup	4.3
Carrageenan	0.6
CMC – 50 000 with low DS of 0.6[a)]	0.1
Potassium chloride, glucose, dextrose	1.44
Nitrite salts	8.57
Flavourings	0.08
Sodium ascorbate	0.16
Sum	100

a) CMC is pre-blended with the dried sugar syrup and then added to the salty brine to ensure that the viscosity development is suppressed or delayed.

10.9.6.3 Sausage Casings

CMC is an important processing aid in the preparation and optimization of sausage casings. An overview of the functions of CMC and recommendations for use are given in Table 10.52.

Table 10.52 Functions of CMC in sausage casings with recommendations.

Application	Function of CMC	Recommendation	CMC type[a)]	Concentration (%)
Collagen casings	• Processing aid during extrusion; • part of the plasticizer solution; • lubricant; • increases the elasticity	CMC grades with low viscosity and minimal water retention capacity	CMC 30; CMC 100	1.5–3
Cellulose-based casings	• Peeling aid for easy peelable sausages (Frankfurt/Strasbourg type); • prevents casing and sausage sticking together ('anti-tack'); • film-former/ lubricant; • ingredient of the spray solution for the inner coating	CMC grades with low viscosity and minimal water retention capacity	CMC 30, CMC 100	1.0–3
Natural casings	• Processing aid during crimping of the casings; • lubricant	CMC with low or medium viscosity	CMC 30, CMC 100, CMC 1000, CMC 2000	0.5–2
	• 'Gluing' the individual pieces of intestine together; • binder/adhesive	CMC with medium viscosity and higher degree of substitution (DS = 0.8–1.0)	CMC 1000 to CMC 30 000	0.2–1

a) The number indicates the viscosity of a 2% CMC solution – measured with a Brookfield LVT at 25 °C (77 °F).

10.9.7
Pet Food and Animal Feed

Carboxymethylcellulose is used as a functional additive and processing aid in animal nutrition, too. Table 10.53 describes in more detail the fields of application for CMC in pet food and animal feed products. Local legislation and acceptance needs to be checked before use.

Table 10.53 Application of cellulose gum in pet food and animal feed.

Application	Function of CMC	Recommendation	CMC type[a)]	Concentration (%)
Pellets	• Improves the structure; • binder for dusts; • lubricant for extrusion; • increased stability, reduced abrasion	CMC grades with low viscosity and minimal water retention capacity	CMC 30, CMC 100	0.05–0.2
Dry food 'Kibbles'	• Coating of the meat pieces with CMC; • addition of water just before consumption; • easy to swallow by forming a sauce/gravy; • slide effect through CMC	High-viscosity CMC	CMC 40 000, CMC 50 000	0.3–0.5
Wet food (canned)	• Thickener; • reduced synaeresis/no or less phase separation	Medium-viscosity CMC	CMC 1000 to CMC 30 000	0.1–0.4
Emulsified meat products	• Increased water uptake; • good water retention, reduced synaeresis; • enhanced shelf life; • potential for protein and fat replacement	High-viscosity CMC with DS 0.8–1.0	CMC 30 000, CMC 40 000	0.1–0.5
Medium-viscosity products (sauces)	• Processing aid to create a higher filling viscosity (the thickness is reduced later through a heating process)	High-viscosity CMC with low DS 0.6–0.7	CMC 50 000 with DS 06	0.3–1
Liquid products (e.g. cat milk)	• Improving the flow properties and consistency; • improved stability; • enhanced durability	CMC with low viscosity and high DS	CMC 30, CMC 100	0.1–1

a) The number indicates the viscosity of a 2% CMC solution – measured with Brookfield LVT at 25 °C (77 °F).

10.9.8 Wine

10.9.8.1 Summary

Tartaric acid is a naturally occurring acid found in grapes.

In wine, it is converted into potassium bitartrate (known as KHT or tartrates). During and after fermentation of wine, tartrates can form insoluble deposits, which will crystallize when wine is exposed to low temperatures. These wine crystals look like little pieces of glass and will not dissolve in the wine. They are harmless but undesirable for consumers.

Winemakers can use physical methods (cold stabilization, electrodialysis) or additives (cream of tartar, *meta*-tartaric acid, mannoproteins) to eliminate tartrate formation from commercial wines. However, these processes can be expensive, time-consuming, and may be ineffective for long-term storage [9].

Sodium carboxymethylcellulose has been shown to be a cost-effective additive for tartrate control. The anionic CMC polymer competes with other ions present in the wine to bond on the surface of crystals, which slows their growth and reduces their size. Cellulose gum is able to suppress the formation of tartar, and thus significantly improves the sensory quality of bottled wines.

Chemical Background

'Tartar' is the collective name for the potassium and calcium salts of L-(+)-tartaric acid[1] that precipitate in the form of small crystals, stick or leave and then settle on the bottom of bottles, on the cork, or at the walls of wine barrels.

The insoluble potassium hydrogen tartrate,[2] potassium bitartrate,[3] or calcium tartrate[4] result from the reaction of tartaric acid with potassium or calcium ions.

Tartaric acid is naturally present in grape juice. By the genesis of alcohol, which is formed during the fermentation of the grapes into wine, the Ca^{2+} and K^+ ions present lead to crystallization of tartaric acid, to the so-called tartar, during the storage of wine in bottles or barrels.

Tartar is colourless and transparent. It is insoluble in ethanol, and very slightly soluble in cold water. The solubility is below $0.4\,g\,l^{-1}$ at 10 °C (50 °F) and decreases with decreasing temperature. The crystals always settle on the walls of wooden barrels or – depending on the type of storage – on the bottom or on corks of bottled wine.

The tartar crystals have no harmful effects on humans. However, they are sensorially noticeable because they have sharp edges, give a sandy feeling in the mouth, and taste sour. The tartar formation is not a 'wine defect', but is seen as a lack of quality for bottled wines.

1) Chemical formula of tartaric acid: $C_4H_6O_6$, HOOC–CHOH–CHOH–COOH, also known as 2,3-dihydroxysuccinic acid or 2,3-dihydroxybutandioic acid.
2) Potassium hydrogen tartrate: $KC_4H_5O_6$, KOOC–$(CHOH)_2$–COOH.
3) Potassium bitartrate: $K_2C_4H_4O_6$, KOOC–$(CHOH)_2$–COOK.
4) Calcium tartrate: $CaC_4H_4O_6$, $^-$OOC–$(CHOH)_2$–COO^-Ca^{2+}.

Therefore, wine producers try to stabilize wines by means of physical or chemical methods so that the formation of crystals is prevented beforehand. Once created, tartar crystals can only be completely removed by filtration.

Common Stabilization Methods

Possible stabilization methods are:

- cold stabilization/wine cooling ($T = -4\,°C$ (25 °F) for 4 weeks):
  - advantage: purely physical treatment,
  - problem: energy intensive process (costs about €2 hl^{-1}) (hl = hectolitre);
- addition of *meta*-tartaric acid:
  - advantage of low cost (about 5–10 cents hl^{-1}),
  - problem: slow degradation of the additive, effective only about 6 to a maximum of 12 months;
- electrodialysis (filtration with ion-selective membranes):
  - costs about €1 hl^{-1},
  - problem: sensory impairment;
- addition of mannoproteins (in the form of autolysed yeast):
  - advantage: natural image,
  - problem: high costs (about €6 h^{-1}), low efficacy.

Table 10.54 presents the common methods for tartaric acid stabilization.

Use of CMC

The use of CMC for the stabilization of the salts of tartaric acid is a novel process that has been used successfully in practice since 2010.

The legal hurdles are cleared. Since August 2009, the use of CMC in wine has been allowed in the EU after the OIV (International Organization of Vine and Wine) assessed it as very positive after intensive investigations.

Table 10.54 Comparison of cellulose gum to other methods of tartaric acid stabilization [9].

	Usage rate ($mg\,l^{-1}$)	Fast processing	Equipment needed	Long-term effectiveness	Costs
Cold stabilization/wine cooling	—	—	Yes	Yes	High
Electrodialysis	—	—	Yes	Yes	Medium
Contact process – cream of tartar	4000	Yes	—	Yes	Medium
meta-Tartaric acid	100	Yes	—	—	Low
Manno-proteins	200	Yes	—	Yes	Very high
CMC	50–100	Yes	—	Yes	Low

Data taken with permission from Reference [9].

This method of CMC supplement provides the following benefits:

- easy, effective, inexpensive way to prevent formation of crystals in wine:
  - low costs (about 10–20 cents hl^{-1}),
  - economically efficient,
  - easy handling;
- low usage rates (40–100 mg l^{-1} of wine);
- long-term effectiveness;
- stability is not impacted by temperature extremes;
- no capital investment or special equipment required;
- no organoleptic influence (no impact on taste and mouthfeel) [9].

The principle of this method is based on the complexation of positively charged potassium ions through the anionic carboxymethyl groups of CMC.

Thus, the K^+-ions are kept in solution and are no longer available for the formation of tartar.

In addition, electrostatic interactions between the CMC molecules and the surfaces of the fine tartrate crystals already formed have been detected.

An example of the reduction of potassium bitartrate (KHT) crystals in white wine using various concentrations of a suitable low-viscosity CMC grade is given below.

The samples were stored at 4 °C (25 °F) for 3 days:

- 520 mg of KHT crystals were formed in the reference sample without CMC;
- 150 mg of KHT crystals were found in the sample stabilized with 40 mg CMC;
- Only 20 mg of KHT crystals occurred in the sample with the addition of 100 mg CMC.

This test series shows that the amount of crystals formed and the concentration of CMC added correlates. A certain amount of cellulosic polymer is necessary to achieve the crystallization control effect and to prevent tartaric crystal formation.

Suitable CMC Grades

To stabilize the tartaric acid or suppress the formation of tartrate crystals, the carboxymethylcellulose used should meet the following criteria:

- low viscosity and short chain length (2% = 20–30 mPa s);
- high degree of substitution (DS $\geq$ 0.9);
- very low fibre content.

Short-chain CMC maintains a constant low viscosity without thickening effect. The high degree of substitution causes good acid stability.

The low fibre content with short fibre lengths is necessary to achieve good (residue-free) filterability.

As the wine producer would incorporate as little extraneous water into the wine as possible, there are two ways to use the CMC:

1) preparation of a concentrated solution of CMC (5% CMC in water) with a traditional CMC;
2) application of a new 'instantized' CMC.

The quantity used is 50–100 mg of CMC per litre of wine (5–10 g hl^{-1}).

When using a concentrated solution of CMC, this concentrate is premixed with the same amount of wine.

The CMC–wine solution is then added to the wine tank before or after the filtration. If the CMC is added before the filtration process, a waiting period of 24 h is to be considered before clarification is carried out.

Owing to these advantages, there is great demand for user-friendly carboxymethylcelluloses for wine stabilization.

A development recently introduced to the market is a so-called 'instantized CMC'. It is a physically modified product that can be stirred directly and lump-free into the wine – without water as a carrier and diluent. The time needed for preparation of the CMC concentrate can thus be saved. In addition, the risk of lump formation is reduced considerably.

10.10 Non-food Applications

10.10.1 Applications for Standard Grades of CMC – Overview

Table 10.55 gives a representative listing of the many applications for sodium carboxymethylcellulose. Many of these applications do not require the use of

Table 10.55 Options to use CMC in technical products [7].

Types of uses	Specific application	Properties utilized
Adhesives	Wallpaper paste	• Water-binding aid, • adhesion, • good open time, • non-straining
	Starch-corrugating adhesive	• Thickener, • water-binding aid, • suspending aid
	Latex adhesives	• Thickener, • water-binding aid

Aerial-drop fluids	Insecticides	• Thickener, • binder • suspending aid
	Drift-control agent	• Thickener
Ceramics	Glazes, porcelain slips, vitreous enamels, refractory mortars	• Binder for green strength, • thickener, • suspending aid
	Welding rod coatings	• Binder, • thickener, • lubricant
Coatings	Foundry core wash	• Binder, • thickener, • suspending aid
	Latex paints, paper coatings	• Rheology control, • suspending aid, • protective colloid
Detergents	Laundry	• Whiteness retention through soil suspension
Lithography	Fountain and gumming solutions	• Hydrophilic protective film
	Water-based inks	• Binder, • rheology control, • suspending aid
Paper and paper products	Internal addition	• High-strength binder, • improves dry strength of paper
	Surface addition	• High-strength binder, • oil-resistant film-former, • provides control of curl and porosity, • provides resistance to oils and greases
	Pigmented coatings	• Thickener, • rheology control, • water-retention aid
Textiles	Laundry and fabric sizes	• Film-former
	Latex-adhesives, backing compounds	• Rheology control, • thickener,
	Printing pastes and dyes	• water binding and holdout
	Warp sizing	• High film-strength, • good adhesion to fibre, • low BOD value
Tobacco	Cigar and cigarette adhesive	• Good wet tack, • high film-strength
	Reconstituted sheet	• High-strength binder, • suspending aid

Data taken with permission from Reference [7].

highly purified CMC grades. Nevertheless, the technical grades of most CMC producers have a high purity of 99% CMC (max. 1% of salt in total); this means an average of only a half-percentage point more salt than food-grades.

10.10.2
Technical and Regulated CMC Applications

Na-CMC is used in various technical and regulated non-food applications [5,10].

1) **Detergent:**
   The ability of CMC to improve detergency was discovered in Germany in 1935.
   CMC is selectively absorbed by cotton fibres through hydrogen bonding.
   The anti-redisposition properties are due to electrostatic repulsion between negatively charged dirt particles and the negative charge of the carboxymethyl groups. 'Tattletale grey' is caused by the dirt particle being redeposited back on the cloth after initial removal by detergent.
   Owing to its anionic charge and its film forming capability Na-CMC acts in detergents as a dirt carrier and prevents secondary deposition on the textile fibre. The amount of cellulose ether is 0.05–0.1 $g\,l^{-1}$ in the wash liquor.
   CMC improves the consistency and wetting in soap, cleansing agents, and detergents.
   However, the advent of wash and wear finished and improved optical brighteners has reduced the competitive advantage of including CMC in the detergent formulation and, consequently, such use has declined in recent years.
2) **Drilling fluids:**
   Sodium carboxymethyl cellulose with low or high viscosity is added to the rinsing fluids in oil and gas drilling. The CMC is supplied to the drilling hopper. It stabilizes the drilling fluids that are constantly carrying the cuttings out of the borehole and sealing the walls of the borehole.
   Oil and gas drilling muds are used to cool the drill bit and to carry the rocks away from the bit face. These 'muds' are aqueous dispersions of clays, weighting agents, and rheology modifiers.
   One rheology modifier is CMC, which serves several functions in the mud system. First, it acts as a suspending agent to keep the clays and weighting agents uniformly suspended in the liquid phase. Second, it serves to provide some 'yield value' to the system to aid carrying the rock cuttings to the surface.
   Although these functions can be performed almost as well by other rheology modifiers, CMC is unmatched by other agents in its ability to disperse clays and to prevent loss of liquid from the drilling fluid into the surrounding rock strata. The property of controlling 'water loss' has resulted in the growing use of CMC in drilling, which is now its largest single application.

3) **Paper industry:**

Large amounts of low-substituted Na-CMC are processed in bulk during the production of printing and writing paper as well as in coloured paper products. In addition, in the production of paperboard, CMC is used to improve the fibre retention, yield of fillers and colours, sizing/gluing, and the strength significantly.

By treatment of paper surfaces with the cellulose ether, the smoothness is increased, the printability is improved, and the dusting is reduced, particularly in offset printing.

The grease-resistant, film-forming characteristics of CMC have made it a useful additive for paper and paperboard manufacture. It can be used alone or in conjunction with starch on the size press to increase dry strength and improve surface characteristics. Size press application aids in controlling viscosity and flow characteristics of the coating colour and suspends and deflocculates pigments. It also binds the coating to the paper more efficiently, providing a more suitable printing surface with decreased dusting.

Recent trends in the paper-coating industry towards the use of less water, to increase running speeds, and to decrease energy consumption have resulted in increased use of CMC.

CMC can be economically applied at the calender stack to increase dry strength and wax pick, decrease porosity, impart greater oil and grease resistance, and improve surface and gloss ink printing characteristics. Paperboard finish is improved because abrasion resistance, stiffness, and smoothness are increased.

Special types of CMC have been developed for addition to the wet end of the papermaking process to promote chemical hydration of the fibres. Greater hydration is reflected in increased dry strength. In Fourdrinier paper or paperboard, increased internal bonding (and, in cylinder machine products, improved ply bonding) are realized without loss of sheet.

CMC in combination with poly[(aminoamide)epichlorohydrin] (PAE) wet strength resin improves wet and dry strength more than PAE resin alone. The improvements result from both increased resin retention and greater effect per unit of retained resin. In wet-end addition, strengths increase with increasing CMC–PAE ratio until the complex becomes anionic and is no longer absorbed by the pulp.

4) **Paints:**

The thickening action and suspending properties of CMC are mainly responsible for its use in emulsion paints. It performs best in anionic systems, such as acrylic emulsions, where its compatibility produces excellent levelling, brushing properties, and pigment dispersion.

Intumescent paints (those that are self-extinguishing or flame-retardant) can be formulated by combining CMC with urea–formaldehyde resin, phosphates, and pigments.

With glue and distemper colours, CMC causes a binding of fillers and pigments, and the colour gets a workable consistency.

Low-viscosity carboxymethylcellulose with up to 4% concentration fulfils this task in dry-wipe-clean interior wall paints.

Washable inner wall paints require the use of medium-viscosity methylcellulose in combination with high polymeric resins such as poly(vinyl acetate) (PVA).

5) **Adhesives and binder:**
Film-forming ability, easy solubility in cold water, and specific adhesion to various surfaces make CMC an excellent adhesive.

A traditional application for cellulose gum is wallpaper paste. It produces a clear film that is non-straining and nearly invisible if traces are left on the surface. CMC is processed in dosages of 2–5%.

Wallpaper is often sold pre-pasted with a dusting of dry CMC powder. The CMC will then go into solution upon being re-wet and can be easily hung in a relatively 'mess free' fashion. For those applications requiring wet tack, gum Arabic, hung glue, dextrin, or D-glucose can be added.

In primer or bonding agents based on latex, CMC is used to stabilize and thicken the water phase.

6) **Cosmetics and pharmaceuticals:**
High amounts of carboxymethylcellulose are processed in the production of toothpaste.

CMC essentially controls the rheology of the dentifrice. The consistency, flow behaviour, and shape of the toothbrush can be set. Additionally, cellulose gum is employed as a suspending agent for mild abrasives and as a thickener.

Low-viscosity CMC grades are used in products with a low content of polyelectrolytes where a stable shape profile is required.

Highly fluoridated toothpastes require highly-substituted CMC types to deliver sufficient CMC solubility and salt stability. The resulting shape stability on the brush and the yield point are comparatively much lower. The cleaning cream tends to flow away from the toothbrush quickly.

Another field of application is denture fixatives in moist semi-firm paste form as adhesive cream or dry as adhesive powder. The ability of CMC to develop wet tack quickly and to dissolve slowly when in a gel state gives huge advantages in dental products. Generally, CMC grades with a high viscosity and very fine particle sizes are used in combination with flavouring additives.

When cellulose gum is used with gelling agents and fillers, a dental impression material can be made.

Cellulose gum is used in laxatives because of its high water-binding capacity. In addition, toxins are physically bonded with the help of CMC – by their incorporation into the swollen polymer – and thereby carried out of the body.

Drugs in the form of syrups, suspensions, and solutions are thickened and stabilized with carboxymethylcellulose. It has been found that CMC delivers the necessary demulcent properties to be beneficial in cough syrup. Eye drops are another example of the application of a very low-viscosity liquid.

Na-CMC is used as a thickening agent for ointments, creams, and lotions for cosmetic and medicinal purposes. The cellulose ether is used as a non-fat ointment base.

Cellulose gum acts as a stabilizer for hand lotions and vitamin–oil emulsions and is widely used as a binder in cosmetic products.

Its grease resistance is utilized in formulating protective hand creams. These protective products based on CMC form a film on the skin that is impermeable to oil and solvents with degreasing effect.

The adhesive qualities of CMC enable it to be readily tableted. Carboxymethylcellulose is insoluble in stomach acid, but soluble in the alkaline fluids of the intestine. This characteristic, in combination with its uniform quality, makes CMC highly suitable for use as an enteric coating for powders and tablets in medicines.

It is a mild but effective laxative. It has been extensively used in geriatric and weight loss formulations because of its ability to retain water without any digestion or irritation of the human intestinal tract.

Its suspending action makes the cellulose derivative valuable for aqueous systems containing calamine (hemimorphites) and antibiotics. In addition, X-ray contrast media contain CMC to stabilize the barium salts used.

In the production of so-called gel pads, the formation of high-viscosity spreadable gels of CMC with trivalent ions is used for medical-technical application. The gels are sealed in plastic bags and can be heated or frozen as often as desired, to then be placed on aching parts of the body.

7) **Ceramics:**

Cellulose gum is used by the ceramics industry to thicken liquids, to suspend solids, and to act as a green strength binder prior to firing. Subsequently, the CMC is burnt out in the kiln.

The most important applications are in glazes for sanitary ware and structural tiles, in porcelain coatings and enamels, in refractory materials such as mortars, in coatings for welding electrodes, as a jigger-body additive, and as a plasticizer for electrical porcelain.

8) **Textiles:**

CMC has received considerable attention in the textile industry because of its readily solubility and excellent film-forming characteristics. Additionally, the high wet adhesion to the fibres and the low biochemical oxygen demand (BOD) value of the CMC are essential for the successful application.

It is particularly useful as a warp size for spun yarns, especially fine cotton yarns, and other fibres. One prime advantage is the ease with which yarns sized with it can be completely separated before weaving.

CMC has replaced starch in many warp-sizing operations because of its advantage in weaving efficiency and in reducing stream pollution at the finishing mill.

Energy savings available from taking advantage of the cold-water solubility of CMC have become more widely recognized. Other advantages are reduced shedding on the loom, elimination of the need for desizing enzymes, and allowable low weave-room humidities.

CMC has also been used successfully for many years as a thickener for textile printing pastes, colorants, latex-containing adhesives, and other adhesive compounds. Cellulose gum controls the rheological properties, binds the water, and affects the absorption resistance. Carboxymethylcellulose is stable to acids and alkalis within the pH range of the printing pastes and can therefore be used with various dyes. It dissolves easily in hot and cold water and is therefore easily removed after printing. Sharp prints can be obtained at high machine speeds because viscosity can be closely controlled.

Fabrics sized with CMC are difficult to soil and easy to clean. This property has generated interest in its utilization as a domestic laundry size and as a spray 'starch' on aerosol packages.

9) **Printing ink:**
The rheological profile of water-based inks is adjusted with CMC, and controlled over their shelf life. The cellulose ether keeps the finest particles in suspension.

In solutions for wetting and sticking together of particles, cellulose gum helps to produce a hydrophilic protective layer with its film-forming properties.

10) **Lithography**
CMC is used as a desensitizing agent for lithographic plates. It imparts desirable ink-repellent properties to the treated areas on the plate. It has better adhesion, is more durable, and necessitates less fountain water to desensitize the plate than gum Arabic. As a result, lithographic plates last longer, production can be increased, and ink colours are brighter and sharper.

11) **Agricultural products – Plant protection:**
Carboxymethylcellulose acts as a protective colloid suspension stabilizer and viscosifier of pesticide emulsions. After application, its adhesive properties function to hold the active ingredients in place without impairing their effectiveness. In addition, in insecticides for plant protection, CMC has the task of binding and thickening the liquid phase.

12) **Metal extrusion:**
CMC is used as lubricant in the extrusion of metals. Thereby, it can significantly reduce the prevailing extremely high frictional forces.

In the manufacture of welding electrodes, the extruder screws can be used up to 30% longer.

13) **Cable industry:**
CMC is used in the sealing of multi-conductor cables. The space between the outer cable insulation of the individual wires and inner sheath is filled to 50% with high-viscosity CMC. If a cable is damaged in wet ground, the cellulose ether swells strongly and seals the cable inner space, so that the ingress of water cannot advance further. Then, only a short part of the cable needs to be replaced.

14) **Tobacco:**
Cellulose gum is used to bind fine tobacco particles to each other to produce new and very thin tobacco films (reconstituted sheets) by means of an extruder.
Moreover, the cellulose derivative has the function of a binding adhesive because of its very high wet tackiness and high film-strength in cigars and cigarettes.
15) **Fire protection:**
CMC also finds use as a thickener for water ad water-chemical systems to produce 'sticky water' for use in fire-fighting.
16) **Soil treatment:**
Cellulose gum improves the structure and air–water relationship of soils. Initial microbial deterioration is overcome by using products of DS greater than 0.7.
17) **Other applications:**
CMC has various other uses. It can be used in:
- auto-polish,
- suspension polymerization,
- moulding extrusion of lead pencils,
- extrusion of coal briquettes,
- lubricants for friction reduction,
- waterproofing of explosives,
- suspending of solids in aqueous gelled explosives,
- emulsions for photographic films,
- clarifying agent for liquids,
- leather dressings,
- embalming fluids.

10.10.3 General Overview for Use of CMC

Table 10.56 shows the application spectrum of CMC in the field of technical and regulated products (without food).

Table 10.56 Technical and regulated application possibilities of Na-CMC [10].

Category/industry	Application	Concentration (%)
Detergents	Washing powder	0.5–1
	Cleaning paste	1–2
	Soap	1
	Detergent and fabric softener	0.5
	Hand cleaning paste	1–2
	Periodic filings	1–3
	Car care products	0.5

(*continued*)

Table 10.56 (*Continued*)

Category/industry	Application	Concentration (%)
Drilling industry	Drilling fluid	0.5–2
	Well construction	0.5–2
Mining	Flotation	0.5–5
	Stocking cartridges	3–5
Textile industry	Smoothing	2–5
	Equipment, floor coverings	2
	Fleece	0.3
Paper industry	Mass	0.2–0.6
	Paper coating	0.5–2
	Surface preparation	0.5
	Wood fibreboards	0.5–5
Wallpaper industry	Fund	0.5–2
	Printing ink	0.5–2
	Dispersion coating	0.5–2
Foundry	Core finishing/smoothing	0.1
	Moulding sand binder	0.1
	Welding wire	0.5
	Rolling mills (drums)	0.1
Ceramic compositions	Porcelain, stoneware:	
	(a) mass	(a) 0.1
	(b) glaze	(b) 0.3
	(c) colour	(b) 0.3
	Enamel	2
Pottery industry	Blank	0.05
	Glaze	0.5
Paints	Distemper	3–4
	Dispersion paints	0.3–0.8
	Synthetic plasters	0.3–0.8
	Mineral paints	1
	Lead and coloured pencils	0.6–1
Adhesives	Wallpaper glue	2–5
	Dispersion adhesives	0.3–1
Construction materials	Gypsum-anhydrite	0.2–0.4
	Bitumen emulsions	0.3–0.5
Films and foils	Tobacco films	15
Cosmetics	Toothpaste	1
	Skin cream	1
	Shampoo	0.5
	Denture adhesive powder	50–100
Pharmacy	Disintegrants	1
	Ointments	1
	Cod liver oil	1
Animal nutrition	Pellets	1–3
	Semi-moist	0.05–0.5
	Canned	0.05–0.5
Agriculture	Plant protection	1
	Seed compounds	1
Cable industry	Multi-conductor cable	Approx. 50%, filling of hollow space

Table 10.57 Nutritional value of carboxymethylcellulose.

Component	Amount (in 100 g)
Energy content	0 kcal
Fat	0 g
Dietary fibre (total)	At least 82 g
Soluble dietary fibre (ß-D-glucose)	At least 82 g
Carbohydrates/sugar	0 g
Protein	0 g
Vitamins (in sum)	0 g
Sodium	6.5–9.5 g
Total salt content	Max. 0.5 g
Sodium chloride	≤0.5 g
Sodium glycolate	≤0.4 g
Water	Max. 8 g
Ash (as sulfate ash)	20–33%

10.11 Nutritional Properties

As it is derived from native cellulose, Na-CMC is also neither absorbed nor metabolized in the human body.

Cellulose gum is considered as harmless and safe, so that FAO and WHO set the ADI (acceptable daily intake) as 'not specified'. In addition, CMC has – as methylcellulose has, too – GRAS ('generally recognized as safe') status, certified by the US FDA (US Food and Drug Administration).

Positive effects of carboxymethylcellulose were found regarding sugar resorption. CMC reduces the speed of absorption of mono- and disaccharides and thus influences the glycemic index (GI).

Table 10.57 contains a list with the most important nutritional information of cellulose gum.

10.12 Legislation

10.12.1 European Union

Approval as Technological Additive

Cellulose gum conforms to the purity criteria of E 466 as mentioned in the Annex of EC directive 2008/84/EC and amendments. In the European Union, sodium carboxymethylcellulose is listed as a 'generally allowed food additive for use in food' in Annex I of the 'Directive on miscellaneous additives'. It may be used as a technological additive to achieve particular effects in almost all food products in unlimited dosage ('quantum satis'). Exceptions are untreated products and those food products that may not be modified by additives. There is an obligation to declare the hydrocolloid used on the label of a food package. Descriptive labelling can be made with 'Thickener' or 'Stabilizer'. Clean labelling without chemical name and E number has been possible since 2006. The use of the term 'cellulose gum' is fully sufficient and offers new possibilities.

An ADI of 'not specified' has been allocated by the Scientific Committee of the EU and the Joint FAO/WHO Expert Committee on Food Additives.

In the *Official Journal of the European Union* (Directive No. 2008/84/EEC), the following purity criteria for CMC are set [1]:

- content: minimum 99.5% (on dry basis);
- total glycolate: maximum of 0.4%, calculated as sodium glycolate;
- sodium: maximum of 12.4% (based on dry matter);
- molecular weight: >17 000 $g\,mol^{-1}$ (DP about 100);
- substitution: 0.2–1.5 carboxymethyl groups ($-CH_2COOH$) per anhydroglucose unit (AGU).

The original target of fixing the lower limit of the molecular weight was to exclude very short CMC molecules that might pass or diffuse through the membranes of the intestinal cells, reach the bloodstream, and 'circle endlessly' there or lead to deposits. These fears have been refuted in part, which is shown and proved by more recent applications in the pharmaceutical field.

The production costs for very low-viscosity CMC grades are comparably high due to the massive leaching losses during purifying. Thus, commercially available products have a degree of polymerization of more than 400 glucose monomers per molecule and molecular weights greater than 50 000 $g\,mol^{-1}$.

At the end of 2003, a revised labelling of carboxymethylcellulose was decided by the EU panel on food law. The written statement can be found in the *Official Journal of the European Union* from 29 January 2004 [11]. After a two-year implementation phase, CMC may be designated as 'Cellulose gum' in the ingredient list of a food product. The name can be translated into all languages. The E-number or the chemical name may be omitted, so that consumer-friendly declaration or labelling is possible, as traditionally done when using pectins or starches.

Special Approval for CMC

In addition, CMC is approved in pasteurized full fat cream. The use is allowed without restriction ('quantum satis') [12].

On the list for 'Infant and Toddler Foods for Special Medical Purposes (FSMP)', E 466 (CMC) is allowed in dietary products for use with inborn disorders in the fatty acid metabolism. The dosage is limited to a maximum 10 g per litre or kg, which corresponds to a concentration of 1% in the product ready for consumption.

Negative List for All Cellulose Derivates 'E 460 and Following'

Like all other cellulose derivatives, CMC may be **not** added in the following food products:

- untreated food;
- honey (in the sense of the 'Honey-Regulation'/Directive 74/409/EEC);
- non-emulsified oils and fats of animal or vegetable origin;
- butter;
- pasteurized and sterilized (including UHT) milk (even full-fat, skimmed, or partially skimmed) and plain pasteurized cream with full fat content;
- unflavoured, with living bacteria fermented, milk products;
- natural mineral water and spring water in terms of 'The mineral and table water regulation or directive 80/777/EEC';
- coffee (excluding flavoured instant coffee) and coffee extracts;
- non-flavoured tea from leaves;
- types of sugar (in the sense of 'The Sugar's Regulation') and Directive 73/473/EWG (i.e. sucrose, glucose, dextrose, and fructose, each in different dosage forms);
- dry pasta, excluding gluten-free pasta and/or pasta intended for a low-protein diet (in terms of 'The diet regulation' or directive 89/389/EEC);
- natural unflavoured buttermilk (excluding sterilized buttermilk);
- cocoa and chocolate products within the meaning of 'The Cocoa Regulation', but not those with reduced calorie content or with no added sugar:
  - Cocoa Regulation: cocoa butter, cocoa powder, chocolate powder, drink chocolate powder, chocolate, milk chocolate – each with a specific composition that is specified in 'The Cocoa Regulation';
- fruit juices and fruit nectars as defined in 'The Fruit Juice Regulation' or Directive 93/77/EEC;
- extra jam and extra jelly as defined in 'The Jam Regulation' or Directive 79/693/EEC;
- jams and jellies within the meaning of 'The Jam Regulation' or similar fruit spreads including low-calorie products;
- dry milk and evaporated milk product as defined in the 'Dairy product Regulation' or Directive 76/118/EEC;
- peeled potatoes;
- unprocessed fruit and vegetables – frozen, deep-frozen, pre-packaged, and cooled;
- unprocessed fish, crustaceans, and molluscs, including frozen – or deep-frozen;
- fruit compote;
- quick-cooking rice;
- sour cream butter;
- ripened cheese;
- mozzarella and whey cheese;
- glass or metal cans of fruit and vegetables;

- preparations of fresh minced meat;
- bread: made from wheat flour, water, yeasts, or sourdough and salt;
- paint courant français;
- fresh pasta;
- beer;
- foie gras;
- UHT goat milk;
- chestnuts in liquids;
- infant start formula and follow-on formulas, complementary baby foods for infants and young children.

10.12.2
Other Countries (Non-EU)

United States of America

Food additives are regulated in the USA by the Food and Drug Administration (FDA) and the US department of Agriculture (USDA). Of the chemical additives, only MC and sodium CMC are generally recognized as safe (GRAS). The specific approval for CMC is listed in the US Code of Federal Regulations (CFR) Title 21 CFR 182.1745. There are other specific approvals for cellulose derivatives such as those that define their use in 'Adhesives and Coatings for Food Use' (21 CFR 175.300) or specific food products such as 'Artificially Sweetened Fruit Jellies' (21 CFR 150.141).

The USDA has regulatory authority over meat products in the USA. The approvals for cellulose ethers in meats are different to those for other food products. For red meat, only sodium CMC is approved as an extender or stabilizer in baked pies. Methylcellulose is approved as an extender or stabilizer in meat and vegetable patties. Both are according to 9 CFR 318.7. For poultry products, both CMC and MC are approved as extenders and stabilizers according to 9 CFR 381.147 [4].

CMC meets the monograph requirements set forth in the Food Chemicals Codex, current edition. FAO/WHO defined the ADI value as 'not specified'.

Canada

In Canada, carboxymethylcellulose is listed in the Food and Drugs Act and Regulations as food additive that may be used as emulsifying, gelling, stabilizing, and thickening agent.

Australia

In Australia, sodium CMC has been approved for inclusion in the Food Standards Regulations under Section A 10, group 1, and referencing modifying agents [4].

Japan

In Japan, sodium and calcium CMC have approval for food use. The materials must meet specifications for the 'Japanese Pharmacopeia' or the 'Japanese Food Codex' [4].

Other Countries and Areas

In other countries and areas, the legal status for cellulose gum has to be checked before the formulation of new food systems or the modification of existing products.

As the product has been intensely examined in Europe, North America, Australia, and Japan, and widely accepted, these successful approvals will support faster and successful implementation in other markets.

References

1 European Union (2008) Commission directive 2008/84/EC of 27 August 2008, laying down specific purity criteria on food additives other than colours and sweeteners. *Official Journal of the European Union*, **51**, L253, 94–102. Available at http://eur-lex.europa.eu/LexUriServ/LexUriServ.do?uri=OJ:L:2008:253:0001:0175:EN:PDF.

2 Ashland Inc. (2012) Aqualon™ sodium carboxymethylcellulose – physical and chemical properties (PRO-250-10H). Information from the library at http://www.ashland.com/products/aqualon-cellulose-gum.

3 Belitz, H.-D., Grosch, W., and Schieberle, P. (2009) *Food Chemistry*, 4th edn, Springer-Verlag, pp. 294–297.

4 Stephen, A.M., Philips, G.O., and Williams, P.A. (eds) (2006) *Polysaccharides and Their Applications*, 2nd edn, Taylor & Francis Group, Boca Raton, FL, pp. 147–179.

5 Whistler, R.L. and BeMiller, J.N. (eds) (1993) *Industrial Gums, Polysaccharides and Their Derivatives*, 3rd edn, Academic Press Inc., San Diego, chapters 18–20, pp. 475–578.

6 Murray, J.C.F. (2000) Cellulosics, in *Handbook of Hydrocolloids* (eds G.O. Phillips and P.A. Williams), Woodhead Publishing Limited, Cambridge, pp. 219–230.

7 Ashland Inc. (2012) Aqualon™ sodium carboxymethylcellulose – recipes and recommendations. Formulations from the library at http://www.ashland.com/products/aqualon-cellulose-gum.

8 Ashland Inc. (2012) Masa flour for tortillas with Aqualon™ 7H4F M cellulose gum (Formula 01-1020A). Formulation from the library at http://www.ashland.com/products/aqualon-cellulose-gum.

9 Ashland Inc. (2012) Aquacel™ WS – Cellulose gum for stabilization of tartrates in white wine. Product information from the library at http://www.ashland.com/products/aquacel-cellulose-gum.

10 Balser, K. and Iseringhausen, M. (1975) Cellulose ether, *Ullmann's Encyclopedia of Industrial Chemistry* (eds E. Bartholomé *et al.*), vol. **9**, 4th edn, Verlag Chemie, Weinheim, pp. 192–212.

11 European Union (2004) From 29 January 2004 to the changed declaration of sodium carboxymethyl cellulose (Na-CMC) as cellulose gum (Directive 2003/114/EC of the European Parliament and of the Council from 22 December 2003 amending Directive 95/2/EC on food additives other than colours and sweeteners). Amendment by M6 on 29 January 2004, page 11. *Official Journal of the European Union*. http://eur-lex.europa.eu/LexUriServ/LexUriServ.do?uri=CONSLEG:1995L0002:20060815:EN:PDF.

12 European Parliament and Council (1995) Directive Number 95/2/EC of the European Parliament and the Committee for food additives other than colours and sweeteners (List of food products in which only a limited number of food additives is permitted). http://ec.europa.eu/food/fs/sfp/addit_flavor/flav11_en.pdf.

13 Herrick, F.W., Casebier, R.L., Hamilton, J. K. and Sandberg, K.R. (1983) Microfibrillated cellulose: morphology and accessibility, in *Proceedings of the Ninth Cellulose Conference* (ed. A. Sarko), Applied Polymer Symposia, vol. **37**, John Wiley & Sons, Inc., New York, pp. 797–813, ISBN 0-471-88132-5.

11
Crosslinked Sodium Carboxymethylcellulose

Synonyms: crosslinked carboxymethylcellulose, crosslinked sodium-CMC, crosslinked Na-CMC, crosslinked cellulose gum, modified cellulose gum, croscarmellose

E number: E 468
EINECS: not listed
CAS number: 74 811-65-7

Related hydrocolloids: cellulose, Na-CMC

11.1
Manufacturing Process

Crosslinked sodium carboxymethylcellulose (croscarmellose) is prepared by the reaction of Na-CMC with acids. The raw material is usually a high-viscosity sodium CMC with a low degree of substitution. Bridging units are produced between the straight chains of the cellulose, so that a dense network is created.

11.2
Chemistry

Chemically speaking, crosslinked sodium carboxymethylcellulose is the sodium salt of a thermally crosslinked, partially O-carboxymethylated cellulose. Figure 11.1 shows how croscarmellose is produced by intra- and intermolecular esterification of carboxymethyl and hydroxyl groups with the formation of water molecules.

The general chemical formula for croscarmellose is:

$$C_6H_7O_2(OR^1)(OR^2)(OR^3)$$

Cellulose and Cellulose Derivatives in the Food Industry: Fundamentals and Applications, First Edition.
Tanja Wüstenberg.

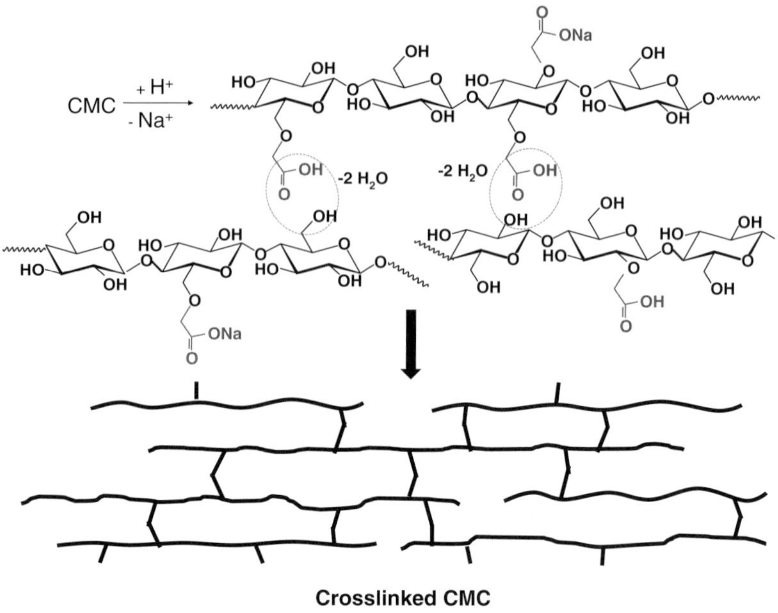

Figure 11.1 Schematic representation of crosslinked CMC (croscarmellose).

where R^1, R^2, and R^3 can be:

H
$CH_2COO^-Na^+$
CH_2COOH
CH_2COO^- 'CMC' (ester).

The degree of substitution is between 0.2 and 1.5 carboxymethyl groups per anhydroglucose unit [1].

11.3 Rheology

11.3.1 Dissolution Behaviour

Croscarmellose is not soluble in water. The proportion of water-soluble components is less than 10% [1]. By networking, the solubility in water is strongly reduced, but the swelling capability is not restricted. Croscarmellose can absorb many times its own weight of water. The highly swollen substance does not feel wet or sticky. Application is limited to tablets in which the crosslinked Na-CMC is used as disintegrant.

11.3.2
Gelation and Behaviour with Other Ingredients

Crosslinked sodium CMC shows no gelation in aqueous systems. Owing to the insolubility in water, there is usually only little to no interaction with other components of the formulation. Under certain circumstances, the croscarmellose rapidly absorbs the available water to swell. This can affect the solubility of more inert ingredients. After the explosion-like disintegration of the tablet, all ingredients (active pharmaceutical ingredients, vitamins, or minerals) are released. The croscarmellose as excipient then lies inert in the human stomach or in the drinking glass.

11.4
Stability

Crosslinked sodium CMC is highly stable due to its insolubility in water. The traditional stress factors (shear force, pH value, heat) are not discussed further because of the very limited application options in food.

11.5
Rheometry

Crosslinked sodium carboxymethylcellulose can be detected by the following staining methods [1]:

1) A 1-g sample is shaken in 100 ml of methylene blue solution ($4\,\mathrm{mg\,kg^{-1}}$) first and then kept until settled. The substance absorbs the methylene blue and settles as a blue fibrous mass.
2) A 1-g sample is shaken in 50 ml of water; 1 ml of the mixture is poured into a test tube, 1 ml of water is added and then 0.05 ml of a freshly prepared solution of α-naphthol in methanol ($40\,\mathrm{g\,l^{-1}}$) is added. The test tube is inclined and 2 ml of sulfuric acid is added carefully down the side, so that a lower layer is formed. The interface turns red–purple.

The explosive force of the croscarmellose can be measured with a corresponding disintegration tester.

11.6
Synergies with Other Hydrocolloids

There are no significant synergies between croscarmellose and other hydrocolloids.

Upon swelling, the crosslinked CMC consumes a certain amount of water and can impede the hydration and dissolution of other substances because of its very high water-absorption speed.

11.7 Food Applications with Typical Formulations

Because of its insolubility in water, crosslinked Na-CMC is not used in typical liquid or semi-liquid food applications. In the European Union, the application of croscarmellose is restricted to dietary supplements, sweetener tablets, and medicines. The dosage in the so-called 'Food Supplements' is a maximum of $30\,g\,kg^{-1}$.

Crosslinked CMC acts as a disintegrant in tablets (see Section 11.8). The relative concentration is 1% of the tablet core.

11.8 Non-food Applications

Crosslinked Na-CMC is a drug carrier (excipient) in medical formulations. It is highly absorbent and insoluble, like a sponge. This enables the remaining components of the formulation to better come into contact with body fluids and, thus, increases their bioavailability. Croscarmellose is used as a disintegrant in pharmaceutical applications. Chemical crosslinking leads to greatly reduced water solubility, resulting in unrestricted high swelling capacity. Croscarmellose can absorb many times its own weight of water. The result is an improved dissolution profile and excellent release of the active agents, with subsequent increased bioavailability. With the use of croscarmellose, the problem of long-term stability being coupled to the tablet hardness is eliminated or reduced. In combination with a suitable coating, the crosslinked CMC can help, in both pharmaceuticals and food supplements, the tablet to dissolve in the correct 'location' in the body and to ensure that the active substances are released quickly and completely. Thereby, a high efficiency is achieved.

11.9 Nutritional Properties

Croscarmellose is considered as safe. No ADI has been set for human consumption. Owing to the lack of available data, evaluation of crosslinked carboxymethylcellulose (E 468) is based on knowledge of Na-CMC (E 466). As insoluble dietary fibre, croscarmellose is neither digested nor metabolized by the human body. In higher doses, it can give a softer consistency to the bowel content [2].

11.10 Legislation

Crosslinked carboxymethylcellulose has been approved only for use in solid food supplements (with a dose maximum of $30\,g\,kg^{-1}$) and sweetener tablets [2].

The authorization for pharmaceutical products is in accordance with the US FDA, USPh, and EUPh.

References

1 European Union (2008) Commission directive 2008/84/EC of 27 August 2008, laying down specific purity criteria on food additives other than colours and sweeteners. *Official Journal of the European Union*, **51**, L253, 94–102. Available at http://eur-lex.europa.eu/LexUriServ/LexUriServ.do?uri=OJ:L:2008:253:0001:0175:EN:PDF.

2 For food legislation in Germany see: Anon (2000) *Federal Law Gazette*, 1521–1536. Available at http://bundesrecht.juris.de/zzulv_1998/anlage_4_16.html.

12
Enzymatically Hydrolysed Carboxymethylcellulose

Synonyms: with enzymes hydrolysed sodium carboxymethylcellulose, Na-carboxymethylcellulose partially enzymatically hydrolysed, enzymatically hydrolyzed cellulose gum

E number: E 469

Related hydrocolloids: sodium CMC, chemically degraded

12.1
Manufacturing Process

Hydrolysed carboxymethylcellulose is obtained by enzymatic digestion of sodium carboxymethylcellulose with cellulases. The cellulases are produced by the ascomycete (Ascomycetum) *Trichoderma longibrachiatum*. In the past, *Trichoderma reesei* was used to break down the cellulose chain. The molecular structure of the carboxymethylcellulose (CMC) is changed to such an extent that the remaining thickening effect is only very weak. The content of mono- and disaccharides is a maximum of 7.5% [1].

12.2
Chemistry

The general chemical formula of the enzymatically hydrolysed sodium carboxymethylcellulose is:

$$\left[C_6H_7O_2(OH)_x(OCH_2COONa)_y\right]_n$$

where:

- n = degree of polymerization ($n \geq 4$);

Cellulose and Cellulose Derivatives in the Food Industry: Fundamentals and Applications, First Edition.
Tanja Wüstenberg.

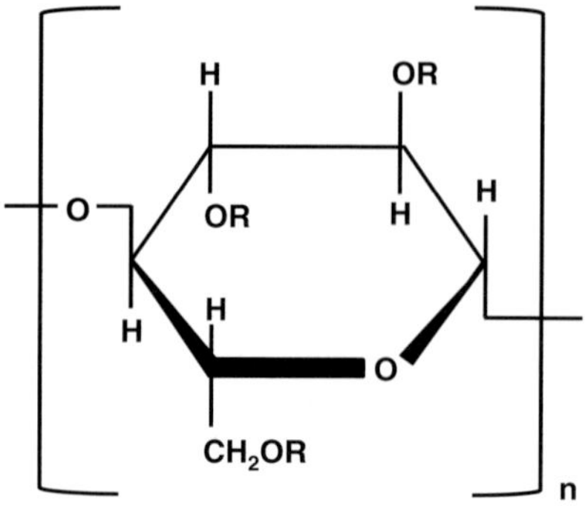

Figure 12.1 Structure formula of enzymatically degraded Na-CMC (R = H, CH_2COONa, or CH_2COOH).

- y = degree of substitution;
- x = number of unsubstituted hydroxyl groups ($x + y = 3$);

$x = 1.50–2.80$ and $y = 0.2–1.50$. Figure 12.1 shows the structural formula. The weight of an anhydroglucose ring is in the following range:

- 178.14, when the degree of substitution (DS) is $y = 0.2$;
- 282.18, when the DS is $y = 1.5$.

The macromolecule has a weight of at least 800 g mol^{-1}, which corresponds to a degree of polymerization of at least four glucose monomers. Figure 12.2 shows a typical molecule of degraded sodium CMC.

The purity is 99.5% based on the dry matter, including the mono- and disaccharides present. The maximum residual salt content (sodium chloride and sodium glycolate in total) is 0.5%.

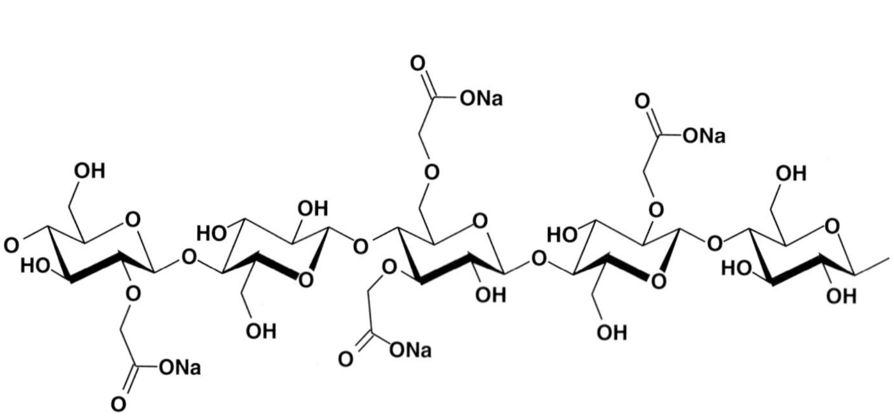

Figure 12.2 Molecular model of an enzyme-hydrolysed Na-CMC with five glucose monomers.

12.3
Rheology

12.3.1
Dissolution Behaviour

Like the original CMC, the enzymatically hydrolysed carboxymethylcellulose is very soluble and is completely dissolved in water at any temperature between 0 and100 °C (32–212 °F). The shorter the cellulose chain, the shorter the hydration time will be. In addition, with higher concentrations, the thickening effect is pronounced only slightly. Degraded CMC is insoluble in pure ethanol, but soluble in mixtures of water and ethanol (up to a maximum of 40% EtOH). All CMC types are – regardless of the chain length of the molecule – not soluble in oil. The presence of minerals, salts, sugars, and other hydrocolloids has an effect on the solubility, depending on the concentration of the individual substances (see Section 10.3.1).

12.3.2
Gelation and Behaviour with Other Ingredients

Enzymatically decomposed carboxymethylcellulose shows no gelation in water. At any temperature and at normal pressures, this short-chain CMC dissolves completely in water in a colloidal form.

Owing to the good solubility in water, its compatibility with other ingredients of polymeric nature (proteins, soluble polysaccharides and dextrins, fats/oils) and with non-polymeric products (mono-, di-, and oligosaccharides, NaCl, phosphates) is good, and only a slight increase in viscosity results. Hydrolysed carboxymethylcellulose shows only moderate protein protection.

12.4
Stability

The influences of temperature, pH value, and shear forces are no longer very strong because the cellulose chains are comparatively short after hydrolytic degradation.

When the temperature is increased from 40 to 100 °C (104 to 212 °F), a slight decrease in viscosity is observed. This reversible heat thinning effect, however, is completely reversible at neutral pH values, and, on cooling, the initial viscosity is restored. By shear stress, there may also be – depending on the viscosity of the system – a reduction in viscosity, which is reversible; after a rest period, the original texture is reached again (thixotropy). At acidic pH values, there is protonation of the negatively charged carboxymethyl groups. The insoluble acid form of carboxymethylcellulose is produced, which is precipitated as solid particles.

This neutralization process is irreversible and leads to a visible flocculation at pH values below 3.5.

Notably, in a neutral medium pre-dissolved CMC is considerably less sensitive to the addition of acid, as a dry solid powder, which is incorporated into an acidic system.

12.5 Rheometry

Enzymatically hydrolysed sodium CMC is characterized and distinguished from other hydrocolloids by the following six characteristics [1]:

1) **Solubility:**
   it is insoluble in 100% ethanol, but completely soluble in cold water.
2) **Foam test:**
   when a 0.1% solution of the sample is shaken vigorously, hydrolysed CMC does not foam.
3) **Precipitation:**
   On addition of 5 ml of a 5% solution of copper sulfate or aluminium sulfate solution to 5 ml of a 0.5% solution of the sample, CMC grades form a precipitate. All other cellulose ethers, gelatine, tragacanth, and locust bean gum show no reaction.
4) **Colour reaction test:**
   Here, 0.5 g of CMC sample is added to 50 ml of water and dissolved with stirring, until a clear solution is achieved. In a test tube, 1 ml of the solution is diluted with 1 ml of water and then mixed with five drops of 1-naphthol-TS. The test tube is then inclined and 2 ml of sulfuric acid is poured carefully along the side, so that a lower layer is formed. The interface is coloured purple.
5) **Viscosity (60% solids):**
   this is a minimum 2500 mPa s (at 25 °C (77 °F)), which corresponds to an average molecular weight of 5000 Da.
6) **Residual enzyme activity:**
   A complete absence of enzymes (cellulases) means that the viscosity of the test solution remains constant. When the viscosity value decreases, however, there are still active enzymes present.

12.6 Synergies with Other Hydrocolloids

Enzymatically hydrolysed carboxymethylcellulose (CMC) is very compatible with most water-soluble thickeners and stabilizers – regardless of their concentration. In hydrocolloid blends, low-viscosity CMC grades are significantly more flexible than high-viscosity CMC products because they have a lower water absorption capacity and, as a consequence, they require less water to develop their full functionality. CMC can be used in combination with non-ionic polymers (such as guar or HPMC (hydroxypropylcellulose)) to achieve a synergistic effect on the viscosity. The mixture with anionic CMC shows a higher thickening factor, as expected theoretically. Negative interactions can be detected between carboxymethylcellulose and highly esterified pectin ('HM pectin'). The similar molecular structure with substituted carboxymethyl groups leads to the same negative charge and thus to electrostatic repulsion of the two substances. From a technological perspective, the combination 'CMC – HM pectin' is not recommended and does not make sense, since both hydrocolloids fulfil the same function (e.g., in the stabilization of acidic dairy drinks).

12.7 Food Applications with Typical Formulations

Owing to the strongly reduced thickening effect, the function of the enzyme-hydrolysed CMC is limited to:

- the stabilization of beverages;
- low-calorie fillings.

The dosage is 0.5–3%.

Carboxymethylcellulose forms a film on the tongue and induces an oily mouthfeel. By the use of CMC, food products obtain a creamy consistency and fat-like melting behaviour. Other possible applications are presented in Section 10.8.

In the food industry, enzymatically hydrolysed CMC (E 469) is not used in significant volumes in Europe or North America. According to new GMO regulations, enzyme-catalysed production processes are strongly prone to criticism. Therefore, product developers prefer to use the original carboxymethylcellulose (E 466) and its chemically degraded low-viscosity variations.

12.8 Non-food Applications

The use of enzymatically degraded CMC corresponds to that of traditional chemically hydrolysed carboxymethylcellulose in technical applications (Section 10.10).

12.9 Nutritional Properties

Enzymatically degraded sodium CMC is assessed as safe. Like the untreated carboxymethylcellulose, the low-viscosity substance is a soluble dietary fibre.

The energy content is a maximum of 30 kcal per 100 g dry matter. An ADI has not been defined.

12.10 Legislation

In the European Union (EU), enzymatically hydrolysed sodium carboxymethylcellulose is generally approved as a technological additive for food products without a maximum limit ('quantum satis'). Exceptions, where E 469 is not allowed, are:

- untreated products;
- those foods that should not be changed by additives;
- special products for baby and infant nutrition.

More information on the excluded food products and on the global legislation of CMC can be found in Section 10.12.

Reference

1 European Union (2008) Commission directive 2008/84/EC of 27 August 2008, laying down specific purity criteria on food additives other than colours and sweeteners. *Official Journal of the European Union*, **51**, L253, 94–102. Available at http://eur-lex.europa.eu/LexUriServ/LexUriServ.do?uri=OJ:L:2008:253:0001:0175:EN:PDF.

13
Nanocellulose

Synonyms: nanocrystalline cellulose (NCC), cellulose nanocrystals (CNC), microfibrillated cellulose (MFC)

Related hydrocolloids: cellulose, microcrystalline cellulose (MCC), powdered cellulose

13.1
Definition and Summary

Nanocellulose, or microfibrillated cellulose (MFC), is composed of nanosized cellulose fibrils with a high aspect ratio (length to width ratio). Typical lateral dimensions are 5–20 nm and the longitudinal dimension is in a wide range from 10 s of nanometres to several microns. It is pseudoplastic and exhibits the property of gels or fluids that are thick (viscous) under normal conditions, but flow (become thin, less viscous) over time when shaken, agitated, or otherwise stressed. This property is known as thixotropy. When the shearing forces are removed, the gel regains much of its original state. The fibrils are isolated from any cellulose containing source, including wood-based fibres (pulp fibres), through high-pressure, high-temperature, and high-velocity impact homogenization (see also Section 13.4 on the 'Manufacturing process') [1].

Nanocellulose can also be obtained from native fibres by acid hydrolysis, giving rise to highly crystalline and rigid nanoparticles (generally referred to as nanowhiskers), which are shorter (100 s to 1000 s nanometres) than the nanofibrils obtained through the homogenization route. The resulting material is known as nanocrystalline cellulose (NCC) [1,2].

13.2
Commercial Products

Microfibrillated cellulose (MFC) is available in three commercial forms:

1) a 4% paste in water;

Cellulose and Cellulose Derivatives in the Food Industry: Fundamentals and Applications, First Edition.
Tanja Wüstenberg.

2) a granular product with 25% MFC and 75% water;
3) a powder with 50% MFC and 50% sucrose.

The paste is readily dispersible in water or a number of water-miscible organic solvents. Medium shear mixing is needed to obtain maximum viscosity and water retention. The granulated and powdered products require high shear mixing to allow the product to swell fully in aqueous systems. Dry blending the MFC with other ingredients before adding to the wet mix is often advantageous. The current commercial status is unclear [3].

13.3 History and Terminology

The terminology microfibrillated cellulose (MFC) or nanocellulose was first used by Turbak, Snyder, and Sandberg in the late 1970s at the ITT Rayonier laboratories in Whippany, New Jersey, USA to describe a product prepared as a gel-type material by passing wood pulp through a Gaulin-type milk homogenizer at high temperatures and high pressures followed by ejection impact against a hard surface.

The terminology (MFC) first appeared publicly in the early 1980s when several patents and publications were issued to ITT Rayonier on this totally new nanocellulose composition [1,4]. In later work, Herrick at Rayonier also published work on making a dry powder form of the gel [5]. Since Rayonier is one of the world's premier producers of purified pulps their business interests have always been:

1) to create new uses and new markets for pulps;
2) never to compete with new or potentially new customers.

Thus, as indicated in the patents issued [6], Rayonier gave free license to whoever wanted to pursue this new use for cellulose. Rayonier, as a company, never pursued scale-up. Rather, Turbak *et al.* pursued:

1) finding new uses for the MFC/nanocellulose; these included using MFC as a thickener and binder in foods, cosmetics, paper formation, textiles, nonwovens, and so on;
2) evaluate swelling and other techniques for lowering the energy requirements for MFC/nanocellulose production [7].

ITT closed the Rayonier Whippany Labs in 1983–84 and further work on making a dry powder form of MFC was carried out by Herrick at the Rayonier laboratories in Shelton, Washington, USA [5]. The field was later taken up in Japan in the mid-1990s by the group of Taniguchi and coworkers and later by Yano and coworkers [8] and a host of major companies (see numerous U.S.

patents issued to P&G, J&J, 3 M, McNeil, etc. using a U.S. patent search under inventor name Turbak).

Today, there are still extensive research and development efforts around the world in this field. There are two pilot plant production facilities in North America – one in Canada manufacturing one ton per day and one just-opened facility in the USA where crystals and fibrils are produced [1,9]. Another pilot plant is located in Sweden, where 100 kg per day can be produced, so that testing material for new development is available on a larger scale.

13.4 Manufacturing Process

13.4.1 Nanocellulose/MFC

Nanocellulose or microfibrillated cellulose can be prepared from any cellulose source material, but wood pulp is normally used. The nanocellulose fibrils are isolated from the wood-based fibres using high-pressure homogenizers. The homogenizers are used to delaminate the cell walls of the fibres and liberate the nanosized fibrils. This process is responsible for the high energy consumption associated with fibre delamination. Values of over 30 MWh per tonne are not uncommon. Pretreatments are sometimes used to address this problem. Examples of such pretreatments are enzymatic or mechanical pretreatment [10] and introduction of charged groups, for example, through carboxymethylation [11] or TEMPO-mediated oxidation. TEMPO is a chemical substance with the official name 2,2,6,6-tetramethylpiperidinyloxyl (sum formula $C_9H_{18}NO$, CAS number 2564-83-2). It is a stabilized radical that is used as oxidant. It has been shown that energy consumption can be greatly decreased by using these pretreatments. The researchers Lindström and Ankerfors have reported values below 1 MWh per tonne. The manufacturing process is visualized in Figure 13.1.

13.4.2 Cellulose Nanowhiskers

Cellulose nanowhiskers, a more crystalline form of nanocellulose, are formed by the acid hydrolysis of native cellulose fibres using a concentrated inorganic salt, commonly sulfuric or hydrochloric acid. The amorphous sections of native cellulose are hydrolysed and, after careful timing, the crystalline sections can be retrieved from the acid solution by centrifugation and washing. Cellulose nanowhiskers are rod-like highly crystalline particles (relative crystallinity index above 75%) with a rectangular cross section. Their dimensions depend on the native cellulose source material and the hydrolysis time and temperature [1].

Nanocellulose obtained from native fibres by acid hydrolysis, the so-called 'nanowhiskers', is shorter (up to 1000 nm) than nanofibrils obtained through the

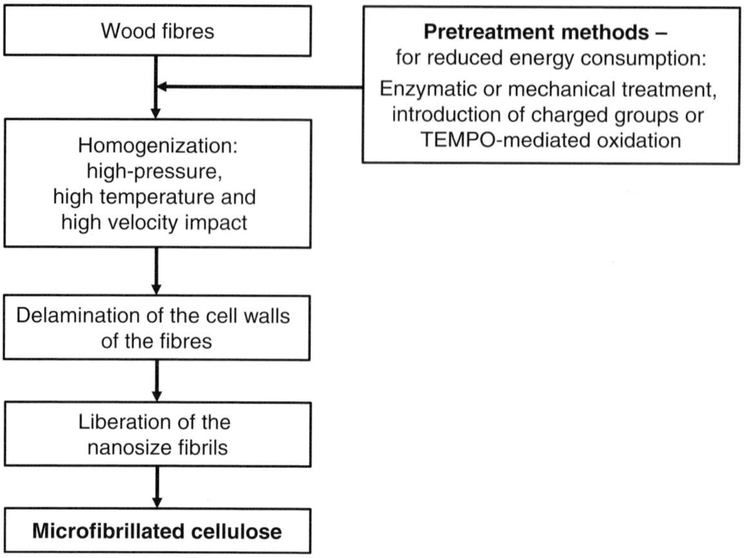

Figure 13.1 Production of nanocellulose/MFC [1].

homogenization route. The resulting material is also known as nanocrystalline cellulose (NCC) [1,2].

Production of NCC starts with 'purified' wood, which has had compounds such as lignin and hemicellulose removed. It is then milled into a pulp and hydrolysed in acid to remove impurities before being separated and concentrated as crystals into a thick paste that can be applied to surfaces as a laminate or processed into strands, forming nanofibrils. These are hard, dense, and tough, and can be forced into different shapes and sizes. When freeze-dried, the material is lightweight, absorbent, and a good insulator. Figure 13.2 illustrates the manufacturing process of cellulose nanowhiskers.

By freeze-drying, the nanofibrils formed turn into a lightweight material with highly absorbent and good insulating properties.

13.4.3 New Biotechnological Production Methods

Recent breakthroughs in nanocellulose production were announced in April 2013 by R. Malcolm Brown Jr., a biology professor at the University of Texas at Austin (www.botany.utexas.edu/mbrown), who presented his team's findings at an American Chemical Society conference in New Orleans [12]. Brown's mass manufacturing method is vastly more efficient and environmentally friendly than using a cellulose source material like wood pulp. His method requires only sunlight, water, and a genetically engineered blue-green algae (a cyanobacteria sourced from vinegar bacterium). Brown's laboratory

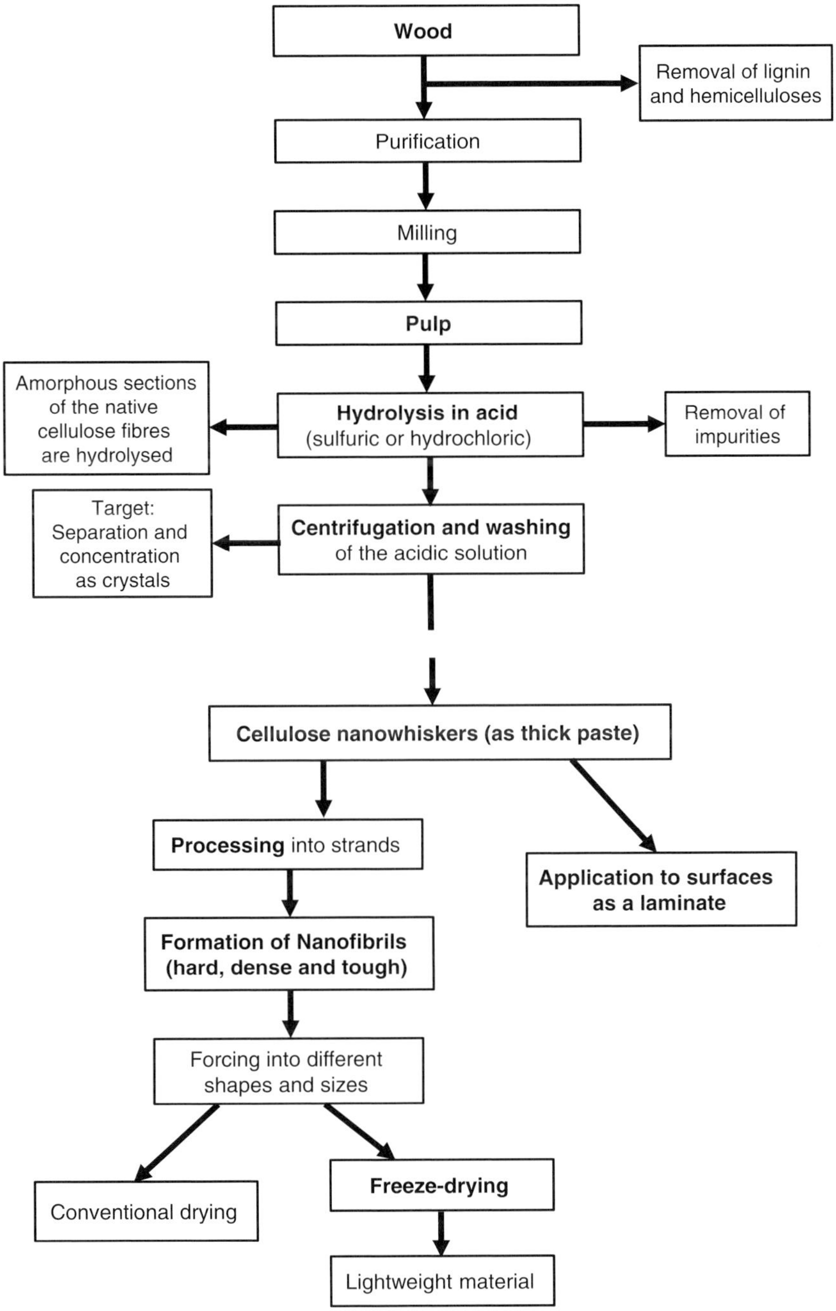

Figure 13.2 Production of highly crystalline nanocellulose or cellulose nanowhiskers [1].

has effectively created organic factories capable of making nanocellulose on a potentially industrial scale. The team is currently working on synthesizing a more complete and stable form of the material, though their progress is already promising. If scaled up, Brown's method could have a dramatic impact across several sectors [13].

Nanocellulose production from cotton fibres, sugarcane bagasse, and other agro-biomass is being carried out at the Central Institute for Research on Cotton Technology in Mumbai, India. India's first pilot plant for production of nanocellulose is established in Mumbai, India and operational since June 2014. It can produce 10 kg Nanocellulose output per day from cotton based cellulose, sugarcane bagasse, agro-biomass and forest biomass. (www.circot.res.in; www.nanocellulose.in) [1].

13.5 Chemistry and Rheometry

13.5.1 Molecular Structure

The chemistry of cellulose is explained in detail in Chapter 3 (see Section 3.4 on 'Chemical composition and structure').

13.5.2 Dimensions and Crystallinity

The ultrastructure of cellulose derived from various sources has been extensively studied. Techniques such as transmission electron microscopy (TEM), scanning electron microscopy (SEM), atomic force microscopy (AFM), wide angle X-ray scattering (WAXS), small incidence angle X-ray diffraction, and solid state ^{13}C cross-polarization magic angle spinning (CP/MAS) nuclear magnetic resonance (NMR) spectroscopy have been used to characterize nanocellulose morphology. These methods have typically been applied for the investigation of dried nanocellulose morphology [14].

Although a combination of microscopic techniques with image analysis can provide information on nanocellulose fibril widths, it is more difficult to determine nanocellulose fibril lengths because of entanglements and difficulties in identifying both ends of individual nanofibrils [15,16].

It has been reported that nanocellulose suspensions may not be homogeneous and that they consist of various structural components, including cellulose nanofibrils and nanofibril bundles [17].

Most methods have typically been applied to the investigation of dried nanocellulose dimensions, although a study has been conducted on the size and size-distribution of enzymatically pretreated nanocellulose fibrils in a suspension using cryo-TEM. The fibrils were found to be rather monodispersed, mostly with a diameter of approximately 5 nm although occasionally thicker fibril bundles were present [10]. Notably, some newly published results indicate that by combining ultrasonication with an 'oxidation pretreatment', cellulose microfibrils with a lateral dimension below 1 nm can be observed by AFM. The lower end of the thickness dimension is around 0.4 nm, which is related to the thickness of a cellulose monolayer sheet [18].

The aggregate widths can be determined by CP/MAS NMR, developed by Innventia AB, Sweden, which also has been demonstrated to work for nanocellulose (enzymatic pretreatment). An average width of 17 nm has been measured by NMR, which corresponds well with SEM and TEM. Using TEM, values of 15 nm have been reported for nanocellulose from carboxymethylated pulp. However, thinner fibrils can also be detected. Wågberg *et al.* have reported fibril widths of 5–15 nm for a nanocellulose with a charge density of about 0.5 meq g^{-1} [11].

The group of Isogai reported fibril widths of 3–5 nm for TEMPO-oxidized cellulose having a charge density of 1.5 meq g^{-1} [19].

The influence of cellulose pulp chemistry on the nanocellulose microstructure has been investigated using AFM to compare the microstructure of two types of nanocellulose prepared at Innventia AB (enzymatically pretreated nanocellulose and carboxymethylated nanocellulose). Owing to the chemistry involved in producing carboxymethylated nanocellulose, it differs significantly from the enzymatically pretreated one. The number of charged groups on the fibril surfaces is very different. The carboxymethylation pretreatment makes the fibrils highly charged and, hence, easier to liberate, which results in smaller and more uniform fibril widths (5–15 nm) compared to the enzymatically pretreated nanocellulose, where the fibril widths were 10–30 nm [20].

The degree of crystallinity and the cellulose crystal structure of nanocellulose were also studied at the same time. The results clearly showed that the nanocellulose exhibited cellulose crystal I organization and that the degree of crystallinity was unchanged by the method used to prepare the nanocellulose. Typical values for the degree of crystallinity were around 63% [1,20].

13.6 Rheology

13.6.1 Dissolution Behaviour

Like microcrystalline cellulose (MCC) and native cellulose, microfibrillated cellulose (MFC) and nanocrystalline cellulose (NCC, CNC) are not soluble in water. They can, however, be dispersed in water using shear forces. Nanocellulose is pseudoplastic. Once dispersed appropriately, it develops high viscosities at low concentrations. It exhibits the property of certain gels or fluids that are thick and highly viscous under normal conditions, but become thin-flowing and less viscous over time when shaken, agitated, or otherwise stressed. This property is known as thixotropy. When the shearing forces are removed the gel regains much of its original state. There are different MFC products in the market:

- Pastes with low cellulose content (approx. 5%) are readily dispersible in water or a number of water-miscible organic solvents. Medium shear mixing is needed to obtain maximum viscosity and water retention.
- Granular products with 25% MFC and 75% water require high shear mixing to allow the product to swell fully in aqueous systems.
- Powders with 50% MFC and 50% sucrose also need high shear forces to be dispersed properly and incorporated homogeneously. Dry blending MFC with other ingredients before adding to the wet mix is often advantageous.

13.6.2 Viscosity and Gelation

The unique rheology of nanocellulose dispersions was recognized by the early investigators. The high viscosity at low nanocellulose concentrations makes nanocellulose very interesting as a non-caloric stabilizer and gelling agent in food applications, the major field explored by early investigators. The dynamic rheological properties were investigated in great detail [10] and revealed that the storage and loss modulus were independent of the angular frequency at all nanocellulose concentrations between 0.125% and 5.9%. The storage modulus values are particularly high (104 Pa at 3% concentration) [10], compared to results for cellulose nanowhiskers (102 Pa at 3% concentration) [21].

There is also a particularly strong concentration dependence as the storage modulus increases by five orders of magnitude if the concentration is increased from 0.125% to 5.9%. Nanocellulose gels are also highly shear thinning (the viscosity is lost upon introduction of shear forces). The shear-thinning behaviour is particularly useful in a range of different coating applications [1,10].

13.7 Stability

Overview

NCC/CNC is a very lightweight, but highly stable material. It is very strong, durable, and tough. Nanocellulose has a high wear and abrasion stability, and it is stable against the damages of light. It can also be incorporated into systems to make light reflective structures that are impermeable to gas. This nanomaterial is produced from natural and renewable ressources; it is recyclable, compostable, and biodegradable [22].

Mechanical Properties

It has long been known that crystalline cellulose has interesting mechanical properties for use in material applications. The tensile strength of crystalline cellulose has been shown to be on the order of 500 MPa, which is similar to that of aluminium. Its stiffness is about 140–220 GPa, which is of the same order as, for instance, that of Kevlar and is better than that of glass fibres, both of which are used commercially to reinforce plastics. Films made from nanocellulose have been shown to have high strength (over 200 MPa), high stiffness (around 20 GPa), and high strain (12%). Its 'strength to weight' ratio is 8 times that of stainless steel [1,9].

Barrier Properties

In semi-crystalline polymers, the crystalline regions are considered to be gas impermeable. Owing to relatively high crystallinity (www.botany.utexas.edu/mbrown), in combination with the ability of the nanofibres to form a dense network held together by strong inter-fibrillar bonds (high cohesive energy density), it has been suggested that nanocellulose might act as a barrier material [19,23,24].

Although the number of reported oxygen permeability values is limited, they attribute high oxygen barrier properties to nanocellulose films. One study reported an oxygen permeability of 0.0006 (cm^3 μm)/(m^2 day kPa) for an approximately 5 μm thin nanocellulose film at 23 °C (73 °F) and 0% RH [23].

In a related study, a more than 700-fold decrease in oxygen permeability of a polylactide (PLA) film was observed when a nanocellulose layer was added to the PLA surface [19].

The influence of nanocellulose film density and porosity on film oxygen permeability has been explored recently [25].

Some authors have reported significant porosity in nanocellulose films [26–28], which seems to be in contradiction with high oxygen barrier properties, whereas Aulin *et al.* [23] measured a nanocellulose film density close to that of crystalline cellulose (cellulose I_β crystal structure, 1.63 g cm^{-3}) [29], indicating a very dense film with a porosity close to zero.

Changing the surface functionality of the cellulose nanoparticle can also affect the permeability of nanocellulose films. Films constituted of negatively charged cellulose nanowhiskers could effectively reduce permeation of negatively charged ions, while leaving neutral ions virtually unaffected. Positively charged ions were found to accumulate in the membrane [1,30].

Foams

Nanocellulose can also be used to make aerogels/foams, either homogeneously or in composite formulations. Nanocellulose-based foams are being studied for packaging applications in order to replace polystyrene-based foams. Svagan *et al.* have shown that nanocellulose has the ability to reinforce starch foams by using a freeze-drying technique [31].

The advantage of using nanocellulose instead of wood-based pulp fibres is that the nanofibrils can reinforce the thin cells in the starch foam. Moreover, it is possible to prepare pure nanocellulose aerogels by applying various freeze-drying and super critical CO_2 drying techniques. Aerogels and foams can be used as porous templates [32,33].

Tough ultrahigh porosity foams prepared from cellulose I nanofibril suspensions have been studied by Sehaqui *et al.* A wide range of mechanical properties, including compression, was obtained by controlling the density and nanofibril interaction in the foams [34]. Cellulose nanowhiskers could also be made to gel in water under low power sonication, giving rise to aerogels with the highest reported surface area (>600 $m^2 g^{-1}$) and lowest shrinkage during drying (6.5%) of cellulose aerogels [33].

In another study by Aulin *et al.* [35], the formation of structured porous aerogels of nanocellulose by freeze-drying was demonstrated. The density and surface texture of the aerogels was tuned by selecting the concentration of the nanocellulose dispersions before freeze-drying. Chemical vapour deposition of a fluorinated silane was used to uniformly coat the aerogel to tune the wetting properties towards nonpolar liquids/oils. The authors demonstrated that it is possible to switch the wettability behaviour of the cellulose surfaces between super-wetting and super-repellent, using different scales of roughness and porosity created by the freeze-drying technique and change of concentration of the nanocellulose dispersion. Structured porous cellulose foams can, however, also be obtained by utilizing the freeze-drying technique on cellulose generated by *Gluconobacter* strains of bacteria, which biosynthesize open porous networks of cellulose fibres with relatively large amounts of nanofibrils dispersed inside. Olsson *et al.* [36] have demonstrated that these networks can be further impregnated with metal hydroxide/oxide precursors, which can readily be transformed into grafted magnetic nanoparticles along the cellulose nanofibres. The magnetic cellulose foam may allow several novel applications of nanocellulose, and the first remotely actuated magnetic super sponges, absorbing 1 g of water within a 60-mg cellulose aerogel foam, were reported. Notably, these highly porous foams (>98% air) can be compressed into strong magnetic nanopapers, which may find use as functional membranes in various applications [1].

Surface Modification

The surface modification of nanocellulose is currently receiving much attention [37]. Nanocellulose displays a high concentration of hydroxyl groups at the surface, which can react and be reacted. However, hydrogen bonding strongly affects the reactivity of the surface hydroxyl groups. In addition, impurities at the surface of nanocellulose such as glucosidic and lignin fragments need to be removed before surface modification to obtain acceptable reproducibility between different batches [1,38].

13.8 Synergies with Other Substances

13.8.1 Reason for Interactions and Options

Cellulose nanofibres isolated from wood pulp are several micrometers long, but only a few nanometres thick. The nanofibres are strongly interlinked with each other and have an extremely large surface area over which they can interact chemically and physically with substances such as water, as well as inorganic, organic, and polymeric compounds.

Cellulose nanofibres can thus be utilized as a robust and highly reactive raw material for industrial use. The product is made of renewable resources and is also biodegradable. Therefore, it is suitable for reinforcing (bio-)polymers. It is a promising environmentally friendly lightweight material for use in the automotive industry, as well as a membrane or filter material in packaging applications or for biomedical products (www.empa.ch).

13.8.2 Chemical Modification

The nanocellulose isolated from pulp is initially present as an aqueous suspension.

During drying, the cellulosic material keratinizes and the cellulose fibres 'glue' together. Thereby, cellulose loses its outstanding mechanical properties. Researchers from the EMPA institute (Swiss Federal Laboratories for Materials Science) in Switzerland recently developed a method by which nanocellulose can be dried without being clumped and keratinized. It is based on the chemical modification of cellulose by a method that is industrially easily realizable, and even acceptable for food applications, and which prevents the cellulose fibrils aggregating one to another and sticking together.

The result is impressive: The dried nanocellulose powder has – after redispersion in water – the same properties as non-modified cellulose that has not previously been dried. This nanocellulose powder is an attractive alternative to conventional cellulosic suspensions for the synthesis of organic nanocomposite materials. Traditional cellulose suspensions contain more than 90% water, which dramatically raises the cost of transport and increases the risk of degradation by bacteria or fungi. In addition, aqueous cellulose suspensions are difficult to process, since in the course of chemical processes the solvents usually must be replaced (www.empa.ch).

13.8.3
Bioactivity

General

Nanocellulose can be 'activated' by integrating bioactive molecules and substances into the fibrous cellulose network. This new biological material offers many new opportunities, especially in human biomedical applications.

Possible applications for nanocellulose and bioactive NCC are:

- cartilage replacement;
- repair of intervertebral disc;
- treatment of hernias, including inguinal hernias;
- substitute for small blood vessels: the fibrous structure makes nanocellulose suitable for use as flexible tubes.

Cartilage Replacement [39]

Bioactive nanocellulose is a novel biomaterial that can serve as a scaffold for the growth of new cartilage tissue.

Background: In the joints cartilage acts as a shock absorber for the bones. The pressure-elastic or fibrous supporting tissue heads off compressive and shear forces that act during all movements to the bones. If the cartilage substance is damaged by wear, metabolic disorders, or injuries and limited in its function, these forces are applied directly to the bone, which is very painful. To help patients with osteoarthritis, diseases such as osteochondritis dissecans, or after sports injuries, orthopaedic surgeons are seeking methods to repair or replace, for example, cartilage in the knee, ankle, or between the vertebrae. The idea of a research project at the University of Jena, Germany, is to introduce a framework structure of nanocellulose in the joint, wherein the cartilage cells be can stored or located upon and can thus form new functional cartilage. Bacterially synthesized nanocellulose is, owing to its high water content, a good candidate to replace the cartilage matrix. The real highlight, however, is integrated bioactive molecules. These substances are meant to entice the cartilage cells present in the joint, facilitate their growth into the matrix, and stimulate matrix formation. These catalysts for cartilage formation are specific protein molecules. The bioactive molecules in combination with the nanocellulose matrix are expected to deliver decisive advantages over already established cartilage replacement procedures. The researchers want to implant the new cartilage matrix in a cell-free way and thus spare the patient the hitherto customary previous intervention, were healthy cartilage cells were taken out beforehand. So far, cells obtained from the medical intervention are cultured in the laboratory on the new matrix and then subsequently re-implemented with it into the human body.

13.9 Food Applications

Nanocellulose can be used as a low-calorie replacement for today's carbohydrate additives used as thickeners, flavour carriers, and suspension stabilizers in a wide variety of food products. It is useful for producing fillings, crushes, chips, wafers, soups, gravies, puddings, and so on. Food applications were recognized early on as a highly interesting application field for nanocellulose due to the rheological behaviour of the nanocellulose gel [1].

13.10 Non-food Applications

NCC is an advanced lightweight material derived from wood fibre that improves the strength, durability, and toughness of products. It can reduce damage caused by wear, abrasion, and light. This high-value nanomaterial can also be incorporated into systems to make light reflective structures (tunable from ultraviolet to infrared) that are impermeable to gas and stable over time. NCC can transform the performance of existing products as well as help create new, unique, and improved products. Natural, renewable, recyclable, compostable, and biodegradable, NCC also has the potential to become a contributor to the green economy in numerous industrial sectors in the coming decades [22].

The properties of nanocellulose such as the mechanical properties, the film-forming properties, viscosity and so on make it an interesting material for many applications:

1) **Paper and paperboard:**
   The potential of nanocellulose applications in the area of paper and paperboard manufacture is obvious. Nanocelluloses are expected to enhance the fibre–fibre bond strength and, hence, have a strong reinforcement effect on paper materials. Nanocellulose may be useful as a barrier in grease-proof type papers and as a wet-end additive to enhance retention and dry and wet strength in commodity-type paper and board products [1].
2) **Composite:**
   In the search for novel polymeric materials with specific desired properties, materials scientists have developed high-performance composite materials (composites), in which nanocellulose fibres are embedded in polymers. These new materials are used as lightweight composites with similar mechanical properties as steel as well as nanoporous 'bio'-foams to replace conventional insulation foams (www.empa.ch). As described above, the properties of nanocellulose make it an interesting material for reinforcing plastics. Nanocellulose has been reported to improve the mechanical properties of, for example, thermosetting resins, starch-based matrixes, soy

protein, rubber latex, and polylactide. Composite applications may be for use as coatings and films, paints, foams, and packaging [1].

3) **Hygiene and absorbent products:** Different applications in this field include [1]:
   - super water absorbent (e.g. material for incontinence pads);
   - nanocellulose used together with super absorbent polymers;
   - use of nanocellulose in tissue, nonwoven products, or absorbent structures;
   - use as antimicrobial films.
4) **Emulsion and dispersion:**
   Apart from the numerous applications in the area of food additives, the general area of emulsion and dispersion applications in other fields has also received some attention [40,41]. Oil in water applications were recognized early on. The area of non-settling suspensions for pumping sand and coal as well as paints and drilling muds was also explored by early investigators [1].
5) **Oil recovery:**
   Hydrocarbon fracturing of oil-bearing formations is a potentially interesting and large-scale application. Nanocellulose has been suggested for use in oil recovery applications as a fracturing fluid. Drilling muds based on nanocellulose have also been suggested [1].
6) **Medical, cosmetic, and pharmaceutical:**
   The use of nanocellulose in cosmetics and pharmaceuticals was also recognized early on. A wide range of high-end applications have been suggested [1]:
   - freeze-dried nanocellulose aerogels used in sanitary napkins, tampons, diapers, or as wound dressing;
   - the use of nanocellulose as a composite coating agent in cosmetics, for example for hair, eyelashes, eyebrows, or nails;
   - a dry solid nanocellulose composition in the form of tablets for treating intestinal orders;
   - nanocellulose films for screening biological compounds and nucleic acids encoding a biological compound;
   - filter medium partly based on nanocellulose for leukocyte-free blood transfusion;
   - a bucodental formulation, consisting of nanocellulose and a polyhydroxylated organic compound;
   - powdered nanocellulose has also been suggested as an excipient in pharmaceutical compositions;
   - nanocellulose in compositions of a photoreactive noxious substance purging agent;
   - elastic cryo-structured gels for potential biomedical and biotechnological applications [42];
   - bioactive nanocellulose is a novel biomaterial that could serve as a scaffold for the growth of new cartilage tissue [39].

7) **Other applications** [1]:
   - activation of the dissolution of cellulose in different solvents;
   - regenerated cellulose products, such as fibres, films, cellulose derivatives;
   - tobacco filter additive;
   - organometallic modified nanocellulose in battery separators;
   - reinforcement of conductive materials;
   - loud-speaker membranes;
   - high-flux membranes;
   - flexible electronic displays [9];
   - computer components [9];
   - lightweight body armour and ballistic glass [9].

13.11 Nutritional Properties and Toxicity

Biochemically, the nutritional properties of nanocellulose particles are comparable to those of traditional cellulose fibres. The human body does not have enzymes to degrade the β-glycosidic linkages between the individual anhydroglucose monomers. A major difference, however, is that nanosize-cellulose (MFC, CNC, NCC) can pass through the barrier membranes and enter the bloodstream. The intravenous injection of short-chain carboxymethylcellulose (Na-CMC) as an ingredient of pharmaceutical products has been practiced for some years already and assessed as harmless. To what extent the permanent inclusion of high levels of nanoparticulate artificial cellulose in the human body is innoxious and, therefore, tolerable is very controversially discussed at present. Several scientific studies have been started to examine the short-, mid-, and long-term effects. Numerous sources for cellulose nanoparticles are possible:

- undesired release during nanocellulose production, during manufacture of products composed of NCC, during disposal and recycling;
- use of nanocellulose-containing products;
- delivery by direct consumption of food products;
- resulting from undesired transfer out of food packaging material;
- release by abrasion from coatings of technical products;
- ageing and wear of construction material (composites).

Toxicity

According to the results of toxicity studies, nanocellulose is not cytotoxic (see also Section 13.12).

13.12 Safety Aspects

Health, safety, and environmental aspects of nanocellulose have been evaluated recently. Processing of nanocellulose does not cause significant exposure to fine particles during friction grinding or spray drying. No evidence of inflammatory effects or cytotoxicity on mouse or human macrophages can be observed after exposure to nanocellulose. The results of toxicity studies suggest that nanocellulose is not cytotoxic and does not cause any effects on the inflammatory system in macrophages. In addition, nanocellulose is not acutely toxic to *Vibrio fischeri* (a Gram-negative facultative anaerobe bacterium that lives in the oceans of the world) in environmentally relevant concentrations [1,43].

13.12.1 Conclusion

According to current knowledge, there are no immediate safety concerns related to wood-based nanocelluloses [44].

13.13 Regulatory Aspects in Europe

Nanotechnology is still a quite new application area and despite efforts within REACH (Registration, Evaluation, Authorization & restriction of Chemicals) and by national authorities in Europe there are no specific regulations for the production of microfibrillated cellulose (MFC), nanocrystalline cellulose (NCC), or cellulose nanocrystals (CNC). Thus, corporate social responsibility is called for and considerations of environmental, health, and safety (EHS) issues should already be incorporated in the early stages of MFC/NCC/CNC based product development [44].

References

1 Anon (2013) Nanocellulose From Wikipedia the free encyclopedia. http://en.wikipedia.org/wiki/Nanocellulose (accessed on 13 April 2013).

2 Peng, B.L., Dhar, N., Liu, H.L. and Tam, K. C. (2011) Chemistry and applications of nanocrystalline cellulose and its derivatives: a nanotechnology perspective. *The Canadian Journal of Chemical Engineering*, **89** (5), 1191–1206.

3 Stephen, A.M., Philips, G.O., and Williams, P.A. (eds) (2006) *Polysaccharides and Their Applications*, 2nd edn, Taylor & Francis Group, Boca Raton, FL, pp. 147–179.

4 Turbak, A.F., Snyder, F.W., and Sandberg, K.R. (1983) Microfibrillated cellulose, a new cellulose product: properties, uses and commercial potential, in *Proceedings of the Ninth Cellulose Conference* (ed. A. Sarko), Applied Polymer Symposia, vol. 37, John Wiley & Sons, Inc, New York, pp. 815–827. ISBN 0-471-88132-5.

5 Herrick, F.W., Casebier, R.L., Hamilton, J. K., and Sandberg, K.R. (1983) Microfibrillated cellulose: morphology and accessibility, in *Proceedings of the Ninth Cellulose Conference* (ed. A. Sarko), Applied Polymer Symposia, vol. **37**, John Wiley & Sons, Inc., New York, pp. 797–813. ISBN 0-471-88132-5.

6 (a) Turbak, A.F., Snyder, F.W., and Sandberg, K.R. (1982) U.S. patent 4,341,807; (b) Turbak, A.F., Snyder, F.W.,

and Sandberg, K.R. (1983) U.S. patent 4,374,702; (c) Turbak, A.F., Snyder, F.W., and Sandberg, K.R. (1983) U.S. patent 4,378,381; (d) Turbak, A.F., Snyder, F.W., and Sandberg, K.R. (1984) U.S. patent 4,452,721; (e) Turbak, A.F., Snyder, F.W., and Sandberg, K.R. (1984) U.S. patent U.S. patent 4,452,722; (f) Turbak, A.F., Snyder, F.W., and Sandberg, K.R. (1984) U.S. patent 4,464,287; (g) Turbak, A.F., Snyder, F.W., and Sandberg, K.R. (1984) U.S. patent 4,483,743; (h) Turbak, A.F., Snyder, F.W., and Sandberg, K.R. (1984) U.S. patent 4,487,634; (i) Turbak, A.F., Snyder, F.W., and Sandberg, K.R. (1985) U.S. patent 4,500,546.

7 Turbak, A.F., Snyder, F.W., and Sandberg, K.R. (1984) Microfibrillated cellulose – a new composition of commercial significance, in *1984 Nonwovens Symposium, Myrtle Beach, SC, April 16–19*, TAPPI Press, Atlanta, pp. 115–124.

8 Berglund, L. (2005) Cellulose-based nanocomposites, in *Natural Fibers, Biopolymers and Biocomposites* (eds A.K. Mohanty, M. Misra, and L. Drzal), CRC Press, Boca Raton, FL, pp. 807–832. ISBN 978-0-8493-1741-5.

9 Ferguson, W. (2012) Why wood pulp is world's new wonder material, New Scientist, 23 August.

10 Pääkkö, M., Ankerfors, M., Kosonen, H., Nykänen, A., Ahola, S., Österberg, M., Ruokolainen, J., Laine, J., Larsson, P.T., Ikkala, O. and Lindström, T. (2007) Enzymatic hydrolysis combined with mechanical shearing and high-pressure homogenization for nanoscale cellulose fibrils and strong gels. *Biomacromolecules*, **8** (6), 1934–1941.

11 Wågberg, L., Decher, G., Norgren, M., Lindström, T., Ankerfors, M. and Axnäs, K. (2008) The build-up of polyelectrolyte multilayers of microfibrillated cellulose and cationic polyelectrolytes. *Langmuir*, **24** (3), 784–795.

12 Brown, R.M. (7 April 2013) Engineering algae to make 'wonder material' nanocellulose for biofuels and more. American Chemical Society, http://www.newswise.com/articles/engineering-algae-to-make-the-wonder-material-nanocellulose-for-biofuels-and-more (accessed on 13 January 2014).

13 Toor, A. (2013) Can we grow a stronger-than-steel 'wonder material' to save the world? http://www.theverge.com/2013/4/8/4195982/nanocellulose-wonder-material-produced-from-algae-solar-energy (accessed on 13 January 2014).

14 Siró, I. and Plackett, D. (2010) Microfibrillated cellulose and new nanocomposite materials: a review. *Cellulose*, **17** (3), 459–494.

15 Chinga-Carrasco, G., Yu, Y., and Diserud, O. (2011) Quantitative electron microscopy of cellulose nanofibril structures from eucalyptus and Pinus radiata kraft pulp fibers. *Microscopy and Microanalysis*, **17** (4), 563–571.

16 Chinga-Carrasco, G., Miettinen, A., Luengo Hendriks, C.-L., Gamstedt, E.K., and Kataja, M. (2011) Structural characterisation of kraft pulp fibres and their nanofibrillated materials for biodegradable composite applications, in *Nanocomposites and Polymers with Analytical Methods* (ed. J. Cuppletti), InTech, ISBN 978-953-307-352-1. doi: 10.5772/21580.

17 Chinga-Carrasco, G. (2011) Cellulose fibres, nanofibrils and microfibrils: The morphological sequence of MFC components from a plant physiology and fibre technology point of view. *Nanoscale Research Letters*, **6**, 417.

18 Qingqing, L. and Renneckar, S. (2011) Supramolecular structure characterization of molecularly thin cellulose I nanoparticles. *Biomacromolecules*, **12** (3), 650–659.

19 Fukuzumi, H., Saito, T., Iwata, T., Kumamoto, Y., and Isogai, A. (2009) Transparent and high gas barrier films of cellulose nanofibers prepared by TEMPO-mediated oxidation . *Biomacromolecules*, **10** (1), 162–165.

20 Aulin, C., Ahola, S., Josefsson, P., Nishino, T., Hirose, Y., Österberg, M., and Wågberg, L. (2009) Nanoscale cellulose films with different crystallinities and mesostructures – their surface properties and interaction with water. *Langmuir*, **25** (13), 7675–7685.

21 Tatsumi, D., Ishioka, S., and Matsumoto, T. (2002) Effect of fiber concentration and axial ratio on the rheological properties of cellulose fiber suspensions. *Journal of the Society of Rheology (Japan)*, **30** (1), 27–32.
22 CelluForce (2012) CelluForce to present at Bioplastek Forum 2012 in Arlington. Press release 23 March.
23 Aulin, C., Gällstedt, M., and Lindström, T. (2010) Oxygen and oil barrier properties of microfibrillated cellulose films and coatings. *Cellulose*, **17** (3), 559–574.
24 Syverud, K. and Stenius, P. (2009) Strength and barrier properties of MFC films. *Cellulose*, **16** (1), 75–85.
25 Chinga-Carrasco, G. and Syverud, K. (2012) On the structure and oxygen transmission rate of biodegradable cellulose nanobarriers. *Nanoscale Research Letters*, **7**, 192.
26 Henriksson, M. and Berglund, L. (2007) Structure and properties of cellulose nanocomposite films containing melamine formaldehyde. *Journal of Applied Polymer Science*, **106** (4), 2817–2824.
27 Svagan, A.J., Azizi Samir, M.A.S., and Berglund, L.A. (2007) Biomimetic polysaccharide nanocomposites of high cellulose content and high toughness. *Biomacromolecules*, **8** (8), 2556–2563.
28 Henriksson, M., Berglund, L.A., Isaksson, P., Lindström, T., and Nishino, T. (2008) Cellulose nanopaper structures of high toughness. *Biomacromolecules*, **9** (6), 1579–1585.
29 Diddens, I., Murphy, B., Krisch, M., and Müller, M. (2008) Anisotropic elastic properties of cellulose measured using inelastic x-ray scattering. *Macromolecules*, **41** (24), 9755–9759.
30 Thielemans, W., Warbey, C.A., and Walsh, D.A. (2009) Permselective nanostructured membranes based on cellulose nanowhiskers. *Green Chemistry*, **11** (4), 531–537.
31 Svagan, A.J., Azizi Samir, M.A.S., and Berglund, L.A. (2008) Biomimetic foams of high mechanical performance based on nanostructured cell walls reinforced by native nanofibrils. *Advanced Materials*, **20** (7), 1263–1269.
32 Pääkkö, M., Vapaavuori, J., Silvennoinen, R., Kosonen, H., Ankerfors, M., Lindström, T., Berglund, L.A., and Ikkala, O. (2008) Long and entangled native cellulose I nanofibers allow flexible aerogels and hierarchically templates for functionalities. *Soft Matter*, **4** (12), 2492–2499.
33 Heath, L. and Thielemans, W. (2010) Cellulose nanowhisker aerogels. *Green Chemistry*, **12** (8), 1448–1453.
34 Sehaqui, H., Salajková, M., Zhou, Q., and Berglund, L.A. (2010) Mechanical performance tailoring of tough ultra-high porosity foams prepared from cellulose I nanofiber suspensions. *Soft Matter*, **6** (8), 1824–1832.
35 Aulin, C., Netrval, J., Wågberg, L., and Lindström, T. (2010) Aerogels from nanofibrillated cellulose with tunable oleophobicity. *Soft Matter*, **6** (Advance publication), 3298.
36 Olsson, R.T., Azizi Samir, M.A.S., Salazar-Alvarez, G., Belova, L., Ström, V., Berglund, L.A., Ikkala, O., Nogués, J., and Gedde, U.W. (2010) Making flexible magnetic aerogels and stiff magnetic nanopaper using cellulose nanofibrils as templates. *Nature Nanotechnology*, **5** (8), 584–588.
37 Eichhorn, S.J., Dufresne, A., Aranguren, M., Marcovich, N.E., Capadona, J.R., Rowan, S.J., Weder, C., Thielemans, W., Roman, M., Renneckar, S., Gindl, W., Veigel, S., Keckes, J., Yano, H., Abe, K., Nogi, M., Nakagaito, A.N., Mangalam, A., Simonsen, J., Benight, A.S., Bismarck, A., Berglund, L.A., and Peijs, T. (2010) Review: current international research into cellulose nanofibres and nanocomposites. *Journal of Materials Science*, **45**, 1–33.
38 Labet, M. and Thielemans, W. (2011) Improving the reproducibility of chemical reactions on the surface of cellulose nanocrystals: ROP of e-caprolactone as a case study. *Cellulose*, **18** (3), 607–617.
39 University Hospital Jena (2008) Cellulose in the knee. http://www.uni-jena.de/PM081002_Bioimplantate.print (accessed 25 March 2014).
40 Xhanari, K., Syverud, K., and Stenius, P. (2011) Emulsions stabilized by microfibrillated cellulose: the effect of hydrophobization, concentration and o/w ratio. *Dispersion Science and Technology*, **32** (3), 447–452.

41 Lif, A., Stenstad, P., Syverud, K., Nydén, M., and Holmberg, K. (2010) Fischer-Tropsch diesel emulsions stabilised by microfibrillated cellulose. *Colloid and Interface Science*, **352** (2), 585–592.

42 Syverud, K., Kirsebom, H., Hajizadeh, S., and Chinga-Carrasco, G. (2011) Cross-linking cellulose nanofibrils for potential elastic cryo-structured gels. *Nanoscale Research Letters*, **6**, 626.

43 Vartiainen, J., Pöhler, T., Sirola, K., Pylkkänen, L., Alenius, H., Hokkinen, J., Tapper, U., Lahtinen, P., Kapanen, A., Putkisto, K., Hiekkataipale, K., Eronen, P., Ruokolainen, J., and Laukkanen, A. (2011) Health and environmental safety aspects of friction grinding and spray drying of microfibrillated cellulose. *Cellulose*, **18** (3), 775–786.

44 Pitkänen, M. and Kangas, H. (2011) Characterization and safety aspects of micro- and nanocelluloses. Internal VTT Research Report VTT-04793-11, 96 pp.

Index

Cellulose and Cellulose Derivatives in the Food Industry: Fundamentals and Applications, First Edition.
Tanja Wüstenberg.